Lecture Notes in Computer Science   7248

Commenced Publication in 1973
Founding and Former Series Editors:
Gerhard Goos, Juris Hartmanis, and Jan va

Cecilia Di Chio et al. (Eds.)

Applications of Evolutionary Computation

EvoApplications 2012: EvoCOMNET, EvoCOMPLEX,
EvoFIN, EvoGAMES, EvoHOT, EvoIASP, EvoNUM,
EvoPAR, EvoRISK, EvoSTIM, and EvoSTOC
Málaga, Spain, April 11-13, 2012, Proceedings

 Springer

Volume Editors

see next page

Cover illustration:
"Chair No. 17" by The Painting Fool (www.thepaintingfool.com)

ISSN 0302-9743   e-ISSN 1611-3349
ISBN 978-3-642-29177-7   e-ISBN 978-3-642-29178-4
DOI 10.1007/978-3-642-29178-4
Springer Heidelberg Dordrecht London New York

Library of Congress Control Number: 2012934050

CR Subject Classification (1998): F.1, D.2, C.2, I.4, I.2.6, J.5

LNCS Sublibrary: SL 1 – Theoretical Computer Science and General Issues

Typesetting: Camera-ready by author, data conversion by Scientific Publishing Services, Chennai, India

Printed on acid-free paper

Springer is part of Springer Science+Business Media (www.springer.com)

Volume Editors

Cecilia Di Chio
cdichio@gmail.com

Alexandros Agapitos
University College Dublin, Ireland
alexandros.agapitos@ucd.ie

Stefano Cagnoni
Dept. of Computer Engineering
University of Parma, Italy
cagnoni@ce.unipr.it

Carlos Cotta
Dept. Lenguajes y Ciencias
de la Computación
University of Málaga, Spain
ccottap@lcc.uma.es

F. Fernández de Vega
University of Extremadura, Spain
fcofdez@unex.es

Gianni A. Di Caro
"Dalle Molle" Institute for
Artificial Intelligence (IDSIA)
Lugano, Switzerland
gianni@idsia.ch

Rolf Drechsler
Cyber-Physical Systems
DFKI Bremen, Germany
rolf.drechsler@dfki.de

Anikó Ekárt
Computer Science
Aston University, Birmingham, UK
ekarta@aston.ac.uk

Anna I. Esparcia-Alcázar
S2 Grupo, Spain
aesparcia@s2grupo.es

Muddassar Farooq
National University of Computer
and Emerging Sciences
Islamabad, Pakistan
muddassar.farooq@nu.edu.pk

William B. Landgon
University College London, UK
w.langdon@cs.ucl.ac.uk

Juan-J. Merelo-Guervós
Departamento de Arquitectura
y Tecnología de Computadores
Universidad de Granada, Spain
jmerelo@geneura.ugr.es

Mike Preuss
TU Dortmund University, Germany
mike.preuss@tu-dortmund.de

Hendrik Richter
Faculty of Electrical Engineering
and Information Technology
HTWK Leipzig University of Applied
Sciences, Germany
richter@eit.htwk-leipzig.de

Sara Silva
INESC-ID Lisboa, Portugal
sara@kdbio.inesc-id.pt

Anabela Simões
Coimbra Institute of Engineering,
Coimbra Polytechnic
Coimbra, Portugal
abs@isec.pt

Giovanni Squillero
Politecnico di Torino, Italy
giovanni.squillero@polito.it

Ernesto Tarantino
Institute for High Performance
Computing and Networking, Italy
ernesto.tarantino@na.icar.cnr.it

Andrea G. B. Tettamanzi
Università degli Studi di Milano, Italy
andrea.tettamanzi@unimi.it

Julian Togelius
Center for Computer Games Research
IT University of Copenhagen
Denmark
juto@itu.dk

Neil Urquhart
Centre for Emergent Computing
Edinburgh Napier University, UK
n.urquhart@napier.ac.uk

A. Şima Uyar
Dept. of Computer Engineering
Istanbul Technical University, Turkey
etaner@itu.edu.tr

Georgios N. Yannakakis
Center for Computer Games Research
IT University of Copenhagen
Denmark
yannakakis@itu.dk

Preface

The field of evolutionary computation (EC) brings together researchers who aim to solve a wide range of problems using nature-inspired techniques and methods. The essential operators of natural evolution and genetics (namely, reproduction, variation and selection) are used to tackle problems in many areas, ranging from optimization to planning, from design to classification, from simulation to control.

All the papers in this volume represent carefully chosen, state-of-the-art examples of applications of EC. They are intended to provide inspiration and guideline to researchers and professionals willing to use an EC approach to answer their own questions.

This was the 15th year that the EvoApplications conference, as one of the main events of the Evo* family (it originally started in 1998 as EvoWorkshops), provided a professional (and social) platform to researchers willing to discuss the varied aspects of applications of EC.

EvoApplications, year after year, evolves and adapts itself in order to accommodate newly emergent topics. Moreover, in this 2012 edition of Evo*, we saw the EvoMusArt event become a conference in its own right, joining EuroGP (a conference since 2000), EvoCOP (2004), EvoBIO (2007) and EvoApplications (2010) in what is described as "Europe's premier co-located events in the field of EC."

EVO* was held during April 11–13, 2012 in the beautiful city of Málaga, Spain. Evo* 2012 included in addition to EvoApplications: EuroGP, the main European event dedicated to genetic programming; EvoCOP, the main European conference on EC in combinatorial optimization; EvoBIO, the main European conference on EC and related techniques in bioinformatics and computational biology; EvoMusArt, the main European conference on evolutionary and biologically inspired music, sound, art and design. The proceedings for all of these events are also available in the LNCS series (volumes 7244, 7245, 7246 and 7247).

The central aim of the EVO* events is to provide researchers, as well as people from industry, students, and interested newcomers, with an opportunity to present new results, discuss current developments and applications, or just become acquainted with the world of EC. Moreover, it encourages and reinforces possible synergies and interactions between members of all scientific communities that may benefit from EC techniques.

EvoApplications 2012 consisted of the following 11 tracks:

- *EvoCOMNET*, track on nature-inspired techniques for telecommunication networks and other parallel and distributed systems
- *EvoCOMPLEX*, track on algorithms and complex systems
- *EvoFIN*, track on evolutionary and natural computation in finance and economics

- *EvoGAMES*, track on bio-inspired algorithms in games
- *EvoHOT*, track on bio-inspired heuristics for design automation
- *EvoIASP*, track on EC in image analysis and signal processing
- *EvoNUM*, track on bio-inspired algorithms for continuous parameter optimization
- *EvoPAR*, track on parallel implementation of evolutionary algorithms
- *EvoRISK*, track on computational intelligence for risk management, security and defence applications
- *EvoSTIM*, track on nature-inspired techniques in scheduling, planning and timetabling
- *EvoSTOC*, track on evolutionary algorithms in stochastic and dynamic environments

EvoCOMNET addresses the application of EC techniques to problems in distributed and connected systems such as telecommunication and computer networks, distribution and logistic networks, interpersonal and interorganizational networks, etc. To address the challenges of these systems, this track promotes the study and the application of strategies inspired by the observation of biological and evolutionary processes, that usually show the highly desirable characteristics of being distributed, adaptive, scalable, and robust.

EvoCOMPLEX covers all aspects of the interaction of evolutionary algorithms (and metaheuristics in general) with complex systems. Complex systems are ubiquitous in physics, economics, sociology, biology, computer science, and many other scientific areas. Typically, a complex system is composed of smaller aggregated components, whose interaction and interconnectedness are non-trivial. This leads to emergent properties of the system, not anticipated by its isolated components. Furthermore, when the system behavior is studied from a temporal perspective, self-organization patterns typically arise.

EvoFIN is the only European event specifically dedicated to the applications of EC, and related natural computing methodologies, to finance and economics. Financial environments are typically hard, being dynamic, high-dimensional, noisy and co-evolutionary. These environments serve as an interesting test bed for novel evolutionary methodologies.

EvoGAMES aims to focus the scientific developments in computational intelligence techniques that may be of practical value for utilization in existing or future games. Recently, games, and especially video games, have become an important commercial factor within the software industry, providing an excellent test bed for application of a wide range of computational intelligence methods.

EvoHOT focuses on all bio-inspired heuristics applied to electronic design automation. The track's goal is to show the latest developments, industrial experiences, and successful attempts to *evolve rather than* design new solutions. EvoHOT 2012 allows one both to peek into the problems that will be faced in the next generation of electronics, and to demonstrate innovative solutions to classical CAD problems, such as fault tolerance and test.

EvoIASP, the longest-running of all EvoApplications which celebrated its 14th edition this year, has been the first international event solely dedicated to the applications of EC to image analysis and signal processing in complex domains of high industrial and social relevance.

EvoNUM aims at applications of bio-inspired algorithms, and cross-fertilization between these and more classical numerical optimization algorithms, to continuous optimization problems in engineering. It deals with theoretical aspects and engineering applications where continuous parameters or functions have to be optimized, in fields such as control, chemistry, agriculture, electricity, building and construction, energy, aerospace engineering, and design optimization.

EvoPAR covers all aspects of the application of parallel and distributed systems to EC as well as the application of evolutionary algorithms for improving parallel architectures and distributed computing infrastructures. EvoPAR focuses on the application and improvement of distributed infrastructures, such as grid and cloud computing, peer-to-peer (P2P) system, as well as parallel architectures, GPUs, manycores, etc. in cooperation with evolutionary algorithms.

Recent events involving both natural disasters and man-made attacks have emphasized the importance of solving challenging problems in risk management, security and defence. EvoRISK seeks both theoretical developments and applications of computational intelligence to subjects such as cyber crime, IT security, resilient and self-healing systems, risk management, critical infrastructure protection (CIP), military, counter-terrorism and other defence-related aspects, disaster relief and humanitarian logistics, and real-world applications of these subjects.

EvoSTIM presents an opportunity for EC researchers in the inter-related areas of planning, scheduling and timetabling to come together, present their latest research and discuss current developments and applications.

EvoSTOC addresses the application of EC in stochastic and dynamic environments. This includes optimization problems with changing, noisy, and/or approximated fitness functions and optimization problems that require robust solutions. These topics recently gained increasing attention in the EC community and EvoSTOC was the first event that provided a platform to present and discuss the latest research in this field.

Continuing in the tradition of adapting the list of events to the needs and demands of the researchers working in the field of EC, two new tracks were introduced: EvoPAR (track on parallel implementation of evolutionary algorithms) and EvoRISK (track on computational intelligence for risk management, security and defence applications).

The number of submissions to EvoApplications 2012 was again fairly high, accumulating 90 entries (compared to 162 in 2011 and 191 in 2010 – bearing in mind that these numbers included submissions for EvoMusArt). The following table shows relevant statistics for EvoApplications 2012, where the statistics for the 2011 edition are also reported.

	2012			Previous edition		
	Submissions	Accept	Ratio	Submissions	Accept	Ratio
EvoCOMNET	6	4	67%	15	8	53%
EvoCOMPLEX	13	9	69%	11	5	45%
EvoFIN	9	6	67%	8	6	75%
EvoGAMES	13	9	69%	17	11	65%
EvoHOT	2	1	50%	7	5	71%
EvoIASP	13	7	54%	19	7	37%
EvoMUSART	-	-	-	43	24	56%
EvoNUM	12	4	33%	9	5	56%
EvoPAR	10	8	80%	-	-	-
EvoRISK	2	1	50%	-	-	-
EvoSTIM	3	2	67%	9	4	44%
EvoSTOC	7	3	43%	8	5	63%
Total	90	54	60%	162	87	54%

As for previous years, accepted papers were split into oral presentations and posters. And similarly to last year, the paper length for these two categories was the same for all the tracks. The low acceptance rate of 60% for EvoApplications 2012 is an indicator of the high quality of the articles presented at the events, showing the liveliness of the scientific movement in the corresponding fields.

Many people helped make EvoApplications a success. We would like to thank the following institutions:

- The University of Málaga, and particularly the School of Computer Science with its director Prof. José M. Troya, and the School of Telecommunications with its director Prof. Antonio Puerta
- The Málaga Convention Bureau
- The Institute for Informatics and Digital Innovation at Edinburgh Napier University, UK, for administrative help and event coordination

Even with an excellent support and location, an event like EVO* would not have been feasible without authors submitting their work, members of the Program Committees dedicating energy in reviewing those papers, and an audience. All these people deserve our gratitude.

Finally, we are grateful to all those involved in the preparation of the event, especially Jennifer Willies for her unfaltering dedication to the coordination of the event over the years. Without her support, running such a type of conference with a large number of different organizers and different opinions would be unmanageable. Further thanks to the local organizer Carlos Cotta (University of Málaga, Spain) for making the organization of such an event possible and successful. Last but surely not least, we want to specially acknowledge Penousal Machado (University of Coimbra, Portugal) for his hard work as Publicity Chair and Webmaster, and Marc Schoenauer (INRIA, France) for his continuous help in setting up and maintaining the MyReview management software.

April 2012

Cecilia Di Chio
Alexandros Agapitos
Stefano Cagnoni
Carlos Cotta
F. Fernández de Vega
Gianni Di Caro
Rolf Drechsler
Anikó Ekárt
Anna I Esparcia-Alcázar
Muddassar Farooq
William B. Langdon
Juan-J Merelo-Guervós

Mike Preuss
Hendrik Richter
Sara Silva
Anabela Simões
Giovanni Squillero
Ernesto Tarantino
Andrea G.B. Tettamanzi
Julian Togelius
Neil Urqhart
A. Şima Uyar
Georgios N. Yannakakis

Organization

EvoApplications 2012 was part of EVO* 2012, Europe's premier co-located events in the field of evolutionary computing, that included the conferences EuroGP 2012, EvoCOP 2012, EvoBIO 2012 and EvoMusArt 2012.

Organizing Committee

EvoApplications Chair

Cecilia Di Chio   UK

Local Chair

Carlos Cotta   University of Málaga, Spain

Publicity Chair

Penousal Machado   University of Coimbra, Portugal

EvoCOMNET Co-chairs

Gianni A. Di Caro   IDSIA, Switzerland
Muddassar Farooq   National University of Computer and Emerging
                                     Sciences, Pakistan
Ernesto Tarantino   Institute for High Performance Computing and
                                     Networking, Italy

EvoCOMPLEX Co-chairs

Carlos Cotta   University of Málaga, Spain
Juan-J. Merelo-Guervós   Universidad de Granada, Spain

EvoFIN Co-chairs

Andrea G.B. Tettamanzi   Università degli Studi di Milano, Italy
Alexandros Agapitos   University College Dublin, Ireland

EvoGAMES Co-chairs

Mike Preuss   TU Dortmund University, Germany
Julian Togelius   IT University of Copenhagen, Denmark
Georgios N. Yannakakis   IT University of Copenhagen, Denmark

EvoHOT Co-chairs

Giovanni Squillero   Politecnico di Torino, Italy
Rolf Drechsler   Cyber-Physical Systems, DFKI Bremen,
                                     Germany

EvoIASP Chair

Stefano Cagnoni   University of Parma, Italy

EvoNUM Co-chairs

Anna I Esparcia-Alcázar   S2 Grupo, Spain
Anikó Ekárt   Aston University, UK

EvoPAR Co-chairs

F. Fernández de Vega   University of Extremadura, Spain
William B. Langdon   University College London, UK

EvoRISK Co-chairs

Anna I Esparcia-Alcázar   S2 Grupo, Spain
Sara Silva   INESC-ID Lisboa, Portugal

EvoSTIM Co-chairs

A. Şima Uyar   Istanbul Technical University, Turkey
Neil Urquhart   Edinburgh Napier University, UK

EvoSTOC Co-chairs

Hendrik Richter   HTWK Leipzig University of Applied Sciences,
                                       Germany
Anabela Simões   Coimbra Institute of Engineering,
                                       Coimbra Polytechnic, Portugal

Program Committees

EvoCOMNET Program Committee

Özgür B. Akan   Middle East Technical University, Turkey
Qing Anyong   National University of Singapore, Singapore
Payman Arabshahi   University of Washington, USA
Mehmet E. Aydin   University of Bedfordshire, UK
Alexandre Caminada   University of Technology Belfort-Montbéliard,
                                       France
Iacopo Carreras   CREATE-NET, Italy
Frederick Ducatelle   IDSIA, Switzerland
Luca Gambardella   IDSIA, Switzerland
Kenji Leibnitz   Osaka University, Japan
Domenico Maisto   ICAR CNR, Italy
Roberto Montemanni   IDSIA, Switzerland
Enrico Natalizio   INRIA Lille, France
Conor Ryan   University of Limerick, Ireland
Muhammad Saleem   National University of Computer and Emerging
                                       Technologies, Pakistan
Chien-Chung Shen   University of Delaware, USA

Jun Suzuki	University of Massachusetts, USA
Tony White	Carleton University, Canada
Lidia Yamamoto	University of Basel, Switzerland
Nur Zincir-Heywood	Dalhousie University, Canada

EvoCOMPLEX Program Committee

Antonio Córdoba	Universidad de Sevilla, Spain
Carlos Cotta	Universidad de Málaga, Spain
Jordi Delgado	Universitat Politècnica de Catalunya, Spain
Albert Díaz-Guilera	University of Barcelona, Spain
Marc Ebner	University of Tübingen, Germany
Carlos Fernandes	University of Granada, Spain
José E. Gallardo	Universidad de Málaga, Spain
María Isabel García Arenas	University of Granada, Spain
Carlos Gershenson	UNAM, Mexico
Anca Gog	Babes-Bolyai University, Romania
Márk Jelasity	University of Szeged, Hungary
Juan Luis Jiménez	University of Luxembourg, Luxembourg
Antonio J. Fernández-Leiva	University of Málaga, Spain
Juan-J Merelo-Guervós	Universidad de Granada, Spain
Antonio Nebro	University of Málaga, Spain
Joshua L. Payne	University of Vermont, USA
Katya Rodríguez-Vázquez	UNAM, Mexico
Robert Schaefer	AGH University of Science and Technology, Poland
Marco Tomassini	Université de Lausanne, Switzerland
Alberto Tonda	Politecnico di Torino, Italy
Leonardo Vanneschi	University of Milano-Bicocca, Italy

EvoFIN Program Committee

Alexandros Agapitos	University College Dublin, Ireland
Jonathan Arriaga	Instituto Tecnológico y de Estudios Superiores de Monterrey, Mexico
Antonia Azzini	Università degli Studi di Milano, Italy
Carlos Cotta	Universidad de Málaga, Spain
Wei Cui	University College Dublin, Ireland
Mauro Dragoni	Fondazione Bruno Kessler, Italy
José Ignacio Hidalgo	Universidad Complutense de Madrid, Spain
Ronald Hochreiter	Vienna University of Economics and Business, Austria
Serafin Martinez Jaramillo	Bank of Mexico, Mexico
Piotr Lipinski	University of Wroclaw, Poland
Michael Mayo	University of Waikato, New Zealand
José Pinto	Instituto Superior Técnico, Portugal
Andrea Tettamanzi	Università degli Studi di Milano, Italy
Nikolaos Thomaidis	University of the Aegean, Greece

EvoGAMES Program Committee

Phillipa Avery	University of Nevada, USA
Wolfgang Banzhaf	Memorial University of Newfoundland, Canada
Luigi Barone	University of Western Australia, Australia
Robin Baumgarten	Imperial College London, UK
Paolo Burelli	IT-Universitetet i København, Denmark
Simon Colton	Imperial College London, UK
Ernesto Costa	Universidade de Coimbra, Portugal
Marc Ebner	University of Tübingen, Germany
Anna Esparcia Alcázar	S2 Grupo, Spain
F. Fernández de Vega	Universidad de Extremadura, Spain
Antonio J. Fernández-Leiva	Universidad de Málaga, Spain
Edgar Galvan-Lopes	University College Dublin, Ireland
Leo Galway	University of Ulster, UK
Johan Hagelbäck	Blekinge Tekniska Högskola, Sweden
John Hallam	University of Southern Denmark
Erin Hastings	University of Central Florida, USA
Philip Hingston	Edith Cowan University, Australia
Stefan Johansson	Blekinge Tekniska Högskola, Sweden
Rilla Khaled	IT-Universitetet i København, Denmark
Krzysztof Krawiec	Poznan University of Technology, Poland
Pier Luca Lanzi	Politecnico di Milano, Italy
Simon Lucas	University of Essex, UK
Rodica Ioana Lung	Babes Bolyai University, Cluj Napoca, Romania
Penousal Machado	Universidade de Coimbra, Portugal
Tobias Mahlmann	IT-Universitetet i København, Denmark
Hector P. Martinez	IT-Universitetet i København, Denmark
Juan-J Merelo-Guervós	Universidad de Granada, Spain
Risto Miikkulainen	University of Texas at Austin, USA
Antonio Mora	Universidad de Granada, Spain
Miguel Nicolau	University College Dublin, Ireland
Steffen Priesterjahn	Wincor Nixdorf, Germany
Jan Quadflieg	TU Dortmund, Germany
Jacob Schrum	University of Texas at Austin, USA
Noor Shaker	IT-Universitetet i København, Denmark
Moshe Sipper	Ben-Gurion University, Israel
Terence Soule	University of Idaho, USA

EvoHOT Program Committee

Varun Aggarwal	Aspiring Minds, Haryana, India
Angan Das	Intel Corporation, USA
Stefano Di Carlo	Politecnico di Torino, Italy
Rolf Drechsler	Cyber-Physical Systems, DFKI Bremen, Germany

Carlos Gershenson	Universidad Nacional Autónoma de México, Mexico
Gregor Papa	Jozef Stefan Institute, Slovenia
E.J. Solteiro Pires	Universidade de Trás-os-Montes e Alto Douro, Portugal
Ernesto Sanchez	Politecnico di Torino, Italy
Lukas Sekanina	Brno University of Technology, Czech Republic
Massimo Schillaci	Dora Tech, Italy
Giovanni Squillero	Politecnico di Torino, Italy
Alberto Tonda	Insitut des Systémes Complexes - Paris Île-de-France (ISC-PIF), France

EvoIASP Program Committee

Antonia Azzini	Università degli Studi di Milano, Italy
Lucia Ballerini	University of Edinburgh, UK
Leonardo Bocchi	University of Florence, Italy
Stefano Cagnoni	University of Parma, Italy
Oscar Cordon	European Center for Soft Computing, Spain
Sergio Damas	European Center for Soft Computing, Spain
Ivanoe De Falco	ICAR - CNR, Italy
Antonio Della Cioppa	University of Salerno, Italy
Laura Dipietro	MIT, USA
Marc Ebner	University of Tübingen, Germany
Francesco Fontanella	University of Cassino, Italy
Şpela Iveković	University of Glasgow, UK
Mario Koeppen	Kyushu Institute of Technology, Japan
Krisztof Krawiec	Poznan University of Technology, Poland
Jean Louchet	INRIA, France
Evelyne Lutton	INRIA, France
Luca Mussi	Henesis srl, Italy
Ferrante Neri	University of Jyväskylä, Finland
Gustavo Olague	CICESE, Mexico
Riccardo Poli	University of Essex, UK
Stephen Smith	University of York, UK
Giovanni Squillero	Politecnico di Torino, Italy
Kiyoshi Tanaka	Shinshu University, Japan
Andy Tyrrell	University of York, UK
Leonardo Vanneschi	University of Milano-Bicocca, Italy
Mengjie Zhang	Victoria University of Wellington, New Zealand

EvoNUM Program Committee

Anne Auger	INRIA, France
Wolfgang Banzhaf	Memorial University of Newfoundland, Canada
Hans-Georg Beyer	Vorarlberg University of Applied Sciences, Austria
Ying-ping Chen	National Chiao Tung University, Taiwan

Marc Ebner	Ernst-Moritz-Universität Greifswald, Germany
F. Fernández de Vega	Universidad de Extremadura, Spain
Nikolaus Hansen	INRIA, France
José Ignacio Hidalgo	Universidad Complutense de Madrid, Spain
Andras Joo	Aston University, UK
William B. Langdon	University College London, UK
Boris Naujoks	Log!n GmbH, Germany
Ferrante Neri	University of Jyväskylä, Finland
Mike Preuss	TU Dortmund University, Germany
Gabriela Ochoa	University of Nottingham, UK
Petr Pošík	Czech Technical University, Czech Republic
Günter Rudolph	University of Dortmund, Germany
Ivo F. Sbalzarini	ETH Zurich, Switzerland
Marc Schoenauer	INRIA, France
P.N. Suganthan	Nanyang Technological University, Singapore
Olivier Teytaud	INRIA, France
A. Şima Uyar	Istanbul Technical University, Turkey
Darrell Whitley	Colorado State University, USA

EvoPAR Program Committee

Pierre Collet	Strasbourg University, France
Gianluigi Folino	L'ICAR-CNR, Cosenza, Italy
Stephane Gobron	EPFL, Switzerland
Simon Harding	IDSIA, Switzerland
Malcolm Heywood	Dalhousie University, Canada
José Ignacio Hidalgo	University Complutense Madrid, Spain
Ogier Maitre	Strasbourg University, France
Juan-J Merelo-Guervós	University of Granada, Spain
Jose Carlos Ribeiro	Polytechnic Institute of Leiria, Portugal
Denis Robilliard	l'Universite du Littoral-Cote d'Opale, France
Marco Tomassini	Lausanne University, Switzerland
Shigeyoshi Tsutsui	Hannan University, Japan
Leonardo Vanneschi	University of Milano-Bicocca, Italy
Garnett Wilson	Afinin Labs, Inc., Canada
Tien-Tsin Wong	The Chinese University of Hong Kong, China
Qizhi Yu	INRIA, France

EvoRISK Program Committee

Hussein Abbass	UNSW@Australian Defence Force Academy, Australia
Robert K. Abercrombie	Oak Ridge National Laboratory, USA
Rami Abielmona	University of Ottawa, Canada
Anas Abou El Kalam	IRIT-INP Toulouse, France
Marco Carvalho	IHMC, USA
Nabendu Chaki	University of Calcutta, India
Sudip Chakraborty	Valdosta State University, USA

Sanem Sariel	Istanbul Technical University, Turkey
Greet Vanden Berghe	Universiteit Brussel, Belgium
Shengxiang Yang	University of Leicester, UK

EvoSTOC Program Committee

Enrique Alba	University of Málaga, Spain
Peter Bosman	Centre for Mathematics and Computer Science, The Netherlands
Juergen Branke	University of Warwick, UK
Tan Kay Chen	National University of Singapore, Singapore
Ernesto Costa	University of Coimbra, Portugal
Kalyanmoy Deb	Indian Institute of Technology Kanpur, India
Andries Engelbrecht	University of Pretoria, South Africa
A. Şima Uyar	Istanbul Technical University, Turkey
Ferrante Neri	University of Jyväskylä, Finland
Hendrik Richter	Leipzig University of Applied Sciences, Germany
Philipp Rohlfshagen	University of Essex, UK
Briseida Sarasola	University of Málaga, Spain
Anabela Simões	Coimbra Institute of Engineering, Coimbra Polytechnic, Coimbra, Portugal
Ke Tang	University of Science and Technology of China, China
Renato Tinós	Universidade de São Paulo, Brazil
Krzysztof Trojanowski	Polish Academy of Science, Poland
Shengxiang Yang	Brunel University, UK

Sponsoring Institutions

- University of Málaga – the School of Computer Science and the School of Telecommunications, Málaga, Spain
- The Málaga Convention Bureau
- The Institute for Informatics and Digital Innovation at Edinburgh Napier University, UK

Table of Contents

EvoFIN Contributions

EvoGAMES Contributions

EvoHOT Contributions

EvoIASP Contributions

EvoNUM Contributions

EvoPAR Contributions

EvoRISK Contributions

EvoSTIM Contributions

EvoSTOC Contributions

Optimizing Energy Consumption in Heterogeneous Wireless Sensor Networks by Means of Evolutionary Algorithms

José Manuel Lanza-Gutiérrez, Juan Antonio Gómez-Pulido,
Miguel A. Vega-Rodríguez, and Juan Manuel Sánchez-Pérez

Dep. of Technologies of Computers and Communications, University of Extremadura,
Polytechnic School, Campus Universitario s/n, 10003 Cáceres, Spain
{jmlanza,jangomez,mavega,sanperez}@unex.es

Abstract. The use of wireless sensor networks has been increased substantially. One of the main inconveniences of this kind of networks is the energy efficiency; for this reason, there are some works trying to solve it. Traditionally, these networks were only composed by sensors, but now auxiliary elements called routers have been included to facilitate communications and reduce energy consumption. In this work, we have studied the inclusion of routers in a previously established traditional wireless sensor network in order to increase its energy efficiency, optimizing lifetime and average energy effort. For this purpose, we have used two multi-objective evolutionary algorithms: NSGA-II and SPEA-2. We have done experiments over various sceneries, checking by means of statically techniques that SPEA-2 offers better results for more complex instances.

Keywords: Heterogeneous wireless sensor networks, multi-objective optimization, evolutionary algorithms, NSGA-II, SPEA-2, energy consumption, lifetime, average energy effort.

1 Introduction

The use of wireless sensor networks (WSNs) has increased substantially in the last years [1-4]. Both, boom of this technology and its versatility, have favored the appearance of applications in civil areas (industrial control, environmental monitoring, intensive agriculture, fire protection systems, and so) and military areas (rescue operations, surveillance, etc.).

An important aspect in the use of WSNs is the energy efficient. Usually, this kind of networks are powered by batteries (both lack of cabling and freedom of positioning are one of its attractions), thus network lifetime depends on amount of information transmitted by sensors, as well as its scope, among others.

The design of an energy-efficient WSN has been established as a NP-hard [5] optimization problem by some authors [6, 7], so it is a suitable problem for being solved by several strategies. Beginning with heuristics, we can cite the contributions of Xiuzhen Cheng et al. [8] (to optimize network lifetime on WSNs, by means of the

C. Di Chio et al. (Eds.): EvoApplications 2012, LNCS 7248, pp. 1–10, 2012.
© Springer-Verlag Berlin Heidelberg 2012

assignment of transmission powers to sensors) and Huang et al. [9] (to minimize power consumption of sensors, using for this purpose various estimation schemes). In addition to heuristics, there are other works that use genetic algorithms for single-objective optimization. For example Ferentinos et al. [10] propose to optimize energetic consumption based on several factors (connectivity, transmission powers, etc.), but using a unique objective function.

Genetic algorithms for multi-objective optimization have been used too. For example, Konstantinidis et al. [11] propose a new multi-objective evolutionary algorithm (MOEA) to optimize coverage and power consumption; and He et al. [12] to optimize reliability coverage and network lifetime.

Nowadays, WSNs are more complex due to the fact that auxiliary elements (routers) have been included in order to minimize communication among sensors [13, 14], increasing both network speed and lifetime of sensors. We can find some references about this topic. Thus, M. Cardei et al. [15] studied sensor position on a pre-established network of routers to optimize both coverage and energetic cost; and Duarte-Melo et al. [16] to optimize lifetime and energy consumption.

In this work, we have studied the inclusion of routers in a previously established homogeneous WSN in order to increase its energy efficiency, optimizing lifetime and average energy effort. As we have said previously, this is a multi-objective NP-hard problem, so we need to use certain techniques to facilitate its resolution, like evolutionary algorithms [17]. We have used two well-known MOEAs: NSGA-II (*Non-dominated Sorting Genetic Algorithm II*) [18] and SPEA-2 (*Strength Pareto Evolutionary Algorithm II*) [19]. In summary, our work shows the following contributions:

1) The problem has been solved by means of evolutionary techniques.

2) We have optimized over two objectives that have not been considered jointly in any paper found: lifetime and average energy effort. In addition, a third non-simultaneous objective was considered: number of routers.

3) The results obtained of both MOEAs have been analyzed in depth using statistical procedures, comparing both heterogeneous and homogeneous WSN for the same instances. Demonstrating both, this conception allows increasing energy efficiency substantially and SPEA-2 provides betters results for more complex instances.

The rest of this paper is organized as follow. In the second section we provide a brief introduction on heterogeneous WSN design. The methodology followed to solve this problem appears in section 3. In the fourth section, we present an evaluation of results using statistical techniques. A comparative with other approaches appear in section 5. Finally, conclusions and future work are left for section 7.

2   Heterogeneous Wireless Sensor Network

In this work, we study the deployment of a heterogeneous WSN as an alternative to traditional homogeneous WSN. This section provides basic design aspects in WSN: first, definition of elements involved in this problem, below fitness functions to determine goodness of solutions obtained (topologies), finally restrictions on topologies.

2.1   Problem Instance Definition

A particular problem instance will be defined by several elements:

- M terminals or sensors that capture physical information about their environment.
- A collector or sink node (C) that collects information provided by routers or sensors.
- N routers that establish network communications and collect information about sensors in its communication radius.
- Width (D_x) and height (D_y) of the scenery (space where network will be placed).
- Communication radius (R_c). It is the capacity of a network element (router, sensor or collector) to establish communications with other elements.
- Sensitivity radius (R_s) of a sensor. It is the terrain portion over which a sensor can obtain information.
- Initial energy (IE). It is the amount of maximum initial energy that each sensor have initially.
- Information packet size (K). It is the packet size send by sensors.
- α is the path lost exponent [11].
- β is the transmission quality parameter [11].
- *amp* is the power amplifier's energy consumption per bit[11].

The definition of this problem is similar to [15]. We can observe in Fig. 1 a representation of the definition of a problem instance.

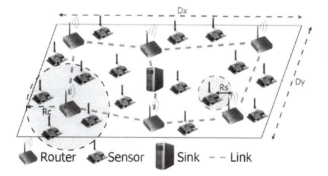

Fig. 1. Definition of a problem instance

2.2   Fitness Functions

The most important energetic factors have been used to deploy the network: lifetime (to maximize) and average energy effort (to minimize). These objectives are simultaneously optimized using the MOEAs. As secondary objective, we define the number of routers (to minimize). This value will be modified along several executions for the same scenery.

Some fitness functions are necessary in order to quantifying the goodness of a solution. Next, we define the fitness functions in detail.

- Lifetime (LT): It is the amount of time units (t.u) that network can provide information of its environment; usually a coverage threshold is used in order to determine whether among of information obtained is enough for the initial size of the network. Initially, all sensors have the same maximum energy charge (IE). Each time unit sensors obtain a measure of its environment and they send it to collector node, whether distance between a sensor and collector is less than R_c, sensor sends the measure to collector node directly, otherwise sensor sends it to other network element (sensor or router) using for this purpose Dijsktra's minimum path [20]. Each time a sensor sends an information packet to other network element consumes an amount of energy proportional to distance between them (2), this amount will be subtracted from its initial energy (IE). When the energy of a sensor is equal to zero, it not will be used again, and then its coverage not will be taken into account. This definition is similar to show in [11, 21].
- Average energy effort (2): It is the average energy consumption in the network lifetime. Whether this value is high, sensors will have greater energy consumption along its lifetime, and the network performance will be decreased faster after crossing the fixed threshold. This definition is based on energy model presented in [11]

$$E_i(t) = k \cdot (r_i(t) + 1) \cdot \beta \cdot d^\alpha(i, e) \cdot amp \tag{1}$$

$$Y_2 = \left(\sum_{j=1}^{LT} \sum_{i=1}^{M} E_i(t)\right)/(LT) \tag{2}$$

$$Y_3 = \left(\sum_{x=1}^{D_x} \sum_{y=1}^{D_y} R_{x,y}\right)/(D_x \cdot D_y) \tag{3}$$

Note that $R_i(t)$ is the number of incoming packets in sensor i at instance t. $d(i,e)$ is the distance between sensor i and element e (router or sensor), following Dijkstra's formulation to collector node.

In order to obtain lifetime fitness function, we use the sensor coverage measure. It is the terrain percentage covered by sensor nodes. There are two possible options [21]. The first one considers that the coverage provided by a sensor is a circumference of radius R_s, so the global coverage will be the intersection of all of them. The second one consists of the use of a boolean matrix of D_x*D_y points over scenery, so for each sensor, the points within its radius will be activated; finally, we have to count the activated points. We have selected the second option, because although the first one is more exact, it is harder. In (3), R represents the boolean matrix and $R_{x,y}$ its position (x,y).

3   Problem Resolution

The design of a heterogeneous WSN is a NP-hard problem as we have mentioned above, so it is necessary to use non-conventional techniques to facilitate its resolution. In this work, we use MOEAs. When a problem is solved by MOEAs, there are some important aspects to tackle: encoding of individuals, crossover and mutation strategies, generation of initial population and description of MOEAs used.

A. Encoding of Individuals

The codification of the individuals is easy. A chromosome is a coordinate list (two dimensions, x and y) of routers.

B. Generation of Initial Population

The initial population is randomly generated only with one restriction: routers must be accessible to collector node. The objective is to start with an adequate population in order to facilitate convergence of MOEAs used.

C. Evaluation of Individuals

In order to evaluate the goodness of an individual, we first study the connectivity among routers, and visibility among sensors and routers. It is possible that there are routers not linked. With this topological information, we obtain all fitness values. Finally, we check whether there are already the same individual into the population; in such case, we will remove it.

D. Crossover and Mutation Strategies

Crossover allows generating new individuals by means of recombination of two previously selected. Mutation allows incorporating random changes in an individual, avoiding local minimums and increasing diversity.

For crossover, we select a crossover point randomly and then, we copy routers from individual 1 until this point, next we copy from individual 2 until the end of the chromosome.

For the mutation of a chromosome, we perform random changes over coordinates of routers. Every time the coordinates of an element are changed, the individual will be evaluated. If this change causes better fitness values will be accepted; in the negative case change will be discarded, back to previous coordinates. The objective is to avoid getting a worse individual than token originally.

The performance of both algorithms is determined by crossover and mutation probabilities. For the crossover, if a randomly generated value is greater than crossover probability, the resulting individual will be a complete-copy of the dominant individual; in other words, crossover will not be performed. For mutation, mutation probability determines that elements will be modified.

E. Multi-objective Evolutionary Algorithms Used

We have used two well-known MOEAs: NSGA-II and SPEA-2.

NSGA-II is characterized by use a methodology which allows sorting the population basing on its dominance (Pareto fronts division), and by using crowding distance to elements in the same front. For more details see [18].

The second one is based on the file concept, an auxiliary population that saves better solutions over generations. Using in this case a strategy that considers for each individual, the number of individual that dominates and which dominates. Also it uses the density concept as a method of fine assignment. For more details see [19].

In both algorithms, we have used the habitual binary tournament [22] to apply the crossover operator. In addition, we have allowed that this selection is not elitist (best individuals do not always win) by means of elitist probability.

4   Experimental Results

The instance data used in this work (Table1) can be obtained in [23]. The instances represent a couple of scenarios of 100x100 and 200x200 meters, in which are placed a set of sensors and a collector node, that we need to study in order to reduce its energy consumption. Both R_c and R_s values are from commercial device MICA2 (Fig. 2) [24], 30 and 15 respectively (in meters). Energy values ($\alpha=2$, $\beta=1$ and amp=100pJ/bit/m^2) are from [11]. The used information packet size is 128kB. Collector node is placed in the scenery center. The coverage threshold used for lifetime is 70%.

Fig. 2. MICA2 wireless sensor device

The number of sensors for both instances is the minimum value to cover all terrain: the area covered for a sensor is $\pi \cdot R_s^2$ and the area of the scenery is $D_x \cdot D_y$, so it is necessary $\left[D_x \cdot D_y\right)/(\pi \cdot R_s^2)\right]$ sensors. Sensor coordinates for both instances have been fixed through a mono-objective evolutionary algorithm that optimizes the coverage. The number of routers used for both instances are variable; this will allow us to observe the different energy performances.

Table 1. Instance data used

Instance	A(m^2)	M	Homogeneous lifetime (t.u)	Homogeneous average energy effort (J)
Inst1	100x100	15	34	0.109
Inst2	200x200	57	9	0.262

For both instances, we have got lifetime and average energy effort (AEE) following homogeneous conception [11, 24]. The use of non-duplicated sensors damages these measures (see table 1), as sensors close to collector node suffer more energy consumption. For this reason, we use routers in order to reduce it.

The strategy for solving the problem by both algorithms (NSGA-II and SPEA-2) is simple. First, we determine the settings that provide the best results. Then, we study if any of them provides a higher performance, using for this purpose statistical tools. To define the best settings, we have set the most common parameters, always over 30 independent runs: crossover, mutation and elitist probabilities, number of evaluations and population size. This methodology is similar to other one proven before [25]:

stating on a default configuration, parameters are adjusted one by one in its optimal value, until all parameters have been adjusted.

To determine the goodness of solutions (Pareto fronts [17]), we have considered the hypervolume metric [26]. This usual metric in MOEAs needs a couple of reference points called *ideal* and *nadir*; these values are maximum and minimum values for tuples {lifetime, AEE}. Maximum values are been defined experimentally for each instance and number of router used (see tables 2 and 3, fields *reference lifetime* and *reference AEE*), minimum values are zero for both.

Table 2. Instance 1. Obtained hypervolumes.

	20.000 evaluations		50.000 evaluations				
Routers	NSGA-II	SPEA-II	NSGA-II	SPEA-2	Reference lifetime	Reference AEE	Statistical study (winner)
4	0.6539	0.6464	**0.7095**	**0.7011**	130	3	=
6	0.7856	0.7795	0.7166	**0.8310**	190	3	S
8	0.7229	0.7193	**0.8004**	0.7798	270	3	N
10	0.8346	0.8354	**0.8310**	**0.8378**	314	3	=

Table 3. Instance 2. Obtained hypervolumes.

	20.000 evaluations		50.000 evaluations				
Routers	NSGA-II	SPEA-II	NSGA-II	SPEA-2	Reference lifetime	Reference AEE	Statistical study (winner)
10	0.7462	0.7462	0.7594	**0.7719**	75	160	S
20	0.7289	0.7453	0.7263	**0.7545**	120	160	S
30	0.8634	0.8702	0.8570	**0.8772**	150	160	S
40	0.7481	0.7528	0.7845	**0.8300**	220	160	S

In tables 3 and 4, we can observe average hypervolume for each instance and routers used; they have been obtained in two different evaluation numbers to see their progressions. In addition, we can see a statistical study in order to determine what algorithm provides better results in each case (field *statistical study*, N means NSGA-II provides better results than SPEA-2, S means the opposite and = both algorithms provide similar results). In this study, we observe that both algorithms provide similar results in the first instance, but SPEA- 2 obtains better results when we use more complex instances (instance 2).

In order to carry out the statistical study for each instance and router used. We have followed procedure shown in Fig. 3 [27]. The first step was to determine if data obtained from these instances (each of them with 30 runs) follow a normal distribution. For this purpose, we used both Shapiro-Wilk [28] and Kolmogorov-Smirnov-Lilliefors [29], obtaining that data did not come from a normal model. To check what algorithm provides betters results, and since we cannot assume a normal distribution, we have used a non-parametric test: Wilcoxon test [30].

If we analyze energy consumption obtained by means of using routers, we can note that network lifetime is increased substantially in comparison with homogeneous conception, for example: in instance 1, with only 4 routers, we have obtained a lifetime above 130 u.t (3 times more). In instance 2, with 20 routers, we have obtained a value above 110 u.t (13 times more). In Fig. 4 we can find relation coverage/lifetime for this second case in comparison with homogeneous conception.

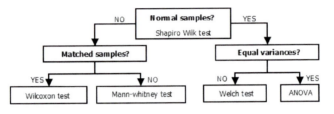

Fig. 3. Statistical procedure

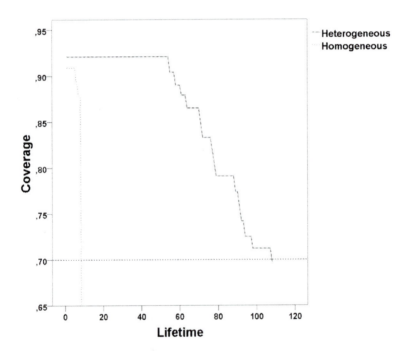

Fig. 4. Coverage/lifetime homogeneous vs. heterogeneous

With these experiment results we can affirm that to use this conception allows to increase lifetime network substantially, without having to increase the number of sensors needlessly.

5   Comparisons with Other Authors

Comparisons with other authors are a complex task, as there are not works with which we can compare our results directly. On the one hand, we can found results from resolution of traditional WSN for energy efficiency [10, 11, 24], but we cannot compare our fitness values with theirs. In these works, redundant sensors are used in order to increase network lifetime, but we use routers to do this task. The main difference is that routers do not have energy limits, and we can use them in order to realize harder communication task, increasing lifetime. On the other hand, we can found works in heterogeneous WSN [15, 16], but their approaches are different from ours.

6   Conclusions and Future Work

In this work, we have tackled the deployment of a heterogeneous WSN optimizing some important energy factors: lifetime and average energy effort. We have used two well-known EAs, NSGA-II and SPEA-2, proving as SPEA-2 provides better results for more complex instances.

As future work, we propose to use more instances and other EAs. In addition, we think that to introduce parallelism would be interesting in order to reduce execution times of algorithms, allowing the use of more complex instances, as well as to study algorithm convergence using a larger number of evaluations.

Acknowledgment. This work has been partially funded by the Ministry of Education and Science and the ERDF (European Regional Development Fund) under the project TIN2008-06491-C04-04 (MSTAR project), and Junta de Extremadura through GR10025 grant provided to group TIC015.

References

[1] Akyildiz, G.I., Su, W., Sankarasubramaniam, Y., Cayirci, E.: A survey on sensor networks. IEEE Communications Magazine, 102–114 (2002)

[2] Vieira, M.A.M., Coelh, C.N., da Silva Jr., D.C.: Survey on wireless sensor network devices. In: Proceedings of IEEE Conference on Emerging Technologies and Factory Automation, ETFA (2003)

[3] Pottie, G.J., Kaiser, W.J.: Wireless integrated network sensors. Commun. ACM 43(5), 51–58 (2000)

[4] Mukherjee, B., Yick, J., Ghosal, D.: Wireless sensor network survey. Comput. Netw. 52(12), 2292–2330 (2008)

[5] Garey, M.R., Johnson, D.S.: Computers and Intractability: A Guide to the Theory of NP-Completeness. Freeman, San Francisco (1979)

[6] Cheng, X., Narahari, B., Simha, R., Cheng, M., Liu, D.: Strong minimum energy topology in wireless sensor networks: Np-completeness and heuristics. IEEE Transactions on Mobile Computing 2(3), 248–256 (2003)

[7] Clementi, A.E.F., Penna, P., Silvestri, R.: Hardness results for the power range assignmet problem in packet radio networks. In: Proceedings of the International Workshop on Approximation Algorithms for Combinatorial Optimization Problems, pp. 197–208. Springer, Heidelberg (1999)

[8] Cheng, X., Narahari, B., Simha, R., Cheng, M.X., Liu, D.: Strong minimum energy topology in wireless sensor networks: np-completeness and heuristics. IEEE Transactions on Mobile Computing 2, 248–256 (2003)

[9] Huang, Y., Hua, Y.: Energy cost for estimation in multihop wireless sensor networks, pp. 2586–2589 (2010)

[10] Ferentinos, K.P., Tsiligiridis, T.A.: Evolutionary energy management and design of wireless sensor networks, pp. 406–417 (2005)

[11] Konstantinidis, A., Yang, K.: Multi-objective energy-efficient dense deployment in Wireless Sensor Networks using a hybrid problem-specific MOEA/D. Applied Soft Computing 11, 4117–4134 (2011)

[12] He J., Xiong, N., Xiao, Y., Pan Y.: A Reliable Energy Efficient Algorithm for Target Coverage in Wireless Sensor Networks, pp. 180–188 (2010)

[13] Heterogeneous Networks with Intel XScale,
http://www.intel.com/research/exploratory/heterogeneous.htm

[14] Yarvis, M.: Exploiting Heterogeneity in Sensor Networks. In: IEEE INFOCOM (2005)

[15] Cardei, M., Pervaiz, M.O., Cardei, I.: Energy-Efficient Range Assignment in Heterogeneous Wireless Sensor Networks, p. 11 (2006)

[16] Duarte-Melo, E.J., Liu, M.: Analysis of energy consumption and lifetime of heterogeneous wireless sensor networks 1, 21–25 (2002)

[17] Deb, K.: Multiobjective optimization using evolutionary algorithms, New York (2001)

[18] Deb, K., Agrawal, S., Pratap, A., Meyarivan, T.: A Fast Elitist Non-dominated Sorting Genetic Al-gorithm for Multi-objective Optimization: NSGA-II (2000)

[19] Zitzler, E., Laumanns, M., Thiele, L.: SPEA2: Improving the strength Pareto evolutionary algorithm. In: EUROGEN (2001)

[20] Cormen, T.: Introduction to algorithms, Cambridge Mass (2001)

[21] Younis, M., Akkaya, K.: Strategies and techniques for node placement in wireless sensor networks: A survey. Ad Hoc Networks 6, 621–655 (2008)

[22] Koza, J.R.: Genetic Programming. MIT Press, Cambridge (1992)

[23] Instance sets for optimization in wireless sensor networks (2011),
http://arco.unex.es/wsnopt

[24] Martins, F.V.C., Carrano, E.G., Wanner, E.F., Takahashi, R.H.C., Mateus, G.R.: A Hybrid Multiob-jective Evolutionary Approach for Improving the Performance of Wireless Sensor Networks. IEEE Sensors Journal 11, 545–554 (2011)

[25] Lanza-Gutiérrez, J.M., Gómez-Pulido, J.A., Vega-Rodríguez, M.A., Sánchez, J.M.: A Multi-objective Network Design for Real Traffic Models of the Internet by Means of a Parallel Framework for Solving NP-hard Problems. In: NABIC IEEE Conference (2011)

[26] Fonseca, C., Knowles, J., Thiele, L., Zitzler, E.: A Tutorial on the Performance Assessment of Stochastic Multiobjective Optimizers. In: EMO (2005)

[27] Ott, L., Longnecker, M.: An introduction to statistical methods and data analysis, Cole Cengage Learning (2008)

[28] Shapiro, S.S., Wilk, M.B.: An analysis of variance test for normality (complete samples). Bio-metrika 52(3 & 4), 591–611 (1965)

[29] Laha, C.: Handbook of Methods of Applied Statistics, pp. 392–394. Wiley J. and Sons (1967)

[30] Wilcoxon, F.: Individual Comparisons by Ranking Methods. Biometrics 1, 80–83 (1967)

Network Protocol Discovery and Analysis via Live Interaction

Patrick LaRoche, A. Nur Zincir-Heywood, and Malcolm I. Heywood

Faculty of Computer Science
Dalhousie University
Halifax, Nova Scotia, Canada
{plaroche,zincir,mheywood}@cs.dal.ca
http://www.cs.dal.ca

Abstract. In this work, we explore the use of evolutionary computing toward protocol analysis. The ability to discover, analyse, and experiment with unknown protocols is paramount within the realm of network security; our approach to this crucial analysis is to interact with a network service, discovering sequences of commands that do not result in error messages. In so doing, our work investigates the real-life responses of a service, allowing for exploration and analysis of the protocol in question. Our system initiates sequences of commands randomly, interacts with and learns from the responses, and modifies its next set of sequences accordingly. Such an exploration results in a set of command sequences that reflect correct uses of the service in testing. These discovered sequences can then be used to identify the service, unforeseen uses of the service, and, most importantly, potential weaknesses.

1 Introduction

In this paper we focus on the analysis of network protocols, presenting a novel approach to do so, by applying an Evolutionary Computing (EC) techniques to study the allowed sequences of a targeted protocol. Our work discovers viable command sequences that represent allowable operations on the protocol implementation under testing (IUT). In exploring these, we then have the ability to analyse the protocol as well as determine specifics to the implementation, gaining insight into the tested service. Our proposed system will identify sequences that are allowable by the specific IUT, not simply what is documented. As such, it will help reveal not only sequences of commands which should be allowable, but also which should not exist that could potentially lead to undesirable consequences. This differs from other work in the field in that we requires only a live interaction with the IUT, no "in hand" binary or detailed knowledge of the targeted implementation. Our system interacts with a live service, adjusts given the responses received, and improves its understanding of the IUT.

This work continues of the principals of previous work in which EC techniques [4, 15] were applied to build variations of known vulnerabilities at the host level [7] as well as network level [11]. By harnessing the exploratory nature of EC

C. Di Chio et al. (Eds.): EvoApplications 2012, LNCS 7248, pp. 11–20, 2012.
© Springer-Verlag Berlin Heidelberg 2012

techniques, our proposed system will explore command sequences of an FTP server in order to investigate the possibility of "learning to communicate with an (un)known protocol". To this end, we performed two sets of experiments, the first giving the proposed system only the correct command set for the FTP protocol. In the second, a larger command set is used (both FTP and SMTP). The second experiment tests our system's ability to not only explore correct sequences of FTP commands, but also its ability to recognize commands that do not belong to the tested protocol, or possibly, find a protocol compliant way to use them.

2   Background

The background work we present in this paper is focused on the analysis of network protocols using learning systems. As such we present related fields, reverse engineering of protocols, evolving communication, syntax based testing as well as the fundamentals of FTP itself.

2.1   Reverse Engineering of Protocols

This work aims to discover the sequences that result in minimal error messages as feedback from a given protocol. Indeed such a system can also be seen as reverse engineering the targeted protocol, albeit requiring limited *a priori* knowledge. This form of protocol analysis is an important tool in network security and protocol verification.

Polygot, a system that uses program binaries for extracting application-level analysis is described in [1]. Unlike Polygot, our only requirement is the ability to interact with the implementation, not the binary itself. This allows our work to be used in situations where the tester does not have an actual instance of the IUT locally.

Also using the binary, Wondracek et al. use observations to create a grammar that describe the possible combinations for the commands of the protocol [18]. The authors input messages into the server binary via a client binary, mark the input as it enters the memory stack of the server and track how that information is used. The client / server relationship is similar to our approach, however, rather than monitoring the internal behaviour directly, we interact with it and observe the output. Hence, we remove the requirement of having the binary of the targeted server.

2.2   Evolving Communication

Application / network protocols are analogous to communication languages between machines. As such, work in the related field of evolving generic agent communication is relevant and forms the basis for our selection of EC towards the domain of protocol analysis. In [3], Froese describes a Genetic Algorithm (GA) that investigates communication in a multi-agent system. In his work, agents are modelled as a neural network with connection weights that are evolved using a GA, where these weights represent the communication between agents (nodes in the neural net). Similarly, Khasteh et al. propose an algorithm that evolves the relationships between agents and elements. In [9] and [10], the authors describe an algorithm that uses a confidence matrix that relates a specific word to a specific element or concept.

In general, the work evolving communication between agents shows promise, but, in most cases, use models where the agents represent some "real world" object trying to communicate with similar objects. In our work, we focus on a less simplistic model where our agents are computer systems, and the language is a network / application protocol where a variety of correct (and incorrect) interactions are possible.

2.3   Syntax Based Testing

In Syntax Based Testing one employs a system specifically designed to test the security of a network protocol or application by focusing on modifications of its protocol syntax, hence directly related to our work. In this type of work, Protocol Data Units (PDUs), are defined by a "frame" of data of a particular length with regards to the IUT [17]. These frames are very protocol specific and require complete knowledge of the protocol in testing. As an example, Tal et al. focus on the OSPF (Open Shortest Path First) routing protocol, a frame in that case being the "Hello" packet, starting at byte 13 and being two bytes long [17]. This level of testing is shown to be successful, however, requires an amount of *a priori* knowledge that our work aims to avoid, as it is not always possible to obtain such knowledge (i.e. we may not know which protocol a botnet is employing).

Another work, Protos [6], specifies a grammar for numerous protocols (HTTP, LDAP, WAP and SNMP), variations are then applied to the PDUs. They considered the IUT to have passed testing if it rejected the anomalous input without any undesired effects, such as illegal memory access or crashing. This approach inspired our work, where we seek to remove *a priori* assumptions made towards finding the variations of the sequence of commands (hence grammar), removing any biases or shortcomings unknowingly introduced by the designer.

Building off of Protos, [17] and [14] gather the protocol information not from knowing the grammar ahead of time and modifying, but from a live network connection. Their system then processes the packets, parses and mutates them in a user-specified manner. The resulting packet is then re-injected into the network stream, measuring the response of the IUT. Similar work with modifying PDUs is demonstrated in [19].

In typical syntax-based testing, one either starts with a defined grammar that one modifies and applies (as in [17]), or one gathers the PDU data from a network stream, modifies and applies (as in [17, 14, 6, 19]). In our work the modifications are motivated by the feedback our system *automatically* derives from the IUT. By using this $real - time$ feedback from the IUT, we seek to gain intelligence in how to modify the inputs to be able to communicate to the IUT based on the actual implementation, not a theoretical or claimed implementation, hence, unique to this field.

2.4 FTP

In this paper, we explore the protocol state of the FTP service for the purpose of selecting and exploring correct sequences of commands. The "correctness" of these sequences are defined in RFC 959 ([16]). In its most basic form, FTP is defined as a communication and file exchange service between a client and a server. FTP was chosen due to its well defined command and response structure, as well as prevalence in today's networks. Our system will interact with an actual FTP server in order to evaluate the "correctness" of any given solution. This "correctness" will be based on the response messages the server will send to our system during live interaction and influences how our system will vary future interactions. The commands that we implement are:

1. RETR - Retrieve a file from the server, store on the client
2. STOR - Store a file from the client, store on the server
3. STOU - Same as store, but name is generated unique to the transfer directory
4. CWD - Change the working directory
5. ABOR - Abort the previous command
6. DELE - Delete a file on the server side
7. RMD - Remove a directory
8. MKD - Make a directory
9. PWD - Print the working directory
10. LIST - List the directory contents
11. QUIT - Disconnect from the server

3 The Model

In this work the evolutionary process is driven by discovering legitimate command sequences. How well a given individual is performing is based on the feedback received from the FTP server using FTP's documented response code system, described in RFC [16]. Our system is based on an EC technique, namely genetic programming (specifically linear GP), as such it evolves a population of individuals, each representing a series of commands. The evaluation of each individual is done by executing these commands against an active FTP server with the response from each contributing to the overall performance (fitness). We are not only interested in finding solutions that discover sequences of commands, but also in promoting "unique" solutions - that is different sequences. As such, we examine an archive of individuals (solutions) [2]. The inclusion of an individual in the archive is based on its uniqueness when compared to others and as such, represents a selection of the best performers that are as unique as possible from each other.

The fitness evaluation is similar to that described in [12,11], where the overall fitness is evaluated based on the proper sequencing of given opcodes. In this work, however, the fitness is directly related to proper sequences of commands, minimizing illegal sequences. Here the fitness metric is a percentage, 100% indicating that a solution has achieved no error messages. Moreover, a command can result in no return code; in such a case, the individual's fitness will not be improved nor diminished as we consider this as a "neutral" response. Furthermore, a secondary metric, "uniqueness", is also used in evaluating an individual. In order to calculate the uniqueness of an individual, we augment the representation of each individual with what we call an individual's *fingerprint*. This *fingerprint* is defined as a vector that is the size of the number of opcodes present in the protocol being targeted, where each entry of the vector represents a counter of how many times that specific opcode was seen in the individual. This fingerprint can then be used to measure the uniqueness of an individual by calculating the euclidean distance between its own fingerprint and other individuals.

3.1   The Archive

Not only are we interested in sequences of commands that result in no error messages, but also in finding a subset of sequences that do so in a variety of methods. To archive this, we implement an archive of solutions based on the work presented in [2]. For an individual to be in the archive, they must be sufficiently unique when compared to individuals already present. The archive is populated with random individuals in the onset of the evolutionary process, at each generation, the individuals selected for the tournament are then compared to the archive; if they are more unique than an individual in the archive, they replace this individual. The result is a subset of the entire population that represents a diverse set of methods for achieving the desired goal.

3.2   Evolutionary Model

The work in this paper employs a page-based linear genetic programming learning model [13]. A population of individual solutions are randomly initialized in both size (number of commands) and content (command sequences) in a linear sequence [4, 5, 15]. At each generational stage, a small subset of the current population is randomly selected (the tournament), search operators are then applied; the best resulting individuals of the tournament then replace the worst and are placed back in the population. At this stage, the best individuals in the tournament are also examined for inclusion into the archive. Table 1 lists the specific model parameters used during the experiments. Similar to [8], the search operators employed in this work are:

- Crossover, Single Point: Single page from parent is swapped with child.
- Swap Selector: Two instructions are selected and swapped; individual length remains fixed.
- Instruction-Wise Mutation: Test for application of mutation for each instruction, if it is to be applied, another instruction from the complete set is chosen with uniform probability and used.

Table 1. Parameters for the Evolutionary Model

Parameter	Value	Parameter	Value
Population	1000	Mutation	0.5 with linear decay
Page Count	100	Swap	0.5
Page Size	6	Crossover	0.9
Tournament Size	4	Stop Criteria	10 000 Tournaments
Archive Size	100		

4   Experiments and Results

Two sets of experiments were performed towards the goal of exploring correct command sequences. The first set of experiments focuses on the FTP command set, rewarding the presence of positive response messages for a given sequence. The second set uses a larger set of commands by augmenting the first with those from SMTP. These experiments are designed to test our hypothesis that the system will be able to learn the relevant commands as was as where the non-protocol specific commands, in this case SMTP commands, can be used, or perhaps not use them at all.

4.1   Results

We present our findings for both sets of experiments, each having been run 30 times, each with different randomly selected initial seeds. Upon completion of 10 000 generations we report both our population and archive fitness (mean and best values). Figures 1 and 2 show all metrics at 1000 generation intervals for the duration of the experiments. In both figures the fitness level of the archive increases as the population does, but not necessarily at the same rate due to the fitness not being directly related to inclusion in the the archive. Table 2 lists the mean and the best fitness for the population and the archive for both sets of experiments. The fitness percentage is based on number of commands and the returned responses from the server in testing, 100% being no negative responses were received, 0% all commands resulted in negative responses from the server.

Table 2. Results

	FTP Only Commands		FTP and SMTP Commands	
	Mean Fitness	Best Fitness	Mean Fitness	Best Fitness
Population	96.642%	97.918%	96.359%	97.436%
Archive	93.755%	97.344%	93.601%	96.783%

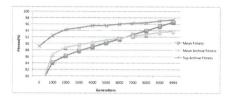

Fig. 1. FTP only Experiment

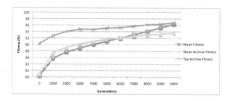

Fig. 2. FTP + SMTP Experiment

4.2   Discussion

Our experiments succeed at accomplishing high performance rates, as well as diverse sets of solutions in the archive. The learning curves of both sets of experiments are consistent, indicating a relatively similar increase in performance over each generation of evaluation. In both cases a few instances in the archive already have achieved a greater fitness then the average population individual at early stages, indicating that some individuals do relatively well early on in the evolutionary process. We can also see, however, that these initial "good individuals" do not represent a "best case" scenario, as the fitness for these do improve over each generation.

Individual Example 1 (from Archive)

1: put RNTO:
2: ls
3: delete MAIL FROM:
4: cd
5: rename FILENAME FILENAME2
6: cd ..
7: cd ..
8: status
9: status
10: mkdir DIR
11: status
12: system
13: ls
14: QUIT

We also remark that at approximately the 6000th generation the general populations fitness surpasses that of the mean archive fitness. This indicates that although the population's mean fitness is improving, it is doing so with a set of individuals that are becoming less diverse. This demonstrates that past 6000 generations the solutions are converging towards a smaller set of representations to achieve high performance.

Given the added complexity of having the larger command set in the second set of experiments, yet still a similar learning curve as the first set, it is worth exploring solutions. In Example 1, we list a sample solution (shortened for brevity). Here we see the individual does have non FTP commands remaining, however, they are placed in locations that are interpreted as arguments to the FTP commands (lines 1 and 3, for example). This results in an FTP server having numerous directories and file names that match SMTP opcode names, demonstrating an important artifact of our learning mechanism; the system learns where it can use non FTP commands such that they do not lower the fitness value, i.e do not produce error messages. The system is demonstrating its ability to minimize error feedback from the server under testing, hence learning to place non compliant commands as arguments.This example solution demonstrates two key features of our work:

1. The solution is human readable, hence execution can be tested by simply executing the individual verbatim on an FTP server.
2. The solution contains FTP commands primarily, SMTP commands (not the protocol of the server being tested) can be seen to exist mainly as arguments to the correct protocol commands.

Both these features are desired, even if the second was not predicted by our initial hypothesis, it nonetheless allows us to determine the IUT by looking at the commands that only exist in the first position per line (i.e not the argument list). Our system also helps identify commands that exist in multiple protocols (such as the case of "QUIT" in the above example, line 14 above), which could be useful in identifying similarities in protocols.

Relevance to our work towards achieving a variety of solutions is the makeup of the archive solutions themselves. In order to aid in this analysis we will focus on 5 common "tasks" of an FTP server (and the related command): 1) store a file (STOR); 2) retrieve a file (RETR); 3) change the working directory (CWD); 4) delete a file (DEL) and 5) rename a file (RNFR). In Table 3 we list the percentages, on average, per individual

Table 3. Precence of specific commands in the archive, per individual

Command	Percentage	Average Count
STOR	5.15%	6.51
RETR	5.03%	6.35
CWD	5.52%	6.98
DEL	5.27%	6.66
RNFR	4.98%	6.29

that these commands appear (as well as the average count), note that the remaining commands make up the remaining percentages. Given these findings, we determine that our solutions are indeed discovering how to perform these tasks against the server, fairly evenly as well. In order to look at the diversity of how these tasks have been accomplished we looked at the "lead in" commands, i.e the commands in the individual directly before the above listed ones. In this case, we show in examples 2 and 3 the two commands directly before the first instance of DEL (delete) from different solution individuals. These examples show how our system is discovering a variety of methods to achieve goals on the server.

Example 2 (from Archive)

1: rmd HELO
2: rmd DIRNAME
3: delete FILENAME1

Example 3 (from Archive)

1: cd
2: status
3: delete FILENAME

5   Conclusion and Future Work

In this work, we present a novel approach to network protocol testing. By using a machine learning based system in order to interact with an IUT, our system learns what commands (and sequences) are applicable with minimal *a priori* information. To test our system, we conducted two sets of experiments, the first with protocol specific commands, focusing on discovering valid sequences, the second with an augmented command set, testing the ability to discover which commands are applicable. Both achieve successful fitness levels, 97.918% in the first set, 97.426% in the second. Moreover, our system discovers diverse solutions, as seen by commands for common tasks having similar representation in the archive. Our system also demonstrated the additional ability to discover the correct use of non protocol appropriate commands as arguments to correct commands in an effort to reduce negative feedback from the IUT, as well as identifying commands that exist in multiple protocols.

Our proposed system reduces *a priori* information required in protocol testing by showing its ability to discover the relevant commands via *live* interaction over a network connection. This removes the requirement of having direct access to the binary of the IUT, of knowing the specific command syntax of the IUT, or for that matter, a complex understanding of the IUT before testing commences. For the purpose of testing this work, we focused on FTP as the IUT, and augmented the command set with SMTP. For future work, we plan to test against other network protocols with larger command sets. We hypothesize our system will remain successful in such cases, however, potentially taking more generations of evolution to achieve similar results. Further experiments will also be made comparing varying implementations of the same protocol towards identifying the differences in design, which will lead to further detailed analysis of the archive of solutions. It is our belief that a system of this manner would not only be valuable in protocol analysis but also in protocol testing and verification.

References

1. Caballero, J., Yin, H., Liang, Z., Song, D.: Polyglot: Automatic extraction of protocol message format using dynamic binary analysis. In: Proceedings of the 14th ACM Conference on Computer and Communications Security, p. 329. ACM (2007)
2. Doucette, J., Heywood, M.I.: Novelty-Based Fitness: An Evaluation under the Santa Fe Trail. In: Esparcia-Alcázar, A.I., Ekárt, A., Silva, S., Dignum, S., Uyar, A.Ş. (eds.) EuroGP 2010. LNCS, vol. 6021, pp. 50–61. Springer, Heidelberg (2010)
3. Froese, T.: Steps toward the evolution of communication in a multi-agent system. In: Symposium for Cybernetics Annual Research Projects, SCARP 2003. Citeseer (2003)
4. Heywood, M.I., Nur Zincir-Heywood, A.: Dynamic page based crossover in linear genetic programming. IEEE Transactions on Systems, Man, and Cybernetics: Part B - Cybernetics 32(3), 380–388 (2002)
5. Huelsbergen, L.: Toward simulated evolution of machine language iteration. In: Koza, J.R., Goldberg, D.E., Fogel, D.B., Riolo, R.L. (eds.) Proceedings of the First Annual Conference on Genetic Programming 1996, July 28-31, pp. 315–320. Stanford University, MIT Press, CA, USA (1996)

6. Kaksonen, R., Laasko, M., Takanen, A.: Vulnerability analysis of software through syntax testing. University of Oulu, Finland, Tech. Rep. (2000)
7. Gunes Kayacik, H., Heywood, M.I., Nur Zincir-Heywood, A.: Evolving Buffer Overflow Attacks with Detector Feedback. In: Giacobini, M. (ed.) EvoWorkshops 2007. LNCS, vol. 4448, pp. 11–20. Springer, Heidelberg (2007)
8. Gunes Kayacyk, H., Nur Zincir-Heywood, A., Heywood, M.: Evolving successful stack overflow attacks for vulnerability testing. In: 21st Annual Computer Security Applications Conference, ACSAC 2005, pp. 225–234. IEEE Computer Society (December 2005)
9. Khasteh, S.H., Shouraki, S.B., Halavati, R., Khameneh, E.: Evolution of a communication protocol between a group of intelligent agents. In: World Automation Congress, WAC 2006, pp. 1–6. Citeseer (2006)
10. Khasteh, S.H., Shouraki, S.B., Halavati, R., Lesani, M.: Communication Protocol Evolution by Natural Selection. In: 2006 and International Conference on Intelligent Agents, Web Technologies and Internet Commerce, Computational Intelligence for Modelling, Control and Automation, p. 152 (2006)
11. LaRoche, P., Nur Zincir-Heywood, A., Heywood, M.I.: Evolving tcp/ip packets: A case study of port scans. In: CDROM: IEEE Symposium on Computational Intelligence for Security and Defense Applications (2009)
12. LaRoche, P., Nur Zincir-Heywood, A., Heywood, M.I.: Using Code Bloat to Obfuscate Evolved Network Traffic. In: Di Chio, C., Brabazon, A., Di Caro, G.A., Ebner, M., Farooq, M., Fink, A., Grahl, J., Greenfield, G., Machado, P., O'Neill, M., Tarantino, E., Urquhart, N. (eds.) EvoApplications 2010. LNCS, vol. 6025, pp. 101–110. Springer, Heidelberg (2010)
13. LaRoche, P., Nur Zincir-Heywood, A., Heywood, M.I.: Exploring the state space of an application protocol: A case study of smtp. In: 2011 IEEE Symposium on Computational Intelligence in Cyber Security (CICS 2011), pp. 152–159 (April 2011)
14. Marquis, S., Dean, T.R., Knight, S.: Scl: a language for security testing of network applications. In: CASCON 2005: Proceedings of the 2005 Conference of the Centre for Advanced Studies on Collaborative Research, pp. 155–164. IBM Press (2005)
15. Nordin, P.: A compiling genetic programming system that directly manipulates the machine code. In: Kinnear Jr., K.E. (ed.) Advances in Genetic Programming, ch. 14, pp. 311–331. MIT Press (1994)
16. Postel, J., Reynolds, J.: File Transfer Protocol. RFC 959 (Standard), Updated by RFCs 2228, 2640, 2773, 3659, 5797 (October 1985)
17. Tal, O., Knight, S., Dean, T.: Syntax-based vulnerability testing of frame-based network protocols. In: Proc. 2nd Annual Conference on Privacy, Security and Trust (2004)
18. Wondracek, G., Comparetti, P.M., Kruegel, C., Kirda, E., Anna, S.S.S.: Automatic network protocol analysis. In: Proceedings of the 15th Annual Network and Distributed System Security Symposium, NDSS 2008. Citeseer (2008)
19. Xiao, S., Deng, L., Li, S., Wang, X.: Integrated tcp/ip protocol software testing for vulnerability detection. In: 2003 International Conference on Computer Networks and Mobile Computing, ICCNMC 2003, pp. 311–319. IEEE (2003)

Evolutionary Design of Active Free Space Optical Networks Based on Digital Mirror Devices

Steffen Limmer[1], Dietmar Fey[1], Ulrich Lohmann[2], and Jürgen Jahns[2]

[1] University of Erlangen-Nuremberg, Martensstr. 3, 91058 Erlangen, Germany
{steffen.limmer,dietmar.fey}@informatik.uni-erlangen.de
[2] University of Hagen, Universitätsstr. 27/PRG, 58097 Hagen, Germany
{ulrich.lohmann,jahns}@fernuni-hagen.de

Abstract. Optical connections have several advantages compared to conventional electrical connections, especially a higher attainable bandwidth. While long distance optical connections are already established, optical board- and chip-level connections are still a subject of current research. In this paper we describe a new setup for optical board-level connections which is based on free space optics and allows the switching of signals within the optical domain. We describe the evolutionary optimization of design parameters for the proposed setup, done with a memetic evolutionary algorithm and present the optimization results.

Keywords: Evolutionary algorithm, digital mirror device, PIFSO, optical connection, CMA-ES.

1 Introduction

The data rates, required for future internet applications, like streaming video or HD-TV on web, currently drive the research activities in the field of optical communication, because optical technology promises much higher bandwidth than conventional copper based interconnections. A further advantage of optical interconnect technology is the possibility of 3D-interconnections, that means, "light" enables a full three dimensional interconnection scheme, which can help to solve the interconnection bottleneck of complex topologies, like high dimensional crossbar architectures with up to 32 channels.

Industrial realizations of complex optical interconnection schemes as an integrated system are still missing, so this work shows a possible design to build up a robust planar integrated free space optical (PIFSO) [1] approach for a high parallel crossbar interconnection. Based on a combination between free space optics, micro-electro-mechanical systems (MEMS) and a new multi-fiber optical interface device (called fibermatrix), the described system represents a 16×16 optical crossbar-connection with a total data rate of up to $160\ Gb/s$. Yeow et al. already demonstrated the ability of MEMS-based components for switching [2].

In [3] we described the evolutionary design of interconnects for Clos networks in PIFSO technology. We were able to find a layout, allowing the connection

C. Di Chio et al. (Eds.): EvoApplications 2012, LNCS 7248, pp. 21–30, 2012.
© Springer-Verlag Berlin Heidelberg 2012

of 256 inputs with 256 outputs. But these optical connections are passive - the switching must be done in the electronic domain. Thus, in a multi stage network there must be a conversion between the optical and electronic medium for every stage. This leads to a high power consumption.

Now we focus on the design of an active optical network respectively crossbar connection. The idea is to combine the PIFSO approach with digital mirror devices (DMDs) (Figure 1(a)). In- and outputs for light beams are brought onto the surface of a substrate, like SiO_2, with help of so called fibermatrix connectors (Figure 1(b)) which allow an individual positioning. With help of fibers, light is transported to the inputs where it is split up into several beams with the help of splitting devices, like diffractive gratings or refractive microlens arrays (Figure 1(c)). The beams enter the substrate and hit a DMD that is located at the bottom of the substrate. The DMD is a 1 cm^2 sized array of 1000×1000 small mirrors. The mirrors can be set independently in two different positions (respectively two different tilt angles). The objective is to switch incoming beams to outputs by switching the mirrors of the DMD. This requires on one side that the beams of one input can be reflected to the outputs by mirrors with certain tilt angles. On the other side, these beams must not hit any output if the mirrors are set to the opposite tilt angle.

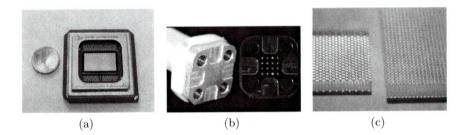

(a)   (b)   (c)

Fig. 1. (a) A digital mirror device. An array of mirrors is located on its surface. These mirrors can be switches independently in two different positions with help of electrostatic fields. (b) Fibermatrix connectors with 16 fibers. It is possible to position the fibers in another layout, than the shown one. (c) Microlens arrays which can be used for beam splitting.

The advantages of this approach is that the PIFSO technology is very robust and well to integrate and to produce [4]. Unlike the conventional electronically solutions, the free space optical approach is free of any crossing problems at such high number of channels. Additionally, DMDs are low-cost mass products (for example used in beamers). Thus, the setup is well suited for the practical implementation.

The question is how to place the in- and outputs and how to split the light beams exactly in order to make it possible to switch as many inputs to as many outputs as possible. This is a difficult task that we were not able to solve analytically. For that reason, we decided to employ an evolutionary algorithm (EA).

2   Optimization Problem

Figure 2 shows one beam of the network. The beam enters the substrate at a point P_1 with the entrance angles β_1 in the x-z plane and γ_1 in the x-y plane. At the point P_2 located at the bottom of the substrate it is reflected by a mirror with a tilt angle α (-12° or +12°) in the y-z plane. This results in new angles β_2 and γ_2. At the point P_3 the beam leaves the substrate.

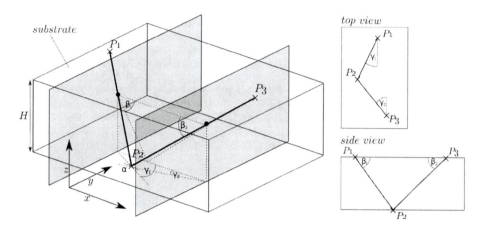

Fig. 2. One beam of the network from three different perspectives. It enters the substrate at point P_1, at P_2 it is reflected by a mirror with tilt angle α and leaves the substrate at point P_3. The gray planes are just indicated for the purpose of illustrating the different angles.

For given P_1 (x_1, y_1, z_1), β_1 and γ_1, the coordinates x_2 and y_2 of P_2 can be calculated as follows:

$$x_2 = x_1 - \frac{H}{\tan(\beta_1)} \tag{1}$$

$$y_2 = y_1 - \frac{H}{\tan(\beta_1)} \cdot \tan(\gamma_1) \tag{2}$$

Where H is the height of the substrate ($z_1 = z_3 = H$ and $z_2 = 0$). Depending on α, the new angles β_2 and γ_2 are calculated as follows:

$$\beta_2 = \arctan(\tan(\beta_1) \cdot (-\cos(2\alpha)) + \tan(\gamma_1) \cdot (-\sin(2\alpha))) \tag{3}$$

$$\gamma_2 = \arctan(\tan(\beta_1) \cdot (-\sin(2\alpha)) + \tan(\gamma_1) \cdot (\cos(2\alpha))) \tag{4}$$

The coordinates x_3 and y_3 of P_3 arise from the following equations:

$$x_3 = x_2 + \frac{H}{\tan(\beta_2)} \tag{5}$$

$$y_3 = y_2 + \frac{H}{\tan(\beta_2)} \cdot \tan(\gamma_2) \tag{6}$$

The given equations can be rearranged to compute β_1 and γ_1 for given in- and output points P_1 and P_3 and a given tilt angle α. For one tilt angle and one in- and output combination there is only one possible beam that is reflected by a mirror with the specified tilt angle and thereby connects the in- with the output. Or to put it differently: for given P_1, P_3, and α the point P_2 is uniquely defined. Since there are only two possible tilt angles (-12° and +12°), there are always only two possible beams that can connect an in-/output pair. That means, for a given in-/output pair there exist only two valid combinations of entrance angles β_1 and γ_1.

To make the splitting of an incoming beam into the multiple beams within the network technically possible, the entrance angles have to meet the following condition: The step sizes between the angles β_1 of the beams of one input have to be as equal as possible. The same holds for the angles γ_1.

That means, if there are n beams $b_1, \ldots, b_n$ to connect a certain input with all the outputs and $(\beta_1^1, \ldots, \beta_1^n)$ and $(\gamma_1^1, \ldots, \gamma_1^n)$ are the sequences of the entrance angles with β_1^i and γ_1^i belonging to beam b_i and $\beta_1^i \leq \beta_1^j$ for $i \leq j$, then the following equations should hold:

$$(\beta_1^2 - \beta_1^1) = (\beta_1^3 - \beta_1^2) = \ldots = (\beta_1^n - \beta_1^{n-1}) \tag{7}$$

$$(\gamma_1^2 - \gamma_1^1) = (\gamma_1^3 - \gamma_1^2) = \ldots = (\gamma_1^n - \gamma_1^{n-1}) \tag{8}$$

Besides this condition, it is necessary to regard some geometrical constraints: Since a beam has a certain diameter, it hits up to 10×10 mirrors (allocating an area of about $0.1 \cdot 0.1\ mm^2$). One mirror should not be hit by two beams, otherwise it would not be possible to switch the two beams independently from each other. So if the points P_2^i and P_2^j on the bottom of the substrate belong to two beams b_i and b_j of the system, then there should be a distance of at least $0.1\ mm$ between these points. Furthermore, it is required for the realization that there is a distance of at least $0.25\ mm$ between all the in-/output positions (this requirement follows from the characteristics of the fibermatrix connectors which are intended to be used).

Hence, the problem is to find a layout of the network that fulfills all the stated conditions. The idea is to find suitable in-/output positions (with distances of at least $0.25\ mm$ among them) that make it possible to connect them with beams that fulfill the restrictions on the entrance angles and the positions P_2. As already mentioned, we were not able to solve this task analytically. Instead, we used an evolutionary algorithm (EA) to search such points. This EA is described in the following section.

3   Optimization Algorithm

An individual of the EA represents a placement of all in- and outputs. The placement of one in- respectively output is encoded as two real numbers specifying the x and y position of the in-/output. For the implementation double precision variables are used for the real values. Thus, an individual for an n×n network consists of 4n double variables.

Before a fitness is assigned to an individual, all beams for the encoded in-/output placement are computed. There are at most two viable beams for a connection of an input with an output: a beam reflected by a mirror with +12° and a beam reflected by a mirror with -12°. For each in-/output combination the two possible beams are computed and one of them is selected for the connection. Of course, if one of the two beams does not hit the mirror array, then it is invalid and will not be selected. If both beams for an in-/output combination are invalid, the combination is marked as invalid - there is no possibility to connect it with a beam. If both beams are valid, the connecting beam is chosen randomly (see Figure 3).

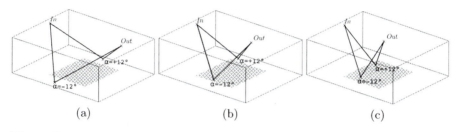

(a)   (b)   (c)

Fig. 3. Illustration of the three possible relations between an in- and an output position. (a) Both beams that can be used to connect the in- with the output do not hit the DMD. Thus, both beams can not be realized and the input can not be connected with the output - the in-/output pair is invalid. (b) The input can be connected with the output by only one beam hitting the DMD (the beam reflected by $\alpha = -12°$). The other beam, reflected by $\alpha = +12°$ does not hit the DMD. (c) The in-/output pair can be connected by two beams hitting the DMD.

The result of the beam computation is used for the fitness assignment: Let $\mathfrak{I}$ be an individual encoding the input positions

$$I_1, ..., I_n = (x_i^1, y_i^1), ..., (x_i^n, y_i^n)$$

and the output positions

$$O_1, ..., O_n = (x_o^1, y_o^1), ..., (x_o^n, y_o^n)$$

of an n×n network. For each in-/output pair (I_k, O_l) a connecting beam b_{kl} with the corresponding entrance angles β_1^{kl} and γ_1^{kl} and the point $P_2^{kl} = (x_2^{kl}, y_2^{kl})$ on the bottom of the substrate is computed in the way described above. For the Individual $\mathfrak{I}$ we define the partial fitness value $f_1(\mathfrak{I})$ as the number of pairs of in- respectively output positions with a distance less than 0.25 mm between them and the partial fitness value $f_2(\mathfrak{I})$ as the number of pairs of points P_2^{kl} with a distance less than 0.1 mm between them. Furthermore we define the third partial fitness value $f_3(\Im)$ as ten times the number of invalid in-/output pairs. An in-/output pair that can not be connected, can be seen as worst case. That is the reason why we use a penalty of 10 in f_3, instead of 1 like in f_1 and f_2. Without loss of generality let us assume that for all inputs I_k, $k = 1, .., n$, the sequence of angles $\beta_1^{k1}, ..., \beta_1^{kn}$ is ordered in increasing order. Then we can define the last partial fitness $f_4(\Im)$ as

Definition 1

$$f_4(\Im) = \sum_{k=1}^{n} \sum_{l=1}^{n-1} (|\Delta_\beta(k) - (\beta_1^{k(l+1)} - \beta_1^{kl})| + |\Delta_\gamma(k) - (\gamma_1^{k(l+1)} - \gamma_1^{kl})|)$$

with

$$\Delta_\beta(k) = \frac{\sum_{l=1}^{n-1} (\beta_1^{k(l+1)} - \beta_1^{kl})}{n-1}$$

and

$$\Delta_\gamma(k) = \frac{\sum_{l=1}^{n-1} (\gamma_1^{k(l+1)} - \gamma_1^{kl})}{n-1}$$

$\Delta_\beta(k)$ and $\Delta_\gamma(k)$ are the average pitches between the entrance angles of the beams of input k. So the smaller $f_4(\Im)$, the better equations 7 and 8 are fulfilled. The total fitness is set to the sum of all partial fitnesses:

Definition 2

$$fit(\Im) = f_1(\Im) + f_2(\Im) + f_3(\Im) + f_4(\Im)$$

The smaller the fitness value, the better is the individual. The best achievable fitness is zero. A fitness value smaller than one, indicates that the partial fitness values f_1, f_2 and f_3 are zero. That means, the in- and outputs can be connected with valid beams and there are no geometrical conflicts.

The evolutionary operators we have used, are derived from the memetic algorithm MA-LSCh-CMA proposed by Molina et al. [6]. For selection, negative assortative mating [7] is used: The first parent individual is selected randomly. Then five other individuals of the population are chosen randomly and from these individuals that one with the highest euclidean distance to the first parent is selected as second parent. As variation operators BLX-0.5 crossover [8] and the mutation operator from the Breeder Genetic Algorithm [5] are used.

The replacement is done in a μ+1 scheme. In every generation *PopSize* times crossover and mutation is done iteratively, leading to one pair of offspring in each iteration. The best of the two produced offspring in terms of fitness replaces the so far worst individual in the population if it is better than that.

After each generation one individual of the generation is improved by local search. CMA-ES (Evolution Strategy with Covariance Matrix Adaptation) [9] is used as the local search strategy.

CMA-ES is an efficient local search algorithm for optimization problems in continuous domains. In 2004, Hansen and Kern demonstrated its competitive performance on eight multimodal test functions [10]. It was employed in a multi-start EA with increasing population size, called G-CMA-ES [11], which yielded the best results in the real-parameter optimization competition at the Congress of Evolutionary Computation 2005 [12]. In our optimization a (μ,λ)-CMA-ES is used. It iteratively creates λ individuals by sampling a multivariate normal distribution with mean value m, global step size σ and covariance matrix C:

$$x_i \sim N(m, \sigma^2 C) \text{ for } i = 1, ..., \lambda$$

After each iteration the new individuals are evaluated and m, σ and C are updated in order to guide the search in a better direction. The initial mean value m is the individual that should be improved by local search. The new m after each iteration is created from the λ current individuals by a weighted recombination of the μ best ones.

Algorithm 1. Overview Optimization Algorithm

Input: G (number of generations)

       P (population size)

Input: Pop (population)

1   **Initialization:**

1.1   Initialize and evaluate Pop. Set $g = 1$.

2   **Global optimization:**

2.1   If $g = G$ then stop, otherwise set $k = 1$.

2.2   Select two parents p1, p2 from Pop with negative assortative mating.

2.3   Create offspring o1 and o2 from p1 and p2 using BLX-crossover and mutation from Breeder Genetic Algorithm.

2.4   Evaluate o1 and o2 with fitness function.

2.5   Insert best individual from {o1,o2} in Pop if it is better than the worst individual in Pop.

2.6   If $k < P$ then set $k = k + 1$ and go to 2.2.

3   **Local improvement:**

3.1   Pick the best individual b from Pop to which CMA-ES was applied less then 30 times or which was improved at least once during the last 30 applications of CMA-ES to it.

3.2   Apply $2 \cdot P/\lambda$ iterations of (μ,λ)-CMA-ES to b.

3.3   Set $g = g + 1$ and go to 2.1.

For the internal parameters of the CMA-ES, like μ and λ, we used values as recommended by Hansen and Kern [10]. In each generation of the used EA CMA-ES is applied to the best individual in the population that fulfills one of the following two conditions:

(a) CMA-ES was applied to the individual less than 30 times.

(b) From the last 30 applications of CMA-ES to the individual, at least one led to an improvement.

In each application $2 \cdot PopSize/\lambda$ iterations of the local search are performed (thus, in the CMA-ES the same number of evaluations is performed like in the rest of the algorithm) and at the end the internal state of the CMA-ES (the step size σ and the covariance matrix C) is stored and reused for the next application of the CMA-ES to the individual, leading to a so called *LS-chain* according to Molina. An overview over the complete optimization algorithm is given in Algorithm 1

We also tried an EA without local search. It was able to find adequate solutions for networks of small dimensions (like 4×4) but not for higher ones. Figure 4 shows how the fitness changes when the position of the first inputs of random initialized individuals for the dimensions 4×4, 8×8 and 16×16 change. One can not only see that the average fitness increases (and so becomes worse) with increasing dimension, but also that the fitness landscape becomes more complex.

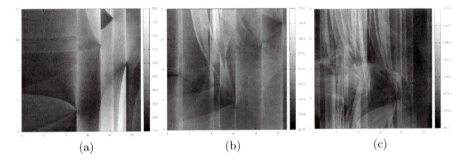

(a)   (b)   (c)

Fig. 4. Fitness as a function of the x and y position of the first input of a random initialized individual. The fitness is indicated by the intensity. (a) Dimension of 4×4. (b) Dimension of 8×8. (c) Dimension of 16×16.

4   Optimization Results

In order to get good optimization results we did 120 runs of the algorithm in parallel on a cluster consisting of 56 AMD Opteron 2.2 GHz Dual-Cores and four AMD Opteron 2.4 GHz Six-Cores. Then the best results of these runs are used as initial populations for further 120 runs and so on, until the results did not further improve. For one run of the algorithm we used a population size of 1000 and computed between 1000 and 4000 generations dependent on the intended dimension of the optical network. For all runs a mutation rate of 0.125 and a crossover rate of 0.9 was used.

Table 1 shows the best found fitness values for the dimensions from 8×8 to 16×16.

Although the results are not perfect in terms of fitness (that means a fitness of 0), they are still satisfying for the realization. For the dimension of 16×16 the maximum variation of a step size between entrance angles of two beams of an

Table 1. Fitness values of the best found individuals for the dimensions from 8×8 to 16×16

Dimension	8×8	9×9	10×10	11×11	12×12	13×13	14×14	15×15	16×16
Fitness	0.00005	0.00011	0.00093	0.00134	0.00775	0.00592	0.00401	0.02286	0.03364

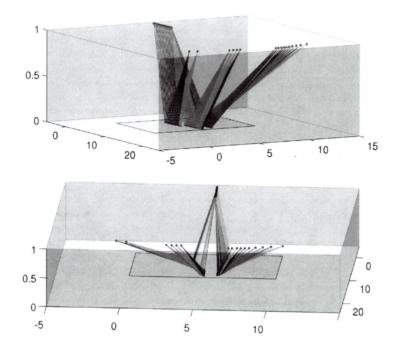

Fig. 5. The evolved setup for dimension 16×16 from two different perspectives. All beams from the inputs to the outputs are drawn in the figure. The inputs are located in the front and the outputs in the rear from the given view point. The specifically marked area at the bottom of the substrate represents the DMD.

input to the average step size for that input is 0.000763°. This is still acceptable. But it is unlikely that the dimension can be increased much more under the given conditions. The geometrical limits are nearly reached. Figure 5 shows a plot of the evaluated system for the dimension of 16×16.

5   Conclusion

Due to the small area of the DMD it was foreseeable that the possible dimension is comparatively small. But it is also possible to use the evolved system as a switch instead as a whole network or stage in a multi stage network. Furthermore, it is conceivable that the reachable dimension can be increased with other preconditions like larger DMDs. This is part of further work. It is important that it could be shown that it is theoretically possible to switch multiple inputs to multiple outputs with the proposed setup. The next step is to verify if it is also practically viable by implementing a prototype under consideration of the achieved optimization results. Furthermore it has to be investigated what is the best way to do the routing.

References

1. Jahns, J., Huang, A.: Planar Integration of Free-Space Optical Components. Applied Optics 28, 1602–1605 (1989)
2. Yeow, T.-W., Law, E., Goldenberg, A.: MEMS Optical Switches. IEEE Communications Magazine 39, 158–163 (2001)
3. Limmer, S., Fey, D., Lohmann, U., Jahns, J.: Evolutionary Optimization of Layouts for High Density Free Space Optical Network Links. In: GECCO 2011, pp. 1635–1642. ACM (2011)
4. Gruber, M.: Multichip Module with Planar-Integrated Free-Space Optical Vector-Matrix-Type Interconnects. Applied Optics 43(2), 463–470 (2004)
5. Mühlenbein, H., Schlierkamp-Voosen, D.: Predictive Models for the Breeder Genetic Algorithm: I. Continuous Parameter Optimization. Evolutionary Computation 1(1), 25–49 (1993)
6. Molina, D., Lozano, M., García Martínez, C., Herrera, F.: Memetic Algorithms for Continuous Optimization Based on Local Search Chains. Evolutionary Computation 18, 27–63 (2010)
7. Fernandes, C., Rosa, A.: A Study on Non-Random Mating and Varying Population Size in Genetic Algorithms Using a Royal Road Function. In: Proceedings of the 2001 Congress on Evolutionary Computation, pp. 60–66. IEEE Press (2001)
8. Eshelman, L.J., Schaffer, J.D.: Real-coded Genetic Algorithms and Interval-chemata. In: Whitley, L.D. (ed.) Foundations of Genetic Algorithms, vol. 2, pp. 187–202. Morgan Kaufmann, San Mateo (1993)
9. Hansen, N., Ostermeier, A.: Completely Derandomized Self-Adaptation in Evolution Strategies. Evolutionary Computation 9, 159–195 (2001)
10. Hansen, N., Kern, S.: Evaluating the CMA Evolution Strategy on Multimodal Test Functions. In: Yao, X., Burke, E.K., Lozano, J.A., Smith, J., Merelo-Guervós, J.J., Bullinaria, J.A., Rowe, J.E., Tiño, P., Kabán, A., Schwefel, H.-P. (eds.) PPSN VIII 2004. LNCS, vol. 3242, pp. 282–291. Springer, Heidelberg (2004)
11. Auger, A., Hansen, N.: Performance Evaluation of an Advanced Local Search Evolutionary Algorithm. In: Proceedings of the IEEE Congress on Evolutionary Computation, CEC 2005, pp. 1777–1784 (2005)
12. García, S., Molina, D., Lozano, M., Herrera, F.: A study on the use of non-parametric tests for analyzing the evolutionary algorithms behaviour: a case study on the CEC 2005 Special Session on Real Parameter Optimization. Journal of Heuristics 15(6), 617–644 (2009)

Frequency Robustness Optimization with Respect to Traffic Distribution for LTE System

Nourredine Tabia[1], Alexandre Gondran[2], Oumaya Baala[1], and Alexandre Caminada[1]

[1] UTBM, SeT, Thierry Mieg, 90010 Belfort, France
[2] ENAC, Edouard Belin, 31055 Toulouse, France
{nourredine.tabia,oumaya.baala,alexandre.caminada}@utbm.fr,
alexandre.gondran@enac.fr

Abstract. The Long Term Evolution (LTE) cellular network is based on Orthogonal Frequency Division Multiple Access (OFDMA) to meet several services and performance requirement. This paper shows the interest of robustness approach due to the uncertainty of traffic distribution while evaluating some antenna parameters. We use a greedy algorithm with different variants to show how a frequency parameter setting can impact the coverage performance indicator based on the SINR metric. The well-known frequency reuse schemes 1x3x3, whereby the entire bandwidth is divided into 3 non-overlapping groups and assigned to 3 co-site sectors within each cell, have been used in our model. Further work must be done on algorithmic approach.

Keywords: LTE, Robustness, SINR, Interference, Frequency, Optimization.

1   Introduction

The Long Term Evolution is a new air-interface designed by the Third Generation Partnership Project (3GPP) [6]. Its goal is to achieve additional substantial leaps in terms of service provisioning and cost reduction. OFDMA has been widely accepted as a promising technology for new generations [9]. This technique provides orthogonality between the channels [7]; it reduces interference and improves the network Quality of Service (QoS). Resource allocation in radio networks essentially depends on the quality of some reference signals received by the user equipment (UE). In LTE, they are the Reference Signal Received Power (RSRP) and the Reference Signal Received Quality (RSRQ) corresponding respectively to Received Signal Code Power (RSCP) and *Ec/No* in (UMTS Universal Mobile Telecommunications System). Each user is assigned a portion of the spectrum depending on RSRP and RSRQ. The more complex optimization of reference signal is the RSRQ which is based on SINR [8] [12]. SINR is an important performance indicator to estimate the achievable throughput from the interference received by the neighboring cluster of first-tier cells. The estimation and optimization of the SINR are well-known problems in radio communication systems such as 802.11, Global System for Mobile Communications (GSM) or UMTS [4], [1], and LTE needs also a good estimation and control of SINR.

C. Di Chio et al. (Eds.): EvoApplications 2012, LNCS 7248, pp. 31–41, 2012.
© Springer-Verlag Berlin Heidelberg 2012

Optimizing antenna parameters configuration is one of main targets. It can significantly improve the coverage and the capacity of the network dealing with the lack of available bandwidth in eNB (evolved Node Base). Several studies have been done in this direction to understand the impact of parameters on antennal QoS offered by the network [12], [10] and [3]. In [3] the authors study the impact of azimuth and tilt inaccuracies on network performance considering three main quality parameters: service coverage, soft handover and the ratio of chip energy to interference *Ec/No*. The approach of simulated annealing or evolutionary computation is used in [10] [1] to study network configuration parameters (CPICH power, down tilt and antenna azimuth) effects toward coverage service in UMTS network. Other approaches for frequency assignment are available at http://fap.zib.de/biblio/. In LTE various combinations of antenna have been studied in term of SINR and throughput performance [12] [2]. However, there is no work on robust optimization for LTE. In this paper, we study the influence of the frequency as preliminary study on some performance metrics (e.g. SINR, coverage) and also, the interest of robust optimization for LTE network. The choice of the robust approach is mainly due to the uncertainty of the traffic distribution and its advantage is to tackle this uncertainty among several traffic scenarios. For this aim, the paper is structured as follows. Section 2 introduces the system model and basic assumptions. Section 3 extends this work and shows the performance metrics and test assumptions. Section 4 presents some results to highlight utility of robust optimization toward the uncertainty of the traffic demands with LTE online optimization. Conclusions are drawn in section 5.

2   Case Study and System Model

2.1   Case Study

The considered network for this study consists of tri-sectors sites in one real city. The service area is a 40kmx20km area with industrial zones. For our model the service area is divided into a grid of equally sized test points. A test point is a 25x25 meters. Due to the very small size of the test point, we assume the same signal propagation conditions within a test point. It is characterized by its number of users and the category of required services for each user (e.g., voice, data). Each sector in the network is equipped with one directive antenna and each antenna is characterized by its parameters: radiation pattern, azimuth, tilt, frequency and output power in downlink. Due to the dynamic aspect of the network and changes in traffic demand, we use the concept of traffic scenarios. A scenario is a given distribution and load of the traffic demand at a given time for each test point. Then the scenarios allow us to compute different situations of network performance to study the robustness problem.

2.2   Basic Assumptions

In this paper, we consider the downlink transmission and illustrate the interference schemes using a theoretical model of seven-cell hexagonal layout as shown in Fig. 1. Three sectors are considered in each site with three eNBs. In Fig. 1 we see the frequency reuse 1x3x3 pattern where one site with three sectors uses three frequency sets. In our real network, cells are not hexagons; the sub-band assignment depends on the azimuth orientation of the sectors. The features of our computational model are the following: 1) Intra-frequency interference is avoided due to the use of OFDMA technique in downlink transmission. In LTE the orthogonality between subcarriers insures that the interference inside the cell can be ignored. 2) The basic resource element in OFDMA is the physical resource block (PRB) which spans both frequency and time dimensions. Here, we do not take into account PRB to estimate the inter-cell interference because it needs huge simulation; we only focus on the frequency sub-band reuse scheme. Two adjacent cells are scrambling each other if they are using the same sub-band to transmit data. It gives a worst but fast estimation of SINR.

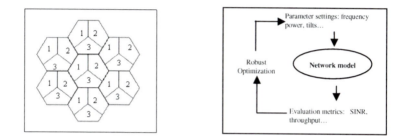

Fig. 1. Reuse 1x3x3 seven-cell hexagonal layout   **Fig. 2.** Online optimization process

2.3   Problem Formulation

The global objective of this study is to propose a methodology to change automatically some antenna parameters settings (power output, tilts, frequency allocation) so that the network will be more responsive to changes in traffic and the environment. The online optimization process is depicted on Fig. 2, where the overall optimization process is presented. The system operator acts on it via the antenna parameters. These parameters are called decision variables of the problem. From the input data, the network model provides an assessment of service quality using endpoints such as the level of interference and the bandwidth required to absorb the flow required by the users. The calculation in this level needs to know the real state of the traffic. Thus the decision variables depend on the considered traffic scenario. The proposed global objective of robust optimization is to minimize the lack of bandwidth of the network for all traffic flow requested. For the deployment of the network, each test point is associated with the eNB according to the quality of the received SINR. The interference model based on SINR is thus calculated as defined in Eq. (1). The detail on the production of this Eq. is given in [11].

$$\gamma_{b,t,s} = \frac{p_{b,t}^{R} f_b}{\sum_{b' \neq b, f_b = f_{b'}} p_{b',t}^{R} f_{b'} \delta_{b,s} \delta_{b',s} + n_0 w} > \gamma^{MIN} \qquad (1)$$

Where, $\gamma_{b,t,s}$ is the SINR received by the test point t and issued from the eNB b in scenario s; f_b and $f_{b'}$ are the frequencies used by eNB b and b' respectively; $\delta_{b,s}$ and $\delta_{b',s}$ are the load factors corresponding to eNBs b and b' in scenario s. The term w represents the total bandwidth used by all eNBs and n_0 is the thermal noise over the considered bandwidth. The terms $p_{b,t}^R$ and $p_{b',t}^R$ are the end power received by UE located in test point t from respectively b and b'. The right part of Eq. (1) γ^{MIN} is the SINR threshold required for the test point to establish a communication; below this value, the users within the given test point are considered as non-covered users. The full problem formulation is given by the following sets of data, parameters and functions. Let: $B=\{1,...,n^B\}$ the set of n^B eNBs of the network; $T=\{1,...,n^{TP}\}$ the set of test points of the map; and $c_{t,s}$ the number of users located on the test point t in scenario s. We need to use the network return to determine the values of the following decision variables.

Decision Variables (parameter settings)

p_b^E is the output power of the eNB b : $p_b^E \in P_b$; t_b^E is the electrical tilt of the eNB b : $t_b^E \in T_b$; $f_{b,n}$ is the variable for carrier assignment to eNB : $f_{b,n} \in \{0;1\}$

Where, P_b and T_b the sets of possible values of the power output and tilt respectively.

Fitness Functions for Robustness: We aim at minimizing one of the following:

Mean: $\displaystyle\sum_{s \in S}(p_s \times \Delta_s)$	Standard deviation: $\displaystyle\sum_{s \in S} p_s \left[\left(\sum_{s' \in S} \Delta_{s'} / n^s - \Delta_s\right)\right]^2$
Absolute robustness: $\displaystyle\sum_{s \in S}(p_s \times \Delta_s)$	Absolute deviation: $\displaystyle\max_{s \in S}\left(p_s\left(\Delta_s^* - \Delta_s\right)\right)$

Where, Δ_s^* is the optimum of Δ_s in the scenario s. This measure therefore requires solving n^s problems in advance. p_s is the probability of using the scenario s.

Where, Δ_s is the lack of bandwidth expressed in Hz required by the network to drain all traffic flow in scenario s.

$$\Delta_s = \max_{b \in S}(\max(0, \Delta_{b,s})) \tag{2}$$

Where, $\Delta_{b,s}$ is the difference between the necessary bandwidth of the eNB b to drain the requested flow of data in the scenario s, and the total available bandwidth w.

$$\Delta_{b,s} = w_{b,s}^S - w \tag{3}$$

The term $w_{b,s}^S$ is the necessary bandwidth to satisfy all the users in scenario s. and w is the total available bandwidth in each eNB. If $\Delta_{b,s} < 0$, then all users associated to the eNB b are satisfied. If $\Delta_{b,s} >= 0$, some users are not satisfied (lack on bandwidth).

Constraints: The main constraints of our model are:

(C1) $\forall s \in S, n_{0,s}^C \leq n_0^C$: the number of non covered users in scenario s should not exceed the threshold n_0^C

(C2) $\forall b \in B, v_b^{MIN} \leq |v_b| \leq v_b^{MAX}$: minimum and maximum of neighbourhood cells for b.

(C3) $\forall t \in T, \displaystyle\sum_{b \in B} u_{b,t} \leq 1$: a test point is associated with exactly one eNB.

Where, $u_{b,t} = \begin{cases} 1 \ if \ t \ is \ associated \quad with \ the \ eNB \ b \\ 0 \ otherwise \end{cases}$

(C4) $\forall b \in B, \displaystyle\sum_{n \in N} f_{b,n} \leq 1$: one eNB b can use only one carrier n.

In the current work, we will not consider all the parameter settings of the antenna. We limit our study to the impact of the frequency parameter on the number of clients not covered by the network in the service area. Other parameters will be added later. The robust approach uses the mean robustness over three different demand scenarios in a traffic day. The proposed evaluation methodology aims to show the effect of the antenna frequency parameter on non covered users with respect to traffic distribution. For the study presented in this paper, the overall problem is reduced to the following.

Decision Variable (frequency assignment): $f_{b,n}$: frequency assignment of the carrier n to the eNB : $f_{b,n} \in \{0,1\}, n = 1,2,3$

Constraints: We keep the constraints (C3) and (C4).

Fitness Function: Let $n_{0,s}^C$ be the number of non covered users in scenario s where $n_{t,s}^C$ is the number of non-covered users in test point t for scenario s.

$$n_{0,s}^C = \sum_{t \in T_0^C} n_{t,s}^C \tag{4}$$

Robustness Function: $f^{Rob} = \displaystyle\sum_{s \in S} n_{0,s}^C \tag{5}$

Where, f^{Rob} is the sum of non-covered users in all scenarios.

3   Assumptions and Performance Metrics

We aim at evaluating the SINR model to identify where the assigned frequency presents a remarkable increase of covered users with respect to traffic distribution.

The main parameters and assumptions we used are those selected by 3GPP for LTE as shown in Table1. Evaluations are performed by a static snapshot of the network level. In addition to Table 1, we assume that the antennas are grouped by site and stored on the basis of an index in ascending order of the x-axis. Two intermediate performance metrics are used for the deployment of the network:

Table 1. Test assumptions for LTE downlink

Parameters	Simulation setting	Parameters	Simulation setting
Network layout	35 sites 88 sectors	TX power range	[36 dBm, 46 dBm]
System frequency	1800 Mhz	Mechanical tilt range	[0°,-6°]
System bandwidth	20 Mhz	Electrical tilt range	[0°,-10°]
Required service/user	2 Mbps	Azimuth range	[0°,360°]
Frequency reuse factor	1x3x3	Horizontal HPBW	+70°
eNB heights range	[17m, 46m]	Vertical HPBW	+10°
UE height	1.5 m	Antenna gain range	[14dBi , 18.9dBi]
Propagation loss model	Hata model [5]	Traffic distribution	Distribution in proportion to UMTS traffic load

Signal-to-Interference Plus Noise Ratio: The SINR, expressed in Eq. (1), is an important indicator to evaluate cellular networks. It is motivated by the fact that it takes into account all the parameters of the antenna, it depends on the traffic distribution of the network, resizes the network and determines which eNB controls each user and also, allows us to estimate the total throughput of the network.

Load Factor: The Load Factor of the sector/cell is the ratio between the total allocated bandwidth to the cell and the maximum bandwidth available in the cell. Let $\delta_{b,s}$ be the load factor, then: $\delta_{b,s} = w_{b,s}^{s} / w$ where, $w_{b,s}^{s}$ is the total allocated bandwidth to the eNB b in the reference scenario s, and w is its maximum available bandwidth. It is worthwhile to mention that load factor is one of the main key indicators. The downlink cell load for a stable network should not exceed 70% [10].

4   Simulation Results and Analysis

The basic network is the city of Belfort described in section 2. The UE are randomly dropped in each cell in proportion to the existing UMTS traffic load. We present a methodology to evaluate robustness taking into account traffic data of the network. A cell is defined as a set of test points of the map which are assigned to the same eNB; a test point is assigned to the eNB which provides the best SINR. The collected traffic data come from a real UMTS network. The tests consider three different scenarios originating from the traffic of one day, as shown in Fig. 3. Three scenarios were selected at different times of the day as follows: : a first scenario at 8am with low traffic and 482 users dropped randomly in the network; a second scenario at 3pm with medium traffic and 1,019 users; and a third scenario at 6pm with high traffic and 1,471 users. We are considering that all users are accessing the network at the same time (saturated traffic condition). Fig. 4.a) and 4.b) show the concentration of the traffic in the 1st and 3rd scenario at 8am and 6pm respectively. The traffic is represented by a colour gradient. The light colour shows a low traffic, while the dark colour shows higher traffic.As we can see, there is more dense traffic at 6pm than at 8am.

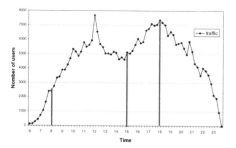

Fig. 3. Example of day traffic with three chosen scenarios

The objective is to test different traffic loads in the day. The program implementing our model is developed in C++. We run the program 10 times to get stable results; different tests are presented in the following to show the interest of robustness in a real network design and traffic load.

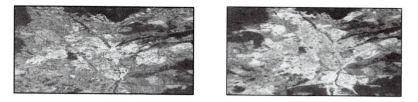

Fig. 4. Example of traffic concentration at: a) 8am. b) 6pm

Our study focuses on the interest of robust optimization for LTE online optimization and shows that optimizing a number of network configurations helps considerably to meet variant services respective to traffic uncertainty. Furthermore, we use a greedy algorithm on frequency assignment where sites and sectors are explored firstly one by one ranked from the input file and secondly randomly. The reuse scheme used here is the 1x3x3 knowing that it presents better results with respect to the number of covered users [11]. We look how the solution of the antenna configuration, especially the frequency parameter, behaves under realistic scenarios.

4.1   Algorithm Description

It can be proved that this problem is NP-hard as it is a graph-colouring problem. For such problems, guaranteeing optimum requires, in the worst case, an enumeration of all possible configurations. The number of possibilities is enormous; for 35 cells, 88 antennas and 3 frequency groups, the number of possibilities is $6^{35} = 1.71 * 10^{27}$. In order to show the interest of robust optimization, we present an algorithm able to quickly find a good solution. An iterative algorithm is used; the purpose is not to find an optimal solution but to get the benefit of robust optimization for 3 scenarios in comparison with local solution based on single scenario. The algorithm with several variants is proposed and measures the performance toward the network coverage.

The robust optimization function takes into account the three scenarios considered above. We run the optimization with different conditions varying: the scenarios of traffic; the initial frequency assignment to the eNBs (deterministic or stochastic per sector from the same site); the sites neighbourhood search to test the permutation of frequency: sites ranked from the input file or randomly chosen during optimization. The algorithm starts with one solution using the scheme 1x3x3. The optimization algorithm is run for each scenario to show the best configuration of the frequency parameter setting with respect to the performance metric given by the Eq. (4). For each explored site, we evaluate the 6 (3! =6) possibilities of permutations for each sector of the site. The algorithm evaluates 6x88 permutations at each iteration. If a frequency permutation improves the evaluation function of the current solution, the algorithm keeps the last modification and goes through the next sector configuration. It stops once the current iteration brings no improvement. This is achieved by the following algorithm which was used for all cases.

Algorithm

Input parameters: Set B of n^B eNBs; Set T of n^T test points; Set S of scenarios: s_1=8am, s_2=3pm, s_3=6pm; Frequency reuse scheme 1x3x3 (3 groups of frequency to assign to eNB)

Variables: Frequency assignment to eNBs

Fitness function: Fitness(F) = Number of outage users for the frequency plan F with $Fitness(F) = n^C_{0,s}(F)$ in s for non robust

optimization and $Fitness(F) = f^{Rob}(F)$ in s_1, s_2 and s_3 for robust optimization.

Algorithm:
```
Initialize F // F is the initial frequency plan
F*:= F // F* is the current best frequency plan
Repeat
Improve:=False
   For each site b of the network // Testing all the sites
      For each permutation j:=1..6//Testing all the permutations on b
         Generate the new frequency plan F from F*
         If Fitness(F) < Fitness(F*) //Evaluation of the current configuration
            F*:= F // Store the new best solution of frequency plan
            Improve:=True
         End IF
      End For
   End For
Until Improve=False // Stopping criteria if there is no improvement.
```

4.2 Results with Non Robust and Robust Optimization

The results of optimization are shown in the Fig. 5. We emphasize that for the non robust (each scenario tackled alone) and for the robust optimization (the 3 scenarios together) we use the same algorithm but in case of robustness the evaluation function is the Eq. (5) and takes into account the configuration of the frequency considering all the scenarios simultaneously. It means that, for each frequency of the network, we evaluate the non-covered users in the three scenarios. So, we run the same algorithm 4 times (one run for each optimization), we use the evaluation function Eq. (4) to optimize the 3 scenarios separately; then we use the evaluation function Eq. (5) for the robust optimization using the 3 scenarios at the same time. The x-axis represents the starting solution and the optimization of scenarios 1, 2 and 3 separately and the robust optimization at the end. The y-axis shows the number of users in outage for scenarios 1, 2, 3 and totally. We can note that scenario1 optimization has the smallest number of non-covered users when evaluating $s1$ (4 users) comparing to the other cases (8, 7, 9). The same analysis can be done for the scenario s_3 and it is a different for s_2 but not far away from the best one. After 20mn runs there is no guarantee on the solution quality. We observe that the result of the robust optimization is a trade off between the three scenarios, the best for s_1 and s_2 but not the best for s_3. Finally, the fitness function value $f^{Rob}=22$ of non-covered users for all cases corresponds to the global best solution, while in other situations, starting solution and non robust cases, the global function values are 43, 25, 27 and 29 respectively from left to right part of the Fig. 5a. The robust optimization does a better compromise between all scenarios. This result shows how the robust approach is important for our study.

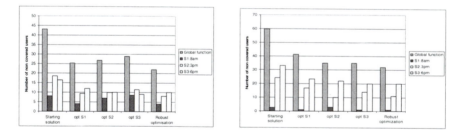

Fig. 5. The non robust and robust optimization. (a) Test 1 with deterministic initial frequency plan and site ranked from input file. (b) Test 3 with random initial frequency plan and sites randomly processed.

Different variants of the algorithm have been tested by varying several parameters. We run the program 20 minutes for each optimization in Test 2 and Test 3, and keep the best solution for the considered fitness function. In Test 1 (Fig. 5a), the initial frequency plan is deterministically assigned and the sites are processed from their rank in the input file. In Test 2, the initial frequency plan is deterministically assigned and the sites are randomly processed during optimization. The results are similar to the test 1 so we do not plot it. In Test 3 (Fig. 5b) the initial frequency plan is randomly assigned to the co-site sectors and the sites are randomly processed during optimization. The Fig. 5 plots one execution but the ten executions provide results that are in the same direction, the robust optimization gives the best trade off between the different scenarios and this is the requirement for LTE online optimization.

5   Conclusion and Perspectives

In this paper, the impact of frequency parameter setting and the interest of robust approach with respect of traffic distribution in LTE downlink system have been discussed and simulations were presented. The analysis has been carried out using model radiation pattern 1x3x3 and simple model of system performance. With respect to coverage, our case study demonstrates the benefit of the robustness approach and how the frequency parameter setting affects the coverage of the macro-cellular scenario. We aim in further studies at analyzing the influence of the load factor throughout the capacity performance metric to show the overloaded cells which represent the bottlenecks of the network and the impact of other antenna parameters like tilts. Variable Neighbourhood Search and Tabu Search are under development for robust optimization. <u>Thanks to Orange Labs for the cooperation on this work.</u>

References

1. Altman, Z., Picard, J.M., Ben Jamaa, S., Fourestie, B., Caminada, A., Dony, T., Morlier, J.F., Mourniac, S.: New challenges in automatic cell planning of UMTS networks. In: 56th IEEE Vehicular Technology Conference, pp. 951–954 (2002)
2. Athley, F., Johansson, M.: Impact of electrical and mechanical antenna tilt on LTE downlink system performance. In: IEEE Vehicular Technology Conference, Ericsson Res., Ericsson AB, Goteborg, Sweden (2010)
3. Didan, I., Kurochkin, A.: The impacts of antenna azimuth and tilt installation accuracy on UMTS network performance. Bechtel Corporation (2006)
4. Gondran, A., Baala, O., Caminada, A., Mabed, H.: Interference management in IEEE 802.11 frequency assignment. In: IEEE Vehicular Technology Conference (VTC Spring), Singapore, pp. 2238–2242 (2008)
5. Hata, M.: Empirical formula for propagation loss in land mobile radio services. IEEE Transactions on Vehicular Technology 29(3) (1980)
6. Holma, H., Toskala, A.: LTE for UMTS OFDMA and SC-FDMA based radio access. John Wiley & sons Ltd. Edition (2009)
7. Mao, X., Maaref, A., Teo, K.: Adaptive soft frequency reuse for inter-cell interference coordination in SC-FDMA based 3GPP LTE uplinks. In: IEEE Global Telecommunications Conference (GlobeCom), New Orleans LO, USA (2008)
8. Rahman, M., Yanikomeroglu, H.: Enhancing cell edge performance: A downlink dynamic interference avoidance scheme with inter-cell coordination. IEEE Transaction on Wireless Telecommunication 9(4), 1414–1425 (2010)
9. Rahman, M., Yanikomeroglu, H.: Interference avoidance with dynamic inter-cell coordination for downlink LTE system. In: IEEE Wireless Communication and Networking Conference, WCNC (2009)

10. Siomina, I., Varbrand, P., Yuan, D.: Automated optimization of service coverage and base station antenna configuration in UMTS networks. IEEE Wireless Communications 13(6), 16–25 (2006)
11. Tabia, N., Gondran, A., Baala, B., Caminada, A.: Interference model and evaluation in LTE networks. In: IFIP&IEEE Wireless and Mobile Networking Conference, Toulouse, France (2011)
12. Yilmaz, O.N.C., Hamalainen, S., Hamalainen, J.: System level analysis of vertical sectorisation for 3GPP LTE. In: IEEE 6th International Symposium on Wireless Communication System, CSWCS (2009)
13. Yilmaz, O.N.C., Hamalainen, S.: Comparison of remote electrical and mechanical antenna downtilt performance for 3GPP LTE. In: IEEE 70th Vehicular Conference Fall, VTC 2009 FALL (2009)

Small-World Optimization Applied to Job Scheduling on Grid Environments from a Multi-Objective Perspective

María Arsuaga-Ríos[1], Francisco Prieto-Castrillo[1],
and Miguel A. Vega-Rodríguez[2]

[1] Extremadura Research Center for Advanced Technologies (CETA-CIEMAT),
Trujillo, Spain
{maria.arsuaga,francisco.prieto}@ciemat.es
http://www.ceta-ciemat.es/
[2] ARCO Research Group, University of Extremadura, Dept. Technologies of
Computers and Communications, Escuela Politecnica,
Campus Universitario s/n, 10003, Cáceres, Spain
mavega@unex.es
http://arco.unex.es/

Abstract. Grid scheduling techniques are widely studied in the related literature to fulfill scientist requirements of deadline or budget for their experiments. Due to the conflictive nature of these requirements - minimum response time usually implies expensive resources - a multi-objective approach is implemented to solve this problem. In this paper, we present the Multi-Objective Small World Optimization (MOSWO) as a multi-objective adaptation from algorithms based on the small world phenomenon. This novel algorithm exploits the so-called small-world effect from complex networks, to optimize the job scheduling on Grid environments. Our algorithm has been compared with the well-known multi-objective algorithm Non-dominated Sorting Genetic Algorithm-II (NSGA-II) to evaluate the multi-objective properties and prove its reliability. Moreover, MOSWO has been compared with real schedulers, the Workload Management System (WMS) from gLite and the Deadline Budget Constraint (DBC) from Nimrod-G, improving their results.

Keywords: Grid computing, scheduling, multi-objective, small world phenomenon.

1 Introduction

Grid computing is a distributed computation paradigm build upon the Internet infrastructure. Its aim is to share distributed resources for processing and storing data from different types of users in a transparent way. The scheduling of experiments is normally composed for a set of jobs and their owners usually need to ensure a maximum execution time or cost in order to fulfill the deadlines and budgets for their projects.

C. Di Chio et al. (Eds.): EvoApplications 2012, LNCS 7248, pp. 42–51, 2012.

Real Grid schedulers such as the *Workload Management System* (WMS)[1] from the most used European middleware Lightweight Middleware for Grid Computing (gLite)[2], implemented during the Enabling Grids for E-sciencE (EGEE)[3] project, does not take into account these requirements. However, the scheduler *Deadline Budget Constraint* (DBC) from the Nimrod-G [1] tries to keep the deadline and budget for an experiment but it does not ensure that all the jobs will execute successfully. Currently, new algorithms are emerging to consider not just one of these requirements; time or cost, but also both of them at the same time with the same importance([11], [15], [16], [17]). Usually, these algorithms are based on genetic algorithms. Genetic algorithms are inspired by biological phenomena and they have reported a good performance in multi-objective approaches as it is the case of the well-known *Non-dominated Sorting Genetic Algorithm-II*(NSGA-II)[2]. But, the test environments used for the cited studies lack of real data and their experiments do not exhibit complex dependencies between jobs and transferred data process.

In this research, we have considered a novel optimization algorithm based on the small world phenomenon. Milgram found this phenomenon during sociological research consisting on tracking the shortest paths in EEUU social networks through a simple experiment [8]. Milgram's experiment originated the famous term "six degrees of separation". Lately, Watts and Strogatz ([14][9]) reported a mathematical model of small world fostering research on the small world by computer scientist. Small-world phenomenon became an active field in complex problems due to its interdisciplinary combining sociology, mathematics and computer science ([3], [13], [6]). The key of the small world phenomenon depends on local connections and few long-range connections to find the shortest paths in the network, which offers an efficient searching strategy [5]. Resource distribution of Grid computing is sometimes compared with the Internet network or social networks due to its scale-free complex topology. Therefore, we develop research of the small world phenomenon to optimize the job-scheduling problem on Grid environments due to the good results found by currently emerging optimization algorithms based on this phenomenon. The approach of this work is the modification of the Tabu Small World Optimization (TSWO) [7] algorithm to support multi-objective optimization and its application to solve the job-scheduling problem to optimize execution time and cost. This proposed algorithm is called Multi-Objective Small World Optimization (MOSWO) and it is compared with the standard multi-objective and also genetic algorithm (NSGA-II). Moreover, a comparison with real-schedulers such as the WMS and DBC is carried out to prove the improvement offered by the MOSWO approach.

This paper is structured as follows. Section 2 defines the problem statement. Section 3 presents the MOSWO algorithm approach. Section 4 describes briefly the NSGA-II applied to this problem. Section 5 shows the experimental results. Finally, Section 6 summarizes the conclusions.

---

[1] http://web.infn.it/gLiteWMS/
[2] http://glite.cern.ch/
[3] http://www.eu-egee.org/

2   Problem Statement

Job scheduling in Grid environments is a challenging task due to their complexity topology from heterogeneous and distributed resources. Desirably, jobs are managed in order to sent themselves to the most suitable resources satisfying the user needs. We have considered two critical and conflictive requirements often demanded by the scientists, the execution time and cost of an experiment (set of jobs). These requirements are conflictive each other because faster resources are often more expensive than the slower ones. By this reason a multi-objective approach is required to tackle this problem. A general multi-objective optimization problem (MOP) [4] includes a set of n parameters (decision variables) and a set of k objective functions. The objective functions are functions from the decision variables. Hence, a MOP could be defined as: $Optimize\ y = f(x) = (f_1(x), f_2(x), ..., f_k(x))$, where $x = (x_1, x_1, ..., x_n) \in X$ is the decision vector and $y = (y_1, y_2, ..., y_n) \in Y$ the objective vector. The decision space is denoted by X and the space objective is defined by Y. A multi-objective problem does not return one solution but also a set of them. The set of optimum solutions is called Pareto optimal set and the point set, defined by the Pareto optimal set in the value space of the objective functions, is known as Pareto front. For a given MOP and Pareto optimal set P*, the Pareto front (PF) is defined as: $PF^* := f = (f_1(x), f_2(x), ..., f_k)|x \epsilon P^*$. The solutions included in the same Pareto front dominates all the points not matched in all the objectives by other solution. By this reason Pareto front consists just in non-dominated solutions. A solution dominates other if and only if, it is at least as good as the other in all the objectives and it is better in at least one of them. Our problem needs to minimize two objectives (time and cost). Given a set of jobs $J = \{J_i\}$, $i = 1,..,m$ and a set of grid resources $R = \{R_j\}$, $j = 1,..,n$ the fitness functions are defined as:

$$Min\ F = (F_1, F_2) \tag{1}$$

$$F_1 = max\ time\,(J_i,\ f_j(J_i)) \tag{2}$$

$$F_2 = \sum cost\,(J_i,\ f_j(J_i)) \tag{3}$$

where $f_j(J_i)$ denotes the job J_i allocation on the resource R_j. The objective function F_1 renders the completion time for the experiment (set of jobs) and the object function F_2 returns the resource cost for processing the experiment. Complex workflows that follow a DAG model have been considered, the experiments used are composed of dependent jobs. Dependent jobs have direct relation to the requirements that are going to be optimized (time and cost) since an experiment (or workflow) is modeled by a weighted directed acyclic graph (DAG) $JG = (V, E, l, d)$, where V is a set of nodes and E is a set of edges. On one hand, each node $j \in V$ represents a job and has assigned a constant length measured in MI (Million of instructions), this length is denoted by $l(j)$. On the other hand, each edge $(j \rightarrow j') \in E$ from j to j' denotes the dependency between the job j' regarding the job j. Job j' could not be executed until job j has been executed successfully. The data transferred length $d(j \rightarrow j')$ between the jobs is specified and measured in bytes.

3 MOSWO: Multi-Objective Small World Optimization

MOSWO is an adaptation of the novel Tabu Small World Optimization (TSWO) [7] algorithm to optimize more than one objective. TSWO is also based on the Small World Algorithm (SWA) [3] but using a Tabu search for the local search operator. SWA implements the solution space of optimization problem as a small world network and takes the optimization procedure as finding the shortest path from a candidate solution to an optimal solution. In the research of TSWO, a good performance for decimal encoding in contrast with the binary used by SWA is showed. Moreover, TSWO and SWA have demonstrated respectively their goodness in comparison with genetic algorithms (GAs) ([3], [6], [7]). MOSWO algorithm is inspired by the Small World phenomenon and it applied the Tabu search as a local search operator with a multi-objective perspective. The main steps of MOSWO are shown in Algorithm 1. This algorithm

---

Algorithm 1. MOSWO pseudocode

INPUT: Population Size, Mutation Probability
OUTPUT: Set of Solutions
 1: Initialize population of solutions;
 2: Evaluate population (Time and Cost);
 3: **while** not time limit (2 minutes) **do**
 4:     Multi-Objective Random Long-Range Operator;
 5:     Multi-Objective Local Shortcuts Search Operator;
 6:     Select Set of Best Solutions (Pareto Front);
 7:     Generate New Population;
 8: **end while**

---

requires two parameters: population size and mutation probability. MOSWO considers the population size as the size set of nodes, which represents the candidate solutions. Two vectors define the nodes: allocation and order vector. On one hand, the allocation vector indicates the assignment between jobs and resources meaning, in which resources the jobs will be executed. On the other hand, order vector denotes the order of execution of the jobs that compound the workflow. Execution time and cost are also attributes of each node according to previous vectors. This representation is based on the work [11]. The algorithm starts with a random initialization of the nodes and their evaluation using two well-known operators for multi-objective problems: *Classification of Pareto front* and *Crowding distance*[2]. Then, main operators in small world optimization algorithms are executed per node: a random long-range search operator Γ and local search operator Ψ.

Multi-objective Γ operator is applied to the node N_i according to mutation probability and if and only if the node is not in the set of best solutions. Next, allocation and order vectors are modified to obtain N_i'. The process of Γ operator applied to allocation and order vectors is similar. For allocation vector, μ and ν jobs (assigned previously to a resource) are selected randomly, $1 \leq \mu$

$< \nu \leq m$, where m is the number of jobs that compound the experiment and hence, it is also the length of the allocation and order vectors. The corresponding resources from μ and ν jobs are substituted. In order vector are also substituted if this change does not fail to fulfill the dependency constraint between jobs. When N_i' is calculated, multi-objective Ψ operator is executed to obtain the best set of neighbours $Best\ (\xi(N_i'))$ using the Tabu search. Tabu search updates a list, called Tabu list, using the *Least Recent Used* (*LRU*) strategy, to store the studied nodes in order to not repeat their study again. Therefore, the operator checks if the node N_i' are in the Tabu list before generating its neighbours. Then, multi-objective Ψ operator generates the set of neighbours from N_i' modifying allocation and order vector. Due to performance, the algorithm generates a limited number of neighbours; this size is the total number of jobs. Hence, allocation vector is across modifying one resource assignation sequentially. This modification consists on increasing the resource identification by one. Order vector uses the same technique but always considering the dependencies among jobs. Once the neighbours are generated, multi-objective Ψ obtains the First Pareto Front from the neighbours set using the *Classification of Pareto front* operator. Then, the front is crossed in order to check if its nodes are in the Tabu list. If one node is in the list it is delete from the front. Otherwise, it is included into the Tabu list in order to not study so far.

Finally, the resulting front is built from all first fronts calculated per node in the current iteration. These fronts are joined with the current population. The *Classification of Pareto front* and also the *Crowding distance* operator sort this new node set. Next, an improvement is applied to this algorithm in order to add further diversity and to avoid the stagnation: The worst node of the set is modified using heuristics from the problem. Two different heuristics are applied per vector for each node. The heuristic for the order vector consists of modifying the vector comparing it with another order vector calculated from a greedy algorithm. This greedy algorithm provides the best order of execution for the workflow according to the number of dependencies. Top positions of these vectors are assigned to the jobs that have more jobs depending on them without forgetting the precedence constraint. Allocation vector heuristic calculates its processing time *MI/MIPS* per each job, where *MI* (*Million Instructions*) denotes the job length (in instructions) and *MIPS* (*Million Instructions per Second*) indicates the speed of the execution job in the resource assigned in the allocation vector. Also, the total time execution of the workflow is calculated considering the dependencies between jobs that, assuming those that have not dependencies between them can be executed in parallel. Therefore, each job is assigned to the resource that reduces the total execution time calculated previously. Moreover, resources are sorted following the value of *processing speed/cost*. The overhead time is also taken into account by the competitive jobs (without dependencies between them) that run sequentially in the same resource. Before finalizing the iteration, population is reduced to its previous size choosing the best nodes that compose the set (old population and neighbours) to obtain the new population and also the set best nodes from it (*First Pareto Front*).

4   NSGA-II: Non-dominated Sort Genetic Algorithm

NSGA-II [2](Non-dominated Sorting Genetic Algorithm) is a well-known multi-objective algorithm that is usually compared with novel multi-objective approaches. The main steps of NSGA-II are showed in Algorithm 2. NSGA-II

Algorithm 2. NSGA-II pseudocode

INPUT: Population Size, Crossover and Mutation Probability
OUTPUT: Set of Solutions

1: Initialize population of solutions;
2: Evaluate population (Time and Cost);
3: **while** not time limit (2 minutes) **do**
4:     Binary Tournament Selection;
5:     Crossover;
6:     Mutation;
7:     Select Set of Best Solutions (Pareto Front);
8:     Generate New Population;
9: **end while**

---

requires the typical parameters of a genetic algorithm (GA): population size, crossover probability and mutation probability. In case of genetic algorithms, a candidate solution is usually called individual. These individuals are codified through the same way that the nodes from MOSWO and the initial population is generated with the same process. Multi-objective operators are applied during the execution of this algorithm: *Classification of Pareto fronts* and *Crowding distance* to sort, select and compare the individuals. *Binary Tournament Selection* process selects the parents that will be crossing. Two tournaments are executed, one per parent, selecting the individual located in the first front. In case that individuals are in the same front, the individual with high *Crowding distance* is used to break the tie.

Typical *Crossover* and *Mutation* operators in genetic algorithms are duplicated in this problem due to the concurrency of vectors in each individual: allocation and order vector. Therefore, this algorithm has two crossover and two mutation operators. *Crossover* operators are based on the work [11]. The allocation crossover swaps the parent vectors from a random position to create two individuals. A similar crossover uses the order vector but keeping the precedent constraints. To do this, a method checks if the new individuals have duplicated order positions of the jobs, the duplicated positions remain at the missing positions. Next, *Mutation* operators are applied per vector using the mutation probability. Each job from the allocation vector is mutated assigning a new resource from the available resource list. The order allocation vector is different due to the constraints from the DAG model. Although each position of job is mutated regarding the probability mutation, the operator identifies the latest order position from its parents of the current job. Also, the first order position from its child jobs is identified. Therefore, the new position will be selected randomly between the latest position parent and the first position child. Finally, new population is made up from the individuals located in the first Pareto fronts until its original size is achieved, and the resulting set of solutions for that iteration is the one composed by the first Pareto front.

5   Test Environment and Experiments

The test environment is configured and implemented in GridSim[4] to simulate real grid behaviour. On one hand, a grid topology is constructed from two testbeds to complete all the information from a real grid. EU Data Grid [10] testbed provides the topology connections and WWG [1] testbed a complete and real resource data. On the other hand, three different and parallel numerical computation workflows are launched such as Parallel Gaussian Algorithm, Parallel Gauss-Jordan Algorithm and Parallel LU decomposition [12]. All of these workflows follow a DAG Model with their respective lengths in MI (Million of Instructions) and the input/output sizes in bytes. The experiments have been divided in two studies to enforce the feasibility of the proposed algorithm (MOSWO). Due to the stochastic nature of multi-objective metaheuristics, each experiment performed in our study includes 30 independent executions. The parameter settings for each multi-objective algorithm are: for MOSWO (population size = 100, mutation probability = 0.25) and for NSGAII (population size = 100, crossover probability = 0.9, mutation probability = 0.1).

The first study evaluates the multi-objective properties of MOSWO and NSGA-II in order to compare them. Hypervolume indicator [18] is showed in Table 1 and Table 2 for each algorithm. The obtained results demonstrate that MOSWO hypervolume is always better than the NSGA-II and also, the fact of obtaining more than 50% of hypervolume is a good point taking into account the volume covered by its founded solutions. Moreover, the standard deviation of MOSWO has better values than NSGA-II giving therefore more reliability than this last one algorithm. Set coverage metrics [18] is used to determine the percentage of solution dominance of each algorithm regarding the other. Each cell gives the fraction of non-dominated solutions evolved by algorithm B, which are covered by the non-dominated points achieved by algorithm A [18]. Table 3 shows that MOSWO solutions dominate almost all the solutions found by NSGA-II with more than 80%. Furthermore, Pareto fronts returned by MOSWO (Figure 1) contain more solutions than NSGA-II, giving to the user more alternatives inside its requirements (budget and deadline).

The second study is related to the comparison between MOSWO with the real schedulers: WMS and DBC. These schedulers have been implemented in the GridSim simulator. WMS has been executed using two options applying two different scheduling strategies. The option 1 sorts the resources according to their time response and the option 2 sorts the resources regarding their number of free CPUs. Results show that MOSWO always executes successfully all the jobs that compound the workflows within the indicated deadline. However, the schedulers

---

[4] http://www.buyya.com/gridsim/

Table 1. MOSWO properties per each workflow

Workflows	Average Hypervolume %	Standard Deviation of the Hypervolume	Reference Point (Time, Cost)
Gaussian	53.18	0.71	(1000, 10000)
Gauss-Jordan	54.45	0.34	(1200, 22000)
LU	52.71	0.70	(1200, 22000)

Table 2. NSGA-II properties per each workflow

Workflows	Average Hypervolume %	Standard Deviation of the Hypervolume	Reference Point (Time, Cost)
Gaussian	46.40	1.09	(1000, 10000)
Gauss-Jordan	47.17	0.71	(1200, 22000)
LU	48.03	0.93	(1200, 22000)

Table 3. Set coverage comparison of MOSWO and NSGA-II per each workflow

Coverage A ≥ B				
Algorithm		Workflows		Average
A	B	Gaussian Gauss-Jordan	LU	
NSGA-II MOSWO		88.88% 83.33%	88.88%	87.03%
MOSWO NSGA-II		0% 16.66%	0%	5.55%

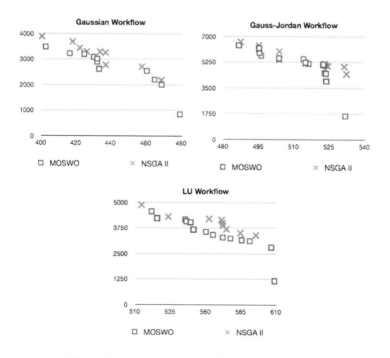

Fig. 1. Pareto fronts per worklow and algorithm

Table 4. Deadline Restriction to check the jobs executed successfully

Workflows	Constraint	WMS (Option 1)			WMS (Option 2)			DBC			MOSWO		
	Deadline	Time	Cost	Jobs	Time	Cost	Jobs	Time	Cost	Jobs	Time	Cost	Jobs
	500	482.68	3434.82	12	442.46	3931.93	12	480.82	850.00	12	479.12	850.00	12
Gaussian	450	401.03	2998.46	9	442.46	3931.93	12	450.58	942.87	10	433.38	2611.49	12
	400	400.01	2566.17	7	417.00	3683.08	10	400.77	903.61	9	400.65	3477.20	12
	600	534.70	6405.24	15	492.18	7366.07	15	533.41	1593.74	15	531.71	1593.74	15
Gauss-Jordan	550	534.70	6405.24	15	492.18	7366.07	15	533.41	1593.74	15	531.71	1593.74	15
	500	427.57	5552.48	13	492.18	7366.07	15	500.08	1487.48	14	496.08	5682.85	15
	650	612.29	4684.63	14	561.46	5367.45	14	610.46	1164.00	14	608.76	1164.00	14
LU	600	585.78	4468.43	13	561.46	5367.45	14	596.66	1422.98	14	591.16	3117.85	14
	550	504.12	3925.96	10	537.14	5123.55	13	550.00	1689.82	12	549.22	4028.32	14

WMS with its two options and DBC do not always achieve the total execution
(see Table 4). In fact, in the cases that real schedulers (WMS and DBC) achieve
the total execution, MOSWO usually offers better values of execution time or
cost than the others schedulers.

6   Conclusions

In this research a new multi-objective optimization algorithm is proposed based
on the small world phenomenon. MOSWO allows the optimization of two objectives (time and cost execution) of workflows with dependent jobs on a Grid environment. MOSWO is compared with real schedulers as DBC and WMS and also
with a well-known multi-objective algorithm, NSGA-II. In all the cases MOSWO
shows better results due to its multi-objective qualities and it also offers a good
range of solutions for decision support. In future works, this algorithm will be
compared with other multi-objective approaches.

Acknowledgment. CETA-CIEMAT acknowledges the support received from
the European Regional Development Fund through its Operational Program,
Knowledge-based Economy.

References

1. Buyya, R., Murshed, M., Abramson, D.: A deadline and budget constrained cost-
   time optimisation algorithm for scheduling task farming applications on global
   grids. In: Int. Conf. on Parallel and Distributed Processing Techniques and Applications, Las Vegas, Nevada, USA, pp. 2183–2189 (2002)
2. Deb, K., Pratap, A., Agarwal, S., Meyarivan, T.: A fast elitist multi-objective
   genetic algorithm: Nsga-ii. IEEE Transactions on Evolutionary Computation 6,
   182–197 (2000)
3. Du, H., Wu, X., Zhuang, J.: Small-World Optimization Algorithm for Function
   Optimization. In: Jiao, L., Wang, L., Gao, X.-b., Liu, J., Wu, F. (eds.) ICNC 2006.
   LNCS, vol. 4222, pp. 264–273. Springer, Heidelberg (2006)

4. Khare, V., Yao, X., Deb, K.: Evolutionary Multi-Criterion Optimization, vol. 2632. Springer, Heidelberg (2003)
5. Kleinberg, J.: The small-world phenomenon: an algorithm perspective. In: Proceedings of the Thirty-Second Annual ACM Symposium on Theory of Computing, STOC 2000, pp. 163–170. ACM, New York (2000)
6. Li, X., Zhang, J., Wang, S., Li, M., Li, K.: A small world algorithm for high-dimensional function optimization. In: Proceedings of the 8th IEEE International Conference on Computational Intelligence in Robotics and Automation, CIRA 2009, pp. 55–59. IEEE Press, Piscataway (2009)
7. Mao, W., Yan, G., Dong, L., Hu, D.: Model selection for least squares support vector regressions based on small-world strategy. Expert Syst. Appl. 38, 3227–3237 (2011)
8. Milgram, S.: The small world problem. Psychology Today 2, 60–67 (1967)
9. Strogatz, S.H.: Exploring complex networks. Nature 410(6825), 268–276 (2001)
10. Sulistio, A., Poduval, G., Buyya, R., Tham, C.: On incorporating differentiated levels of network service into gridsim. Future Gener. Comput. Syst. 23(4), 606–615 (2007)
11. Talukder, A.K.M.K.A., Kirley, M., Buyya, R.: Multiobjective differential evolution for scheduling workflow applications on global grids. Concurr. Comput.: Pract. Exper. 21(13), 1742–1756 (2009)
12. Tsuchiya, T., Osada, T., Kikuno, T.: Genetics-based multiprocessor scheduling using task duplication. Microprocessors and Microsystems 22(3-4), 197–207 (1998)
13. Wang, X., Cai, S., Huang, M.: A Small-World Optimization Algorithm Based and ABC Supported QoS Unicast Routing Scheme. In: Li, K., Jesshope, C., Jin, H., Gaudiot, J.-L. (eds.) NPC 2007. LNCS, vol. 4672, pp. 242–249. Springer, Heidelberg (2007)
14. Watts, D.J., Strogatz, S.H.: Collective dynamics of 'small-world' networks. Nature 393(6684), 440–442 (1998)
15. Ye, G., Rao, R., Li, M.: A multiobjective resources scheduling approach based on genetic algorithms in grid environment. In: International Conference on Grid and Cooperative Computing Workshops, pp. 504–509 (2006)
16. Yu, J., Kirley, M., Buyya, R.: Multi-objective planning for workflow execution on grids. In: GRID 2007: Proceedings of the 8th IEEE/ACM International Conference on Grid Computing, pp. 10–17. IEEE Computer Society, Washington, DC, USA (2007)
17. Zeng, B., Wei, J., Wang, W., Wang, P.: Cooperative Grid Jobs Scheduling with Multi-objective Genetic Algorithm. In: Stojmenovic, I., Thulasiram, R.K., Yang, L.T., Jia, W., Guo, M., de Mello, R.F. (eds.) ISPA 2007. LNCS, vol. 4742, pp. 545–555. Springer, Heidelberg (2007)
18. Zitzler, E., Thiele, L.: Multiobjective Optimization Using Evolutionary Algorithms - A Comparative Case Study. In: Eiben, A.E., Bäck, T., Schoenauer, M., Schwefel, H.-P. (eds.) PPSN 1998. LNCS, vol. 1498, pp. 292–304. Springer, Heidelberg (1998)

Testing Diversity-Enhancing Migration Policies for Hybrid On-Line Evolution of Robot Controllers*

Pablo García-Sánchez[1], A.E. Eiben[2], Evert Haasdijk[2],
Berend Weel[2], and Juan-Julián Merelo-Guervós[1]

[1] Dept. of Computer Architecture and Technology, University of Granada, Spain
[2] Dept. of Computer Science, Vrije Universiteit Amsterdam, The Netherlands
pgarcia@atc.ugr.es

Abstract. We investigate on-line on-board evolution of robot controllers based on the so-called hybrid approach (island-based). Inherently to this approach each robot hosts a population (island) of evolving controllers and exchanges controllers with other robots at certain times. We compare different exchange (migration) policies in order to optimize this evolutionary system and compare the best hybrid setup with the encapsulated and distributed alternatives. We conclude that adding a difference-based migrant selection scheme increases the performance.

1 Introduction

Evolutionary robotics concerns itself with evolutionary algorithms to optimise robot controllers [7]. Traditionally, robot controllers evolve in an off-line fashion, through an evolutionary algorithm running on some computer searching through the space of controllers and only calling on the actual robots when a fitness evaluation is required. To distinguish various options regarding the evolutionary system Eiben et al. proposed a naming scheme based on *when*, *where* and *how* this evolution occurs [3]. The resulting taxonomy distinguishes between design time and run-time evolution (off-line vs. on-line) as well as between evolution inside or outside the robots themselves (on-board vs. off-board). In a system comprising of multiple robots, there are three options regarding the *'how'*:

Encapsulated: A population of genotypes encoding controllers evolves inside each robot independently, without communication with other robots.

Distributed: Each robot carries a single genotype and reproduction requires the exchange of genotypes with other robots. The evolving population is formed by the combined genotypes of all the robots.

* This work was supported in part by Spanish Projects EvOrq (TIC-3903), CEI BioTIC GENIL (CEB09-0010), MICINN CEI Program (PYR-2010-13) and FPU research grant AP2009-2942 and the European Union FET Proactive Initiative: Pervasive Adaptation funding the SYMBRION project under grant agreement 216342. The authors wish to thank Selmar Smit and Luis Pineda for their help and fruitful discussions.

C. Di Chio et al. (Eds.): EvoApplications 2012, LNCS 7248, pp. 52–62, 2012.
© Springer-Verlag Berlin Heidelberg 2012

Hybrid: Each robot has its own locally evolving population and there is exchange of genotypes between robots. In terms of parallel evolutionary algorithms, this can be seen as an island-model evolutionary algorithm with migration.

In this paper we investigate aspects of the hybrid approach: we test the effects of the *migration policy* (migration of the best, random, or most different individual), the *admission policy* (always accept the migrant, or accept only after re-evaluation) and the *island topology* (ring vs. fully connected). Furthermore, we look into these effects for different numbers of robots (4, 16 or 36).

Specifically, our research questions are:

- Using the hybrid approach (island model), which is the best combination of migration policy, admission policy, and island topology?
- Is this combination better than the encapsulated and distributed alternatives?

The rest of the work is structured as follows: after the state of the art, we present the developed algorithms and experimental setting. Then, the results of the experiments are shown (Section 4), followed by conclusions and suggestions for future work.

2   State of the Art

Migration among otherwise reproductively isolated populations has been proven to leverage the inherent parallelism in evolutionary algorithms, not only by obtaining speed-ups, but also by increasing the quality of results, since the reproduction restrictions inherent in the division of the population into islands is a good mechanism to preserve population diversity, as shown in, for instance, [2].

To improve population diversity in an island model evolutionary algorithm, the MultiKulti algorithm [1] takes the genotypic differences of individuals when selecting migrants into account. It is based in the idea that the inflow of migrants that differ from the rest of an island's population increases diversity and thus improves performance. An island requests a migrant from one of its neighbours by sending a genotype that represents the population. This can either be the the best individual (based in the assumption that when a population tends to converge after a few generations, the best is a fair representation of the whole population) or a *consensus sequence* (the most frequent allele in each position of the genotype using binary genomes). In answer, an island selects the most different genotype in either its whole population or the top individuals (the elite). In their experiments, the islands were connected in a ring topology, with migration taking place asynchronously. Results of the experiments performed in [1] show that MultiKulti policies outperform classic migration policies (send the best or random individuals from the population), especially with a low number of individuals but larger number of islands. It is shown to be better to send the consensus as a representation and that sending the most different of a well-chosen elite (those with the best fitness) is better than sending the most different overall.

On-line evolutionary robotics has been studied in works like [8], where genetic programming was used to evolve a robot in real time, and [11], where several robots evolve at the same time, exchanging parts of their genotypes when within communication range. [5] compares an encapsulated and a distributed version; the latter is implemented as a variant of EVAG [6], where each robot has one active solution (genotype) a cache of genotypes that are active in neighbouring robots. Parents are selected through a binary tournament in each robot's cache. If the new solution (candidate) is better than the active, it replaces the active solution. The work compares this algorithm with a panmictic version, where parents are selected (again using binary tournament) from the combined active solutions of all robots.

One of the peculiarities of evolutionary robotics, particularly on-line, is that the fitness evaluations are very noisy [4]. The conclusions in [1], however, are based on experiments with noiseless fitness functions, so we cannot take these conclusions for granted in on-line evolutionary robotics and we have to test the MultiKulti algorithm in our particular setting.

3   Algorithms and Experimental Setup

We carried out our experiments with e-puck like robots simulated in the RoboRobo simulator[1]. The robot is controlled by an artificial neural net with 9 inputs (corresponding to the robot's distance sensors and a bias node), 2 outputs (wheel speeds). Genetically, this was represented as a vector coding the network's 18 weights. All algorithms were evaluated using the *Fast Forward* task and next fitness function:

$$f = \sum_{t=0}^{\tau}(v_t \cdot (1 - v_r)) \tag{1}$$

where v_t and v_r are the translational and the rotational speed, respectively. v_t is normalised between -1 (full speed reverse) and 1 (full speed forward), v_r between 0 (movement in a straight line) and 1 (maximum rotation). Whenever a robot touches an obstacle, $v_t = 0$, so the fitness increment during collisions is 0. There is more information about this function in [5]. This fitness is noisy: a controller configuration can produce different fitness values depending on the robot's position in the arena when evaluation starts. The robots are placed in a small maze-like arena (Fig. 2). To ensure a fair comparison across different numbers of robots, each robot is placed in a separate instance of the arena to avoid physical interaction between robots. Robots can communicate across arenas instances.

In our experiments, we compare three algorithms:

Encapsulated Evolutionary Algorithm. The encapsulated algorithm we use is the $\mu+1$ on-line algorithm presented in [4]. Here, each robot runs a stand-alone evolutionary algorithm with a local population of μ individuals. In each cycle, one new solution (controller) is created and evaluated. This solution replaces

---

[1] `http://www.lri.fr/~bredeche/roborobo/`

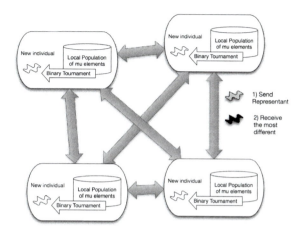

Fig. 1. Migration mechanism: each robot has a local population and in each migration cycle request a different type individual from others robots' populations. If MultiKulti is used, then a message is sent (gray genotype) to receive the most different (black genotype).

the worst individual of the population if it has higher fitness. To combat the effects of noisy evaluations, an existing solution can be re-evaluated, instead of generating and testing a new one, depending on the re-evaluation rate ρ.

Distributed Evolutionary Algorithm. As a benchmark distributed algorithm we use the panmictic algorithm presented in [5]. Here, a single controller is present in each robot. New controllers are created using the controllers of two robots as parents. In each iteration, a robot randomly selects two others to create a new chromosome by crossover and mutation. If the new chromosome is better, it replaces the actual one.

Hybrid Evolutionary Algorithm. This algorithm is an adaptation of the $\mu+1$ on-line algorithm that includes a migration mechanism to exchange genotypes among robots (every robot is an island) as shown in Fig. 1. We test two migrant acceptance mechanisms: a migrant can be added to the local population either regardless of its fitness (to give it a chance to be selected) or only if it is better than the worst.[2]

Each experiment lasts for 50,000 evaluation steps. In on-line evolution, the robots train on the job: this means that the *robot's* performance is not (only) determined by the best individual it stores at any one time, but by the joint performance of all the candidate controllers it considers over a period. Therefore, we assess the algorithms' performance using the average of the last 10% evaluations over all robots.

---

[2] Source code of the presented algorithms is available in http://atc.ugr.es/~pgarcia, under a GNU/GPL license.

As stated in [9], an algorithm's parameters should be tuned to obtain (approximately) the best possible parameter settings and so ensure a fair comparison between the best possible instances of the algorithms. We used Bonesa [10] to tune the parameters for the algorithms we investigate in the following configurations:

- Number of robots: executions with 4, 16 and 36 robots have been performed.
- Migrant selection: select the *Best, random* or *most different* (MultiKulti) individual as a migrant.
- Admission policy: when a new migrant arrives, it is evaluated and accepted only if is better than the worst (*no-replacement*) or accepted regardless, always replacing the worst of the population (*replacement*).
- Topology: migration can move between neighbours and the islands are arranged in a *ring* or in a *random* topology, which is rewired after every evaluation.

We conducted Bonesa runs for each possible combination of these configurations to tune the settings for canonical parameters (e.g., mutation step size, crossover rate) and the following more specific parameters:

Along the canonical GA parameters (like mutation or crossover rate) the MultiKulti parameters to study are the next:

- Migration rate: likelihood of migration occurring per evaluation cycle.
- Best rate: probability of representing the population with the best individual or with a consensus sequence (average of genes). This parameter applies only for MultiKulti instances.
- Elite percentage: the size of the elite group to select the migrant from (if 1, receive the most different of all the population). This parameter applies only for MultiKulti instances.

Population size μ was fixed to 10 individuals to isolate the interactions between the other parameters. Figure 3 lists all tuned parameters and their ranges. For the final analysis, we ran 50 iterations of each configuration with the parameters set to those reported as optimal by Bonesa.

4   Results and Analysis

4.1   Comparing Migration Configurations

The first question we asked ourselves was "which is the best combination of migration policy, admission policy, and island topology?" To answer this question, we analyse the results as reported by Bonesa for each of the configurations we considered. Table 1 shows the best parameters obtained for all configurations with 4, 16 and 36 robots. We discuss the results in the following four paragraphs, each discussing the results for one combination of admission policy and island topology.

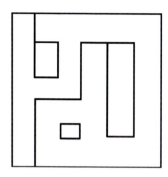

Parameter Name	Range
Evaluation steps	300-600
Mutation step size	0.1-10
μ	10
Re-evaluation rate	0-1
Crossover rate	0-1
mutation rate	0-1
migration rate	0-1
elite percentage	0-1
best Rate	0-1

Fig. 2. Arena used in the experiments   **Fig. 3.** Parameters to tune

Replacement Admission Policy and Panmictic Topology. In all cases, the re-evaluation, crossover, mutation and migration rates are very high. Also, EliteSize is almost 1 everywhere: the migrant is selected from almost the whole population. It also turns out that is better to send a consensus sequence rather than the best individual as a representative of the population (bestRate has low values). There is no clear trend for migration rate.

Replacement Admission Policy and Ring Topology. Changing the island topology to a ring arrangement, three settings change materially: as can be see in Table 1 for MultiKulti with 4 and 16 robots, the migration rate is much lower, but for 36 robots it remains very high. Also, but only for 4 robots, BestRate is higher (send the best individual as representative, not the consensus sequence).

No-replacement Admission Policy and Panmictic Topology. When changing the replacement policy a remarkable decrease can be seen in the migration rate and, more importantly, the re-evaluation rate across the board. For 4 robots, EliteSize is much lower than in all three other combinations of admission policy and topology.

No-replacement Admission Policy and Ring Topology. Apart from lower migration rates for most of the policies and a drop in EliteSize for 16 robots, Bonesa reports similar values for this combination of admission policy and topology and the previous one. For 4 robots, EliteSize again has a high value.

Comparing Performance. Figures 4a, 4b and 4c show box plots summarising 50 repeats of each configuration, grouped by number of robots.

Although in terms of performance levels there is no clear trend it is clear that the admission policy does have an appreciable impact: choosing the no-replacement admission policy always leads to a marked decrease in performance variation, with an increase of minimum performance. So we can conclude that evaluating an immigrant and only admitting it if it outperforms the worst individual in the population leads to more consistent performance with fewer very poor results.

Table 1. Parameters obtained with Bonesa for all admission policies and topology configurations

Replacement admission policy and panmictic topology								
	4 ROBOTS			16 ROBOTS			36 ROBOTS		
	MK	RANDOM	BEST	MK	RANDOM	BEST	MK	RANDOM	BEST
evolutionSteps	345	310	312	310	306	425	538	561	584
stepSize	9.038	9.874	5.38	8.804	8.786	9.199	4.842	8.096	9.684
reEvaluation	0.868	0.72	0.739	0.619	0.812	0.949	0.964	0.751	0.777
Crossover	0.926	0.816	0.929	0.017	0.879	0.917	0.963	0.915	0.941
Mutation	0.943	0.977	0.936	0.98	0.839	0.909	0.937	0.923	0.938
Migration	0.809	0.989	0.958	0.987	0.499	0.993	0.956	0.988	0.567
EliteSize	0.849	-	-	0.988	-	-	0.995	-	
BestRate	0.04	-	-	0.192	-	-	0.181	-	

Replacement admission policy and ring topology								
	4 ROBOTS			16 ROBOTS			36 ROBOTS		
	MK	RANDOM	BEST	MK	RANDOM	BEST	MK	RANDOM	BEST
evolutionSteps	304	319	312	304	311	372	554	589	573
stepSize	9.29	8.149	8.769	7.008	7.37	9.953	9.465	9.307	9.94
reEvaluation	0.868	0.749	0.792	0.953	0.721	0.861	0.935	0.705	0.939
Crossover	0.999	0.983	0.96	0.83	0.955	0.455	0.996	0.848	0.991
Mutation	0.986	0.952	0.691	0.914	0.809	0.889	0.971	0.777	0.98
Migration	0.597	0.892	0.974	0.559	0.624	0.996	0.988	0.816	0.955
EliteSize	0.49	-	-	0.93	-	-	0.827	-	
BestRate	0.862	-	-	0.172	-	-	0.145	-	

No-replacement admission policy and panmictic topology								
	4 ROBOTS			16 ROBOTS			36 ROBOTS		
	MK	RANDOM	BEST	MK	RANDOM	BEST	MK	RANDOM	BEST
evolutionSteps	305	304	308	302	304	306	567	362	516
stepSize	9.895	4.04	9.547	9.731	9.146	9.8	8.832	7.526	9.988
reEvaluation	0.385	0.039	0.489	0.449	0.291	0.692	0.048	0.344	0.528
Crossover	0.828	1	0.934	0.847	0.945	0.671	0.31	0.822	0.963
Mutation	0.976	0.927	0.899	0.849	0.958	0.969	0.921	0.986	0.879
Migration	0.577	0.788	0.72	0.658	0.757	0.577	0.835	0.753	0.7
EliteSize	0.279	-	-	0.716	-	-	0.911	-	-
BestRate	0.198	-	-	0.703	-	-	0.013	-	-

No-replacement admission policy and ring topology								
	4 ROBOTS			16 ROBOTS			36 ROBOTS		
	MK	RANDOM	BEST	MK	RANDOM	BEST	MK	RANDOM	BEST
evolutionSteps	325	302	323	314	303	306	600	375	581
stepSize	9.821	9.726	9.731	9.925	8.44	9.329	9.661	9.686	9.493
reEvaluation	0.045	0.007	0.53	0.044	0.332	0.505	0.752	0.317	0.396
Crossover	0.311	0.51	0.933	0.286	0.986	0.867	0.963	0.992	0.952
Mutation	0.983	0.805	0.873	0.751	0.964	0.889	0.93	0.869	0.913
Migration	0.533	0.517	0.554	0.662	0.706	0.685	0.59	0.71	0.698
EliteSize	0.772	-	-	0.952	-	-	0.413	-	-
BestRate	0.018	-	-	0.624	-	-	0.061	-	-

Table 2. Parameters obtained with Bonesa for the encapsulated, distributed and hybrid algorithms

	4 ROBOTS			16 ROBOTS			36 ROBOTS		
	$\mu+1$	Distr	MK	$\mu+1$	Distr	MK	$\mu+1$	Distr	MK
evolutionSteps	308	303	305	308	301	302	308	583	567
stepSize	9.615	4.306	9.895	9.615	5.621	9.731	9.615	8.197	8.832
reEvaluation	0.091	0.647	0.385	0.091	0.558	0.449	0.091	0.002	0.048
Crossosver	0.19	0.399	0.828	0.19	0.122	0.847	0.19	0.1	0.31
Mutation	0.978	0.908	0.976	0.978	0.86	0.849	0.978	0.606	0.921
Migration	-	-	0.577	-	-	0.658	-	-	0.835
EliteSize	-	-	0.279	-	-	0.716	-	-	0.911
BestRate	-	-	0.198	-	-	0.703	-	-	0.013

Combined with the no-replacement admission policy, MultiKulti is either the best or at a par with the best migrant selection scheme, especially as the number of robots increases.

Finally, the ring topology shows a slight, but not always significant, drop in performance. This may be explained by the fact that in a ring topology, good solutions spread over the islands at a much slower rate than in a randomly connected topology.

Selecting migrants randomly seems always to lead to a smaller spread in performance than either selecting the best or the most different. Vis a vis the MultiKulti algorithm at least, this makes sense because this specifically aims at increasing population diversity, so a larger variation in performance is to be expected.

To conclude, we select a configuration with no-replacement admission policy, MultiKulti migrant selection and a random island topology to compare with the encapsulated and distributed algorithms.

4.2 Comparing Encapsulated, Distributed and Hybrid On-Line Evolution

The second question we asked ourselves is whether the optimal hybrid instance we selected in the previous section outperforms its encapsulated and distributed counterparts. Table 2 shows the settings that Bonesa reported as optimal for the three algorithms that we compare for groups of 4, 16 and 36 robots. Note that the optimal parameters for $\mu + 1$ have only been calculated once: since there is no interaction between robots, $\mu + 1$'s performance and settings are independent of the number of robots. Running 50 repeats with these settings resulted in performances as reported in Figure 4d.

Even for as small a number of robots as 4, the distributed and hybrid algorithms both significantly outperform the encapsulated algorithm. The difference between the algorithms that share the population across robots is only significant for 36 robots, but even there not material. The difference with the encapsulated algorithm may lie in the exploitation of evolution's inherent parallelism, but we

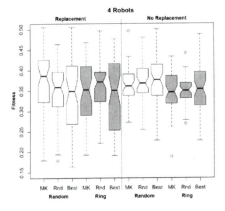

(a) Box plot of all hybrid configurations with 4 robots.

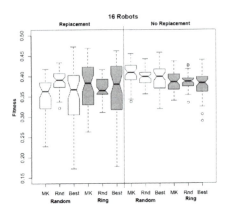

(b) Box plot of all hybrid configurations with 16 robots.

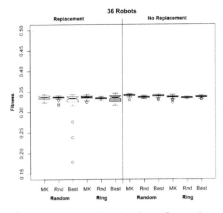

(c) Box plot of all hybrid configurations with 36 robots.

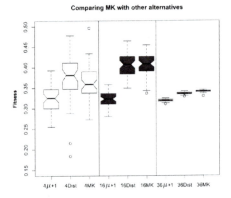

(d) Box plot comparing $\mu + 1$, distributed and multikulti migration with replacement for 4, 16 and 36 robots.

Fig. 4. Box plots of executing each algorithm with the best parameters obtained with Bonesa 50 times

think this is also due to the increased diversity that stems from dividing the total population across islands. This would explain the large benefit of communication even for small numbers of robots, where the distributed algorithm actually has a smaller total population than the individual robots with $\mu + 1$.

Set to the best found parameters, the hybrid algorithm causes much less communication overhead than the distributed algorithm: the latter shares genotypes among all robots at every evaluation, while the hybrid algorithm has comparatively low migration rates (0.577, 0.658 and 0.835). This reduction of communication cost comes at no significant loss of performance, and even a significant gain for 36 robots.

5   Conclusions and Future Work

In this paper, we compared combinations of migrant selection schemes, migrant admission policies and island topologies in a hybrid algorithm for on-line, on-board Evolutionary Robotics. Results show that the migrant admission policy –which determines when a migrant is admitted into the population– is more important in performance than migrant selection or the island topology. But the most important finding is that adding migration between robots significantly and materially increases performance. We have demonstrated that adding a difference-based migrant selection scheme (MultiKulti) leads to optimal or at least near-optimal performance compared to another migration mechanisms. This migration mechanism can compete with the on-line distributed algorithm, where only an individual per robot exist, even with a lower number of data transmissions. Our aim is to continue exploring other techniques, like a self-adaptive migration mechanism to ask for new migrants when the population stagnates and perform new tests for new tasks other than the Fast-Forward. New experiments with different number of individuals in the local population also will be carried out. Also, further investigation will be performed in swarming and cooperation techniques among robots, with different communication mechanisms.

References

1. Araujo, L., Merelo, J.J.: Diversity through multiculturality: Assessing migrant choice policies in an island model. IEEE Trans. Evolutionary Computation 15(4), 456–469 (2011)
2. Cantú-Paz, E.: Migration policies, selection pressure, and parallel evolutionary algorithms. Journal of Heuristics 7(4), 311–334 (2001)
3. Eiben, A.E., Haasdijk, E., Bredeche, N.: Embodied, on-line, on-board evolution for autonomous robotics. In: Levi, P., Kernbach, S. (eds.) Symbiotic Multi-Robot Organisms: Reliability, Adaptability, Evolution, vol. 10, pp. 361–382. Springer, Heidelberg (2010)
4. Haasdijk, E., Eiben, A.E., Karafotias, G.: On-line evolution of robot controllers by an encapsulated evolution strategy. In: Proceedings of the 2010 IEEE Congress on Evolutionary Computation. IEEE Computational Intelligence Society, IEEE Press, Barcelona, Spain (2010)
5. Huijsman, R.-J., Haasdijk, E., Eiben, A.E.: An on-line, on-board distributed algorithm for evolutionary robotics. In: Proceedings of the Biennial International Conference on Artificial Evolution, EA 2011 (2011) (to appear)
6. Laredo, J.L.J., Eiben, A.E., van Steen, M., Merelo, J.J.: Evag: a scalable peer-to-peer evolutionary algorithm. Genetic Programming and Evolvable Machines 11(2), 227–246 (2010)
7. Nolfi, S., Floreano, D.: Evolutionary Robotics: The Biology, Intelligence, and Technology of Self-Organizing Machines. MIT (2000)

8. Nordin, P., Banzhaf, W.: An on-line method to evolve behavior and to control a miniature robot in real time with genetic programming. Adaptive Behavior 5(2), 107–140 (1997)
9. Smit, S.K., Eiben, A.E.: Comparing parameter tuning methods for evolutionary algorithms. In: IEEE Congress on Evolutionary Computation, CEC 2009, pp. 399–406. IEEE (2009)
10. Smit, S.K., Eiben, A.E.: Multi-problem parameter tuning. In: Proceedings of the Biennial International Conference on Artificial Evolution, EA 2011 (2011) (to appear)
11. Watson, R.A., Ficici, S.G., Pollack, J.B.: Embodied evolution: Distributing an evolutionary algorithm in a population of robots. Robotics and Autonomous Systems 39(1), 1–18 (2002)

Evolutionary Optimization of Pheromone-Based Stigmergic Communication

Tüze Kuyucu, Ivan Tanev, and Katsunori Shimohara

Information Systems Design
Doshisha University, Kyotanabe, Japan
{tkuyucu,itanev,kshimoha}@mail.doshisha.ac.jp

Abstract. Pheromone-based stigmergic communication is well suited for the coordination of swarm of robots in the exploration of unknown areas. We introduce a guided probabilistic exploration of an unknown environment by combining random movement and stigmergic guidance. Pheromone-based stigmergic communication among simple entities features various complexities that have significant effects on the overall swarm coordination, but are poorly understood. We propose a genetic algorithm for the optimization of parameters related to pheromone-based stigmergic communication. As a result, we achieve human-competitive tuning and obtain a better understanding of these parameters.

1 Introduction

Members of the animal phyla utilize pheromones as a form of communication; from bacteria to mammals, no matter how capable an organism is of achieving direct communication and complex behaviours, pheromones play an important role in establishing an indirect communication mechanism for a variety of purposes [13]. Pheromones serve a number of functions for living organisms, including aggregation, attraction, alarm propagation, territorial marking and group decision making [13].

Pheromones provide a stigmergic medium of communication, which influence the future actions of a single or a group of individuals via changes made to the environment. Stigmergy allows the history of an individual's actions to be tracked without the need to construct a model of the environment within the individual's own memory; giving rise to the emergence of higher complexity behaviours from a group of simple individuals.

In this work, we are interested in achieving optimum exploratory behaviour via a large number of real robots in unknown environments. Such environments include areas of high devastation (e.g. earthquake or tsunami stricken areas) or distant and dangerous missions. Exploring an unknown area quickly is a mission-critical objective in rescue operations. Such operations can face a list of limitations, such as the lack of a terrain map, the failure of previously established communication networks and lack of reliable GPS tracking. In such missions, the first task is to search the area in question as quickly as possible and locate targets. The robots would be required to be capable of various functionalities other than area exploration, therefore it is desired that the integration to a swarm and the ability to search are seamless and do not consume a large amount of the robot's resources. Utilizing a real stigmergic communication would be an

C. Di Chio et al. (Eds.): EvoApplications 2012, LNCS 7248, pp. 63–72, 2012.

efficient method of achieving such emergent behaviour with low overhead. Robots that utilize stigmergic trails for communication have been shown to effectively coordinate and quickly explore a given terrain [1,12].

Beckers et al. successfully built a group of mobile robots that operate with a very simple algorithm and work collectively to achieve a foraging behaviour by using the objects gathered as sources of stigmergic information [1]. Russell developed short-lived navigational markers for small robots where heat trails were used to replace pheromone trails [10]. Svennebring and Koenig studied terrain-covering robots that used a black marker to form trails on the floor [12]. Ferranti et al demonstrated the use of the exploration of unknown environments via agents that make use of "tags" (such as motes or RFIDs) as stigmergic markers [3]. Purnamadjaja and Russel showed that aggregation with real robots using real chemicals is possible [8], who also later on use real chemicals to establish two way communication between robots [9].

Given that the hardware implementation of stigmergic coordination of robots is still in its early stages, it makes sense to investigate and improve pheromone-based algorithms in simulations. By understanding the optimal conditions required for pheromone-based coordination, the hardware implementations can then also be better directed. There are various works that utilize simulations and report notable improvements in pheromone-based exploration algorithms. However, most of these works focus on improving the overall performance of a given algorithm via design changes [4] or the optimization of a customized algorithm-specific parameters [11]. Due to the lack of a solid understanding of how parameters (such as diffusion rate, sampling frequency, etc.) that relate to the shaping of the stigmergic information via individual agents, there is no known formalization that can guide the optimal tuning of a stigmergy-based robot exploration algorithm. Consequently, it has been reported in many cases that optimising the various aspects of the stigmergy-based algorithms via artificial evolution creates human competitive results [2,6,11].

Furthermore, it is common to make some unrealistic assumptions in order to simplify the simulation environment; such as the ability of the individual robots to execute perfect discreet movements of fixed distances, and the ability to divide a given environment into a grid to allow discreet movement of robots [2,3].

The work presented here aims to demonstrate the use of pheromone-based communication in achieving efficient exploration of unknown environments via realistic robots. We utilize a simulation environment (Webots) that closely models the physics of two wheel differential robots and their interactions with the real world. In this work we also propose a Genetic Algorithm (GA) for the optimization of the parameters related to the characteristics of stigmergic markers (pheromones) and their sampling in order to achieve the best possible performance of the exploration. Our goal is not only to optimize the performance of the algorithm used but also to obtain an insight into the effects of parameters related to the sharing and storage of stigmergic information, and to the systematic and probabilistic guidance of robots in unknown environments.

We design an algorithm that is similar to the "trail avoidance" algorithm described in [12], but in our case the goal is to achieve a quick "survey" of an area by visiting key locations as quickly as possible, instead of visiting all the physical locations in the environment. In Section 2, we will first describe the pheromone-based algorithm we use in our experiments. We then detail the simulation experiments we have executed to demonstrate the performance of our algorithm in Section 3.1, followed by experiments with the optimization of these parameters with a GA in Section 3.2. We then conclude our work and provide some future directions in Section 4.

2   Pheromone-Based Stigmergic Coordination

As in biology, our model uses the environment as a medium of communication by leaving traces of pheromones. These pheromones are deposited by the robots and are also detected by them when they are within the proximity of the pheromone. Our objective is to develop a simple algorithm that can utilize pheromones to benefit from the existing physical properties of a real environment in achieving complex collective behaviour within a large group of simple homogeneous robots.

The aim of the pheromone-based control is to provide an indirect communication mechanism among a group of homogeneous robots. With the availability of various sensors, a range of environmental markers (such as chemicals, metals, heat sources, electronic tags) can be used as a way of encoding information in the environment. We use a three-layered subsumption architecture for the controllers of the robots: random walk ("exploration") is the lowest priority layer in this architecture, while the pheromone-based coordination is the middle layer, and the higher priority layer implements the wall avoidance behaviour (see Figure 1(a) for an illustration of the architecture). The adopted subsumption architecture provides us with a modular and scalable design that leaves the possibility to add higher-level control mechanisms for future development.

The algorithm used for the pheromone-based control is similar to the "trail avoidance behaviour" in [12], and it is detailed in Algorithm 1. Key differences in the algorithms include the ability of the algorithm presented here to use variable concentrations in depositing pheromones, ignoring pheromones below a certain concentration, evaporation and diffusion of pheromones, and a bias towards forward movement. In our case we have a bias towards making smaller turns when there are same levels of pheromone concentrations in more than one direction. There is even a stronger bias towards moving straight, and the robots choose to go straight even if the pheromone concentration in the forward facing direction is higher (than the lowest pheromone concentration) by an amount less than 1%. It is already shown that a biased movement can be more efficient for some exploration scenarios [4]. In our algorithm, the robots only check for the pheromone levels in 5 directions (NORTH, NORTH EAST, NORTH WEST, EAST, WEST; NORTH being the face forward direction), instead of all 8 directions in order to encourage forward movement.

The layer of the pheromone-based behaviour is used to take guiding decisions for a robot during exploration. On the other hand, the wall avoidance layer is always executed by default during the movement of a robot. The wall avoiding is accomplished according to the equations 1 and 2 (when an obstacle is detected by the sensors) for controlling the rotational speed (radians) of the left and right wheels ($v_{leftWheel}$ and $v_{rightWheel}$).

$$v_{leftWheel} = \frac{MS}{2} * \left(s_4 + \frac{MS}{2} * s_3 - s_1 - \frac{MS}{2} * s_2 \right) \tag{1}$$

$$V_{righttWheel} = \frac{MS}{2} * \left(s_1 + \frac{MS}{2} * s_2 - s_4 - \frac{MS}{2} * s_3\right) \tag{2}$$

The robot has four infra-red sensors: the values s_1, s_2, s_3 and s_4 range from 0.0 to 1.0, and the corresponding bearings of the sensors are $46.3^o, 17.2^o, -17.2^o, -46.3^o$ with respect to the forward facing direction of the robot. The constant MS in equations 1 and 2 denotes the maximum speed. The MS in our experiments is 2π radians per second, which is around $7cm/s$. The robot used in simulations is a differential wheeled robot that is roughly spherical in shape, and 7cm in diameter (occupies an area of $38.5cm^2$, similar to a commercially available e-puck robot).

Algorithm 1 provides the pseudo-code for the pheromone-based control, which is executed periodically, and the frequency of its execution can be adjusted by the experimenter. The pheromone deposition is carried out every time the algorithm is executed, after the robot has moved an arbitrary distance. The actual distance travelled by the robot is not regulated in order to provide a realistic simulation environment: we assume that depositing chemicals in precise intervals in a real environment is impractical.

In order to simulate both the presence and the diffusion of pheromones, we augmented the separately encoded diffusion layer over the simulated environment. This layer actively communicates with the robots in order to carry out the updates of levels of the pheromones. The diffusion of pheromones is carried out in three dimensions, i.e. the diffusion pattern from a source of pheromone is in the shape of a dome. By modifying the value of the diffusion constant we can model the markers that diffuse at different rates, including those markers that never diffuse (as proposed in [12]).

$$\Delta Y_{pa} = (Y_{paq} - Y_{qa})/DC \tag{3}$$

$$\Delta Y_{pa} = Y_{pa}/(DC * EC) \tag{4}$$

The amount of chemical that diffuses in the x-y plane is described by Equation 3: the flow (ΔY_{paq}) of chemical a from position p to position q is calculated as a difference in their concentrations Y. DC denotes the diffusion constant. In order to simulate the diffusion of chemicals, we partitioned the map into a grid of cells in the x-y plane and modelled the diffusion of chemicals from one cell to another every second using the Equation 3. The amount of chemical that evaporates into the air is described by Equation 4: the flow (ΔY_{pa}) of chemical a from position p to air is calculated by using the diffusion constant as well as another constant EC; the evaporation constant.

Algorithm 1. The Pheromone-based robot control.

1: $LPC =$ Lowest pheromone concentration
2: $NPC =$ Pheromone concentration in front of the robot
3: $HPC =$ Highest pheromone concentration
4: *Deposit pheromone*
5: **if**(HPC< DetectableConcentration) **then** *Go straight*
6: **else if**($LPC > 0.99 \times NPC$) **then** *Go straight*
7: **else** *Turn towards the LPC*

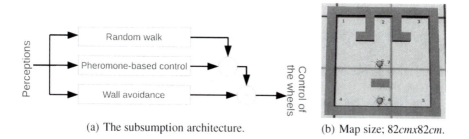

(a) The subsumption architecture.   (b) Map size; 82cmx82cm.

Fig. 1. The subsumption architecture utilized in this paper (a), and the experimental map with two robots to scale (b), respectively

Other than the parameters that determine the density of the chemicals (i.e. diffusion constant and evaporation constant), there are few other parameters that determine both the presence of chemicals in the explored map and consequently the behaviour of the robots. These parameters are: (i) *Production Rate:* the amount of pheromone, deposited every time a robot decides to leave a marker in the environment, (ii) *Detectable concentration:* the minimum amount of pheromone level required to trigger a decision based on nearby pheromone levels (as detailed in Algorithm 1), (iii) *Rate of sampling:* defines the frequency of execution of the control algorithm of the robots (in between robot movements), (iv) *Wall avoiding distance:* the threshold of the distance to a wall, at which the wall avoiding behaviour of the robots is engaged (as elaborated in equations 1 and 2). (v) *Resolution of the diffusion simulation map:* determines the grid size of the virtual map used for the diffusion of chemicals.

In the first set of experiments we set these parameters manually by using a common sense on how they relate to each other and via trial-and-error-correcting runs.

3   Experiments

For all the experiments presented in this section, except when stated otherwise, we use the manually set values for the parameters mentioned in Section 2. The values of these parameters are shown in Table 1. Travel time, which determines the *"rate of sampling"* listed in Section 2, has a minimum value of $\frac{32}{1000}$ since this is the time required (in seconds) for all the physics calculations required by each time step in Webots simulator.

Table 1. The manually set parameters that control the emergent pheromone communication, and the accepted range of each parameter during optimization runs

	Travel Time(s)	Detec. Conc.	Prod. Rate	Diff. Const	Evap. Const	Wall avoid dist(cm)
Man. Set Val.	1.0	2.0	10.0	100.0	8.0	4
Range	$\frac{32}{1000} \to 8$	$0.1 \to 100$	$1 \to 255$	$1 \to \infty$	$0.1 \to 100$	$0 \to \infty$

3.1   Exploration of Environment with Obstacles

Obstacle-free environments are commonly used in the experiments for the exploration of unknown environments by realistically simulated or real robots [7,12]. However, as demonstrated in [3], often the robot control algorithms do not perform well even when simple unconnected obstacles are introduced in the environment.

In this section, we consider a simple map that was initially modelled in the work on exploration of unknown environments, conducted by Kuremoto et al. [5]. This map (shown in Figure 1(b)) is referred to as "complicated" by the authors; due to the orientation of the featured obstacles, which create traps for robots that can easily get stuck at narrow corridors or small rooms, and unconnected obstacles. Here we present the results of our experiments on the same map in order to verify the performance of the proposed pheromone-based algorithm. Furthermore, in the next subsection we employ the same map of the environment in order to optimize the values of the parameters of the control algorithms of the robots (as listed in Table 1), and to attempt to understand their effects in the overall behaviour of the robots.

We study three experimental cases in this map as follows: **(i)** *No pheromones:* two robots with controllers featuring a wall avoiding (as detailed in Section 2) and a random walk (exploring) behaviours, **(ii)** *Pheromones without diffusion:* two robots with controllers featuring a pheromone-based behaviour. No diffusions of pheromones are simulated in this case (i.e. the diffusion constant is ∞), **(iii)** *Pheromones with diffusion:* two robots with the same controllers as considered in the previous case, with the only difference that the diffusion of the pheromones is modelled in the environment.

Each experimental case is tested for 10 runs until 7 key points on the map are visited by at least one robot (shown in Figure 1(b)). The considered experimental task differs slightly from the task described by Kuremoto et al. In the latter case, (i) the robots are aware about the target locations that have to be visited and (ii) there are only three key locations that need to be visited (locations marked as 1, 2 and 5 in Figure 1(b)).

Table 2. The results of runs exploring the environment shown in Figure 1(b)

	Pheromones with Diff.	Pheromones No Diff.	No Pheromones
Avg. time (s)	126	305	2690
Success Rate	100%	100%	50%

The experimental results are summarised in Table 2. The time to finish a run is averaged over 10 runs for each experimental case. The lack of pheromones in the first of the considered cases implies an unguided exploration. However, even in this simple case the robots still managed to complete the task of exploring the whole map 50% of the time. On the other hand, the use of pheromones as well as their diffusion, significantly improved the performance of the team of robots, with the latter being able to explore the key locations in a reasonable amount of time. Kuremoto et al report that an earlier model, which they present improvements on, is unable to explore this map due to the earlier model's inability to circumvent the obstacles efficiently [5]. We can not directly compare our model with the "improved model" due to a lack of quantitative data.

3.2   Optimizing the Parameters

The pheromone-based stigmergic communication among simple entities for coordination and cooperation of their group-wise actions features various complexities that are directly affected by the pheromones. Often, the emergent behaviour might be greatly altered by small changes to the parameters responsible of pheromone-based communication. The complexities, involved in the design of the various aspects of both the morphology and the behaviour of the entities, motivated a number of researchers to use simulated evolution as a heuristic approach for the design of the optimized conduct of agents that results in a desired group-wise, emergent behaviour. For example, Sauter et al. used GA to tune the control-specific parameters to achieve an efficient use of pheromones in guiding a team of robots to locate targets and paths [11]. The authors stated that the evolved parameters, "consistently outperformed the best hand-tuned parameters that took skilled programmers over a month to develop." Panait and Luke evolved genetic programs to control agents in foraging tasks in an obstacle-free environment [6]. Similar to the ant-colonies in nature, their approach used pheromones as trails between the food sources and the nest. The authors reported that the ability to deposit the maximum amount of pheromone provides a quicker foraging behaviour among agents. Recently, Connelly et al. [2], also showed that utilizing simulated evolution to optimize the cooperative use of pheromones results in a better agent-based behaviour in locating targets. We can generalize that the use of evolutionary approaches offer a fast, automated optimization of the exploration algorithms utilizing pheromone-based communication. In this section we propose GA for automated evolution of the optimal values of control parameters of robots. In addition, by analyzing the evolved optimal behaviour of robots exploring an unknown area, we will try to understand the emergent use of pheromone-based communication.

It is recognized that there is no established formal model that describes adequately the influence of the control parameters of the robots on the overall performance of the team. Therefore, we set the values of these parameters (listed in Table 1) via informed guessing and trial-and-error correcting runs. For example, we choose a travel time of 1*second* as a good solution to the exploration-exploitation dilemma: indeed, it allows the robot to have some occasional probabilistic movements (exploration), rather than being completely obsessed by the pheromone-based guidance (exploitation), determined by eventual lower values. On the other hand, using a longer travel time usually makes the movement random, resulting in a worsened overall performance of the team of robots. This is consonant with the observations, reported by Svennebring and Koening [12]. Some of the remaining parameters in our initial experiments, set by the same trial-and-error approach, are the diffusion constant (set to 100), the evaporation constant (set to 8), and the wall avoiding distance (4*cm*).

While well-performing team of robots did emerge for the selected values of control parameters, neither the degree of optimality of these values nor the way to incrementally improve them is evident. Thus, an automated mechanism for evaluation of these values of parameters, and corresponding rules for incremental optimization of the intermediate values of parameters (e.g. based on various models of learning or evolution of species in the nature) are needed. The proposed approach of employing GA implies that the sets of values of all six control parameters (Table 1), that govern the behaviour of robots communicating via pheromones, are represented as linear chromosomes and automatically tuned by a computer system via simulated evolution through selection and survival of the fittest, in a way similar to the evolution of species in the nature. The main attributes of the utilized GA are as follows: *population size* is 200 chromosomes, 4 chromosomes are picked every generation as *elite* individuals, a *mutation rate* of 4% is used, *selection process* involves 10% binary tournament selection and 90% crossover, an *evaluation run* lasts a maximum of 200 seconds, and the *termination criteria* is reaching the total number of generations of 40. We have chosen to set the aforementioned parameters as described relying on our experience with previous experimental runs.

$$Fitness = No_{goals} \times Time_{max} - No_{goalsReached} \times (Time_{max} - Time_{termination} + 1) \quad (5)$$

The experimental environment is as described in Section 3.1; Figure 1(b) is also used for the runs here. The fitness values assigned to each parameter set after a run is calculated as shown in Equation 5. $Time_{max}$ is the maximum evaluation period (which is 200 seconds), and $Time_{termination}$ is the amount of time required for the bots to reach all the goals (this value is set to $Time_{max}$ if the robots are unable to reach all the goals). We evolved the values of the six control parameters in 10 independent runs of GA, and as a result obtained various well-performing combinations of these values. The results are shown in Figure 2(a).

From 10 evolutionary runs of GA, 7 runs evolved parameter settings that result in a performance of the robots that is superior to the manually adjusted parameters. The correspondence of two of the evolved optimal parameters and the manually set parameters are shown in Table 3. As Table 3 illustrates, the pheromone production rate is set to higher values and the diffusion constants are set to lower values (i.e. high diffusion rates) in the evolved parameters. These settings make sure that the released pheromones diffuse out covering as much space as possible and forming pheromone gradients all over the environment. Using these evolved parameters for multiple reruns, we could observe that the diffusing pheromones would actually provide a better guidance for the robots with ambiguous cases (i.e. areas with same pheromone concentrations) for the robots encountering pheromones. The obstacle distance for engaging the wall avoidance behaviour is also significantly lower for the evolved parameters than the manually adjusted. When manually adjusting this value, we were cautious not to set it too low because otherwise the robots could run into obstacles or worse get stuck in the corners, as it happened in some of the evolved cases. However, setting this value high prevents the robots getting close to walls, and entering narrow corridors. Evolution was able to adjust it better, and the quick formation of pheromone gradients due to high pheromone production rates and diffusion rates also meant that robots getting too close to corners would easily find their way out with the help of the pheromone-based control.

Table 3. The best two evolved sets of parameters and the corresponding manually adjusted values

	Travel Time(s)	Detec. Conc.	Prod. Rate	Diff. Const	Evap. Const	Wall avoid dist(cm)
Manually Set Value	1.0	2.0	10.0	100.0	8.0	4
Evolved 1	1.19	36.6	94.0	86.5	45.9	0.53
Evolved 1	2.0	0.5	227.0	20.9	4.7	0.61

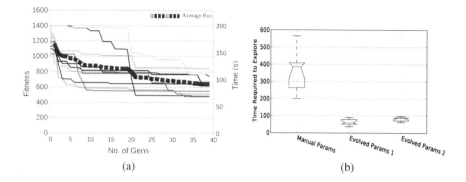

Fig. 2. The fitness convergence characteristics of 10 independent evolutionary runs of GA are plotted in Figure 2(a): the fitness value is proportional to the trial time required for the robots explore the experimental map. The actual corresponding time of the trial is shown in the right hand ordinate.The box and whisker plots of the results from runs utilizing the parameters shown in Table 3 in the exploration of the environment shown in Figure 1(b) is displayed above in Figure 2(b): the data for each case represents 10 runs.

4   Conclusions

In this work we provided a simple algorithm that is aimed at providing real robots with quick and effective exploration capabilities of completely unknown environments via stigmergic communication. We experimentally show that in an environment with multiple obstacles, the proposed pheromone-based exploration algorithm performs well.

Also, we demonstrated that by using GA to automatically tune the values of parameters of pheromone-based communication in a multi-robot system, we could obtain such values of parameters that compared to the manually tuned values of parameters, yield a much better performance of the team of robots exploring an unknown environment. By using simulated evolution, we showed that higher diffusion rate of pheromones as well as a larger amount of deposited pheromone result in significantly improved exploration capabilities. Even obtained in a different context, these findings are consonant with the results reported in [6], and [4]. From the optimized parameters, it was also observed that a robot movement of ≈ 1.5 seconds in between pheromone-based guidance is optimal in balancing exploration vs exploitation behaviours. Finally, the obtained results indicate that using too cautious, manually tuned wall avoiding behaviour that is engaged too far away from the walls might be detrimental to the overall performance. Using GA to optimise the sensitivity of the wall avoiding behaviour provided a better tuning than we could have manually done.

The future directions of this work include: **(i)** investigating the optimal values of evolved parameters for different maps (different complexity, size, and layout) and larger number of robots, **(ii)** expanding the model described here by using attractive pheromones in order to achieve different behaviours such as foraging and collective retrieval of oversized objects is one of the future goals, **(iii)** improving the pheromone-based algorithm in order to achieve a well-scalable coordination in teams comprising very large number of robots, and **(iv)** symbolic regression (via genetic programming) of the multiple sets of evolved optimal values of parameters in order to infer the mathematical model (if any) that describes their relationship.

The presented research was supported (in part) by the Japan Society for the Promotion of Science (JSPS)

References

1. Beckers, R., Holl, O.E., Deneubourg, J.L.: From local actions to global tasks: Stigmergy and collective robotics. In: Artificial Life IV (1996), pp. 181–189. MIT Press (1994)
2. Connelly, B.D., McKinley, P.K., Beckmann, B.E.: Evolving cooperative pheromone usage in digital organisms. In: IEEE Symposium on Artificial Life, ALife 2009, March 3-April 2, pp. 184–191 (2009)
3. Ferranti, E., Trigoni, N., Levene, M.: Rapid exploration of unknown areas through dynamic deployment of mobile and stationary sensor nodes. Autonomous Agents and Multi-Agent Systems 19, 210–243 (2009)
4. Fu, J.G.M., Ang, M.H.: Probabilistic ants (pants) in multi-agent patrolling. In: IEEE/ASME International Conference on Advanced Intelligent Mechatronics, AIM 2009 (July 2009)
5. Kuremoto, T., Obayashi, M., Kobayashi, K., Feng, L.B.: An improved internal model of autonomous robots by a psychological approach. Cognitive Computation, 1–9 (2011)
6. Panait, L.A., Luke, S.: Learning ant foraging behaviors. In: Proceedings of the Ninth International Conference on the Simulation and Synthesis of Living Systems ALIFE9 (2004)
7. Payton, D., Estkowski, R., Howard, M.: Pheromone Robotics and the Logic of Virtual Pheromones. In: Şahin, E., Spears, W.M. (eds.) Swarm Robotics 2004. LNCS, vol. 3342, pp. 45–57. Springer, Heidelberg (2005)
8. Purnamadjaja, A.H., Russel, R.A.: Congregation behaviour in a robot swarm using pheromone communication. In: Australasian Conference on Robotics and Automation (2005)
9. Purnamadjaja, A.H., Russel, R.A.: Bi-directional pheromone communication between robots. Robotica 28, 69–79 (2010)
10. Russell, R.A.: Heat trails as short-lived navigational markers for mobile robots. In: Proceedings of IEEE International Conference on Robotics and Automation, vol. 4, pp. 3534–3539 (April 1997)
11. Sauter, J.A., Matthews, R., Parunak, H.V.D., Brueckner, S.: Evolving adaptive pheromone path planning mechanisms. In: Proceedings of the First International Joint Conference on Autonomous Agents and Multiagent Systems: Part 1, AAMAS 2002, pp. 434–440. ACM, New York (2002)
12. Svennebring, J., Koenig, S.: Building terrain-covering ant robots: A feasibility study. Auton. Robots 16, 313–332 (2004)
13. Wyatt, T.D.: Pheromones and Animal Behaviour: Communication by Smell and Taste. Cambridge University Press (2003)

Hyperparameter Tuning in Bandit-Based Adaptive Operator Selection

Maciej Pacula, Jason Ansel, Saman Amarasinghe, and Una-May O'Reilly

CSAIL, Massachusetts Institute of Technology, Cambridge, MA 02139, USA
{mpacula,jansel,saman,unamay}@csail.mit.edu

Abstract. We are using bandit-based adaptive operator selection while autotuning parallel computer programs. The autotuning, which uses evolutionary algorithm-based stochastic sampling, takes place over an extended duration and occurs *in situ* as programs execute. The environment or context during tuning is either largely *static* in one scenario or *dynamic* in another. We rely upon adaptive operator selection to dynamically generate worthy test configurations of the program. In this paper, we study how the choice of hyperparameters, which control the trade-off between exploration and exploitation, affects the effectiveness of adaptive operator selection which in turn affects the performance of the autotuner. We show that while the optimal assignment of hyperparameters varies greatly between different benchmarks, there exists a single assignment, for a context, of hyperparameters that performs well regardless of the program being tuned.

1 Introduction

We are developing an autotuning technique, called SiblingRivalry, based upon an evolutionary algorithm (EA) which tunes poly-algorithms to run efficiently when written in a new programming language we have designed. The autotuner runs in two different kinds of computing environments: static or dynamic. In either environment, multiple execution times and accuracy of results will vary to different degrees. Using special software infrastructure, the online technique, embedded and running in the run-time system, is able to continuously test candidate poly-algorithm configurations in parallel with the best configuration to date whenever a program is invoked. The technique generates a candidate configuration by selecting one of a set of specific mutation operators that have been derived for it during the program's compilation. If it finds a better configuration, it makes a substitution and continues. We call this process "racing". The technique needs to generate candidate configurations that both explore poly-algorithm space and exploit its knowledge of its best configuration.

The choice of which mutation operator to use is vital in optimizing the overall performance of the autotuner, both in time to converge to efficient programs and their answer quality. Some mutation operators will have large effects on program performance, while others will have little or no effect. If the evolutionary algorithm spends too much time exploring different mutation operators,

C. Di Chio et al. (Eds.): EvoApplications 2012, LNCS 7248, pp. 73–82, 2012.
© Springer-Verlag Berlin Heidelberg 2012

convergence will be slow. If the evolutionary algorithm spends too much time trying to exploit mutation operators that have yielded performance gains in the past, it may not find configurations that can only be reached through mutation operators that are not sufficiently tested. Noise in program performance due to execution complicates the picture and make optimal mutation operator selection imperative.

To address this challenge, SiblingRivalry uses what we call "bandit-based adaptive operator selection". Its underlying algorithm is the Upper Confidence Bound (UCB) algorithm, which is a technique inspired by a provably optimal solution to the Multi-Armed Bandit (MAB) problem. This technique introduces two hyperparameters: W - the length of the history window, and C - the balance point between exploration and exploitation. UCB is only optimal if these hyperparamters are set by an oracle or through some other search technique. In practice, a user of this technique must either use a fixed, non-optimal assignment of these hyperparameters, or perform a search over hyperparameters whenever the search space changes. Unfortunately, in practice, finding good values of these hyperparameters may be more expensive that the actual search itself. While [5] addresses the robustness of hyperparameters in empirical academic study, in this paper, we present a practically motivated, real world study on setting hyperparameters. We define evaluation metrics that can be used in score functions that appropriately gauge the autotuner's performance in either a static or dynamic environment and use them to ask:

- How much does the optimal assignment of hyperparameters vary when tuning different programs in two classes of environments - static or dynamic?
- Does there exist a single "robust" assignment of hyperparameters for a context that performs close to optimal across all benchmarks?

The paper proceeds as follows: in Section 2 we provide a necessarily brief description of our programming language and its autotuner. Section 3 reviews related work. Section 4 describes the UCB algorithm and the hyper parameters. Section 5 describes our evaluation metrics and scoring functions for tuning the hyperparameters. Section 6 provides experimental results. Section 7 concludes.

2   PetaBricks and Its Autotuner

PetaBricks is a language designed specifically to allow the programmer to expose both explicit and implicit choices to an integrated autotuning system [1,2]. The goal of the PetaBricks autotuner is to, on each machine, find a program that satisfies the user's accuracy requirements while minimizing execution time. Accuracy is a programmer-defined metric, while execution time is measured by running the program on the given hardware. Given a program, execution platform and input size, the autotuner must identify an ideal configuration which is a set of algorithmic choice and cutoff selectors, synthetic functions for input size transforms and a set of discrete tunable parameters. The autotuner is an evolutionary algorithm which uses a program-specific set of mutation operators.

These mutation operators, generated by the compiler, each target a specific single or a set of tunable variables of the program that collectively form the genome. For example, one mutation operator can randomly change the scheduling policy for a specific parallel region of code. Another set of mutation operators can randomly add, remove, or change nodes (one mutation operator for each action) in a decision tree used to dynamically switch between entirely different algorithms provided by the user.

3 Related Work and Discussion

In the context of methods in evolutionary algorithms that provide parameter adjustment or configuration, the taxonomy of Eiben [4] distinguishes between offline "parameter tuning" and online "parameter control". Operator selection is similar to parameter control because it is online. However, it differs from parameter control because the means of choosing among a set of operators contrasts to refining a scalar parameter value.

Adaptive methods, in contrast to self-adaptive methods, explicitly use isolated feedback about past performance of an operator to guide how a parameter is updated. An adaptive operator strategy has two components: operator credit assignment and an operator selection rule. The *credit assignment* component assigns a weight to an operator based on its past performance. An operator's performance is generally measured in terms related to the objective quality of the candidate solutions it has generated. The *operator selection rule* is a procedure for choosing one operator among the eligible set based upon the weight of each. There are three popular adaptive methods: probability matching, adaptive pursuit and multi-armed bandit. Fialho has authored (in collaboration with assorted others) a large body of work on adaptive operation selection, see, for example, [5,6]. The strategy we implement is multi-armed bandit with AUC credit assignment. This strategy is comparison-based and hence invariant to the scale of the fitness function which can vary significantly between PetaBricks programs. The invariance is important to the feasibility of hyperparameter selection on a general, rather than a per-program, basis.

There is one evolutionary algorithm, differential evolution [10], that takes a comparison-based approach to search like our autotuner. However, differential evolution compares a parent to its offspring, while our algorithm is not always competing parent and offspring. The current best solution is one contestant in the competition and its competitor is not necessarily its offspring. Differential evolution also uses a method different from applying program-dependent mutation operators to generate its offspring.

4 Adaptive Operator Selection

Selecting optimal mutators online, while a program executes numerous times over an extended duration, can be viewed as an instance of the Multi-Armed Bandit problem (MAB), with the caveats described in [8]. We would like to explore the efficacy of all mutators so that we can make an informed selection of one of them. The MAB resolves the need to optimally balance exploration and exploitation in a way that maximizes the cumulative outcome of the system.

In the general case, each variation operator corresponds to one of N arms, where selecting i-th arm results in a reward with probability p_i, and no reward with probability $1 - p_i$. A MAB algorithm decides when to select each arm in order to maximize the cumulative reward over time [8]. A simple and provably optimal MAB algorithm is the *Upper Confidence Bound (UCB)* algorithm, originally proposed by Auer *et al.* [3]. The empirical performance of the UCB algorithm has been evaluated on a number of standard GA benchmarks, and has been shown to be superior to alternative adaptive operator selection techniques such as Probability Matching [8].

The UCB algorithm selects operators according to the following formula:

$$\text{Select } \underset{i}{\arg\max} \left(\hat{q}_{i,t} + C \sqrt{\frac{2 \log \sum_k n_{k,t}}{n_{i,t}}} \right) \tag{1}$$

where $\hat{q}_{i,t}$ denotes the empirical quality of the i-th operator at time t (exploitation term), $n_{i,t}$ the number of times the operator has been used so far during a sliding time window of length W (the right term corresponding to the exploration term), and C is a user defined constant that controls the balance between exploration and exploitation [3,8]. To avoid dividing by zero in the denominator, we initially cycle through and apply each operator once before using the UCB formula, ensuring $n_{i,t} \geq 1$.

Our PetaBricks autotuner uses the *Area Under the Receiving Operator Curve (AUC)* to compute the empirical quality of an operator. AUC is a comparison-based credit assignment strategy devised by Fialho et al. in [7]. Instead of relying on absolute average delta fitness, this method ranks candidates generated by a mutator i, and uses the rankings to define the mutator's *Receiving Operator Curve*, the area under which is used as the empirical quality term $\hat{q}_{i,t}$ (Equation 1). To extend this method to variable accuracy, we use the following strategy: If the last candidate's accuracy is below the target, candidates are ranked by accuracy. Otherwise, candidates are ranked by throughput (inverse of time).

5   Tuning the Tuner

The hyperparameters C (exploration/exploitation trade-off) and W (window size) can have a significant impact on the efficacy of SiblingRivalry. For example, if C is set too high, it might dominate the exploitation term and all operators will be applied approximately uniformly, regardless of their past performance. If, on the other hand, C is set too low, it will be dominated by the exploitation term $\hat{q}_{i,t}$ and new, possibly better operators will rarely be applied in favor of operators which made only marginal improvements in the past.

The problem is further complicated by the fact that the optimal balance between exploration and exploitation is highly problem-dependent [5]. For example, programs with a lot of algorithmic choices are likely to benefit from a high exploration rate. This is because algorithmic changes create discontinuities in the program's fitness, and operator weights calculated for a given set of algorithms will not be accurate when those algorithms suddenly change. When such changes occur, exploration should become the dominant behavior. For other programs, e.g. those where only a few mutators improve performance, sacrificing exploration in favor of exploitation might be optimal. This is especially true for programs with few algorithmic choices - once the optimal algorithmic choices have been made, the autotuner should focus on adjusting cutoffs and tunables using an exploitative strategy with a comparatively low C.

The optimal value of C is also closely tied to the optimal value of W, which controls the size of the history window. The autotuner looks at operator applications in the past W races, and uses the outcome of those applications to assign a quality score to each operator. This is based on the assumption that an operator's past performance is a predictor of its future performance, which may not always be true. For example, changes in algorithms can create discontinuities in the fitness landscape, making past operator performance largely irrelevant. However, if W is large, this past performance will still be taken into account for quite some time. In such situations, a small W might be preferred.

Furthermore, optimal values of C and W are not independent. Due to the way $\hat{q}_{i,t}$ is computed, the value of the exploitation term grows with W. Thus by changing W, which superficially controls only the size of the history window, one might accidentally alter the exploration/exploitation balance. For this reason, C and W should be tuned together.

5.1   Evaluation Metrics

Because there is no single metric that will suffice to evaluate performance under different hyperparameter values, we use three separate metrics to evaluate SiblingRivalry on a given benchmark program with different hyperparameters:

1. **Mean throughput**: the number of requests processed per second, averaged over the entire duration of the run. Equal to the average number of races per second.
2. **Best candidate throughput**: inverse of the runtime of the fastest candidate found during the duration of the run. For variable accuracy benchmarks, only candidates that met the accuracy target are considered.
3. **Time to convergence**: number of races until a candidate has been found that has a throughput within 5% of the best candidate for the given run. For variable accuracy benchmarks, only candidates that met the accuracy target are considered.

To enable a fair comparison between SiblingRivalry's performance under different hyperparameter values, we define a single objective metric for each scenario that combines one or more of the metrics outlined above. We call this metric the *score function* f_b for each benchmark b, and its output the *score*.

We consider two classes of execution contexts: static and dynamic. In the static context, the program's execution environment is mostly unchanging. In this setting, the user cares mostly about the quality of the best candidate. Convergence time is of little concern, as the autotuner only has to learn once and then adapt very infrequently. For the sake of comparison, we assume in this scenario the user assigns a weight of 80% to the best candidate's throughput, and only 20% to the convergence time. Hence the score function for the static context:

$$f_b(C, W) = 0.8 \times \text{best\_throughput}_b(C, W) + 0.2 \times \text{convergence\_time}_b^{-1}(C, W)$$

In the dynamic context, the user cares both about average throughput and the convergence time. The convergence time is a major consideration since execution conditions change often in a dynamic system and necessitate frequent adaptation. Ideally, the autotuner would converge very quickly to a very fast configuration. However, the user is willing sacrifice some of the speed for improved convergence time. We can capture this notion using the following score function:

$$f_b(C, W) = 0.5 \times \text{mean\_throughput}_b(C, W) + 0.5 \times \text{convergence\_time}_b^{-1}(C, W)$$

We normalize throughput and convergence time with respect to their best measured values for the benchmark, so that the computed scores assume values in the range $[0, 1]$, from worst to best. Note that those are theoretical bounds: in practice it is often impossible to simultaneously maximize both throughput and convergence time.

6   Experimental Results

We evaluated the hyperparameter sensitivity of SiblingRivalry by running the autotuner on a set of four benchmarks: Sort, Bin Packing, Image Compression and Poisson. We used twenty different combinations of C and W for each benchmark: $(C, W) = [0.01, 0.1, 0.5, 5, 50] \times [5, 50, 100, 500]$.

For each run, we measured the metrics described in Section 5.1 and used them to compute score function values. Due to space constraints, we focus on the resulting scores rather than individual metrics (we refer the curious reader to [9] for an in-depth analysis of the latter). We performed all tests on the Xeon8 and AMD48 systems (see Table 1). The reported numbers for Xeon8 have been averaged over 30 runs, and the numbers for AMD48 over 20 runs. The benchmarks are described in more detail in [2].

Table 1. Specifications of the test systems used

Acronym	Processor Type	Operating System	Processors
Xeon8	Intel Xeon X5460 3.16GHz	Debian 5.0	2 (×4 cores)
AMD48	AMD Opteron 6168 1.9GHz	Debian 5.0	4 (×12 cores)

	static context				dynamic context			
	Xeon8		AMD48		Xeon8		AMD48	
	C	W	C	W	C	W	C	W
Sort	50.00	5	5.00	5	5.00	5	5.00	5
Bin Packing	0.01	5	0.10	5	5.00	500	5.00	500
Poisson	50.00	500	50.00	500	0.01	500	5.00	5
Image Compression	0.10	100	50.00	50	0.01	100	50.00	50

(a) Best performing values of the hyperparameters C and W over an empirical sample.

	static context		dynamic context	
	Xeon8	AMD48	Xeon8	AMD48
Sort	0.8921	0.8453	0.9039	0.9173
Bin Packing	0.8368	0.8470	0.9002	0.9137
Poisson	0.8002	0.8039	0.8792	0.6285
Image Compression	0.9538	0.9897	0.9403	0.9778

(b) Scores of the best performing hyperparameters.

Fig. 1. Best performing hyperparameters and associated score function values under static and dynamic autotuning scenarios

Figures 2 and 3 show select scores as a function of C and W on the Xeon8 amd AMD48 systems for benchmarks in both static and dynamic scenarios. All benchmarks except Image Compression show moderate to high sensitivity to hyperparameter values, with Bin Packing performance ranging from as low as 0.1028 at $(C, W) = (0.01, 5)$ to as high as 0.9002 at $(C, W) = (5, 500)$ in the dynamic scenario on the Xeon8. On average, the dynamic context was harder to autotune with a mean score of 0.6181 as opposed to static system's 0.6919 (Figure 4). This result confirms the intuition that maintaining a high average throughput while minimizing convergence time is generally more difficult than finding a very high-throughput candidate after a longer autotuning process.

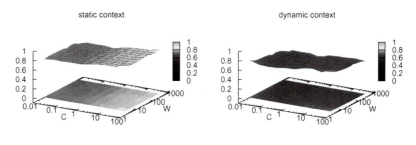

(a) Sort on Xeon8

Fig. 2. Scores for the Sort benchmark as a function of C and W. The colored rectangle is a plane projection of the 3D surface and is shown for clarity.

The optimal hyperparameter values for each benchmark ranged considerably and depended on both the scenario and the architecture (Table 1). Sort tended to perform best with a moderate C and a low W, underlining the importance of exploration in the autotuning process of this benchmark. Bin Packing in the static context favored a balance between exploration and exploitation of a small number of recently tried operators. In the dynamic context Bin Packing performed best with much longer history windows (optimal $W = 500$) and with only a moderate exploration term $C = 5$. This is expected as Bin Packing in the dynamic context is comparatively difficult to autotune and hence benefits from a long history of operator performance. Poisson was another "difficult" benchmark, and as a result performed better with long histories ($W = 500$ for almost all architecures and contexts). In the static scenario it performed best with a high $C = 50$, confirming the authors' intuition that exploration is favorable if we are given more time to converge. In the dynamic context exploration was favored less (optimal $C = 0.01$ for the Xeon8 and $C = 5$ for the AMD48). In the case of Image Compression, many hyperparameters performed close to optimum suggesting that it is an easy benchmark to tune. Medium W were preferred across architectures and scenarios, with $W = 100$ and $W = 50$ for the static and dynamic contexts, respectively. Image Compression on AMD48 favored a higher $C = 50$ for both scenarios, as opposed to the low $C = 0.1$ and $C = 0.01$ for the static and dynamic contexts on the Xeon8. This result suggests exploitation of a limited number of well-performing operators on the Xeon8, as opposed to a more explorative behavior on the AMD48. We suspect this is due to a much higher parallelism of the AMD48 architecture, where as parallelism increases different operators become effective.

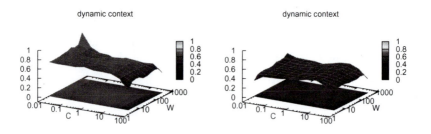

(a) Poisson on Xeon8 (left) and AMD48 (right)

Fig. 3. Measured scores for the Poisson benchmark on each architecture

6.1 Hyperparameter Robustness

Our results demonstrate that autotuning performance can vary significantly depending on the selection of hyperparameter values. However, in a real-world setting the user cannot afford to run expensive experiments to determine which values work best for their particular program and architecture. For this reason,

Table 2. Benchmark scores for the globally optimal values of hyperparameters normalized with respect to the best score for the given benchmark and scenario. The optimal hyperparameters were $C = 5$, $W = 5$ for the static context, and $C = 5$, $W = 100$ for the dynamic context. Mean normalized scores were 88.32% and 82.45% for the static and dynamic contexts, respectively.

	static context		dynamic context	
	Xeon8	**AMD48**	**Xeon8**	**AMD48**
Sort	95.71%	100%	74.16%	61.12%
Bin Packing	85.61%	94.72%	67.42%	88.74%
Poisson	70.64%	71.09%	90.77%	96.07%
Image Compression	92.44%	96.35%	89.92%	91.42%

we performed an empirical investigation whether there exists a single assignment of C and W that works well across programs and architectures.

We used the score functions from Section 5.1 to find hyperparameters that maximized the mean score on all the benchmarks. We found that the hyperparameters $(C, W) = (5, 5)$ for the static context and $(C, W) = (5, 100)$ for the dynamic context maximized this score. The results are shown in Table 2. For the sake of illustration, we normalized each score with respect to the optimum for the given benchmark and scenario (Table 1(b)).

Despite fixing hyperparameter values across benchmarks, we measured a mean normalized score of 88.32% for the static and 82.45% for the dynamic context, which means that we only sacrificed less than 20% of the performance by not tuning hyperparameters on a per-benchmark and per-architecture basis. This result shows that the hyperparameters we found are likely to generalize to other benchmarks, thus providing sensible defaults and removing the need to optimize them on a per-program basis. They also align with our results for individual benchmarks (Figure 1), where we found that exploration (moderate to high C, low W) is beneficial if we can afford the extra convergence time (static context), whereas exploitation (low to moderate C, high W) is preferred if average throughput and low convergence time are of interest (dynamic context).

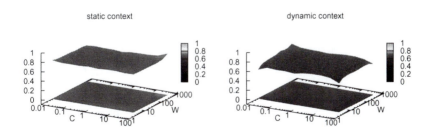

Fig. 4. Scores for the static and dynamic scenarios averaged over the Sort, Bin Packing, Poisson and Image Compression benchmarks and the Xeon8 and AMD48 architectures. The mean scores across all benchmarks, architectures and hyperparameter values were 0.6919 for the static and 0.6181 for the dynamic contexts.

7   Conclusions

We performed a detailed experimental investigation of hyperparameter effect on the performance of the PetaBricks autotuner, a real-world online evolutionary algorithm that uses adaptive operator selection. We evaluated four benchmarks with respect to three metrics which we combined into a performance indicator called the score function, and demonstrated that optimal hyperparameter values differ significantly between benchmarks. We also showed how two possible autotuning scenarios can affect the optimal hyperparameter values. We further demonstrated that a single choice of hyperparameters across many benchmarks is possible, with only a small performance degradation. Such a choice provides sensible defaults for autotuning, removing the need for the user to tune hyperparameters per-program, and thus making our approach feasible in a real-world setting.

References

1. Ansel, J., Chan, C., Wong, Y.L., Olszewski, M., Zhao, Q., Edelman, A., Amarasinghe, S.: Petabricks: A language and compiler for algorithmic choice. In: ACM SIGPLAN Conference on Programming Language Design and Implementation, Dublin, Ireland (June 2009)
2. Ansel, J., Wong, Y.L., Chan, C., Olszewski, M., Edelman, A., Amarasinghe, S.: Language and compiler support for auto-tuning variable-accuracy algorithms. In: International Symposium on Code Generation and Optimization, Chamonix, France (April 2011)
3. Auer, P., Cesa-Bianchi, N., Fischer, P.: Finite-time analysis of the multiarmed bandit problem. Mach. Learn. 47, 235–256 (2002)
4. Eiben, A., Hinterding, R., Michalewicz, Z.: Parameter control in evolutionary algorithms. IEEE Transactions on Evolutionary Computation 3(2), 124–141 (1999)
5. Fialho, Á.: Adaptive Operator Selection for Optimization. PhD thesis, Université Paris-Sud XI, Orsay, France (December 2010)
6. Fialho, Á., Da Costa, L., Schoenauer, M., Sebag, M.: Analyzing bandit-based adaptive operator selection mechanisms. Annals of Mathematics and Artificial Intelligence – Special Issue on Learning and Intelligent Optimization (2010)
7. Fialho, Á., Ros, R., Schoenauer, M., Sebag, M.: Comparison-Based Adaptive Strategy Selection with Bandits in Differential Evolution. In: Schaefer, R., Cotta, C., Kołodziej, J., Rudolph, G. (eds.) PPSN XI. LNCS, vol. 6238, pp. 194–203. Springer, Heidelberg (2010)
8. Maturana, J., Fialho, Á., Saubion, F., Schoenauer, M., Sebag, M.: Extreme com-pass and dynamic multi-armed bandits for adaptive operator selection. In: CEC 2009: Proc. IEEE International Conference on Evolutionary Computation, pp. 365–372. IEEE Press (May 2009)
9. Pacula, M.: Evolutionary algorithms for compiler-enabled program autotuning. Master's thesis, Massachusetts Institute of Technology, Cambridge, MA (2011)
10. Price, K., Storn, R.M., Lampinen, J.A.: Differential Evolution: A Practical Approach to Global Optimization. Natural Computing Series. Springer-Verlag New York, Inc., Secaucus (2005)

Analyzing Dynamic Fitness Landscapes of the Targeting Problem of Chaotic Systems

Hendrik Richter

HTWK Leipzig University of Applied Sciences
Faculty of Electrical Engineering & Information Technology
Postfach 30 11 66, D–04251 Leipzig, Germany
richter@eit.htwk-leipzig.de

Abstract. Targeting is a control concept using fundamental properties of chaotic systems. Calculating the targeting control can be related to solving a dynamic optimization problem for which a dynamic fitness landscape can be formulated. We define the dynamic fitness landscape for the targeting problem and analyze numerically its properties. In particular, we are interested in the modality of the landscape and its fractal characteristics.

1 Introduction

Chaotic behavior is defined by time evolutions that depend highly sensitively on tiny perturbations to the initial conditions and/or parameter values. This poses fundamental limits to the long–term prediction of the trajectory. Targeting is a control concept for chaotic systems that explicitly takes advantage of this property [20,21]. It allows to steer the system towards a target on the chaotic attractor by using a small control input only [11,4,17]. Because of the sensitivity of chaotic dynamics, calculating the control input that actually drives the system closest to the target requires solving an dynamic optimization problem. A popular choice for providing such a solution are methods of evolutionary computation [5,13,7], which in turn can be based on the theoretical concept of fitness landscapes [12,22,23]. A fitness landscape gives a model of the optimization problem and allows to describe how the evolutionary algorithm interacts with it. Hence, the concept permits conclusions on how difficult the problem is for an evolutionary algorithm and what solution behavior can be expected.

In the paper we consider dynamic fitness landscapes for the targeting problem. We first describe the connection between the targeting problem and optimal control. In the Sec. 3, we briefly review dynamic fitness landscapes and show how the targeting problem can be formulated as a dynamic fitness landscape. Following this, we define, calculate and analyze the landscapes numerically in Sec. 4. The paper closes with concluding remarks and a discussion about open question.

C. Di Chio et al. (Eds.): EvoApplications 2012, LNCS 7248, pp. 83–92, 2012.

2   Optimal Control and Targeting

An optimal control problem can be formulated as to consist of a discrete–time dynamical system

$$x(k+1) = f(x(k), u(k)) \tag{1}$$

with the state variable $x(k) \in \mathbb{R}^n$, the control input $u(k) \in \mathbb{R}^m$, the discrete time variable k of a time set $\mathbb{N}_0$, and an equation describing how the next state $x(k+1)$ is generated from the current state $x(k)$ and current input $u(k)$, $f : \mathbb{R}^n \times \mathbb{R}^m \to \mathbb{R}^n$. We intend to find a control input $u(k)$ that drives the system (1) from an initial state $x(0)$ to a final state $x(T)$. Moreover, amongst all control input sequences

$$u = (u(0), u(1), u(2), \dots, u(T-1)) \tag{2}$$

that actually drive the system from $x(0)$ to $x(T)$, we are supposed to find the one u_S that minimizes a cost function J, that is we are looking for the control input with minimal cost

$$J_S = \min_{u(0),u(1),\dots,u(T-1)} J\left(x(0), u(0), u(1), \dots, u(T-1)\right) \tag{3}$$

and hence $u_S = arg\, J_S$ is the actual solution.

Optimal control problems arise in a huge variety of engineering problems [1,2]. We here consider a special kind of optimal control problems, the so–called targeting problem of chaotic systems. An interesting property of a nonlinear dynamical system (1) is that it may exhibit chaotic behavior for certain initial states $x(0)$ and constant control inputs $u(k) = \bar{u} = const$. Chaotic behavior means that the system trajectory is highly sensitive against tiny perturbations of the initial state $x(0)$ and/or the control input $\bar{u}$. Applying such tiny perturbations results in exponential divergence of nearby trajectories. In other words, chaos implies that the system is locally instable (in the sense of Lyapunov), but globally settles on a bounded and closed subset of the state space $\mathbb{R}^n$. No trajectory starting from this subset escapes towards infinity. The subspace built by the chaotic trajectory is the chaotic attractor A_R, which has but for exceptional cases a fractal dimension. This is linked with orbit density in such a sense that the trajectory comes arbitrarily close to all points embedded in the chaotic attractor.

Targeting addresses and employs these properties of chaotic systems, which is why it can only be applied here. The targeting problem poses the following question: Is it possible to steer a chaotic system by a bounded control input

$$U = \{u \in \mathbb{R}^m | \|u(k) - \bar{u}\| \leq \eta\} \tag{4}$$

with $\eta > 0$ being small from any initial state $x(0)$ on the chaotic attractor A_R within T time steps to the neighborhood of any target point $\bar{x}$ on the chaotic attractor. That is we intend to achieve $\|x(T) - \bar{x}\| \leq \epsilon$ with $\epsilon > 0$ being a small constant.

To specify and restrict the control sequence (2) even more, targeting of chaotic systems can be achieved by only using the control input at the initial time $k = 0$:

$$u = (u(0), \bar{u}, \ldots, \bar{u}). \tag{5}$$

Observing (5), we denote the multiple application of (1) by $f\left(f(x(0), u(0), \bar{u}\right) = f^2(x(0), u(0))$ and so on, and can write the cost function of the targeting problem and the corresponding optimization problem

$$J_S = \min_{u(0) \in U} \left\| f^T(x(0), u(0)) - \bar{x} \right\|. \tag{6}$$

We interpret this cost function of the targeting problem as a fitness function over the search space U. Hence, the function (6) in connection with the state equation (1) and the control input (5) also defines an optimal control problem and consequently the optimization problem to be solved. We show next that it also constitutes a dynamic fitness landscape.

3   Dynamic Fitness Landscapes

In the following we draw a connection between the targeting problem (6) and dynamic fitness landscapes. We start with asking what the essence of a dynamic optimization problem is. Therefore, in turn, we first look at a static optimization problem. It consists of an objective function (frequently equated with a fitness function in evolutionary computation) $F(s)$ defined over a search space S with $s \in S$. The search space might need a neighborhood structure $n(s)$ that gives every $s \in S$ a set of neighbors, if it is not intrinsic to S. Optimization means to find the lowest (or highest) value of $F(s)$ and its coordinates among all $s \in S$:

$$F_S = \min_{s \in S} F(s), \tag{7}$$

with the location $s_S = arg\ F_S$. A fitness function together with a search space and a neighborhood structure builds a static fitness landscape [12,23].

The static optimization problem (7) can be thought of as becoming dynamic by solving it not just once, but somehow modified for a second time. The modification might be an alteration of the fitness function, or the search space, or the neighborhood structure, or a combination of them. If we restrict ourselves to a modified fitness function $F^*(s)$ (for the other modifications the following applies likewise), we may write the modified problem as

$$F_S^* = \min_{s \in S} F^*(s). \tag{8}$$

To rewrite the two static problems (7) and (8) as one dynamic problem, we introduce the time variable $i \in \mathbb{N}_0$ and define the dynamic fitness function $F(s, i)$ where we set

$$F(s, 0) = F(s), \quad F(s, 1) = F^*(s).$$

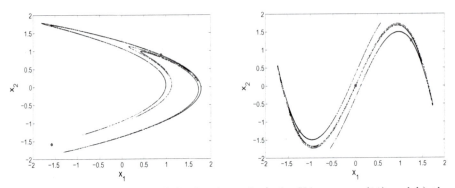

Fig. 1. Chaotic attractors and fixed points of: a) the Hénon map (11) and b) the Holmes map (12)

We may carry on with doing so for the next modification of $F(s)$ to obtain $F(s,2)$, and so on. Hence, a dynamic optimization problem is

$$F_S(i) = \min_{s \in S} \quad F(s,i), \quad \forall\, i \geq 0 \tag{9}$$

with the solution trajectory $s_S(i) = arg\, F_S(i)$. A dynamic fitness function (9) together with a search space, a neighborhood structure and an evolution law describing how $F(s,i+1)$ is generated builds a dynamic fitness landscape [18,19].

Looking at the cost function of the targeting problem (6), we see that in our interpretation it equates with the fitness function (7), which is clearly static. This is in line with evolutionary approaches to solve optimal control problems, see for instance [16,3,6,14], which have essentially in common to tackle static cost functions. Of course, the term $f^T(x(0), u(0))$ depends on the discrete time variable k of the dynamical system (1), but in which way this time variable is in relation to the time variable i of the dynamic problem (9) needs to be specified. One way to do so is to ask how the problem (6) evolves if we were to vary the final target time T. Note that by varying T a problem is addressed that has some similarity to questions asked and principles involved in dynamic programming. In both cases one thing we want to know is how the optimal control policy looks like if we were to wait for one more time step.

In summary, the dynamic fitness landscape of the targeting problem is

$$J(u(0), i) = \left\| f^i(x(0), u(0)) - \bar{x} \right\| \tag{10}$$

with search space U according to (4) (and hence an inherent neighborhood structure) and $J(u(0), i+1) = \left\| f^{i+1}(x(0), u(0)) - \bar{x} \right\|$.

4   Numerical Studies

For studying the dynamic fitness landscapes for the targeting problem numerically, we consider two discrete–time dynamical system according to (1), the

Hénon map [8]

$$x(k+1) = \left(u(k) - x_1(k)^2 + 0.3x_2(k), \qquad x_1(k) \right)^T \tag{11}$$

and the Holmes map [9]

$$x(k+1) = \left(x_2(k), \qquad -0.2x_1(k) + u(k)x_2(k) - x_2(k)^3 \right)^T. \tag{12}$$

Both systems show chaotic behavior, map (11) for $u(k) = \bar{u} = 1.4$ and map (12) for $u(k) = \bar{u} = 2.77$. The systems have slightly different chaoticity with Lyapunov exponent $\lambda = 0.38$ for (11) and $\lambda = 0.59$ for (12). Both Lyapunov exponents are positive, indicating first and foremost that the dynamics is chaotic, but also that the behavior is locally instable. As target points we use the systems' fixed points, which are

$$\bar{x}^{(1,2)} = \frac{1}{2} \left(-0.7 \pm \sqrt{0.49 + 4\bar{u}}, -0.7 \pm \sqrt{0.49 + 4\bar{u}} \right)^T$$

for the Hénon map and

$$\bar{x}^{(1,2)} = \pm \left(\sqrt{\bar{u} - 1.2}, \sqrt{\bar{u} - 1.2} \right)^T, \qquad \bar{x}^{(3)} = (0,0)^T.$$

for the Holmes map. Fig. 1 shows the chaotic attractors of both systems and the location of the fixed points. We notice that for the Hénon map the fixed points are asymmetric and only the point $\bar{x}^{(1)}$ is embedded in the chaotic attractor. The other point lays outside of it and is hence not generally accessible by targeting. For the Holmes map all three fixed points are embedded and $\bar{x}^{(1,2)}$ are symmetric to the line $x_2 = x_1$.

As result of a first experiment, we depict the dynamic fitness landscape for the targeting problem using the Hénon system (11) with $\bar{x}^{(1)}$. As in this case the search space is one–dimensional, we can view its dynamics in a 2D plot, see Fig. 2. On the horizontal dimension the timely evolution of the landscape is shown, in the vertical dimension is the spatial evolution. The values of the fitness (10) are given as a color code according to the colorbar next to the graph. Fig 2a shows the landscape for the relevant parameter range $1.36 \leq u(0) \leq 1.44$, which is $\eta = 0.04$ according to (4). Outside of this parameter range, targeting is not robustly observable. For larger values, the Hénon map tends to become instable, for smaller values it is not chaotic. For the numerical study, the search space with the given parameter range is divided into $par = 3200$ partitions. We vary the target time for $1 \leq i \leq 40$. It can be seen that targeting is not generally possible for small target times. Of course, for $i = 4$ we get close to the target, but this is mainly due to the initial state used $x(0) = (-1.0, 1.5)^T$ for the given example that the system's trajectory reaches the vicinity of the target point. Also, this happens for a large range of control inputs $u(0)$. The real targeting effect (control that is sensitive to the initial state and allows to reach the target) can be observed for approximately $i > 15$. Values of $u(0)$ that create trajectories that come close to the target point are next to values with contrary

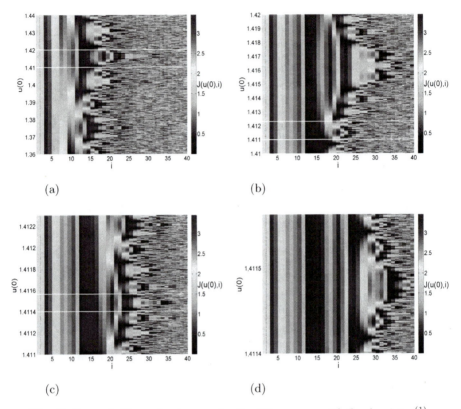

Fig. 2. Dynamic fitness landscapes for the Hénon map with fixed point $\bar{x}^{(1)}$

results. This also has effect on the dynamic fitness landscape. For small values of i it has a small modality or is even unimodal. Also the spatial ruggedness is low. Moreover, also temporal ruggedness (that is how does the landscape at i predicts the landscape at $i + 1$) appears to be lower compared to higher values of i. For these values the landscape has an increased modality and also spatial and temporal ruggedness. For increasing i further, this becomes more prominent. For about $i > 25$, the targeting effect is clearly visible. Input values have a very sensitive effect on fitness.

At the first glance, the landscape appear to be fractal, which is a characteristics reported for static fitness landscapes [10,24]. Fractal here means that the fitness landscape has self–similar structures on all spatial scales. It also has the consequence that the number of local mimima increases as we go from one scale to the next smaller one. To study if it is fractal, we look at the small scale structure of the landscape and zoom into it, see Fig. 2b,c,d. Fig 2b enlarges the region within white lines in Fig. 1a, Fig. 2c is the same for Fig. 2b, and so on. The number of partitions is kept. We notice that for the scale $1.36 \leq u(0) \leq 1.44$, see Fig. 2a, the graph suggests a self–similar structure for approximately $i > 25$, indicating that no regular color pattern are visible. All colors and hence all levels

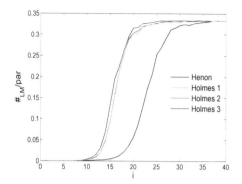

Fig. 3. Modality of the dynamic fitness landscapes for the Hénon map with fixed point $\bar{x}^{(1)}$ and the Holmes map with fixed points $\bar{x}^{(1)}$ (Holmes 1), $\bar{x}^{(2)}$ (Holmes 2), and $\bar{x}^{(3)}$ (Holmes 3)

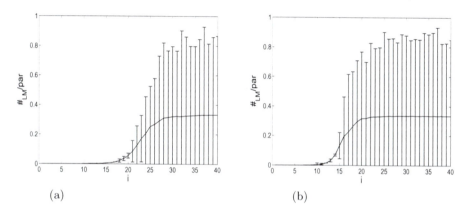

(a)   (b)

Fig. 4. Modality of the dynamic fitness landscapes and the 95% confidence intervals for 100 samples of $x(0)$. a) Hénon map with $\bar{x}^{(1)}$, b) Holmes map with $\bar{x}^{(3)}$.

of fitness can appear next to each other. In a second experiment we look at how the number of local optima depends on the landscape dynamics. We consider both systems, and all fixed points reachable by targeting. The results are shown in the Figs. 3 and 4. The number of local minima is calculated numerically. As in the example the search space is one–dimensional, and because of the numerical approach considered here, the search space consist of a connected set of partitions. We can sort the $u(0)$ according to their size by $u_j(0) < u_{j+1}(0) < u_{j+2}(0)$, with $j = 1, 2, \ldots, par - 2$. Here, par is the number of partitions, with $par = 3200$ in the given experiment. A $u_{j+1}(0)$ is a local minimum if the condition

$$J(u_j(0), i) \geq J(u_{j+1}(0), i) \leq J(u_{j+2}(0), i) \tag{13}$$

is met. For a given $x(0)$ we denote the number by $\#_{LM}(x(0))$. As the actual number of local minima might vary for different initial conditions $x(0)$, we

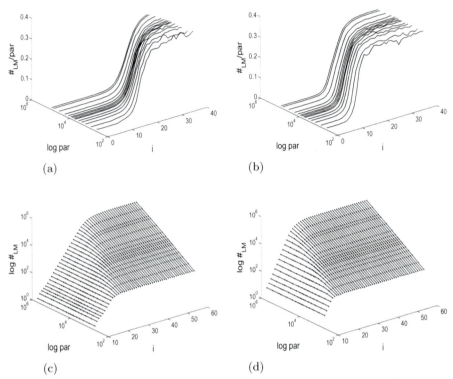

Fig. 5. The normalized and absolute modality over logarithmic number of partitions and targeting time i. a) and c) Hénon map with $\bar{x}^{(1)}$, b) and d) Holmes map with $\bar{x}^{(3)}$

calculate the mean value $\#_{LM} = \langle \#_{LM}(x(0)) \rangle$ over 500 samples of $x(0)$ on the chaotic attractor. Fig. 3 shows the normalized value $\frac{\#_{LM}}{par}$. It can bee seen that for approximately $i > 10$ the number of local minima increases. This happens for the Holmes map for smaller value of i that for the Hénon map. This is most likely because the Holmes map is slightly more chaotic than the Hénon map, indicated by the difference in the Lyapunov exponent. For the Holmes map, the three fixed points show similar results, with the point $\bar{x}^{(3)} = (0,0)^T$ slightly before the other two symmetric points. Fig. 4 gives the 95% confidence intervals over the 500 samples of initial states. We see that not only the average number of local minima increase for i becoming larger, but also the variance, which underlines the stochastic aspects in chaotic dynamics. The variance for the Holmes map is slightly larger than for the Hénon map, which is again to be attributed to the effect of the former being slightly more chaotic that the latter.

Next, we further analyze the small scale structure of the landscape. We therefore vary the number of partitions par and note the normalized number of local minima $\frac{\#_{LM}}{par}$ we obtain, see Fig 5 a,b. It can be seen that for the number of partitions getting larger the increase in the normalized number of minima is slightly postponed, compare to Fig. 3. But the final value that we obtain is the same. This strongly hints at a fractal relationship between $\#_{LM}$ and par.

A fractal relationship would require to establish a fractal dimensionality D where there should apply

$$\log(\#_{LM}) = -D\log(par) \tag{14}$$

with $D > 0$, see e.g. [15]. In Fig. 5 c,d the number of local minima $\#_{LM}$ and numer of partitions par are given on logarithmic scale over the target time i. In the logarithmic plots we notice a linear relationship for larger values of i that support the characteristics (14). Again, for Holmes map chaos this happens for smaller values of i than for Hénon map chaos. In conclusion, the results suggest fractality for the dynamic fitness landscape of the targeting problem.

5 Conclusion

Targeting is a control concept for chaotic systems that explicitly takes advantage of fundamental properties of deterministic chaos, namely orbit density and sensitive dependence on initial conditions and parameter values. Calculation the control input requires solving a dynamic optimization problem. Recently, several approaches using methods of evolutionary computation have been suggested [5,13,7]. In the paper, we defined, calculated and analyzed the dynamic fitness landscape underlying the targeting problem. In particular, we have shown evidence that the landscape has fractal characteristics, and modality of the landscape depends on the target time. Based on the preliminary results presented here, further experiments with larger target times and a wider and finer range of the control input should be conducted. Also, the spatial distribution of the fractal and granular subspaces of the dynamic fitness landscape should be examined.

References

1. Bertsekas, D.P.: Dynamic Programming and Optimal Control, vol. 1. Athena Scientific, Belmont (2005)
2. Bertsekas, D.P.: Dynamic Programming and Optimal Control, vol. 2. Athena Scientific, Belmont (2007)
3. Bobbin, J., Yao, X.: Solving optimal control problems with a cost on changing control by evolutionary algorithms. In: Bäck, T., Michalewicz, Z., Yao, X. (eds.) Proc. 1997 IEEE International Conference on Evolutionary Computation (ICEC 1997), pp. 331–336. IEEE Press, Piscataway (1997)
4. Bollt, E.M.: Targeting control of chaotic systems. In: Chen, G., Yu, X., Hill, D.J. (eds.) Chaos and Bifurcations Control: Theory and Applications, pp. 1–25. Springer, Berlin (2003)
5. Cai, X., Cui, Z.: Using stochastic dynamic step length particle swarm optimization to direct orbits of chaotic systems. In: Sun, F., Wang, Y., Lu, J., Zhang, B., Kinsner, W., Zadeh, L.A. (eds.) Proc. 9th IEEE Int. Conf. on Cognitive Informatics (ICCI 2010), pp. 194–198. IEEE Press (2010)

6. Fleming, P.J., Purshouse, R.C.: Evolutionary algorithms in control systems engineering: A survey. Control Engineering Practice 10, 1223–1241 (2002)
7. Gao, W.F., Liu, S.Y., Jiang, F.: An improved artificial bee colony algorithm for directing orbits of chaotic systems. Applied Mathematics and Computation 218, 3868–3879 (2011)
8. Hénon, M.: A two-dimensional mapping with a strange attractor. Commun. Math. Phys. 50, 69–77 (1976)
9. Holmes, P.J.: A nonlinear oscillator with a strange attractor. Philos. Trans. R. Soc. London A 292, 419–448 (1979)
10. Hoshino, T., Mitsumoto, D., Nagano, T.: Fractal fitness landscape and loss of robustness in evolutionary robot navigation. Autonomous Robots 5, 199–213 (1998)
11. Iplikci, S., Denizhan, Y.: Targeting in dissipative chaotic systems: A survey. Chaos 12, 995–1005 (2002)
12. Kallel, L., Naudts, B., Reeves, C.R.: Properties of fitness functions and search landscapes. In: Kallel, L., Naudts, B., Rogers, A. (eds.) Theoretical Aspects of Evolutionary Computing, pp. 177–208. Springer, Heidelberg (2001)
13. Liu, B., Wang, L., Jin, Y.H., Tang, F., Huang, D.X.: Directing orbits of chaotic systems by particle swarm optimization. Chaos, Solitons & Fractals 29, 454–461 (2006)
14. Lopez Cruz, I.L., Van Willigenburg, L.G., Van Straten, G.: Efficient Differential Evolution algorithms for multimodal optimal control problems. Applied Soft Computing 3, 97–122 (2003)
15. Mandelbrot, B.B.: The Fractal Geometry of Nature. Freeman, New York (1983)
16. Michalewicz, Z., Janikow, C.Z., Krawczyk, J.B.: A modified genetic algorithm for optimal control problems. Computers and Mathematics with Applications 23, 83–94 (1992)
17. Paskota, M., Mees, A.I., Teo, K.L.: Geometry of targeting of chaotic systems. Int. J. Bifurcation and Chaos 5, 1167–1173 (1995)
18. Richter, H.: Coupled map lattices as spatio–temporal fitness functions: Landscape measures and evolutionary optimization. Physica D237, 167–186 (2008)
19. Richter, H.: Evolutionary Optimization and Dynamic Fitness Landscapes: From Reaction–Diffusion Systems to Chaotic CML. In: Zelinka, I., Celikovsky, S., Richter, H., Chen, G. (eds.) Evolutionary Algorithms and Chaotic Systems. SCI, vol. 267, pp. 409–446. Springer, Heidelberg (2010)
20. Shinbrot, T., Grebogi, C., Ott, E., Yorke, J.A.: Using chaos to direct trajectories to targets. Phys. Rev. Lett. 65, 3215–3218 (1990)
21. Shinbrot, T., Grebogi, C., Ott, E., Yorke, J.A.: Using small perturbations to control chaos. Nature 363, 411–417 (1993)
22. Smith, T., Husbands, P., Layzell, P., O'Shea, M.: Fitness landscapes and evolvability. Evolut. Comput. 10, 1–34 (2002)
23. Stadler, P.F., Stephens, C.R.: Landscapes and effective fitness. Comm. Theor. Biol. 8, 389–431 (2003)
24. Weinberger, E.D., Stadler, P.F.: Why some fitness landscapes are fractal. J. Theor. Biol. 163, 255–275 (1993)

Self-organization and Specialization in Multiagent Systems through Open-Ended Natural Evolution

Pedro Trueba, Abraham Prieto, Francisco Bellas, Pilar Caamaño, and Richard J. Duro

Integrated Group for Engineering Research, Universidade da Coruña, Spain
{pedro.trueba,abprieto,fran,pcsobrino,richard}@udc.es
http://www.gii.udc.es

Abstract. This paper deals with the problem of autonomously organizing the behavior of a multiagent system through a distributed approach based on open-ended natural evolution. We computationally simulate life-like dynamics and their evolution from the definition of local and low level interactions, as used in Artificial Life simulations, in a distributed evolutionary algorithm called ASiCo (Asynchronous Situated Coevolution). In this algorithm, the agents that make up the population are situated in the environment and interact in an open-ended fashion, leading to emergent states or solutions. The aim of this paper is to analyze the capabilities of ASiCo for obtaining specialization in the multiagent system if required by the task. Furthermore, we want to study such specialization under changing conditions to show the intrinsic self-organization of this type of algorithm. The particular task selected here is multi-robot collective gathering, due to the suitability of ASiCo for its application to real robotic systems.

1 Open-Ended Natural Evolution

An open-ended natural evolution strategy is a complex system made up of a set of elements, both active (agents) and passive, that interact in the environment either physically or in some other more information intensive manner (i.e. communications). This interaction is not driven by a preset synchronization mechanism as in other algorithms, but by the result of the behaviors of the active agents in the environment and their interactions with other active and passive elements within it. The key parameter in these systems is energy, which is used to regulate the population dynamics and the efficiency of the individuals through the association of the success of their performance of tasks and the variations in their energy levels.

The evolution process in this strategy is clearly asynchronous as agents mate as a result of their individual behaviors, for instance when they meet and have a high enough energy level, and not all of the agents mate at the same time. It is also local, as the mating or information transfer process only depends on the agents that are interacting and not on the whole population as in traditional evolution. It is situated and embodied as the agents are the individuals that make up the population and evolve while "living" within the environment. Finally, it takes place in an open-ended fashion without and explicit stopping criterion. Intrinsic to this type of strategy are self-organization and adaptation to the environment where the agents "live" as usual in complex systems.

C. Di Chio et al. (Eds.): EvoApplications 2012, LNCS 7248, pp. 93–102, 2012.

Open-ended natural evolution has been mainly used in a bottom-up approach, with the main objective of analysing the properties of the emergent system [1]. But through the establishment of an adequate energetic regulation it is possible to guide evolution towards a fixed objective and let the evolution process obtain the behavior of the agents to achieve such objective in a collective manner.

A few years ago [2], we implemented such an open-ended natural evolution algorithm called ASiCo (Asynchronous Situated Coevolution) to be applied in the case of distributed optimization problems that could be solved with a multiagent approach. It has provided successful results in different application domains like routing [2], cleaning [3] or shipping freight [4] problems, showing the validity of the approach. For a detailed explanation of ASiCo working we recommend [3], but in next section we will briefly present its main elements and working to clarify the paper objective.

1.1   ASiCo

Asynchronous Situated Co-evolution (ASiCo) performs a coevolutionary process where the individuals are situated in a scenario and their interactions, including repro-duction, are local and depend on spatial and/or temporal coincidence, making the algorithm intrinsically decentralized. Consequently, ASiCo is highly suitable for distributed and dynamic problems. In this algorithm the *scenario* must be the problem itself instead of the usual representation of it. Furthermore, unlike in traditional evolutionary algorithms, where each individual represents a solution, the whole population makes up the solution in ASiCo. Fig. 1 displays the pseudocode of the algorithm.

```
Do Initialization:
        For each element ∈ Scenario
                For each parameter ∈ element
                        Random generation within parameter range
    :End
    While Scenario active:
        For each interaction ∈ Interaction set
            if(interaction(i).preconditions == TRUE)
                Interaction(i).modifyparameters
        For each individual
            if(elimination.preconditions == TRUE)
                    individual(j).elimination
            if(mating.preconditions == TRUE)
                    new individual = individual(j).mating(partner)
    :End
```

Fig. 1. ASiCo pseudocode

The rules that define the interactions among *elements* (individuals and Environment components) and between elements and the scenario are called the *Interaction Set* and, as usual in complex systems, they imply cost in terms of energy and sometimes may result in an energetic gain. The *energy flow strategy* represents the rules that regulate energy variations and transmission between the individuals and the scenario and vice versa. ASiCo uses the *principled evaluation function selection* procedure developed by Agogino and Tumer [5], a formal procedure to obtain the individual utility function from the global function, in order to define the energy flow strategy towards the desired task objective. This way, we are able to use open-ended natural evolution in a top-down fashion.

A very important element of ASiCo is the reproductive strategy. In the case of using fixed size populations it has been named Embryo Based Reproduction (EBR). The background idea in EBR is that each agent carries, in addition to its own parameters, another set of parameters corresponding to its embryo and an associated pre-utility value for the embryo that estimates its utility. Thus, when a new robot is introduced, its embryo is generated as a mutation of the parent genotype with half of its energy. During the life of an agent, the embryo is modified whenever the robot meets another robot and evaluates it positively (accepted candidate), meaning that the average of the utility of the two parents is higher than the pre-utility of the current embryo and the affinity criteria are met. Finally, when the parent dies because it ran out of energy or time or for whatever other reason, the embryo substitutes the parent. This way, it is ensured that the size of the population remains constant and that the process takes place in an asynchronous and decentralized manner.

1.2   Problem Definition

After presenting the open-ended natural evolution particularized to multiagent systems through the ASiCo algorithm, in this section we are going to define the specific problem we have faced in this paper.

Self-organization is an intrinsic property of open-ended natural systems. In fact, once a stable state is achieved, if something changes in the environment or in the task definition, the population will adapt to the new conditions or become extinct. An important aspect of self-organization in multiagent systems is specialization, that is, the emergence of heterogeneous individuals if required by the task. It has been shown that heterogeneous systems are more appropriate in those tasks that can be naturally decomposed into a set of complementary sub-tasks like collective gathering, collective communication or multi-agent computer games [6].

Evolution in ASiCo is situated, so genetic exchange depends on the spatial coincidence of individuals. Consequently, if a task is divided into subtasks that are spatially separated, species can emerge easily. But if the subtasks do not imply such a spatial separation, the reproduction operators must be carefully implemented because it could easily appear a bias towards homogeneous populations, limiting the emergence of specialization.

In ASiCo, EBR uses a Bipolar Crossover to exchange genetic information between individuals. To avoid the aforementioned bias, it is based on the idea that an offspring must have a larger probability of being similar to one of its parents than of being a 50% crossover between them. Consequently, once the two parent chromosomes have been selected ($C_A[g_{A1}, g_{A2}, g_{A3}, \ldots, g_{AM}]$, $C_B[g_{B1}, g_{B2}, g_{B3}, \ldots, g_{BM}]$), one of them is chosen randomly as the base chromosome. Each gene of the offspring is then generated using a Gaussian-like probability function centered on the corresponding gene of the base chromosome and with an amplitude based on the deviation ($D_i = |g_{Aj}-g_{Bj}|$) between the value of the corresponding genes in both parents:

$$g_{Nj} = N(g_{Aj}, |g_{Aj} - g_{Bj}|)$$

where $N()$ is a normal distribution.

Mutation also follows a Gaussian-like function but, in this case, it is not D_i dependent. With this, when the population has converged to a stable state, D_i tends to zero and there is almost no change on the genetic code of new generations. Only mutation makes some slight variations that allow a "fine tuning" of the solution.

Thus, the objective of this paper is to analyze if Bipolar Crossover and EBR allow for emergent specialization in ASiCo and under which conditions. To do it, we will study a problem where the cooperation among different species or types of agents is necessary to achieve the goal. Specifically, we have designed a multi-robot experiment due to the suitability of this type of approach to the robotics field, as will be explained in the next sub-section.

1.3   Evolution in Multi-robot Systems

Obtaining coordinated behaviors in multi-robot systems is a complex problem that has been faced using evolutionary algorithms for decades [7][8]. The typical approach consists on evolving the robot controller offline using a simulator and then transfer the best result to the real platforms. This approach is easy to implement, the computational cost is not a restriction and implies a low risk of robot damage. But its main drawback is that, if one seeks for real autonomous robots, it is not affordable to contemplate all the possible environmental conditions and tasks. Thus, it is necessary to carry out the evolution in real time, and here is where it is highly suitable an open-ended strategy.

This approach is not new, and Watson et al. proposed in 2002 [9] the Embodied Evolution (EE) methodology, where each real robot embodies an individual of the population and an open-ended evolution is carried out in real time until a solution is achieved. Different authors have continued the EE approach in robotics, although with two clearly different perspectives. In the original EE methodology, each robot executed a single controller and the genetic exchange was performed in a local and asynchronous manner depending on the robots encounters in the space. This is what has been named *distributed evolution* [10], and ASiCo follows this approach. It has been demonstrated in several examples that auto-organization and adaptation to dynamic environments is possible using this strategy [3][11][12]. The other perspective starts from the premise that distributed evolution requires a large number of robots in the team to avoid premature convergence of the solution due to the fact that the team size is the population size, and consequently it cannot be applied in small robot teams. As a consequence, the authors in [13][14] propose an EE variation where each robot carries a whole population of controllers and an independent evolutionary algorithm runs on it. This approach has been called *encapsulated evolution* [10] and it has provided successful results in tasks with a small number of real robots [13][14] (even with only one). Encapsulated evolution is out of the scope of this paper because ASiCo is an intrinsically distributed algorithm in which the agents or robots must be simple elements and the power comes from the interactions among them, so a small number of robots is counterproductive.

None of the previous works belonging to the *distributed evolution* approach have studied the emergence of specialization of this type of open-ended natural evolution strategies in multi-robot systems. That is, different authors have obtained solutions to multi-robot tasks without any consideration about the heterogeneity of the team, which, as commented in the previous sub-section, is very a relevant practical aspect. As a consequence, in the next section we will present the particular experiment carried out to study specialization in ASiCo and the main results obtained will be commented in detail.

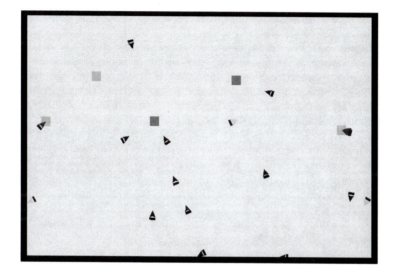

Fig. 2. ASiCo scenario for the specialization experiment. Squares represent blocks and triangles represent robots. Black robots are explorers, dark blue robots are gatherers and light blue robots are hybrid.

2   Self-organization and Specialization Analysis

2.1   Task Description

The experiment consists in a collective gathering task carried out by a team of simulated autonomous robots. The ASiCo scenario with the basic elements, robots and blocks, is displayed in Fig. 2. The team size is fixed to 20 robots. The task can be divided in two smaller subtasks that must be accomplished sequentially. First, the robots must activate some blocks (grey blocks in Fig. 2) that come up randomly in the scenario. The activation occurs if a robot touches a grey block. Once a block is activated (blue blocks in Fig. 2) it is ready for its gathering. The robots have to pair activated blocks. The only condition for this pairing is that both blocks must be activated. When two activated blocks are paired, they are automatically transported to a collection area outside the scenario. When this happens, two new deactivated blocks appear randomly in the arena so, consequently, the number of blocks in the scenario is a constant. As we can see, this experiment implies two clear species, explorer and gatherer robots, besides other hybrid combinations of them that could emerge.

To perform the two subtasks, the robots are endowed with a set of sensors and actuators, namely: a block sensor that provides the distance and angle to their closest block, a movement actuator to move through the scenario, an activation device to activate grey blocks and a gripper to grab the activated blocks. For the sake of simplicity, the robots have the low-level behaviors innate on them, so they know how to reach one block or another. The ASiCo objective is to adjust the optimum number of robots for each subtask according to the energetic conditions established, as we will explain later.

Consequently, the genotype of the robots is very simple, and it consists in a single continuous parameter that represents the probability of being an explorer, a gatherer or a hybrid robot. As usual in robotic applications, hybrid configurations can execute several subtasks but it can also be associated to an efficiency penalty motivated by a higher hardware complexity. In this experiment, we have implemented this penalty by applying a reduction in the detection range of the block sensor. Specifically, we have introduced a c coefficient that ranges from hybrid configurations with no penalty at all ($c=0$), which allow for "super-robots" able to accomplish both tasks in the same way, to a highly penalized hybrid configuration ($c=100$), leading to a robot with poor performance in both (it has to be extremely near of the block to detect it). In the middle, we have several combinations of hybrid robots, but in this paper they will not be considered.

The block detection sensor ranges are calculated using the following expressions, for the detection of single blocks:

$$r_i^{sb} = \beta R_{max}, \text{with } \beta = \begin{cases} 1, & g_i < L \\ \left(\dfrac{g_i - L}{1 - 2L}\right)^c, & L \le g_i < 1 - L \\ 0, & g_i > 1 - L \end{cases}$$

For the detection of activated blocks:

$$r_{ab}^i = \gamma R_{max}, \text{with } \gamma = \begin{cases} 1, & g_i > 1 - L \\ \left(\dfrac{1 - 2L - g_i}{1 - 2L}\right)^c, & L \le g_i < 1 - L \\ 0, & g_i < L \end{cases}$$

Being:

r_i^{sb}, r_{ab}^i: vision range of the single and activated blocks respectively for the i-th robot.

R_{max}: maximum detection distance of the sensors.

g_i: gene responsible of the task activation for the i-th robot

c: hybridization penalty coefficient

L: portion of the range of the gene used to activate the behavior associated to each subtask.

Regarding the energy flow strategy in ASiCo, the individual utility is associated to each of the particular steps required to perform the whole task, and it is introduced in the population as an energy reward. Specifically, when a robot activates a single block, it receives one energy unit, when a robot moves an activated block towards another activated block that is isolated in the scenario, it receives two energy units, and when a robot moves an activated block to another activated block that is being transported by another robot, each of the involved robots receives one energy unit. As in every implementation of ASico, every time step the robots lose a specific amount of energy which makes efficient robots live longer and, consequently, their genetic code can be transmitted with a higher probability.

2.2   Self-organized Specialization

In a basic initial configuration both tasks are balanced so, ideally, they will require half of the population each. Additionally, the energy used to reward the actions of the robots is equally distributed for both subtasks. For testing the specialization and the adaptability of ASiCo, we modify some conditions during the period of evolution, which affects the optimal solution. In this particular case, we present the results obtained when modifying the penalty parameter of the hybrid configuration during evolution, from a high value (c = 100) in the first 700.000 iterations, which makes hybrid configurations very inefficient, to a low value (c = 0) from 700.000 to 900.000 iterations, where hybrid genotypes imply "super-robots", and returning to a high penalization phase again up to the final iteration 1.600.000. We must clarify here that 100.000 iterations in this experiment took about 1 minute of real time to complete.

Fig. 3 shows the evolution of the global utility of the population in this case (dark line). It corresponds to the time elapsed between consecutive collected blocks, the lower the better, and it was obtained as the average value of 30 independent runs of ASiCo. To have a reference value of a successful utility level, in Fig. 3 we have included a dotted line that corresponds to the utility level reached by a "manual" solution (10 explorers/10 gatherers for c=100 and 20 hybrids/0 explorers/0 gatherers for c=0). This figure must be analyzed together with Fig. 4 that corresponds to the same runs and that displays the average number of individuals of each species throughout evolution.

In these two figures we can observe that, after an initial period of 100.000 iterations where the species balance is inadequate and the utility value is high, ASiCo obtains a stable utility value near the reference value in the phase with c=100, while it reaches a species configuration with almost no hybrid robots and with a balanced number of explorer and gatherer robots, as expected.

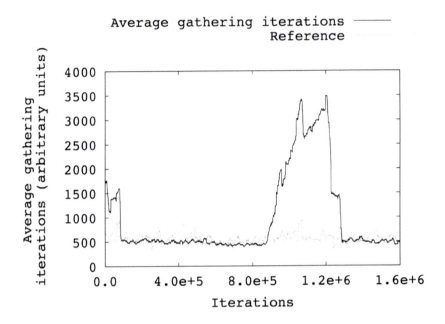

Fig. 3. Global utility evolution

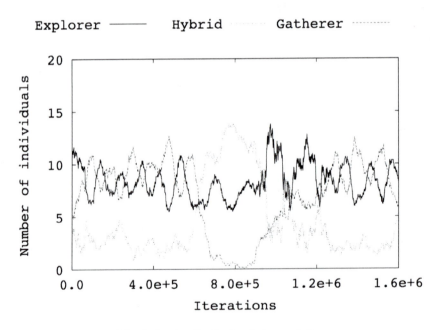

Fig. 4. Species distribution during evolution

During the second phase (c=0) between iterations 700.000 and 900.000, the global utility displayed in Fig. 3 slightly improves and, what is more significant, the distribution of species in the population changes completely. Now, most of the robots are hybrid as expected (see Fig. 4). Anyway, there are a few explorers that remain in the population. Analyzing the simulation in detail, we have noticed that hybrid robots are mainly performing the gathering task, so there is a chance for explorer robots without influence in the global utility. But as hybrid ones can accomplish both tasks, the number of explorers is smaller than in the previous phase. In this case, gatherer robots are basically unnecessary. This type of unexpected equilibrium state is one of the most interesting properties of an open-ended evolutionary approach like ASiCo, mainly due to the knowledge of the problem features it provides to researchers.

Finally, we set c=100 again penalizing the hybrid robots until the evolution finishes. At this point the genetic code of the individuals is not randomly spread throughout its range, but it is concentrated in the range that corresponds to hybrid configurations. Therefore there is a low genetic diversity and the readjustment of the groups is much more difficult as we notice in the utility evolution of Fig. 3. The utility worsens quickly and it takes 300.000 iterations to start improving again, but it finally reaches the reference value. Regarding the species in this last phase, Fig. 4 shows how the original balance with a low number of hybrid robots and a higher and similar value of explorer and gatherer robots is achieved.

3   Conclusions

In this work we have studied the emergence of specialization in multi-agent systems that are coordinated using open-ended natural evolution strategies. This type of strategies involves a situated evolution, which could have a bias towards homogeneous populations if the mating policy is not regulated properly. Specifically, here we have focused our study in the Asynchronous Situated Coevolution (ASiCo) algorithm, to conclude that the Embryo Based Reproduction and the Bipolar Crossover that make up the algorithm mating strategy, do not introduce the commented bias in the emergence of specialization. A high level task of multi-robot collective gathering was employed to carry out this analysis. The task was defined so as to be regulated in order to require or not the division of labor within the ideal population. We have studied variations in the employment of hybrid configurations, which can be penalized as is found in real robotic situations, that change the optimal group of individuals. With this, we aimed to show that the algorithm is not enforced to create species in every case but it adapts the final population, which represents the solution to the task, to its requirements.

Acknowledgements. This work was partially funded by the Xunta de Galicia and European Regional Development Funds through projects 09DPI012166PR and 10DPI005CT.

References

1. Floreano, D., Mattiussi, C.: Bio-Inspired Artificial Intelligence: Theories, Methods, and Technologies. MIT Press (2008)
2. Prieto, A., Bellas, F., Caamaño, P., Duro, R.J.: A Complex Systems Based Tool for Collective Robot Behavior Emergence and Analysis. In: Corchado, E., Abraham, A., Pedrycz, W. (eds.) HAIS 2008. LNCS (LNAI), vol. 5271, pp. 633–640. Springer, Heidelberg (2008)
3. Duro, R.J., Bellas, F., Prieto, A., Paz-López, A.: Social Learning for Collaboration through ASiCo based Neuroevolution. Journal of Intelligent and Fuzzy Systems 22, 125–139 (2011)
4. Prieto, A., Bellas, F., Caamaño, P., Duro, R.J.: Solving a Heterogeneous Fleet Vehicle Routing Problem with Time Windows through the Asynchronous Situated Coevolution Algorithm. In: Kampis, G., Karsai, I., Szathmáry, E. (eds.) ECAL 2009, Part II. LNCS, vol. 5778, pp. 200–207. Springer, Heidelberg (2011)
5. Agogino, A., Tumer, K.: Efficient evaluation functions for evolving coordination. Evolutionary Computation 16(2), 257–288 (2008)
6. Nitschke, G., Schut, M., Eiben, A.: Collective Neuro-Evolution for Evolving Specialized Sensor Resolutions in a Multi-Rover Task. Evolutionary Intelligence 3(1), 13–29 (2010)
7. Baldassarre, G., Nolfi, S., Parisi, D.: Evolving mobile robots able to display collective behavior. Artificial Life 9(1), 255–267 (2003)
8. Bryant, B., Miikkulainen, R.: Neuro-evolution for adaptive teams. In: Proceedings of the Congress on Evolutionary Computation, pp. 2194–2201 (2003)
9. Watson, R., Ficici, S., Pollack, J.: Embodied evolution: Distributing an evolutionary algorithm in a population of robots. Robot. and Auton. Syst. 39(1), 1–18 (2002)
10. Eiben, A.E., Haasdijk, E., Bredeche, N.: Embodied, On-line, On-board Evolution for Autonomous Robotics. In: Symbiotic Multi-Robot Organisms: Reliability, Adaptability, Evolution, pp. 361–382. Springer, Heidelberg (2010)
11. Karafotias, G., Haasdijk, E., Eiben, A.E.: An algorithm for distributed on-line, on-board evolutionary robotics. In: Proceedings of the 13th Annual Conference on Genetic and Evolutionary Computation (2011)
12. Bredeche, N., Montanier, J.M., Liu, W., Winfield, A.: Environment-driven Distributed Evolutionary Adaptation in a Population of Autonomous Robotic Agents. Mathematical and Computer Modelling of Dynamical Systems (2011)
13. Elfwing, S., Uchibe, E., Doya, K., Christensen, H.: Darwinian embodied evolution of the learning ability for survival. Adaptive Behavior - Animals, Animats, Software Agents, Robots, Adaptive Systems 19(2), 101–120 (2011)
14. Montanier, J.-M., Bredeche, N.: Embedded Evolutionary Robotics: The (1+1)-Restart-Online Adaptation Algorithm. In: Doncieux, S., Bredèche, N., Mouret, J.-B. (eds.) New Horizons in Evolutionary Robotics. SCI, vol. 341, pp. 155–169. Springer, Heidelberg (2011)
15. Knudson, M., Tumer, K.: Coevolution of heterogeneous multi-robot teams. In: Proceedings of the 12th Annual Conference on Genetic and Evolutionary Computation, pp. 127–134. ACM, New York (2010)

An Empirical Tool for Analysing the Collective Behaviour of Population-Based Algorithms

Mikdam Turkey and Riccardo Poli

School of Computer Science & Electronic Engineering
University of Essex
Wivenhoe Park
Colchester, Essex, CO4 3SQ
{mturkey,rpoli}@essex.ac.uk

Abstract. Understanding the emergent collective behaviour (and the properties associated with it) of population-based algorithms is an important prerequisite for making technically sound choices of algorithms and also for designing new algorithms for specific applications. In this paper, we present an empirical approach to analyse and quantify the collective emergent behaviour of populations. In particular, our long term objective is to understand and characterise the notions of exploration and exploitation and to make it possible to characterise and compare algorithms based on such notions. The proposed approach uses self-organising maps as a tool to track the population dynamics and extract features that describe a population "functionality" and "structure".

Keywords: Collective behaviour analysis, population dynamics, population-based algorithms, exploration, exploitation, emergent properties.

1 Introduction

A population-based search algorithm operates on a set of interacting individuals — the population. The algorithm uses interaction mechanisms to control the movement of individuals around the search space and redistribute the density of the population in different regions according to information obtained via individuals previously generated in such regions. This bias in directing the movement of the population leads the algorithm to explore promising areas of the search space with more intensity and/or to acquire information about new regions. As the interaction mechanisms operate on moving, creating and/or eliminating individuals, the algorithm shows an emergent behaviour which represents the collective behaviour of the population as a whole and describes its dynamics.

Understanding the emergent collective behaviour of population-based algorithms is important. For example, identifying collective properties related to an algorithm's dynamics allows the comparison of algorithms in terms of behaviours, not just performance (e.g., number of fitness evaluations to reach an optimum). So, this is also a prerequisite for making technically sound choices of algorithms and also for designing new algorithms for specific applications.

C. Di Chio et al. (Eds.): EvoApplications 2012, LNCS 7248, pp. 103–113, 2012.

Various approaches have aimed at developing a better understanding of the collective behaviour of population-based algorithms. Some of them attacked the issue by theoretically modelling the dynamics of the algorithm. For example, within evolutionary computation theory, approaches include: the schema theory [5], which was one of the first and, today, best developed theories of the dynamics of search algorithms (e.g., see [11,10] for some recent results), Markov chain formulations [12], Walsh-function-based analyses [2] and statistical-mechanical formulations [9]. All of these have seen some successes at mathematically modelling evolutionary algorithms.

An alternative approach to theory is to practically study the dynamic behaviour of an algorithm with respect to a specific problem by defining and analysing suitable empirical measures. A number of empirical measures have been proposed to assess the *performance* of evolutionary algorithms. For example, semi-empirical measures of problem difficulty, such as the fitness-distance correlation [6] and the negative slope coefficient [8], have been proposed to characterised what makes a problem easy or hard for evolutionary algorithms. However, *much less has been done to empirically assess properties of an algorithm's emergent/collective behaviour.* This is a major gap, since the quantification of such proprieties of the dynamic behaviour of algorithms would have numerous benefits. It would help, for example, develop a better understanding about the population evolutionary phenomena. It would also make it possible to compare search algorithms based on high-level aspects of functionality and structure.

Looking at what has been done in this area, we find that researchers have concentrated on two emergent features of search algorithms: exploration and exploitation. Exploration refers to behaviour resulting in the discovery of new good regions in the search space, while exploitation refers to the behaviour of exploring previously discovered good regions [5,4]. Existing approaches have focused on controlling the explorative/exploitative behaviour by tuning the parameters of search algorithm prior to a run or changing them dynamically throughout the run based on features of the the population such as fitness or diversity . However, there is no precise definition in the literature of the notions of exploration and exploitation, no precise characterisation of the distinction between them, and no numerical quantification of them.

In this paper we present an algorithm-independent approach to analyse the collective dynamic behaviour of a population. Our approach uses Self-Organising Maps (SOMs) [7] to track population movement and to mine information about the emergent collective behaviour of the algorithm as it operates in solving a problem. Our aim is to use the information to calculate measures that are related (and may eventually lead to quantifying) exploration and exploitation.[1]

In related work [3], an approach was proposed to measure exploration and exploitation of an evolutionary algorithm. The approach uses an ancestry tree as a data structure for recording information about the process of creating individuals by different genetic operators. The information is recorded throughout

---

[1] SOMs had previously been used in evolutionary algorithm to improve their performance by enhancing the search strategy and avoiding genetic drift [1].

the run and used to calculate the exploration/exploitation rate of the algorithm based on the number of ancestry trees resulting after splitting them into small trees based on a distance threshold. While this approach depends on mutual distance between individuals in quantifying exploration and exploitation, other approaches use fitness growth to assess the evolvability of genetic operators, or fitness distributions to predict the distribution of successive generation (e.g., [8,9]). Both fitness and distances can be useful to represent explorative and exploitative search. This is why, to assess emergent properties, in this paper we use a variety measures involving distances, fitnesses or both.

The paper is organised as follows. Sec. 2 introduces self-organising maps and provides details about how our approach makes use of them. Sec. 3 introduces the proposed measures of emergent properties of collective behaviour, while Sec. 4 presents experimental results obtained applying these measures to different evolutionary algorithms. We conclude with discussion and future work.

2 Self-Organising Maps and Population Dynamics

A Self-Organising Map [7] is an artificial neural network that can be trained using unsupervised learning. After training, SOMs can be used for mapping (classifying) or visualising high dimensional data. SOMs consist of a set of nodes (or neurons) arranged, usually, in a two-dimensional grid. A node, i, has an associated vector, m_i, of the same dimension of the input space and is connected to its nearest neighbours. The training is done by feeding a SOM with a large number of training samples drawn from the data space. Each time a sample, x, is fed into a SOM, the best matching node (BMN) is identified as the node whose vector has the smallest distance from the input sample, i.e., $bmn = \arg\min_i \|x - m_i\|$. Then m_{BMN} is updated by moving it slightly in the direction of x. The change to the BMN's vector results in changing the vectors of its neighbours as well.

In this work, we use SOMs in two stages. The first one is the training stage. In each training iteration the SOM is fed individuals randomly selected from the initial population. In this phase the learning rate and the neighbourhood radius are decreased over time as standard in SOMs. The resulting SOM provides a 2–D representation of the search space (as represented by the initial random population). The second stage is where we use the SOM for tracking the population dynamics. In this stage, which is effectively *another training stage*, we use a fixed learning rate and neighbourhood radius, and the SOM is trained by the newly create individuals produced in a run of the algorithm. In addition to the change that new individuals bring to the node vectors of the SOM, more information about those individuals is recorded.

In the proposed approach, the node grid is viewed as a matrix of centroids representing the population distribution, where the collective dynamic behaviour is captured by the analysing changes introduced to the node vectors by new individuals as the population moves in the search space. We use a grid of $n \times n$ nodes. Each node is represented by the following tuple

$$C_r^t = \langle m^{C_r^t}, d^{C_r^t}, f^{C_r^t}, f_{best}^{C_r^t}, f_{best\text{-}so\text{-}far}^{C_r^t}, h^{C_r^t} \rangle$$

where in C_r^t, t is time and $r \in \{1 \ldots n\}^2$ is the position in the grid.

The elements of the tuple are as follows: $m^{C_r^t} \in \mathbb{R}^D$ represents the node vector, D being the search space dimension; $d^{C_r^t}$ is the sum of the distances between $m^{C_r^t}$ and input individuals for which C_r^t is identified as BMN for the period between two sampling points $t - 1$ and t (*hit distance*); $f^{C_r^t}$ is the sum of the fitnesses of all the individuals for which C_r^t is identified as BMN (in the period between two sampling points $t - 1$ and t); $f_{best}^{C_r^t}$ is the best fitness value of an individual for which C_r^t is identified as BMN between sampling points $t - 1$ and t; $f_{best\text{-}so\text{-}far}^{C_r^t}$ is the best fitness value of an individual for which C_r^t is identified as BMN since time $t = 0$; $h^{C_r^t}$ (*hits counter*) is the number of individuals where C_r^t is identified as a BMN between sampling points $t - 1$ and t. A node vector, m^{C^r}, is updated using the following function:

$$l(C_r, \alpha, \sigma, x) = m^{C^r} + \alpha e^{\left(-\frac{\|r - r_{bmn}\|^2}{2\sigma^2}\right)} \left(x - m^{C^r}\right)$$

where x is the input individual vector, α is the learning rate, σ is the neighbourhood radius and r_{bmn} is the index of the BMN.

During the training stage, the learning rate is set to $\alpha = 0.07$ and the neighbourhood radius is set to $\sigma = n/2$. Each learning iteration consists of feeding the SOM with N (typically the size of population) individuals selected randomly form the initial population. For iteration c, the learning rate is computed as $\alpha_c = \alpha \exp\left(-c/\#iterations\right)$ and the neighbourhood radius is computed as $\sigma_c = \sigma \exp\left(-\frac{c}{\#iterations/log(\sigma)}\right)$. The purpose of the training stage is to capture the distribution of the initial population and to create a topologically consistent grid. After the training stage we use fixed values for learning rate and neighbourhood radius, such that $\alpha = 0.07$ and $\sigma = 1$.

3   Extracting Properties of Collective Behaviour

After randomly generating initial SOM vectors, the SOM is trained using the initial population. During this stage we only update the node vectors without recording any information about individuals. The task of the second stage is to track the dynamic collective behaviour of population by detecting changes introduced by newly created individuals and to record information about them. Every time the search algorithm creates τ individuals, we send them as input to the SOM. Information about the collective behaviour is then extracted and analysed. In the proposed approach we use $\tau = 100$, that means the analysis process work on detecting the collective dynamic behaviour every 100 new points in the search space have been examined by the algorithm. The first step in the second stage is to identify *activity regions* in SOM.

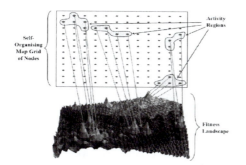

Fig. 1. SOM's Activity Regions and Fitness Landscape

An activity region can be defined as any set of adjacent grid nodes that have been hit by (matched) at least one individual between two sampling points. Formally that is defined as $ActiveN^t = \{C_r^t | h^{C_r^t} > 0\}$. Furthermore, let $Neigh(C_r^t)$ be a function that returns a set of all immediate neighbours of a node that is a member of $ActiveN^t$. Formally:

$$Neigh(C_r^t) = \left\{ C_{r'}^t \in ActiveN^t \mid 0 < \|r - r'\| < \sqrt{2} \text{ and } \|m^{C_r^t} - m^{C_{r'}^t}\| < \omega \right\}$$

where $\omega = 0.05 \times SearchSpaceD$ is a distance threshold, where $SearchSpaceD$ is the largest distance between any two points in search space. The function above returns the empty set $\emptyset$ in case C_r^t has no neighbours. An *activity region*, $Region(C_r^t)$ for $C_r^t \in ActiveN^t$, is the set of all the nodes that are either direct neighbours of node C_r or are neighbours of its neighbours. If C_r has no connected neighbour, then $Region(C_r^t) = \{C_r^t\}$.

Then the set of all activity regions (set of sets) is defined as follows:

$$ActivityRegions^t = \bigcup_{x \in ActiveN^t} Region(x)$$

Figure 1 depicts the distribution of node vectors in the fitness landscape and the identification of activity regions based on neighbour definition. Then the activity regions are divided into $GrowthRegions^t$ and $NonGrowthRegions^t$, as follows:

$$GrowthRegions^t = \left\{ R \mid X^t \in R \text{ and } f_{best\text{-}so\text{-}far}^{X^t} > f_{best\text{-}so\text{-}far}^{X^{t-1}} \text{ where} \right.$$

$$\left. f_{best\text{-}so\text{-}far}^{X^t} = \max_{w \in R} f_{best\text{-}so\text{-}far}^w \right\}$$

$$NonGrowthRegions^t = ActivityRegions^t \setminus GrowthRegions^t$$

We need also to define the best region $BestRegion^t \in ActivityRegions^t$,

$$BestRegion^t = \arg \max_{x \in ActivityRegions^t} f_{best\text{-}so\text{-}far}^x$$

Monitoring changes that occurred on the regions above and analysing the information recorded by nodes belonging to them reveals details of the dynamics and the collective behaviour of an algorithm. Many features that describe aspects of the emergent behaviour can be extracted from the SOM by looking at region changes over two sampling points and analysing the recorded information. The features of emergent collective behaviour can be classified into four types of features as follows: **i)** Fitness-related features: fitness averages or growth in different SOM regions provide an insight about the level of achievement of each region. Furthermore, regions are actually divided according to current or past fitness values; **ii)** Distance-related features: two distance measures can be calculated from the SOM: the sum of the distances between the node's vector and input individuals ($d^{C_r^t}$) and the displacement distance for each node vector. The former can provide information on local search properties, while the later gives information on the population search bias and direction; **iii)** Activity-related features: the number of individuals that have been associated with a certain node (hits counter), the number of nodes in one region and the number of regions. All of these can give indications of the extent of population activities; and **iv)** Correlated features: they analyse the correlation between two different types of features over time, such as the change of fitness and node vector displacement, or the amount of activity. In the paper we explore some measures for the emergent collective behaviour using some of the features outlined above.

In table 1, regions are categorised based on change in regions status over two sampling point $t-1$ and t.

Table 1. Categorising regions based on status change

Description	Regions Sets Definition
Activity led to Activity	$AA^t = \{R \mid R \in NonGrowthRegions^{t-1} \text{ and } R \in NonGrowthRegions^t\}$
Activity led to Growth	$AG^t = \{R \mid R \in NonGrowthRegions^{t-1} \text{ and } R \in GrowthRegions^t\}$
Activity led to Nothing	$AN^t = \{R \mid R \in NonGrowthRegions^{t-1} \text{ and } R \notin ActivityRegions^t\}$
Growth led to Activity	$GA^t = \{R \mid R \in NonGrowthRegions^{t-1} \text{ and } R \in GrowthRegions^t\}$
Growth led to Growth	$GG^t = \{R \mid R \in GrowthRegions^{t-1} \text{ and } R \in GrowthRegions^t\}$
Growth led to Nothing	$GN^t = \{R \mid R \in GrowthRegions^{t-1} \text{ and } R \notin ActivityRegions^t\}$
New Regions	$New^t = \{R \mid R \notin ActivityRegions^{t-1} \text{ and } R \in ActivityRegions^t\}$

Then the ratios of each region can be calculated as follows:

$$AARatio^t = |AA^t|/|NonGrowthRegions^{t-1}|$$

$$AGRatio^t = |AG^t|/|NonGrowthRegions^{t-1}|$$

$$ANRatio^t = |AN^t|/|NonGrowthRegions^{t-1}|$$

$$GGRatio^t = |AG^t|/|GrowthRegions^{t-1}|$$

$$GARatio^t = |AG^t|/|GrowthRegions^{t-1}|$$

$$GNRatio^t = |AG^t|/|GrowthRegions^{t-1}|$$

$$NewRatio^t = |New^t|/|ActivityRegions^t|$$

The activity rate in set of regions S^t is calculated as the number of hits a set of regions receives at time t and defined as:

$$ActivityRate(S^t) = \frac{1}{\tau}\left(\sum_{C \in u(S^t)} h^C\right)$$

where $u(S^t) = \cup_{\lambda \in S^t}\lambda$ represents all the nodes within the set of regions S^t. The previous equation can be applied to any region and it is used to measure the amount of focus an algorithm gives to region(s) or to direct the search to new areas. This can also be used to find the rate of the activities in the best region, interpreted a set, i.e., $ActivityRate(\{BestRegion^t\})$.

The change in node vector positions (*vector displacement*) reveals information about the bias of algorithm and studying this change over time for different regions along with the hit distance can help clarify the collective behaviour of populations in local areas of the search space. Vector displacements and hit distances are defined as follows:

$$Displacement(S^t) = \frac{1}{|u(S^t)|}\left(\sum_{C^t \in u(S^t)} \|m^{C^t} - m^{C^{t-1}}\|\right)$$

$$HitDistance(S^t) = \left(\sum_{C^t \in u(S^t)} d^{C^t}\right)/\left(\sum_{C \in u(S^t)} h^C\right)$$

Vector displacement and hit distance of the best region can be calculated as $Displacement(\{BestRegion^t\})$ and $HitDistance(\{BestRegion^t\})$, respectively.

Finally, the size ratio of a region, the number of active nodes of a certain region relative to the total number of active nodes, can be defined as follows:

$$SizeRatio(S^t) = \left(\sum_{R^t \in S^t} |R^t|\right)/|ActiveN^t| \tag{1}$$

4   Experimental Results

To practically test the approach outlined above, we used a basic EA of 100 real-valued individuals using uniform mutation with 0.2 mutation probability and 1.0 mutation step size. The algorithm uses a intermediate arithmetic recombination with 0.75 probability. We used tournament selection for both mating and survival selections with tournament sizes 4 and 2, respectively. In addition to the

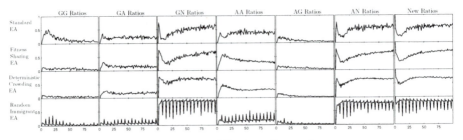

Fig. 2. Ratios of Regions Categories

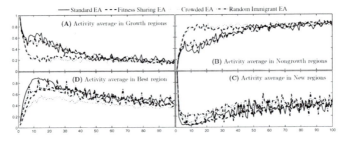

Fig. 3. Activity rates in different population regions

standard form the EA, three different techniques have been used: *fitness sharing, deterministic crowding and random immigrants*. We applied these algorithms on a 5-dimensional problem consisting of the sum of 100 *Sphere, Ranstrigin and Griewank* functions. These function were randomly generated, shifted, rotated and combined. In our experiments, we conducted 100 runs for each of the four algorithms on a single randomly generated fitness landscape.

The measures that we have designed above are aimed at mining emergent properties of the collective behaviour by extracting features that describe the structure and the functionality of population as it moves around the search space. Figure 2 gives indications on how the algorithms handle different regions of the search space and how they respond to discovering new *useful information*. The plots also illustrate how the algorithms distribute their "attention". *GGRatio* and *GARatio* give information on how often the algorithm follows promising information which, of course, gives an indication on the exploitation behaviour, while *AGRatio* gives an indication of how successful are activities that led to local growth in fitness. On the other hand, *ANRatio* shows the rate of the activities that have resulted in discovering nothing and have not been followed. The two previous ratios plus *NewRatio* give indication on explorative behaviour.

Standard EA tends to exploit the search space, while fitness sharing and deterministic crowding EAs give less focus on exploitation. The random immigrant EA tries to exploit useful information but the phase of generating random immigrants (individuals) distracts this activity and pushes the algorithm toward more exploration.

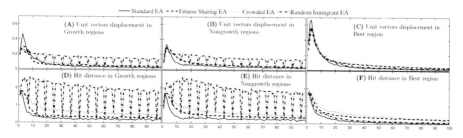

Fig. 4. Displacement and hit distance of unit vectors

Figure 3 shows how each algorithm distributes its activity as measured by the number of individuals created in each region. While Standard and random immigrant EAs concentrate most of their individuals in growth regions or the best region discovered so far and give less attention to non-growth regions and new regions, fitness sharing and deterministic crowding EAs do exactly the opposite. However, they all, more and less, converge to the same fashion of behaviour.

Another aspect of collective behaviour can be noticed by tracking the change caused by the population moving the SOM vectors. Significant change signifies concentration and bias in search. The less these changes, the less explorative the population becomes. This is because when the population concentrates on a specific region of the search space for some time, the SOM comes to equilibrium and that indicates that the population has stopped moving. The hit distance conveys, more and less, the same message. If the hit distance is high, that indicates that the population is exploring and not working on the same area of the search space. The less the distance, the less the population explores. The random immigrant EA has the highest rates of displacement and hit distance because of the unbiased nature of generating random individuals. The deterministic crowding EA shows high rates of hit distance, although it has the same displacement as fitness sharing EA. That means a deterministic crowding EA tends to explore more. The standard EA starts with the highest displacement and drops down to the lowest value due to the exploitative nature of the algorithm.

Figure 5 describes the structural features of population in terms of the size of regions as defined by Eqn. 1. It gives an indication on how the population distributes itself among the regions. From the figure we can notice that growth regions in algorithms using a niching technique occupy less area than for other algorithms that tend initially to use large areas of activities for growth regions

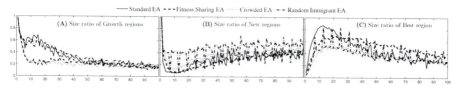

Fig. 5. Size rations of different regions of population

and shrink down as they converge. The larger an area an algorithm allocates for the best region, the more exploitative the behaviour it exhibits.

5   Discussion and Future Work

In this paper, we presented an algorithm-independent approach to evaluate the collective behaviour of population-based search algorithm. The approach uses self-organising maps to track population dynamics and extract properties characterising the collective behaviour. Our aim is to develop measurements to identify emergent properties attributed to functionality and structure of population in real-time as it explores the search space.The proposed approach can help in the characterisation of algorithms and in the production on taxonomies based on features related to population dynamic.

In the paper we applied our proposed measures to the issue of understanding the exploitation and exploration behaviour of algorithms. We hope in future work to be able to develop measures to quantify these behaviours based on the properties presented here. More features and measures can be mined by using the proposed approach, such as the correlation between fitness and displacement or hit distance. Also, the proposed approach can be applied to different types of algorithm such as dynamic optimisation, multi-objective, or multi-modal algorithms, and it can be used to identify preferred behavioural features.

Further work is required to explore the relation between SOM grid size and the population size and/or search space size, and how the size of SOM grid effects the quality of tracking the population dynamics.

Acknowledgement. This work was supported by EPSRC Doctoral Training Account (DTA) Research Studentship.

References

1. Amor, H., Rettinger, A.: Intelligent exploration for genetic algorithms: using self-organizing maps in evolutionary computation. In: Proceedings of the 2005 Genetic and Evolutionary Computation Conference (GECCO 2005), pp. 1531–1538. ACM (2005)
2. Bethke, A.: Genetic algorithms as function optimizers. Doctoral dissertation, University of Michigan (1981)
3. Crepinsek, M., Mernik, M., Liu, S.: Analysis of exploration and exploitation in evolutionary algorithms by ancestry trees. International Journal of Innovative Computing and Applications 3(1), 11–19 (2011)
4. Eiben, A., Schippers, C.: On evolutionary exploration and exploitation. Fundamenta Informaticae 35(1), 35–50 (1998)
5. Holland, J.H.: Adaptation in natural and artificial systems: an introductory analysis with applications to biology, control, and artificial intelligence. The MIT Press (1992)
6. Jones, T., Forrest, S.: Fitness distance correlation as a measure of problem difficulty for genetic algorithms. In: Eshelman, L.J. (ed.) ICGA, pp. 184–192. Morgan Kaufmann (1995)

7. Kohonen, T.: Self-organizing maps. Springer series in information sciences. Springer, Heidelberg (2001)
8. Poli, R., Vanneschi, L.: Fitness-proportional negative slope coefficient as a hardness measure for genetic algorithms. In: Proceedings of the 9th Annual Conference on Genetic and Evolutionary Computation, GECCO 2007, pp. 1335–1342. ACM (2007)
9. Shapiro, J., Prügel-Bennett, A., Rattray, M.: A Statistical Mechanical Formulation of the Dynamics of Genetic Algorithms. In: Fogarty, T.C. (ed.) AISB-WS 1994. LNCS, vol. 865, pp. 17–27. Springer, Heidelberg (1994)
10. Stephens, C., Poli, R.: Coarse graining in an evolutionary algorithm with recombination, duplication and inversion. In: The 2005 IEEE Congress on Evolutionary Computation, vol. 2, pp. 1683–1690. IEEE (2005)
11. Stephens, C., Poli, R.: Coarse-grained dynamics for generalized recombination. IEEE Transactions on Evolutionary Computation 11, 541–557 (2007)
12. Vose, M., Liepins, G.: Punctuated equilibria in genetic search. Complex Systems 5, 31–44 (1991)

Sales Potential Optimization on Directed Social Networks: A Quasi-Parallel Genetic Algorithm Approach

Crown Guan Wang and Kwok Yip Szeto[*]

Department of Physics, The Hong Kong University of Science and Technology
Clear Water Bay, Hong Kong, HKSAR, China
phszeto@ust.hk

Abstract. New node centrality measurement for directed networks called the Sales Potential is introduced with the property that nodes with high Sales Potential have small in-degree and high second-shell in-degree. Such nodes are of great importance in online marketing strategies for sales agents and IT security in social networks. We propose an optimization problem that aims at finding a finite set of nodes, so that their collective Sales Potential is maximized. This problem can be efficiently solved with a Quasi-parallel Genetic Algorithm defined on a given topology of sub-populations. We find that the algorithm with a small number of sub-populations gives results with higher quality than one with a large number of sub-populations, though at the price of slower convergence.

Keywords: Multi-Agent marketing, Social Networks, Security of networks, Parallel Genetic Algorithm.

1   Introduction

There are many complex systems that can be studied from the point of view of complex networks, with examples ranging from biological systems and communication to social networks. In complex networks, objects are represented by nodes and the interactions or the relationships between the objects are represented by edges. Recent research on complex networks has revealed many interesting properties such as the power law distribution and the preferential attachment rule [1]. One popular topic of research concerns the importance of a vertex within a complex network. This importance is often formulated as some measures on the "centrality" of the vertex. Four common measures of centrality used include degree centrality, betweenness, closeness and eigenvector centrality [2].

We propose a new measurement of centrality in directed networks called the Sales Potential of a node. In this paper, we focus on the in-degree for the definition of Sales Potential, but the generalization to out-degree or mixed case can be made similarly. Nodes with high Sales Potential have small in-degree and high second-shell in-degree. The importance of such nodes is often underestimated, because of their low

---

[*] Corresponding author.

C. Di Chio et al. (Eds.): EvoApplications 2012, LNCS 7248, pp. 114–123, 2012.

in-degree (small number of followers); they are considered easier to reach than nodes with larger degrees. However they have great potential for information propagation, since one can get easy access to many higher degree nodes in the second or higher shells, once one reaches these nodes with high Sales potential. We can now pose an interesting question on the collective Sales Potential of a set of nodes as follow: how should we select the nodes so that the combined Sales Potential is maximized?

This turns out to be an excellent optimization problem solvable by genetic algorithms. One can find many successful applications of genetic algorithms [3,4], such as in the solution of the crypto-arithmetic problem [5], time series forecasting [6], traveling salesman problem [7], function optimization [8], adaptive agents in stock markets [9,10], and airport scheduling [11,12]. While genetic algorithm has been used in many industrial applications, recent developments see the importance of parallel computing with multiple CPUs in computer science. Evolution in parallel sub-populations is a natural extension of the idea of "divide and conquer" to the idea of "divide and search in respective subspace". Our previous work [13,14] has investigated the Quasi-parallel Genetic Algorithm (QPGA) performance relationship with different topologies and chromosome exchange rates. Here, we will employ this new framework of QPGA to solve the optimization problem of collective Sales Potential for a fixed number of nodes selected in the network. We also investigate the relation between its performance and the number of sub-populations in QPGA.

Our paper is constructed in the following way. In Section 2 we define Sales Potential on Directed Social Networks. In Section 3 we review the Mutation Only Genetic Algorithm (MOGA) which we will use for each sub-population in the QPGA. Section 4 introduces the Quasi-parallel Genetic Algorithm we use for solving the optimization problem. Section 5 contains the experiment set-up and results. We finally give some discussion and conclusion in Section 6.

2   Sales Potential on Directed Social Networks

The concept of Sales Potential is inspired from the observation in directed social networks like Twitter that some famous people with massive followers follow some "unknown" individuals with few followers. These "unknown" individuals live with a low profile, but have great influence to some popular individuals. Thus, despite their obscurity in society, they provide great potential for marketing strategists because the barriers to contact them are much lower than those protecting famous people. If one is interested in marketing in a social network, convincing these low profile people with strong influence on their famous friends will be very effective. As a simple example, convincing the classmates of a movie star's daughter to use a certain kind of mobile device may be an effective way to market the product. It can also happen that a node with many edges directed to it acts as the source of information for its neighbor which has a small in-degree. This small in-degree neighbor may actually be the real hidden commander and is of great importance. In Fig.1, we illustrate a directed network and the Sales Potential of the node 0.

We define the Sales Potential (SP) of a single node i as the following

$$SP(i) = \frac{1}{K_{in}(i)}\left(\sum_{m=1}^{K_{in}(i)} K_{in}\left(j_i^1(m)\right)\right) \tag{1}$$

where $K_{in}(i)$ is the in-degree of node i and $j_i^1(m)$ is the m-th nearest (1st shell) in-neighbor of node i. From this definition, we see that a node with high Sales Potential will have a small in-degree but large second-shell in-degree. Generalization of this definition to the out-degree version will be meaningful for information transmitting in a different context. Also, one can extend the above definition to the higher n-th shell with $j_i^n(m)$.

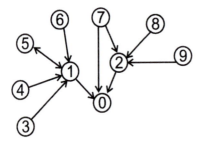

Fig. 1. The Sales Potential in directed network. Node 0 has three nearest in-neighbors: Node 1, 2 and 7, so that the in-degree of Node 0 is 3. There are six second-shell in-neighbors of Node 0: Nodes 3, 4, 5, 6, 8, and 9. Thus, the second shell in-degree of Node 0 is 6. Note that Node 7 has been counted as the nearest in-neighbors and thus we do not count it again in the second shell. The Sales Potential for Node 0 is 6/3=2.

From the perspective of marketing, one can pose the following optimization problem: given a directed complex network, find N nodes that will maximize their combined Sales Potential. We limit ourselves to the task of targeting N nodes because of the limitation in marketing resource. Now, for any two nodes taken at random, there is a possibility that they share some common neighbors. In marketing, we do not want to find a pair of targets that have overlap neighbors, since a similar pair of targets without overlap neighbors will likely have a higher combined Sales Potential. To generalize this observation to many nodes, we define the combined Sales Potential of M nodes as

$$S(i_1...i_M) = \frac{1}{\sum_{k=1}^{M} K_{in}(i_k)}\left(Size\left(\bigcup_{k=1}^{M}\sigma_{i_k}\right)\right) \tag{2}$$

where σ_{i_k} is the set of second shell in-neighbors of the node i_k. The size function, $Size(A)$, returns the cardinality of the set A. In exhaustive search, if we are to select M nodes out of a total of $N>M$ nodes, the number of times we have to calculate Sales Potential in Eq.(2) will be C_M^N. From our experience, this is a very difficult task in computation. However, we may use some topological features of networks to simplify this task. With few exceptions, most real networks obey a generalized Aboav-Weaire law[15,16], which states that the first shell degree of node i is expressible as a linear function in the degree of node i. This law on the degree of the nearest neighbors of a given node in a complex network has recently been generalized and computed for the case of random network, small world network and Barabasi-Albert network [17] and to higher shell of two-dimensional spatial networks [18]. We generalize this notion to directed networks so that we can replace the numerator in Eq.(1) with a linear function of the in-degree.

$$SP(i) = \frac{aK_{in}(i) + b}{K_{in}(i)} = a + \frac{b}{K_{in}(i)} \tag{3}$$

Now, if we assume that the Aboav-Weaire law holds for the network, then in our selection process for nodes with highest sales potential, the maximum $SP(i)$ for the node i obviously occurs at those nodes with small in-degree. This will reduce greatly the number of nodes that we need to investigate, thereby reducing the complexity of the problem. In fact, our problem involves the search of a set of M nodes from a large social network with N nodes, so that the Sales Potential $S(i_1...i_M)$ is maximized. This can be greatly facilitated by starting on the subset of nodes with small $SP(i)$. Therefore, we begin by checking if the given social network obeys the Aboav-Weaire law. If it does, like most cases, we can focus on the subset of nodes with small $K_{in}(i)$. This set will be much smaller than the original network. This effectively reduces the N greatly so that the search space for the optimization problem of finding a set of M nodes is greatly reduced. Despite the reduction in the dimension of the search space by using Aboav-Weaire law, we still have a huge task in finding the M nodes with maximum Sales Potential $S(i_1...i_M)$. To handle this search problem, we will employ our Quasi-Parallel Genetic Algorithm, which has been shown to be very efficient compared to other standard search methods in solving these kind of problems [11-14,19-21].

3   Mutation Matrix

Before we discuss parallel genetic algorithm, we have to state the algorithm used by individual computing node. Traditionally, simple genetic algorithm (SGA) is used, but in this paper we will use a generalization of SGA which permits parameter free adaptation of the mutation and crossover rate. This generalization of SGA is called Mutation Only Genetic Algorithm or MOGA in our publications [19-21] (not to be confused with multi-objective GA). The rational for choosing MOGA instead of SGA is that SGA is a special case of MOGA without adaptive parameter control and in all cases we have investigated so far [11-14,19-21] MOGA is more efficient than SGA. Here we briefly review MOGA. In simple genetic algorithm, we consider a population of N chromosomes, each of length L and binary encoded. We describe the population by a $N \times L$ matrix, with entry $A_{ij}(t), i = 1,...,N; j = 1,...,L$ denoting the value of the j-th locus of the i-th chromosome. The convention is to order the rows of A by the fitness of the chromosomes, $f_i(t) \le f_k(t)$ for $i \ge k$. Traditionally we divide the population of N chromosomes into three groups: (**1**) Survivors who are the fit ones. They form the first N_1 rows of the population matrix $A(t+1)$. Here $N_1 = c_1 N$ with the survival selection ratio $0 < c_1 < 1$. (**2**) The number of children is $N_2 = c_2 N$ and is generated from the fit chromosomes by genetic operators such as mutation. Here $0 < c_2 < 1 - c_1$ is the second parameter of the model. We replace the next N_2 rows in the population matrix $A(t+1)$. (3) The remaining $N_3 = N - N_1 - N_2$ rows are the randomly generated chromosomes to ensure the diversity of the population so that the genetic algorithm continuously explores the solution space. In MOGA, we use a mutation matrix with elements $M_{ij}(t) \equiv a_i(t)b_j(t), i = 1,...,N; j = 1,...,L; \ 0 \le a_i(t), b_j(t) \le 1$ where $a_i(t)$ and $b_j(t)$ are called the row mutation probability and column mutation probability. A MOGA that corresponds to SGA with mutation as the only genetic operator will have a time *independent* mutation matrix with elements $M_{ij}(t) \equiv 0$ for $i = 1,...,N_1$, $M_{ij}(t) \equiv m \in (0,1)$ for $i = N_1 + 1,...,N_2$, and finally we have $M_{ij}(t) \equiv 1$ for $i = N_2 + 1,...,N$. Here m is the time independent mutation rate. Thus, SGA with mutation as the only genetic operator requires at least three parameters: N_1, N_2, and m. In a general MOGA formalism, we will compute the mutation matrix element by the statistics of the fitness and ranking of the chromosome to determine the row mutation probability $a_i(t)$ and the statistics of the loci to determine the column mutation probability $b_j(t)$. Various ingenious methods have been developed to exploit the time dependent statistics on the population matrix A to compute a time dependent mutation matrix M. Generalization to crossover operators under the mutation matrix formalism has also been developed [20]. Here our focus is not on the efficiency of variants of MOGA, but just to use it as an efficient and simple replacement for SGA for individual computation node. We aim at using the parallel implementation of this form of Genetic Algorithms to solve the optimization problem of sales potential for a directed social network.

4   Quasi-Parallel Genetic Algorithms

Difficult computation problems can be solved by addressing the scalability of the algorithm or hardware for the system [22]. One way is to add more power on one machine by upgrading the CPU, which progress has been described well by Moore's law [23] and limited by the physics of the hardware. An alternative approach is to add more computers to existing system, forming a computer cluster. In either approach, we have to address the practical issue of connecting a large number of CPUs in order to handle the increased workload and system demands. Therefore, an efficient parallel computing algorithm should be investigated so that each computing node runs simultaneously to address a specific problem of a generally much more complex task. For parallel genetic algorithm, the population of chromosomes is divided into sub-population consisting equal number of chromosomes and each sub-population evolves on a computing node. The nodes are linked according to a given topology defined by a network. Population on each node can communicate by exchanging information with each other along links on the network. The exchange of information is achieved through the migration of chromosomes among the nodes [24]. The topology of the network of computer nodes is an important feature of parallel genetic algorithms. The links in the network define the how information of solution will be communicated.

Given N_c chromosomes, we divide them into n sub-populations, each with (N_c / n) chromosomes. The rule of communication is defined by a simple model where a sub-population i on node i will gets the **best** M chromosomes from each of its K_i neighbors in the network, so that a total of $M \cdot K$ chromosomes will replace the **worst** $N_i \equiv M \cdot K$ chromosomes in sub-population i. We find that both the topology of the network and the rule of communication contribute to the performance of the parallel genetic algorithms and both affect the exchange rate $R_X = nMK / N_c$ between sub-populations. In practice, the total number N_c of chromosomes is restricted by the computational resource and can be considered a constant here. Thus, there are only three parameters determining the exchange rate R_X. The first one is the degree K, which is determined by the topology of the network. The second one is M, which represents the information flow capacity on each link. In our previous publications [13,14] we have investigated the performance based on the topology and the information flow. Here we address the effect of the third parameter, n, the number of sub-populations. We study the performance based on different number of computing nodes while N_c and R_X are fixed.

5   Experiments

We use the Slashdot0902 from Stanford Large Network Dataset Collection [25,26] as the data set for our experiment on social network. Slashdot is a technology-related news website and has a specific user community. In 2002 they introduced the Slashdot Zoo feature which allows users to tag each other as friends or foes. The Slashdor0902 database contains friend/foe links between the users of Slashdot for the period up to February 2009. It is a directed social network and is suitable for our study on the Sales Potential. We have 82168 nodes in this network. An edge from A to B represents that A tags B as a friend. Our goal is to find the best combination of $M(=5)$ target nodes from the data that gives us the highest combined Sales Potential according to Eq.(2). The choice of M=5 is for computational purpose, as our aim is to illustrate the importance concept of Sales Potential and a method of solving the optimization problem. We leave the engineering application specific to the Slashdot data for future research. To begin, we calculate the Sales Potential of each node according to Eq.(1) and plot the distribution of Sales Potential in Fig.2. We see that many nodes with large potential deviates from the power law obeyed by nodes with smaller potential. This can be a very nice sales list for the salesman and protection list for the department in charge of network security. We also plot the average Sales Potential for each in-degree in Fig.2. For small in-degree the average Sales Potential is quite large, which confirms our previous conjecture resulting from the Aboav-Weaire law. In a separate paper we will report our investigation that verifies the validity of the Aboav-Weaire Law for the in-degree of Slashdot0902 data and many other real social networks. In the present work which concerns the problem of optimization of Sales Potential, we need only to pay attention to the searching problem for a much smaller set of nodes with in-degree less than or equal to 3. This simplification using the Aboav-Weaire law for Slashdot0902 data reduces the original set of 82168 nodes to only 44102 nodes.

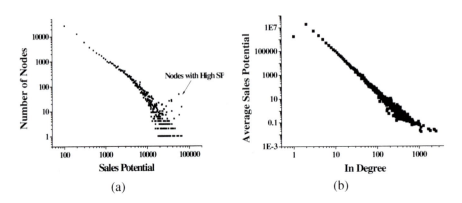

Fig. 2. Sales Potential Distribution for Slashdot0902. (a) shows the number of nodes for each Sales Potential value, and (b) shows the average Sales Potential for each in-degree.

Now, we can use QPGA to search for the optimal set. Without loss of generality, we choose the first 32768 of those nodes according to their original index to define the subspace for the search of $M(=5)$ targeted nodes. The number 32768 is chosen so that we can encode the range of node index into a 15-bit binary string. The chromosome for the 5 targeted nodes is thus the combination of 5 such 15-bit binary numbers, making up a total length of 75-bits. Before the real search we preprocess the dataset of the list of index of the second shell in-neighbors for each one of these 32768 nodes. This greatly reduces the time for the computation of the size of the union in Eq.2 for the combined Sales Potential.

In our previous work [13,14] we have studied the performance of QPGA under different topologies and exchange rates. Here, we expand our investigation to the performance of QPGA as a function of the number of sub-populations. For each sub-population we run the ordinary Mutation Only Genetic Algorithms (MOGA), and the communication topologies are shown in Fig.3.

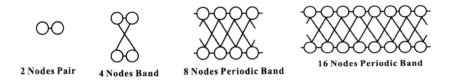

2 Nodes Pair   4 Nodes Band   8 Nodes Periodic Band   16 Nodes Periodic Band

Fig. 3. Topologies for Quasi-parallel Genetic Algorithms Sub-populations

To properly compare the performance for different number of sub-populations, we fix the total number of chromosomes and the exchange rate and perform five sets of experiments listed in Table 1. The band and pair structures are defined in [13] and illustrated in Fig.3. The band structure for the 4-node QPGA cannot be periodic so the number of exchanged chromosome is 8 to ensure the same exchange rate of 0.4. The 2-node QPGA is a pair structure so we exchange 32 chromosomes for each sub-population to fix the exchange rate to be 0.4. In this way we have the same total number of chromosomes and the same exchange rate for all versions of QPGA, so that we have the same time-complexity. Communications are made every five generations and we run each experiment for a total of 500 generations, while each experiment is repeated 100 times, with the average results shown in Fig.4.

Table 1. Experiments of QPGA with a population of N_c chromosomes divided into n sub-populations, each containing N_c / n chromosomes. The topology of QPGA is defined by the degree K of the sub-population. The number of exchange chromosomes is M for each pair of linked sub-populations, while R_X is the exchange rate.

Form	n	N_c / n	N_c	Structure	K	M	R_X
MOGA	1	160	160	N/A	0	N/A	N/A
QPGA	16	10	160	Periodic Band	4	1	4*1/10=0.4
QPGA	8	20	160	Periodic Band	4	2	4*2/20=0.4
QPGA	4	40	160	Band	2	8	2*8/40=0.4
QPGA	2	80	160	Pair	1	32	1*32/80=0.4

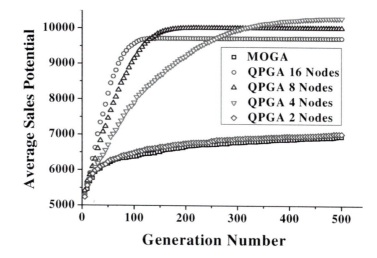

Fig. 4. Average Sales Potential Optimization Results by MOGA and QPGA

We can see that QPGA outperforms the ordinary MOGA. It's interesting to note that if we choose a large number of sub-populations, such as 16, QPGA can get its optimal result in a very short time. If we choose a relatively small number of sub-populations, such as 4, then it takes longer time to get the optimal result, but with higher quality. Depending on the desirable quality of solution and computational resources available, one can choose the appropriate number of sub-populations. We see from these experiments that QPGA is an effective method for solving the combined sales potential problem. Although our objective does not include a detailed analysis of its efficiency on Sales Potential optimization, we nevertheless arrive at a similar conclusion about its performance for other similar optimization problem, such as the knapsack and the Weierstrass function [13].

6   Conclusion

Inspired by the marketing potential of a customer (node) in a complex network such as online sales, we propose a new measurement of centrality through the concept of the Sales Potential of a node. This concept is of special interest for directed networks, as it is relevant for direct marketing as well as protective protocols to safeguard the system against malicious attack. We pose a problem of searching for the best combination of targeted nodes for marketing that gives the maximum combined Sales Potential. This is a well-defined optimization problem and can be efficiently solved by our Quasi-parallel genetic algorithm. We confirm the advantage of using QPGA over ordinary genetic algorithms with a single population. Our study also shows the relationship between performance and the number of sub-populations in QPGA. We can conclude that QPGA with small number of sub-populations gives a slower optimization process but better results than QPGA with large number of sub-populations. This helps us in choosing the population size based on our specific need: to achieve high quality result or fast optimization. We will extend this search through adaptively changing the number of sub-population: begin with a large number of sub-populations to speed up the searching process, and later merge some sub-populations to improve the quality of the results. This hybrid approach may give us a fast QPGA algorithm with results of high quality. At present, the study of Quasi-parallel genetic algorithms is still in the stage of numerical experiments. We expect further theoretical investigation on the relation between QPGA performance and the number of sub-populations, as well as the communication topology and the exchange rate between sub-populations. Finally, we will generalize the concept of Sales Potential to higher shells. We know from the research on small world that essentially the whole network can be reached within roughly six steps, our generalized Sales Potential to a sufficiently high shell number will be similar to searching over the entire network. Thus, there should be an interesting balanced point between the number of shell and the variance of the Sales Potential.

References

1. Barabasi, A.-L., Albert, R.: Emergence of Scaling in Random Networks. Science (5439), 509–511 (1999)
2. Newman, M.E.J.: Networks: An Introduction. Oxford University Press, Oxford (2010)
3. Holland, J.H.: Adaptation in Natural and Artificial Systems. University of Michigan Press, Ann Arbor (1975)
4. Goldberg, D.E.: Genetic algorithms in Search, Optimization, and Machine Learning. Addison-Wesley, Reading (1989)
5. Li, S.P., Szeto, K.Y.: Crytoarithmetic problem using parallel Genetic algorithms. In: Mendl 1999, Brno, Czech (1999)
6. Szeto, K.Y., Cheung, K.H.: Multiple time series prediction using genetic algorithms optimizer. In: Proceedings of the International Symposium on Intelligent Data Engineering and Learning, IDEAL 1998, Hong Kong, pp. 127–133 (1998)
7. Jiang, R., Szeto, K.Y., Luo, Y.P., Hu, D.C.: Distributed parallel genetic algorithms with path splitting scheme for the large traveling salesman problems. In: Shi, Z., Faltings, B., Musen, M. (eds.) Proceedings of Conference on Intelligent Information Processing, 16th World Computer Congress 2000, Beijing, August 21-25, pp. 478–485. Publishing House of Electronic Industry (2000)

8. Szeto, K.Y., Cheung, K.H., Li, S.P.: Effects of dimensionality on parallel genetic algorithms. In: Proceedings of the 4th International Conference on Information System, Analysis and Synthesis, Orlando, Florida, USA, vol. 2, pp. 322–325 (1998)
9. Szeto, K.Y., Fong, L.Y.: How Adaptive Agents in Stock Market Perform in the Presence of Random News: A Genetic Algorithm Approach. In: Leung, K.-S., Chan, L., Meng, H. (eds.) IDEAL 2000. LNCS (LNAI), vol. 1983, pp. 505–510. Springer, Heidelberg (2000)
10. Fong, A.L.Y., Szeto, K.Y.: Rule Extraction in Short Memory Time Series using Genetic algorithms. European Physical Journal B 20, 569–572 (2001)
11. Shiu, K.L., Szeto, K.Y.: Self-adaptive Mutation Only Genetic Algorithm: An Application on the Optimization of Airport Capacity Utilization. In: Fyfe, C., Kim, D., Lee, S.-Y., Yin, H. (eds.) IDEAL 2008. LNCS, vol. 5326, pp. 428–435. Springer, Heidelberg (2008)
12. Chen, C., Guan, W., Szeto, K.Y.: Markov Chains Genetic algorithms for Airport Scheduling. In: Proceedings of the 9th International FLINS Conference on Foundations and Applications of Computational Intelligence (FLINS 2010), pp. 905–910 (August 2010)
13. Wang, G., Wu, D., Szeto, K.Y.: Quasi-Parallel Genetic Algorithms with Different Communication Topologies. In: 2011 IEEE Congress on Evolutionary Computation (CEC), June 5-8, pp. 721–727 (2011)
14. Wang, G., Wu, D., Chen, W., Szeto, K.Y.: Importance of Information Exchange in Quasi-Parallel Genetic Algorithms. In: Krasnogor, N. (ed.) Proceedings of the 13th Annual Conference Companion on Genetic and Evolutionary Computation (GECCO 2011), pp. 127–128. ACM, New York (2011)
15. Aboav, D.A.: Metallography V.3, 383 (1970); ibid, V.13, 43 (1980)
16. Weaire, D.: Metallography. 7, 157 (1974)
17. Ma, C.W., Szeto, K.Y.: Phys. Rev. E 73, 047101 (2006)
18. Szeto, K.Y., Fu, X., Tam, W.Y.: Universal Topological Properties of Two-dimensional Cellular Patterns. Phys. Rev. Lett., 138302-1–138302-3 (2002)
19. Ma, C.W., Szeto, K.Y.: Locus Oriented Adaptive Genetic algorithms: Application to the Zero/One Knapsack Problem. In: Proceeding of the 5th International Conference on Recent Advances in Soft Computing, RASC 2004, Nottingham, UK, pp. 410–415 (2004)
20. Law, N.L., Szeto, K.Y.: Adaptive Genetic algorithms with Mutation and Crossover Matrices. In: Proceeding of the 12th International Joint Conference on Artificial Intelligence (IJCAI 2007), Hyderabad, India, January 6-12. Theme: AI and Its Benefits to Society, vol. II, pp. 2330–2333 (2007)
21. Szeto, K.Y., Zhang, J.: Adaptive Genetic Algorithm and Quasi-parallel Genetic Algorithm: Application to Knapsack Problem. In: Lirkov, I., Margenov, S., Waśniewski, J. (eds.) LSSC 2005. LNCS, vol. 3743, pp. 189–196. Springer, Heidelberg (2006)
22. Bondi, A.B.: Characteristics of scalability and their impact on performance. In: Proceedings of the 2nd International Workshop on Software and Performance, pp. 195–203 (2000)
23. Moore, G.E.: Cramming more components onto integrated circuits. Electronics Magazine, 4 (1965)
24. Cantú-Paz, E.: Efficient and accurate parallel genetic algorithms, pp. 16, 17, 22. Kluwer Academic, USA (2000)
25. http://snap.stanford.edu/data/soc-Slashdot0902.html
26. Leskovec, J., Lang, K., Dasgupta, A., Mahoney, M.: Community Structure in Large Networks: Natural Cluster Sizes and the Absence of Large Well-Defined Clusters. Internet Mathematics 6(1), 29–123 (2009)

The Emergence of Multi-cellular Robot Organisms through On-Line On-Board Evolution

Berend Weel, Evert Haasdijk, and A.E. Eiben

Vrije Universiteit Amsterdam, The Netherlands
{b.weel,e.haasdijk}@vu.nl, gusz@cs.vu.nl

Abstract. We investigate whether a swarm of robots can evolve controllers that cause aggregation into 'multi-cellular' robot organisms without a specific reward to do so. To this end, we create a world where aggregated robots receive more energy than individual ones and enable robots to evolve their controllers on-the-fly, during their lifetime. We perform experiments in six different implementations of the basic idea distinguished by the system of energy distribution and the level of advantage aggregated robots have over individual ones. The results show that 'multi-cellular' robot organisms emerge in all of these cases.

1 Introduction

Swarm-robot systems and (self-)reconfigurable modular robot systems paradigms have been invented to facilitate multi-purpose robot design. Swarm-robot systems use large numbers of autonomous robots which cooperate to perform a task [12]. Similarly, self-reconfigurable modular robot systems use many modules to form a larger, reconfigurable, robot that can tailor its shape to suit a particular task [17]. These two subjects have largely been studied separately, with swarm robotics focusing on cooperating robots which do not assemble into an organism. Research in reconfigurable modular robotics focuses on creating actual modules, finding suitable morphologies for a task, and reconfiguring from one morphology to another.

Recently, a new kind of self-reconfigurable robots has been proposed based on modules that are also capable of autonomous locomotion [9]. In such a system, the modules can form a swarm of autonomous units as well as a 'multi-cellular' robot organism consisting of several physically aggregated units. To date, there has been very little research on self-assembly – the transition from swarm to organism – as emergent, not pre-programmed, behaviour.

The main subject of this paper is emergent self-assembly through evolution. We are interested in the emergence of robot organisms from swarms as a response to environmental circumstances. To this end, we design environments where organisms have an advantage over individual modules and make the robots evolvable. In particular, we implement an on-line and on-board evolutionary mechanism where robot controllers undergo evolution on-the-fly: selection and reproduction of controllers is not performed by an outer entity in an off-line

C. Di Chio et al. (Eds.): EvoApplications 2012, LNCS 7248, pp. 124–134, 2012.

fashion (e.g. a genetic algorithm running on an external computer), but by the robots themselves [7]. One of the premises of our study is that we do not include a specific fitness measure to favour organisms. Rather, we build a system with an implicit environmental pressure towards aggregation by awarding more energy to robots in an aggregated state. The environmental advantage is scalable and we compare the effects of low, medium and high values.

The research questions we seek to answer are the following:

1. Will organisms evolve purely because the environment favours modules that are part of an organism? Or, does the system need a specific user defined fitness to promote aggregation?
2. How does system behavior depend on the level of environmental benefit? Will organisms evolve even if the extra advantage is low?
3. What are the characteristics of the evolved organisms? How large and how old do organisms become?

2   Related Work

Swarm Robotics. As mentioned, our work is situated between swarm robotics and self-reconfigurable modular robot systems. Swarm Robotics [12] is a field that stems from Swarm Intelligence [2], where swarm-robots often have the ability for physical self-assembly. Swarm-bots were created in order to provide a system which was robust towards hardware failures, versatile in performing different tasks, and navigating different environments.

In [3], Şahin categorizes tasks for which to use swarm robots as: tasks that cover a region, tasks that are too dangerous, tasks that scale up or down in time, and tasks that require redundancy. In [12], the self-assembly of s-bots allows for the navigation of crevices and objects too large for a single robot, as well as the transport of objects which are too heavy to be transported by a single robot.

Self-reconfigurable Modular Robot Systems. Self-reconfigurable modular robot systems (SMRSs) were designed with three key motivations: versatility, robustness and low cost. The first two are identical to motivations for swarm-robots, while low cost can be achieved through economy of scales and mass production as these systems use many identical modules. The main advantage advocated is the adaptability of these systems to different tasks, however most of the research in this field is still exploring the basics: module design, locomotion and reconfiguration.

Yim gives an overview of self-reconfigurable modular robot systems in [17], the research is mainly on creation of modules in hardware and showcasing their abilities to reconfigure and lift other modules. Most of these self-reconfigurable modular robot systems are incapable of locomotion as independent modules. In recent years however, a number of SMRSs were developed that incorporate independent mobility as a feature [16], [10], [9].

The SYMBRION/REPLICATOR projects, of which this research is part, develops its own SMRS, exploring two alternatives for hardware, as presented in

[9]. Both versions are independently mobile and so can operate as a swarm, both also have a mechanical docking mechanism allowing the modules to form and control a multi-robot organism.

Self-assembly. The task of multiple robots connecting autonomously is usually called self-assembly, and has been demonstrated in several cases: [5], [18], [14], [16]. Most of these however, are limited to pre-programmed control sequences without any evolution. In self-reconfigurable robots, self-assembly is restricted to the docking of two modules as demonstrated in [16], [10].

The work in this paper is most closely related to that of Groß, Nolfi, Dorigo, and Tuci: [1], [5], [6], in which they explore self-assembly of swarm robots. They evolved Recurrent Neural Networks for the control of s-bots to be capable of Self-Assembly in simulation, they then took the best controllers evolved in this manner and tested them in real s-bots. This research shows it is possible to evolve controllers which create organisms. As it is difficult to evolve controllers in a situation where either robot can grip the other, they use a target robot in most of their research. This target robot, also called a seed, bootstraps the problem of who grips who, by showing which robot should be gripped by the other robot. Furthermore they assign fitness to the s-bots based on whether they succeeded in forming an organism, or if failed the distance between the robots.

On-Line On-Board Evolutionary Algorithms. We use an evolutionary algorithm (EA) as a heuristic optimiser for our robot controller, as do many robotics projects. The field of evolutionary robotics in general is described by Nolfi and Floreano in [13]. Eiben et al. describe a classification system for evolutionary algorithms used in evolutionary robotics [4]. They distinguish evolution based on *when* evolution takes place: off-line or design time vs. on-line or run time. *Where* evolution takes place: on-board or intrinsic vs. off-board or extrinsic. And *how* evolution takes place: encapsulated or centralized vs. distributed.

Whereas most evolutionary robotics research uses offline and extrinsic approaches to evolving controllers. We use an on-line on-board (or intrinsic) hybrid approach, based on EVAG [11] and $(\mu + 1)$ ON-LINE [7]. It is described in detail in [8]. Each robot maintains a population of μ individuals locally, and performs cross-over and mutation to produce offspring. These individuals can be exchanged between robots as part of the parent selection mechanism. The offspring is then instantiated as the controller for evaluation.

We do not include a task in our system other than gathering energy. Nor do we include any type of morphology engineering, or purposeful reconfiguration of an organism. Our goal is to investigate only the very first step: forming an organism under environmental pressure.

3   System Description and Experiments

Simulator. We conduct our experiments with simulated e-puck robots in a simple 2D simulator: RoboRobo[1]. The robots can steer by setting their desired

---

[1] http://www.lri.fr/~bredeche/roborobo/

left and right wheel speeds. Each robot has 8 sensors to detect obstacles (static obstacles as well as other robots), indicated by the lines protruding from their circular bodies in Fig. 1. While a such a simple 2D simulation ignores a lot of the intricacies of robots in the real world, it is still complex enough that creating intentional, meaningful, and effective organisms is not trivial and serves our purpose of investigating organism creation under environmental pressure.

Connections. In our experiments robots can create new organisms, join an already existing organism, and two existing organisms can merge into a larger organism. When working with real robots, creating a physical connection between two robots can be challenging, and movements of joints are noisy because of actuator idiosyncrasies, flexibility of materials used, and sensor noise. We choose to disregard these issues and create a very simple connection mechanism which is rigid the moment a connection is made. The connection is modelled as a magnetic slip-ring, which a robot can set to 'positive', 'negative' or 'neutral'. When robots are close enough, they automatically create a rigid connection if neither of them has the slip-ring on 'negative' and at least one of them has it on the 'positive' setting. The connection remains in place as long as these conditions hold (i.e., neither sets its slip-ring to 'negative' and at least one is set to 'positive').

Environment. The robots start each evaluation cycle with a fixed amount of energy and lose energy at regular intervals to simulate power consumption by actuators. When a robots energy reaches 0 it is deactivated, and is unable to move for the rest of the evaluation cycle. The environment provides energy through a 'feeding ground' from which robots can gather energy by standing on top. The amount of energy gathered during evaluation is the fitness measure for on-line evolution.

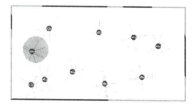

Fig. 1. 10 robots in an arena with a feeding ground which is also a scale, the scale regularly changes position

There are two ways in which the environment provides an advantage to organisms over single robots. The first is that the amount of energy awarded depends on whether or not the robots are part of an organism as described below. The second advantage is that organisms can move faster, by driving in the same direction, although manoeuvring is slightly more complicated. It is important to note that there is no *direct* reward for being part of an organism; the benefit is indirectly defined by the environment. We have implemented the advantages for organisms in two different scenarios: one based on a scale metaphor, and one on a riverbed. We compare these with a separate *baseline* experiment, where organisms have no benefit.

Power Scale. In a rectangular arena without stationary obstacles, there is a single feeding ground (the circle in Fig. 1); the environment awards energy to robots in this feeding ground. This feeding ground acts as a scale: the environment supplies more energy to modules belonging to an organism. For each

Table 1. Power gain formulas for power scale (left) and riverbed (right) scenarios

Logarithmic	$log\left(|O| + 1\right) * 2$	Logarithmic	$log\left(\frac{P}{W} * 2 * |O|\right)$
Linear	$\left(|O| + 1\right) * 2$	Linear	$\frac{P}{W} * 2 * |O|$
Exponential	$exp\left(|O| + 1\right) * 2$	Exponential	$exp\left(\frac{P}{W} * 2 * |O|\right)$

organism on the scale, the gain in awarded energy increases with the organism's size $|O|$. This gain can depend linearly, logarithmically or exponentially on $|O|$ as shown in Table 1. Robots on the scale but not part of the organism do not affect the amount of energy received by the organism.

Riverbed. In this scenario the arena is analogous to a river which pushes the robots downstream. Again, there are no obstacles. Now, the entire arena is a feeding ground, but there is an upstream gradient: the amount of energy awarded increases as a robot finds itself more upstream. To counteract the current that pushes robots down the energy gradient, robots can aggregate into an organism: together they are faster and able to move or stay upriver, and so receive more energy. Again, this gain can increase linearly, logarithmically or exponentially as shown in Table 1. In the formulas used W is the width of the arena and P the position of the centre of the organism.

Baseline. As a baseline, we use an experiment in which being part of an organism holds no benefit. We set up an empty environment where the robots receive a fixed amount of energy at every time-step, regardless of their position in the arena or whether they were part of an organism. There is no current driving the organisms anywhere, and organisms are not able to move faster either. The baseline can be viewed as an extension of either experiment: the power scale the scale now extends to the entire arena and no longer takes the number of robots on it into account. For the riverbed scenario, the current is reduced to 0 and the feeding ground has no gradient.

Controller. The controller consists of a feed-forward artificial neural network that selects one of 4 pre-programmed strategies based on sensory inputs. The neural net has 13 inputs (cf. Table 2), 5 outputs and no hidden nodes. It uses a *tanh* activation function.

Table 2. Neural Network inputs (left) and outputs (right)

8 Distance sensors	Vote for Random Walk
1 Size of the organism	Vote for Wall Avoidance
1 Angle to nearest feeding ground	Vote for Go to feeding ground
1 Distance to nearest feeding ground	Vote for Create organism
1 Energy Level	Magnetic ring setting.
1 Bias node.	

The inputs are normalised: Distance sensors, Organism Size, Distance to nearest feeding ground, and Energy level are normalised between 0 and 1, angle to nearest feeding ground is normalised between -1 and 1.

The output of the neural network, as described in Table 2, is interpreted as follows: the first four outputs each vote for an action, the action with the highest activation level is selected. The fifth output governs the magnetic ring setting: 'negative' if the output is smaller than −0.33, 'positive' if it is bigger than 0.33 and 'neutral' otherwise.

Evolutionary Algorithm and Runs. We use a genome which directly encodes the weights of the neural net using a real-valued vector. It also includes N mutation step sizes for N genes.

Each controller is evaluated for 800 time steps, followed by a 'free' phase of 200 time steps to allow it to get out of bad situations. Each 1000 time steps therefore constitutes 1 generation. At the end of the evaluation cycle the active controller is compared to the local population, and added if it is better than the worst one. A new controller is created using either mutation or crossover.

Mutation is a Gaussian perturbation using self-adaptation. The algorithm uses averaging crossover and parents are selected using a binary tournament on the entire population across all robots (panmictic layout) as described in [8].

At the start of the new generation, control is switched to the new controller, which potentially has a completely different setting for the magnet than the previous one, potentially destroying an organism. We ran the experiments with different reward functions described above using 10 robots. We used this number of robots as we expect to have a similar number of real robots to repeat the experiment with available from the project this research is part of. To ensure good parameter settings, we used the BONESA toolbox[2] [15] to optimise settings for crossover rate, mutation rate, initial mutation step size, re-evaluation rate, and population size. Using the best parameters found, as shown in Table 3, we repeated each experiment 40 times, each run lasting 1000 generations.

Table 3. Parameters of the EA for the Riverbed (R.B.) and Power Scale (P.S.) experiments

	Scaling	re-evaluation rate	crossover rate	population size	mutation rate
R.B.	logarithmic	0.62385	0.03602	3	0.57369
	linear	0.44908	0.04369	4	0.00509
	exponential	0.40806	0.03411	3	0.21407
P.S.	logarithmic	0.50294	0.93562	3	0.07154
	linear	0.54149	0.46759	3	0.06662
	exponential	0.56736	0.99323	3	0.05807
	baseline	0.62385	0.03602	3	0.57369

---

[2] http://sourceforge.net/projects/tuning/

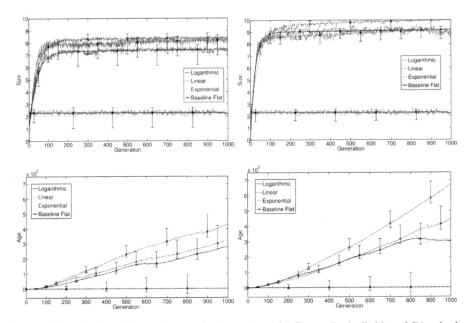

Fig. 2. Sizes (top) and Ages (bottom) of organisms for Power Scale (left) and Riverbed (right) scenarios

4   Results and Analysis

The results we obtained are shown in Fig. 2.[3] They show that random interactions already lead to some organisms, however, under influence of environmental pressure the organisms become much larger and older than without. The amount of pressure did not result in large differences in either organism size or longevity. Even a small amount of environmental advantage suffices to cause significantly bigger and older organisms – the logarithmic reward function that implements the least pressure in our comparison may not even represent the minimum amount of pressure needed.

Comparing results for the two scenarios, the riverbed scenario leads to bigger and older organisms than the power scale scenario – a markedly larger difference than that between logarithmic, linear and exponential benefit. This indicates that the environmental pressure is determined by more than only the reward functions we tested: since there is no quantitative difference between the reward functions in each scenario, the difference in organism longevity and size can only be caused by other, qualitative, differences between the scenarios. This shows that the design of the environment itself can be more important than the specific function that determines the environmental benefit of being in an organism.

---

[3] We used a beta distribution as a basis for the 95% confidence interval for our graphs.

Organism Size. Figure 2 shows the mean organism size of a generation averaged over 40 runs for the power scale experiment on the left and the riverbed experiment on the right. The x-axis is the time measured in generations. The y-axis displays the mean number of robots in an organism. The raw data is plotted as a grey line for each fitness scaling, for each we also show a second order exponential trend line. The bottom line is that of the baseline experiment.

The baseline experiment results in organisms that are not very large, on average approximately 2.3. The other experiments produce much larger organisms: averages between 7 and 9 for power scale, between 8 and 10 for riverbed. These are significantly higher than the baseline, at 99% confidence, as seen by the disjoint confidence intervals between the baseline and other plots. The difference in averages between the logarithmic, linear and exponential experiments are not significant at 99% confidence (overlapping confidence intervals).

In both experiments, environmental benefit positively influences the emergence of organisms, but the level of influence does not seem to differ between the tested reward functions. With logarithmic being the lowest reward tested we can conclude that the minimum pressure lies somewhere between no advantage and logarithmic. Note that in both scenarios the linear reward leads to the highest average. This suggests that there is a sweet-spot somewhere between logarithmic and exponential scaling.

Organism Age. Figure 2 shows the mean organism age for each generation averaged over 40 runs for the power scale experiment on the left and the riverbed experiment on the right. The x-axis is the time measured in generations. The y-axis displays the mean age of an organism in number of 10^5 ticks. The raw data is plotted as a grey line for each fitness scaling, for each we also show a trend line based on the fifth Fourier series. The baseline experiment's results are included in both graphs.

The lines for the three reward functions rise rapidly to values significantly higher than the baseline in both scenarios. In the power scale scenario, the average organism age reaches more than 400.000 ticks, or 400 generations. In the riverbed scenario this goes up to almost 700.000 ticks, or 700 generations for the linear reward.

The lines are rising rapidly and almost monotonously, suggesting that the organisms do not 'die' between generations. The age values in the riverbed scenario are higher than those in the power scale scenario, notable is also that the incline of the plots for riverbed are steeper than the ones for power scale.

These graphs support our earlier conclusion that the reward positively influences the size, and the age of organisms. The steeper and higher graphs of the riverbed scenario lead us to conclude that the environmental pressure is determined by more than just the reward function. It also shows that the reward has a different impact on the size of organisms than it does on the age of organisms. Overall we conclude that, when designing experiments, more effort should go into creating an appropriate environment than into designing the reward.

We also observe that organisms do not disintegrate when switching from one controller to another, from which we can conclude that the evolutionary algorithm converges very quickly (within 50 generations) to values for a positive ring setting.

5   Conclusion and Further Research

We have shown that large organisms emerge in an environment which favours modules that are part of an organism, without the need for a specific fitness function to promote aggregation. Organisms emerge even without environmental pressure by chance, but these are significantly smaller and have a significantly shorter life span.

We tested three reward functions in two separate scenarios. The amount of pressure from the reward functions did not result in large differences. Even the lowest amount of pressure, – logarithmic with respect to organism size – leads to significantly bigger and older organisms. We notice a trend: the linear reward performs slightly better than both logarithmic and exponential, suggesting an optimal setting between logarithmic and exponential. The differences between the scenarios did result in different sizes and organisms. In other words, the environmental pressure is more than just a reward function: our results also show that the design of the environment is more important when designing an experiment than the reward function.

We only used 10 robots for the experiments, this raises the question whether the same results would be obtained when using more robots. More robots would also imply a larger arena, so care should be taken to correctly scale the experiments. Furthermore, we used a controller which had pre-programmed parts that were very solution specific. To alleviate this specificity, further experiments can be executed in which the control is at a lower level, and hence the problem more difficult. Here the use of different controllers, which are more powerful, can be investigated. We observed evidence that the reward function may have an optimum, this could be further investigated to answer questions like: where is the optimum? Is it the same in different scenarios? Does finding the optimum lead to significantly bigger or longer lasting organisms?

We noted evidence that differences in qualitative environmental pressure may be more important than differences in quantitative pressure in their influence on the emergence of organisms, and therefore should be investigated. This research could be part of the upcoming discipline of complexity-engineering: harnessing emergent phenomena to create interesting or useful characteristics in complex systems. Lastly we would like to investigate the 'unlearning' of organism forming behaviour by letting controllers evolve in a changing environment which first favours organisms and over time puts organisms at a disadvantage.

Acknowledgements. This work was made possible by the European Union FET Proactive Initiative: Pervasive Adaptation funding the SYMBRION project under grant agreement 216342. The authors would like to thank Selmar Smit and our partners in the SYMBRION consortium for many inspirational discussions on the topics presented here.

References

1. Bianco, R., Nolfi, S.: Toward open-ended evolutionary robotics: evolving elementary robotic units able to self-assemble and self-reproduce. Connection Science 16(4), 227–248 (2004)
2. Bonabeau, E., Dorigo, M., Theraulaz, G.: Swarm Intelligence: From Natural to Artificial Systems. Oxford University Press (1999)
3. Şahin, E.: Swarm robotics: From sources of inspiration to domains of application. Swarm Robotics, 10–20 (2005)
4. Eiben, A.E., Haasdijk, E., Bredeche, N.: Embodied, on-line, on-board evolution for autonomous robotics. In: Levi, P., Kernbach, S. (eds.) Symbiotic Multi-Robot Organisms: Reliability, Adaptability, Evolution, vol. ch. 5.2, pp. 361–382. Springer, Heidelberg (2010)
5. Groß, R., Bonani, M., Mondada, F., Dorigo, M.: Autonomous self-assembly in swarm-bots. IEEE Transactions on Robotics 22, 1115–1130 (2006)
6. Groß, R., Dorigo, M.: Evolution of solitary and group transport behaviors for autonomous robots capable of self-assembling. Adaptive Behavior 16(5), 285 (2008)
7. Haasdijk, E., Eiben, A.E., Karafotias, G.: On-line evolution of robot controllers by an encapsulated evolution strategy. In: Proceedings of the 2010 IEEE Congress on Evolutionary Computation. IEEE Press (2010)
8. Huijsman, R.J., Haasdijk, E., Eiben, A.: An on-line on-board distributed algorithm for evolutionary robotics. In: Hao, J., Legrand, P., Collet, P., Monmarch, N., Lutton, E., Schoenauer, M. (eds.) Proceedings of Artificial Evolution, 10th International Conference on Evolution Artificielle, EA 2011 (2011)
9. Kernbach, S., Meister, E., Scholz, O., Humza, R., Liedke, J., Rico, L., Jemai, J., Havlik, J., Liu, W.: Evolutionary robotics: The next-generation-platform for on-line and on-board artificial evolution. In: 2009 IEEE Congress on Evolutionary Computation, pp. 1079–1086 (2009)
10. Kutzer, M.D.M., Moses, M.S., Brown, C.Y., Scheidt, D.H., Chirikjian, G.S., Armand, M.: Design of a new independently-mobile reconfigurable modular robot. In: 2010 IEEE International Conference on Robotics and Automation (ICRA), pp. 2758–2764. IEEE (2010)
11. Laredo, J.L.J., Eiben, A.E., Steen, M., Merelo, J.J.: EvAg: a scalable peer-to-peer evolutionary algorithm. Genetic Programming and Evolvable Machines 11(2), 227–246 (2009)
12. Mondada, F., Pettinaro, G.C., Guignard, A., Kwee, I.W., Floreano, D., Deneubourg, J.L., Nolfi, S., Gambardella, L.M., Dorigo, M.: Swarm-bot: A new distributed robotic concept. Autonomous Robots 17(2/3), 193–221 (2004)
13. Nolfi, S., Floreano, D.: Evolutionary Robotics: The Biology, Intelligence, and Technology of Self-Organizing Machines. MIT Press, Cambridge (2000)
14. O'Grady, R., Christensen, A.L., Dorigo, M.: Autonomous Reconfiguration in a Self-assembling Multi-robot System. In: Dorigo, M., Birattari, M., Blum, C., Clerc, M., Stützle, T., Winfield, A.F.T. (eds.) ANTS 2008. LNCS, vol. 5217, pp. 259–266. Springer, Heidelberg (2008)
15. Smit, S.K., Eiben, A.E.: Multi-problem parameter tuning using BONESA. In: Hao, J., Legrand, P., Collet, P., Monmarch, N., Lutton, E., Schoenauer, M. (eds.) Proceedings of Artificial Evolution, 10th International Conference on Evolution Artificielle (EA 2011), pp. 222–233 (2011)

16. Wei, H., Cai, Y., Li, H., Li, D., Wang, T.: Sambot: A self-assembly modular robot for swarm robot. In: 2010 IEEE International Conference on Robotics and Automation (ICRA), pp. 66–71. IEEE (2010)
17. Yim, M., Shen, W.M., Salemi, B., Rus, D., Moll, M., Lipson, H., Klavins, E., Chirikjian, G.S.: Modular self-reconfigurable robot systems [grand challenges of robotics]. IEEE Robotics & Automation Magazine 14(1), 43–52 (2007)
18. Yim, M., Shirmohammadi, B., Sastra, J., Park, M., Dugan, M., Taylor, C.: Towards robotic self-reassembly after explosion. In: 2007 IEEE/RSJ International Conference on Intelligent Robots and Systems, pp. 2767–2772. IEEE (2007)

Evolving Seasonal Forecasting Models with Genetic Programming in the Context of Pricing Weather-Derivatives

Alexandros Agapitos, Michael O'Neill, and Anthony Brabazon

Financial Mathematics and Computation Research Cluster
Natural Computing Research and Applications Group
Complex and Adaptive Systems Laboratory
University College Dublin, Ireland
{alexandros.agapitos,m.oneill,anthony.brabazon}@ucd.ie

Abstract. In this study we evolve seasonal forecasting temperature models, using Genetic Programming (GP), in order to provide an accurate, localised, long-term forecast of a temperature profile as part of the broader process of determining appropriate pricing model for weather-derivatives, financial instruments that allow organisations to protect themselves against the commercial risks posed by weather fluctuations. Two different approaches for time-series modelling are adopted. The first is based on a simple system identification approach whereby the temporal index of the time-series is used as the sole regressor of the evolved model. The second is based on iterated single-step prediction that resembles autoregressive and moving average models in statistical time-series modelling. Empirical results suggest that GP is able to successfully induce seasonal forecasting models, and that autoregressive models compose a more stable unit of evolution in terms of generalisation performance for the three datasets investigated.

1 Introduction

Weather conditions affect the cash flows and profits of businesses in a multitude of ways. For example, energy companies (gas or electric) may sell fewer supplies if a winter is warmer than usual, leisure industry firms such as ski resorts, theme parks, hotels are affected by weather metrics such as temperature, snowfall or rainfall, construction firms can be affected by rainfall, temperatures and wind levels, and agricultural firms can be impacted by weather conditions during the growing or harvesting seasons [3]. Firms in the retail, manufacturing, insurance, transport, and brewing sectors will also have weather "exposure". Less obvious weather exposures include the correlation of events such as the occurrence of plant disease with certain weather conditions (i.e. blight in potatoes and in wheat) [9]. Another interesting example of weather risk is provided by the use of "Frost Day" cover by some of the UK town/county councils whereby a payout is obtained by them if a certain number of frost days (when roads would require gritting - with an associated cost) are exceeded. Putting the above into context,

C. Di Chio et al. (Eds.): EvoApplications 2012, LNCS 7248, pp. 135–144, 2012.

it is estimated that in excess of $1 trillion of activity in the US economy is weather-sensitive [5]. In response to the existence of weather risk, a series of financial products have been developed in order to help organisations manage these risks. *Weather derivatives* are financial products that provide a payout which is related to the occurrence of pre-defined weather events [7].

A key component of the accurate pricing of a weather derivative are forecasts of the expected value of the underlying weather variable and its associated volatility. The goal of this study is to produce predictive models by the means of Genetic Programming [4] (GP) of the stochastic process that describes temperature. Section 2 introduces weather derivatives, motivates the need for seasonal temperature forecasting, and reviews the major statistical and heuristic time-series modelling methods. Section 3 describes the experiment design, Section 4 discusses the empirical findings, and finally Section 5 draws our conclusions.

2 Background

2.1 OTC Weather Derivatives

The earliest weather derivatives were traded over-the-counter (OTC) as individually negotiated contracts. In OTC contracts, one party usually wishes to hedge a weather exposure in order to reduce cash flow volatility. The payout of the contract may be linked to the value of a weather index on the Chicago Mercantile Exchange (CME) or may be custom-designed. The contract will specify the weather metric chosen, the period (a month, a season) over which it will be measured, where it will be measured (often a major weather station at a large airport), the scale of payoffs depending on the actual value of the weather metric and the cost of the contract. The contract may be a simple "swap" where one party agrees to pay the other if the metric exceeds a pre-determined level while the other party agrees to pay if the metric falls below that level.

In the US, many OTC (and all exchange-traded) contracts are based on the concept of a *'degree-day'*. A degree-day is the deviation of a day's average temperature from a reference temperature. Degree days are usually defined as either *'Heating Degree Days'* (HDDs) or *'Cooling Degree Days'* (CDDs). The origin of these terms lies in the energy sector which historically (in the US) used 65 degrees Fahrenheit as a baseline, as this was considered to be the temperature below which heating furnaces would be switched on (a heating day) and above which air-conditioners would be switched on (a cooling day). As a result HDDs and CDDs are defined as

$$HDD = \text{Max } (0, 65''\text{F - average daily temperature}) \tag{1}$$

$$CDD = \text{Max } (0, \text{average daily temperature - } 65''\text{F}) \tag{2}$$

For example, if the average daily temperature for December 20th is $36^\circ F$, then this corresponds to 29 HDDs (65 - 36 = 29). The payoff of a weather future is usually linked to the aggregate number of these in a chosen time period (one HDD or CDD is typically worth $20 per contract). Hence, the payoff to a December contract for HDDs which (for example) trades at 1025 HDDs on 1st December - assuming that there was a total of 1080 HDDs during December - would be $1,100 ($20 * (1080-1025). A comprehensive introduction to weather derivatives is provided by [7].

2.2 Seasonal Forecasting for Pricing a Weather Derivative

A substantial literature exists concerning the pricing of financial derivatives. However, models from this literature cannot be simply applied for pricing of weather derivatives as there are a number of important differences between the two domains. The *underlying* (variable) in a weather derivative (a weather metric) is non-traded and has no intrinsic value in itself (unlike the underlying in a traditional derivative which is typically a traded financial asset such as a share or a bond). It is also notable that changes in weather metrics do not follow a pure random walk as values will typically be quite bounded at specific locations. Standard (arbitrage-free) approaches to derivatives pricing (such as the Black-Scholes option pricing model) are inappropriate as there is no easy way to construct a portfolio of financial assets which replicates the payoff to a weather derivative [6].

One method that is used to price weather risk is *index modelling*. This approach attempts to build a model of the distribution of the underlying weather metric (for example, the number of seasonal cumulative heating degree days), typically using historical data. A wide variety of forecasting approaches such as time-series models, of differing granularity and accuracy, can be employed.

In considering the use of weather forecast information for derivatives pricing, we can distinguish between a number of possible scenarios. In this paper we are focusing on weather derivatives that are traded long before the start of the relevant weather period. In this case we can only use *seasonal forecasting* methods as current short run weather forecasts have no useful information content in predicting the weather than will arise during the weather period. *Seasonal forecasts* are long-term forecasts having a time horizon beyond one month [10]. There are a plethora of methods for producing these forecasts ranging from the use of statistical time-series models based on historic data to the use of complex, coursegrained, simulation models which incorporate ocean and atmospheric data. The following sections briefly review some of the major techniques that fall into the two families of *statistical* and *heuristic* approaches to time-series forecasting.

2.3 Statistical Time-Series Forecasting Methods

Statistical time-series forecasting methods fall into the following five categories: (a) *exponential smoothing methods*; (b) *regression methods*; (c) *autoregressive integrated moving average methods* (ARIMA); (d) *threshold methods*; (e) *generalised autoregressive conditionally heteroskedastic methods* (GARCH). The first three categories can be considered as linear, whereas the last two as non-linear methods.

In *exponential smoothing*, a forecast is given as a weighted moving average of recent time-series observations. The weights assigned decrease exponentially as the observations get older. In *regression*, a forecast is given as a linear combination of one or more explanatory variables. ARIMA models give a forecast as a linear function of past observations and error values between the time-series itself and past observations of explanatory variables. These models are essentially based on a composition of *autoregressive models* (linear prediction formulas that attempt to predict an output of a system based on the previous outputs), and moving average models (linear prediction model based on a *white noise* stationary time-series). For a discussion on smoothing, regression and ARIMA methods see [8]. Linear models cannot capture some featured that commonly occur in real-world data such as asymmetric cycles and outliers.

Threshold methods [8] assume that extant asymmetric cycles are cause by distinct underlying phases of the time-series, and that there is a transition period between these phases. Commonly, the individual phases are given a linear functional form, and the transition period is modelled as an exponential or logistic function. GARCH methods [2] are used to deal with time-series that display non-constant variance of residuals (error values). In these methods, the variance of error values is modelled as a quadratic function of past variance values and past error values.

2.4   Genetic Programming for Time-Series Modelling

In GP-based time-series prediction [1] the task is to induce a model that consists of the best possible approximation of the stochastic process that could have generated an observed time-series. Given *delayed vectors* v, the aim is to induce a model f that maps the vector v to the value x_{t+1}. That is,

$$x_{t+1} = f(v) = f(x_{t-(m-1)\tau}, x_{t-(m-2)\tau}, \dots, x_t) \tag{3}$$

where m is embedding dimension and τ is delay time. The embedding specifies on which historical data in the series the current time value depends. These models are known as *single-step predictors*, and are used to predict to predict one value x_{t+1} of the time series when all inputs $x_{t-m}, \dots, x_{t-2}, x_{t-1}, x_t$ are given. For long-term forecasts, *iterated single-step prediction models* are employed to forecast further than one step in the future. Each predicted output is fed back as input for the next prediction while all other inputs are shifted back one place. The input consists partially of predicted values as opposed to observables from the original time-series. That is,

$$x'_{t+1} = f(x_{t-m}, \dots, x_{t-1}, x_t); m < t$$
$$x'_{t+2} = f(x_{t-m+1}, \dots, x_t, x'_{t+1}); m < t$$
$$\vdots$$
$$x'_{t+k} = f(x_{t-m+k-1}, \dots, x'_{t+k-2}, x'_{t+k-1}); m < t, k \geq 1$$

$$\tag{4}$$

where k is the prediction step.

2.5   Scope of Research

The goal of this study is to produce predictive models of the stochastic process that describes temperature. More specifically, we are interested in modelling aggregate monthly HDDs. The incorporation of this model into a complete pricing model for weather derivatives is left for future work. We also restrict attention to the case where the contract period for the derivative has not yet commenced. Hence, we ignore short-run weather forecasts, and concentrate on seasonal forecasting.

We investigate two families of program representations for time-series modelling. The first is the standard GP technique, genetic symbolic regression (GSR), applied to the forecasting problem in the same way that is applied to symbolic regression problems. The task is to approximate a periodic function, where *temperature* (HDDs) is the dependent variable, and *time* is the sole regressor variable. The second representation allows the induction of iterated single-step predictors that can resemble autoregressive (GP-AR) and autoregressive moving average (GP-ARMA) time-series models as described in Section 2.3.

3   Experiment Design

3.1   Model Data

Three US weather stations were selected: (a) Atlanta (ATL); (b) Dallas, Fort Worth (DEN); (c) La Guardia, New York (DFW). All the weather stations were based at major domestic airports and the information collected included date, maximum daily temperature, minimum daily temperature, and the associated HDDs and CDDs for the day. This data was preprocessed to create new time-series of *monthly* aggregate HDDs and CDDs for each weather station respectively.

There is generally no agreement on the appropriate length of the time-series which should be used in attempts to predict future temperatures. Prior studies have used lengths of twenty to fifty years, and as a compromise this study uses data for each location for the period 01/01/1979 - 31/12/2002. The monthly HDDs data for each location is divided into a *training set* (15 years) that measures the performance during the learning phase, and a *test set* (9 years) that quantifies model generalisation.

3.2   Forecasting Model Representations and Run Parameters

This study investigates the use of two families of seasonal forecast model representations, where the forecasting horizon is set to 6 months. The first is based on standard GP-based symbolic regression (GSR), where *time* serves as the regressor variable (corresponding to a month of a year), and *monthly HDDs* is the regressand variable. Assuming that time t is the start of the forecast, we can obtain a 6-month forecast by executing the program with inputs $\{t+1, \ldots, t+6\}$.

Table 1. Learning algorithm parameters

EA	panmictic, generational, elitist GP with an expression-tree representation
No. of generations	51
Population size	1,000
Tournament size	4
Tree creation	ramped half-and-half (depths of 2 to 6)
Max. tree depth	17
Subtree crossover	30%
Subtree mutation	40%
Point mutation	30%
Fitness function	Root Mean Squared Error (RMSE)

The second representation for evolving seasonal forecasting models is based on the iterated single-step prediction that can emulate autoregressive models, as described in Section 2.3. This method requires that delayed vectors from the monthly HDDs time-series are given as input to the model, with each consecutive model output being added at the end of the delayed input vector, while all other inputs are shifted back one place.

Table 2. Forecasting model representation languages

Forecasting model	Function set	Terminal set
GSR	add, sub, mul, div, exp, log, sqrt, sin, cos	index t corresponding to a month 10 rand. constants in -1.0, ..., 1.0 10 rand. constants in -10.0, ..., 10.0
GP-AR(12)	add, sub, mul, div, exp, log, sqrt	10 rand. constants in -1.0, ..., 1.0 10 rand. constants in -10.0, ..., 10.0 $HDD_{t-1}, ..., HDD_{t-12}$
GP-AR(24)	add, sub, mul, div, exp, log, sqrt	10 rand. constants in -1.0, ..., 1.0 10 rand. constants in -10.0, ..., 10.0 $HDD_{t-1}, ..., HDD_{t-24}$
GP-AR(36)	add, sub, mul, div, exp, log, sqrt	10 rand. constants in -1.0, ..., 1.0 10 rand. constants in -10.0, ..., 10.0 $HDD_{t-1}, ..., HDD_{t-36}$
GP-ARMA(36)	add, sub, mul, div, exp, log, sqrt	10 rand. constants in -1.0, ..., 1.0 10 rand. constants in -10.0, ..., 10.0 $HDD_{t-1}, ..., HDD_{t-36}$ M($HDD_{t-1}, ..., HDD_{t-6}$), SD($HDD_{t-1}, ..., HDD_{t-6}$) M($HDD_{t-1}, ..., HDD_{t-12}$), SD($HDD_{t-1}, ..., HDD_{t-12}$) M($HDD_{t-1}, ..., HDD_{t-18}$), SD($HDD_{t-1}, ..., HDD_{t-18}$) M($HDD_{t-1}, ..., HDD_{t-24}$), SD($HDD_{t-1}, ..., HDD_{t-24}$) M($HDD_{t-1}, ..., HDD_{t-30}$), SD($HDD_{t-1}, ..., HDD_{t-30}$) M($HDD_{t-1}, ..., HDD_{t-36}$), SD($HDD_{t-1}, ..., HDD_{t-36}$)

Table 2 shows the primitive single-type language elements that are being used for forecasting model representation in different experiment configurations. For GSR, the function set contains standard arithmetic operators (protected division) along with e^x, $log(x)$, $\sqrt{x}$, and finally the trigonometric functions of sine and cosine. The terminal set is composed of the index t representing a month, and random constants within specified ranges. GP-AR(12), GP-AR(24), GP-AR(36), all correspond to standard autoregressive models that are implemented as iterated single-step prediction models. The argument in the parentheses specifies the number of past time-series values that are available as input to the model. The function set in this case is similar to that of GSR excluding the trigonometric functions, whereas the terminal set is augmented with historical monthly HDD values. For the final model configuration, GP-ARMA(36), the function set is identical to the one used in the other autoregressive models configurations, however the terminal set contains moving averages, denoted by $M(HDD_{t-1}, \ldots, HDD_{t-\lambda})$, where λ is the time-lag and HDD_{t-1} and $HDD_{t-\lambda}$ represent the bounds of the moving average period. For every moving average, the associated standard deviation for that period is also given as model input, and is denoted by $SD(HDD_{t-1}, \ldots, HDD_{t-\lambda})$. Finally, Table 1 presents the parameters of our learning algorithm.

4   Results

Table 3. Comparison of training and testing RMSE obtained by different forecasting configurations, each experiment was ran for 50 times. Standard error for mean in parentheses. Bold face indicates best performance on test data.

Dataset	Forecasting configuration	Mean Training RMSE	Best Training RMSE	Mean Testing RMSE	Best Testing RMSE
ATL	GSR	140.52 (9.55)	68.82	149.53 (8.53)	**72.73**
	GP-AR(12)	92.44 (0.54)	81.78	111.87 (0.41)	103.60
	GP-AR(24)	91.33 (0.68)	83.33	96.15 (0.51)	91.26
	GP-AR(36)	88.96 (0.81)	77.30	90.38 (0.81)	79.44
	GP-ARMA	85.20 (0.86)	75.84	85.71 (0.82)	74.31
DEN	GSR	165.76 (11.46)	103.09	180.46 (11.74)	**95.23**
	GP-AR(12)	133.18 (0.43)	121.38	126.78 (0.25)	117.19
	GP-AR(24)	130.41 (0.73)	111.48	124.36 (0.66)	110.31
	GP-AR(36)	131.13 (1.08)	114.86	111.41 (0.57)	103.73
	GP-ARMA	126.46 (1.29)	106.18	108.90 (0.64)	101.57
DFW	GSR	118.96 (8.02)	66.49	118.69 (7.20)	66.12
	GP-AR(12)	88.75 (0.66)	80.64	90.37 (0.26)	86.57
	GP-AR(24)	96.14 (0.95)	83.55	85.36 (0.42)	78.24
	GP-AR(36)	89.52 (0.69)	81.12	62.11 (0.43)	55.84
	GP-ARMA	87.09 (0.82)	75.41	60.92 (0.52)	**55.10**

We performed 50 independent evolutionary runs for each forecasting model configuration presented in Table 2. A summary of average and best training and test results obtained by different models is presented in Table 3. The distributions of training and test errors obtained at the end of the evolutionary runs are depicted in Figure 1 for the DFW time-series. Graphs for the other time-series exhibited a similar trend and were omitted due to lack of space. Results suggest that the family of autoregressive moving average models perform better on average than those obtained with standard symbolic regression. A statistical significance difference (unpaired t-test, two-tailed, $p < 0.0001$, degrees of freedom $df = 98$) was found between the average test RMSE for GSR and GP-ARMA in all three datasets. Despite the fact that the ARMA representation space offers a more stable unit for evolution than the essentially free-of-domain-knowledge GSR space, best testing RMSE results indicated that GSR models are better performers in ATL and DEN datasets, as opposed to the DFW dataset, where the best-of-50-runs GP-ARMA model appeared superior.

Given that in time-series modelling it is often practical to assume a *determin-istic* and a *stochastic* part in a series' dynamics, this result can well corroborate on the ability of standard symbolic regression models to effectively capture the deterministic aspect of a time-series, and successfully forecast future values in the case of time-series with a weak stochastic or volatile part. Another inter-esting observation is that there is a difference in the generalisation performance between GP-AR models of different order, suggesting that the higher the order of the AR process the better its performance on seasonal forecasting. Statistical significant differences (unpaired t-test, $p < 0.0001$, $df = 98$) were found in mean test RMSE between GP-AR models of order 12 and those of order 36, in all three datasets.

During the learning process, we monitored the generalisation performance of the best-of-generation individual, and we adopted a model selection strategy whereby the best-generalising individual is designated as the outcome of the run. In the context of *early stopping* for counteracting the marked tendency of model overtraining, Figures 1(g), (h), (i) illustrate the distributions of the generation number where model selection was performed, for the three datasets. It can be seen that GSR models are less prone to overtraining, then follows

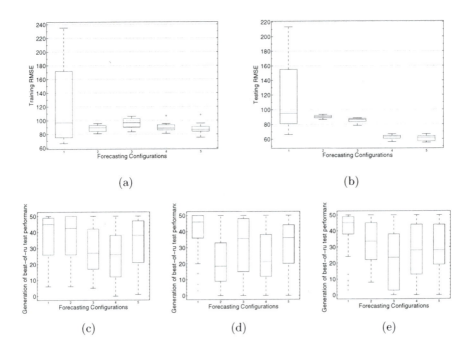

(a) (b)

(c) (d) (e)

Fig. 1. Distribution of best-of-run training and test RMSE accrued from 50 indepen-dent runs. Figures (a), (b) for DFW. Figures (c), (d), (e) show the distribution of generation number where each best-of-run individual on test data was discovered for the cases of ATL, DEN, and DFW respectively.

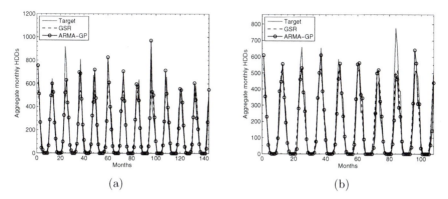

Fig. 2. Target vs. Prediction for best-performing models of GSR and GP-ARMA for the DFW dataset. (a) training data; (b) test data.

GP-ARMA, and finally it can be noted that GP-AR models of high order are the most sensitive to overfitting the training data. Interestingly is the fact that this observation is consistent in all three datasets.

Finally, Figures 2(a), (b) show the target and predicted values from best-performing GSR and GP-ARMA models for the DWF dataset, for training and testing data respectively. Both models achieved a good fit for most of the out-of-sample range. Equation 5 illustrates the parabolic GP-ARMA model that generated the predictions in Figure 2.

$$f(t) = \sqrt{HDD_{t-12} * \left(HDD_{t-36} + \sqrt{HDD_{t-12} * \left(\frac{HDD_{t-26}}{-0.92 + (HDD_{t-7} * log(HDD_{t-21}))} \right)} \right)}$$

(5)

5   Conclusion

This paper adopted a time-series modelling approach to the production of a seasonal weather metric forecast, as part of a general method for pricing weather derivatives. Two GP-based methods for time series modelling were used; the first one is based on standard symbolic regression; the second one is based on autoregressive time-series modelling that is realised via an iterated single-step prediction process and a specially crafted terminal set of past time-series information.

Results are very encouraging, suggesting that GP is able to successfully evolve accurate seasonal forecasting models. More specifically, for two of the three time-series considered in this study, standard symbolic regression was able to capture the deterministic aspect of the modelled data and attained the best test performance, however its overall performance was marked as unstable, producing some very poor-generalising models. On the other hand, the performance of search-based autoregressive moving average models was deemed on average the most stable in out-of-sample data. On a more general note, experiments also revealed a marked tendency of the GP-AR models to overfit the most, with GSR being the most resilient program representation in this problem domain. Whether this is due to a slower learning curve in the case of GSR, or to a very sensitive to overfitting representation in the case on GP-AR models is left to be seen in future work. B

Acknowledgement. This publication has emanated from research conducted with the financial support of Science Foundation Ireland under Grant Number 08/SRC/FM1389.

References

1. Agapitos, A., Dyson, M., Kovalchuk, J., Lucas, S.M.: On the genetic programming of time-series predictors for supply chain management. In: GECCO 2008: Proceedings of the 10th Annual Conference on Genetic and Evolutionary Computation (2008)
2. Bollerslev, T.: Generalised autoregressive conditional heteroskedasticity. Journal of Econometrics 31, 307–327 (1986)
3. Garcia, A., Sturzenegger, F.: Hedging Corporate Revenues with Weather Derivatives: A case study. Master's thesis, Universite de Lausanne (2001)
4. Koza, J.: Genetic Programming: on the programming of computers by means of natural selection. MIT Press, Cambridge (1992)
5. Cao, M., Wei, J.: Equilibrium valuation of weather derivatives, working paper. School of Business, York University, Toronto (2002)
6. Campbell, S., Diebold, F.: Weather forecasting for weather derivatives. Journal of the American Statistical Association 100(469), 6–16 (2005)
7. Jewson, S., Brix, A., Ziehman, C.: Weather Derivative Valuation: The Meteorological, Statistical, Financial and Mathematical Foundations. Cambridge University Press (2005)
8. Makridakis, S., Wheelwright, S., Hyndman, R.: Forcasting: Methods and Applications. Willey, New York (1998)
9. Sprundel, V.: Using weather derivatives for the financial risk management of plant diseases: A study on Phytophthora infestans and Fusarium head blight. Ph.D. thesis, Wageningen University (2011)
10. Weigel, A., Baggenstos, D., Liniger, M.: Probabilistic verification of monthly temperature forecasts. Monthly Weather Review 136, 5162–5182 (2008)

Steepest Ascent Hill Climbing
for Portfolio Selection

Jonathan Arriaga and Manuel Valenzuela-Rendón

Tecnológico de Monterrey, Campus Monterrey,
Av. Eugenio Garza Sada 2501 C.P. 64849 Monterrey, N.L. México
jonathan.arriaga@exatec.itesm.mx, valenzuela@itesm.mx

Abstract. The construction of a portfolio in the financial field is a problem faced by individuals and institutions worldwide. In this paper we present an approach to solve the portfolio selection problem with the Steepest Ascent Hill Climbing algorithm. There are many works reported in the literature that attempt to solve this problem using evolutionary methods. We analyze the quality of the solutions found by a simpler algorithm and show that its performance is similar to a Genetic Algorithm, a more complex method. Real world restrictions such as portfolio value and rounded lots are considered to give a realistic approach to the problem.

Keywords: Portfolio optimization, Hill Climbing.

1 Introduction

The Markowitz mean-variance model was the first approach to solve the portfolio optimization problem, it is a powerful framework for asset allocation under uncertainty. This model defines the optimal portfolios as those that maximize return while minimizing risk; the solution is an efficient frontier, a smooth non-decreasing curve representing the set of Pareto-optimal non-dominated portfolios. Quadratic Programming (QP) optimization has been widely extended and used as a benchmark to compare different approaches to solve the problem. QP solves this model in an efficient and optimal way if all the restrictions given are linear. However, no analytic method exist that solves the problem when considering non-linear constraints present in the real world such as asset cardinality, transaction costs, minimum and maximum weights, among others.

Metaheuristic methods can overcome the limitations of the QP approach since they are not affected by the underlying dynamics of the problem. Some examples of metaheuristic search methods which have been used to solve the portfolio selection problem are genetic algorithms (GA) [1], memetic algorithms [2], simulated annealing, particle swarm optimization [6], differential evolution [11] and tabu search [4,15]. New metaheuristic methods have been developed to solve this problem such as the hybrid local search algorithm introduced in [10] and the Velocity Limited Particle Swarm Optimization presented in [16]. Other approaches have been reported in the literature as the Bayesian portfolio selection [8].

C. Di Chio et al. (Eds.): EvoApplications 2012, LNCS 7248, pp. 145–154, 2012.

An overview of the use of heuristic optimization for the portfolio selection problem can be found in [9].

Different ways to measure risk are widely used such as in [14], where a memetic algorithm is used for the portfolio selection problem while considering Value at Risk (VaR) as the risk measure. The Markowitz mean-variance model assumes a normal distribution, in [12] theoretical return distribution assumptions are compared against empirical distributions using memetic algorithms as the portfolio selection method.

As the problem involves finding an Pareto-optimal frontier, multiobjective (MO) evolutionary algorithms have also been reported in the literature. MO evolutionary algorithm are employed in [5,7] to optimize the Pareto frontier; in [7] robustness of the portfolio is part of the evaluation. A review of MO evolutionary algorithms in economics and finance applications can be found in [3].

All these methods can obtain good approximations to the global optimum solution of the portfolio selection problem. We studied the performance of Steepest Ascent Hill Climbing (SAHC), a local search algorithm and found that can also provide good solutions in spite of its simplicity. Simpler algorithms are preferred over more complex ones due to the ease of implementation, are less prone to raise errors while running, are faster and consume less computer memory if properly coded. It is very efficient regarding memory usage because only one solution is saved through the run; different from population-based methods that load many solutions.

The rest of the paper is organized as follows: Section 2 describes the model for portfolio optimization. Section 3 presents the SAHC algorithm. The setup used for experiments, and how the proposed approach will be compared with a Genetic Algorithm (GA), is presented in section 4. Finally, in section 5 the results of this paper are discussed in conclusion.

2   The Optimization Model

To solve the portfolio optimization problem by means of metaheuristics, we first present the optimization model for the unconstrained and constrained problem.

2.1   Unconstrained Problem

The Markowitz portfolio selection model to find the Unconstrained Efficient Frontier (UEF) is given by:

$$\min \sigma_P^2 \tag{1}$$

$$\max \mu_P \tag{2}$$

subject to

$$\sigma_P^2 = \sum_{i=1}^{N}\sum_{j=1}^{N} w_i w_j \sigma_{ij} \tag{3}$$

$$\mu_P = \sum_{i=1}^{N} \mu_i w_i \tag{4}$$

$$\sum_{i=1}^{N} w_i = 1 \tag{5}$$

where $w_i \in \mathbb{R}_0^+ \forall i$, μ_P is the portfolio expected return, N is the total number of assets considered, μ_i is the mean historical return of asset i, σ_{ij} is the covariance between historical returns of assets i and j, and w_i is the weight of asset i in the portfolio.

With this framework, the identification of the optimal portfolio structure can be defined as the quadratic optimization problem of finding the weights w_i that assure the least portfolio risk σ_P^2 for an expected portfolio return μ_P. This model assumes a market where assets can be traded in any fraction, without taxes or transaction costs and no short sales.

2.2 Constrained Problem

As pointed out in [4] the formulation for the Constrained Efficient Frontier (CEF) problem can formulated by the introduction of zero-one decision variables as:

$$z_i = \begin{cases} 1 \text{ if any of asset } i \text{ is held} \\ 0 \text{ otherwise} \end{cases} \tag{6}$$

where $(i = 1, \ldots, N)$. The objective and constraints are given by:

$$\min \sum_{i=1}^{N} \sum_{j=1}^{N} w_i w_j \sigma_{ij} \tag{7}$$

subject to the same constraints as for the UEF plus

$$\sum_{i=1}^{N} z_i = K \tag{8}$$

$$\varepsilon_i z_i \leq w_i \leq \delta_i z_i, \quad i = 1, \ldots, N \tag{9}$$

$$z_i \in \{0, 1\}, \quad i = 1, \ldots, N \tag{10}$$

where ε_i and δ_i are the minimum and maximum weights allowed in the portfolio for asset i.

Rounded Lots. When purchasing shares, commissions fees are usually charged per transaction and often a percentage of the total price is also considered. To reduce the average commission cost, shares are bought in groups commonly known as *lots*. A group of 100 shares is known as a *rounded lot*, an *odd lot* is a group of less than 100 shares even as small as 1 share. We introduce this restriction to the portfolio optimization problem as follows:

$$l_i = \frac{w_i A}{p_i L} \qquad (11)$$

where $l_i \in \mathbb{Z}^+$, A is the portfolio value in cash with no stock holdings, L is the lot size, p_i and l_i are the price and number of equivalent rounded lots of asset i respectively. The number of shares is given by $q_i = w_i/p_i$.

Thus, we incorporate this restriction in the portfolio selection problem by re-interpreting the weights of the assets in the portfolio. In order to do so, we first calculate the quantity of integer lots that can be bought with the proportion of cash destined to asset i in the portfolio, see Eq. 11. Then the actual weight of the asset in the portfolio is determined with the following equation

$$w_i' = l_i \frac{p_i L}{A} \; . \qquad (12)$$

where w_i' is the weight of asset i after being rounded down to the nearest lot. The remaining cash not invested in any asset is assumed to be risk-free. With the reinterpreted weights of the assets in the portfolio, we proceed to obtain the respective portfolio expected return μ_P and risk σ_P^2.

2.3   Efficient Frontier

With the QP equations 1–5 we can find the efficient frontier by solving them repeatedly, varying the objective μ_P; this efficient frontier represents the set of Pareto-optimal portfolios. Another approach is to trace out the efficient frontier by optimizing the weighted sum of the portfolio's risk and expected return:

$$\max \left[(1 - \lambda)\mu_P - \lambda\sqrt{\sigma_P^2} \right] \qquad (13)$$

where λ ($0 \leq \lambda \leq 1$) is the weighting parameter. Values for the λ parameter can be increased gradually or be selected at random. The resulting portfolio's return and risk have an inverse non-linear relationship with this parameter, so a better selection criteria is needed. We take an adaptive approach to populate the efficient frontier by selecting λ so that the resulting portfolios have risks and expected returns distributed along their respective axes.

First, portfolios are optimized for 10 equally spaced intervals of λ in the range 0-1, always measuring the resulting portfolio risk and expected return. Any of the expected return or risk axes of the efficient frontier plot is selected randomly and the most distant points from each other in that axis are searched; the next value is selected as the mid-point of the corresponding λ values for these points.

This procedure relies on the portfolio optimization algorithm to decide the next λ value. If optimization is not perfect, the distribution of portfolios may be enclosed in a small region of the efficient frontier; with a small probability a random λ value is chosen to overcome this pitfall.

3   Steepest Ascent Hill Climbing for Portfolio Selection

Steepest Ascent Hill Climbing (SAHC) is a subset of metaheuristic optimization methods. SAHC starts off with a random solution and generates new solutions by moving randomly within the current solution's neighborhood. Unlike Hill Climbing, SAHC examines various moves in the neighborhood of a single node. A single solution is handled all the time, so memory usage is minimal and its implementation becomes easier.

Before the algorithm starts the solution structure must be defined, which is composed by s nodes and each of a certain type. For each type, the set of possible values must be defined beforehand. The neighborhood of a node is defined by the values in the set of its type; the algorithm makes moves in a discrete-space neighborhood. The solution nodes can be optimized in random or sequential order; for the experiments we carried out, this setting meant no significant change in the performance. Once the solution structure is defined, the algorithm starts evaluating a number of random solutions, and the best found so far is saved. Then, for each of the nodes, the solution is evaluated with a range of size V of different values from the set defined for the node type. If some of the values enhances the solution fitness, the new solution remains with this change. This process is repeated for G iterations. Below are listed the steps of SAHC:

```
Define set of values for every type of node
Evaluate initial random solutions
for iterations in range(G):
  for node in solution:
    for neighbor in neighbors(node):      # Select V neighbors
      neighbor_fitness = eval(neighbor)
      if neighbor_fitness > best_fitness:
        best_fitness = neighbor_fitness
        best_solution = neighbor
```

For the CEF, K assets compose the portfolio. In this case the solution to be optimized by the heuristic algorithm is composed of K assets and K weights as [4], thus the length of the solution is $s = 2K$ nodes. In order to obtain portfolios with K assets, we specify in the algorithm not to consider the same asset when selecting values for V on the nodes that represent the assets. A minimum weight ε_i is set for each asset i in the solution, with the constraint $\sum_i^K \varepsilon_i < 1$. The remaining fraction of the portfolio is left for the algorithm to be optimized, so weights are normalized to $1 - \sum_i^K \varepsilon_i$. This solution structure is similar to the approaches reported in [4,15].

The way SAHC algorithm moves in the solution space is very similar to the hybrid local search algorithm introduced in [10], which consists of a population of solutions evolving with the simulated annealing method. Instead of evaluating a single change on multiple solutions, SAHC evaluates multiple changes on the same solution; leading to a wider neighborhood exploration and more chances to find or get close to the global optimum.

4   Experimental Setup

We compare the described SAHC with a Genetic Algorithm (GA) to demonstrate the performance of the proposed method. GAs have been widely used in the financial field to solve the portfolio selection problem, see [4,13,15]. Since the portfolio optimization problem is NP-hard following the Markowitz model [10] and no global optimum can be found in a reasonable time, we take the QP approach as the benchmark for both algorithms. Using the solutions derived by QP we take the equivalent number of assets from the weights using Eq. 11 and round down to the nearest integer to obtain rounded lots; actual weights in the portfolio are generated using Eq. 12 we will refer to these solutions as *QP rounded*.

4.1   Data and Parameters

Three stocks sets were considered: the components of the NASDAQ-100 index (100 assets), the FTSE-100 index (100 assets) and the DAX index (30 assets). Data was obtained from the `yahoo.finance.historicaldata` datatable using the Yahoo! Query Language (YQL). Returns were taken in 4-week intervals, 20 samples were considered. The date for the portfolio creation was 1-Nov-2011 with the first date being as for 4-Apr-2010. We assume a lot size of $L = 100$ for all the problems. In order to populate the efficient frontier, we performed optimization with SAHC and GA 1,000 times, selecting a different value for λ each run. All the reported experiments were coded in Python and run in a workstation with Intel Core i5-760 processor @ 2.80GHz and 3.8GB memory.

To perform a fair comparison of SAHC with the GA and QP approaches, we relax the cardinality constraint by taking K as the upper limit of the quantity of assets in the portfolio. As we have no preference or restriction for a certain asset, the ε_i and δ_i values are set to 0 and 1 respectively for all assets, so its allowed weight in the portfolio is $w_i \in [0, 1]$. Assets are allowed to be repeated in the solution, weights for the same asset are added; the total of weights are normalized such that $\sum_i^K w_i = 1$. With this modification, solutions in SAHC and GA will represent portfolios composed from 1 to at most K assets. For SAHC and GA $K \leq 10$ assets in a portfolio is a realistic number, QP rounded portfolios may have more assets.

Settings for the SAHC algorithm were $G = 15$, $V = 10$ and 100 initial evaluations. Values for the selection of assets in the solution were the components of each index and for weights were set 20 evenly spaced values in the range 0-1. For the GA we used 150 generations and 200 individuals, crossover probability was set to $p_c = 1.0$, mutation probability for an individual was $p_m = 10\%$. The solution was coded with 7 bits for the assets and 5 bits for weights in the range 0-1; the solution string for each of the three problems was 120 bits long.

Portfolio Value. With the rounded lots restriction, the A parameter and the stock prices change the difficulty of the problem. To select a proper value for A, we performed experiments using the QP rounded approach. Figure 1 shows the

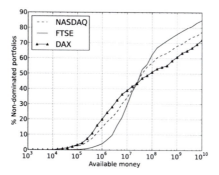

Fig. 1. QP rounded non-dominated portfolios by self with increasing values of A

effect of A on the number of non-dominated portfolios obtained when increased logarithmically from 1,000 to 10,000,000,000 currency units.

When there is a sufficiently large amount of money, rounded lots of different assets are easy to combine; there are more dominated portfolios because the rounding operation means almost no change in the actual weight of assets in the portfolio. Limited cash makes some rounded lots not feasible fewer possible combinations of assets. The worst case is the FTSE problem, its percentage of non-dominated portfolios starts to increase when $A > 100,000$ currency units, below this point rounded lots are not feasible; we selected $A = 1,000,000$ currency units for each market. With the chosen value the FTSE problem is solvable, though hard.

4.2   Results

In figure 2 are shown the non-dominated portfolios of each of the methods considered for each problem. Although no global optimum can be computed for the CEF considering rounded lots, the efficient frontier plots show that the portfolios found by SAHC and GA are very close to the UEF, see figures 2a, 2b and 2c.

Table 1 summarizes the results obtained by using the SAHC and GA approaches to solve the portfolio optimization problem and the performance of QP rounded. FTSE was the hardest problem for both algorithms, even for QP rounded. The reason is because the mean price of its components is 1010.43 GBP, while for the NASDAQ 62.27 USD and for the DAX 52.78 EUR. On the other hand, the DAX was the easiest problem to optimize because is composed by 30 stocks and due to the low prices of its components; only in this problem the GA was able to beat the SAHC, with minor differences. SAHC showed almost the same performance for the NASDAQ and DAX, while performing much better in the FTSE problem, with nearly the double of portfolios generated.

Both methods, the GA and SAHC found portfolios with at most 7 different assets for the NASDAQ and the FTSE problems, while 8 for the DAX; the $K \leq 10$ cardinality limit is a proper value. From these results we can determine that the SAHC did not have major problems while solving the proposed problems.

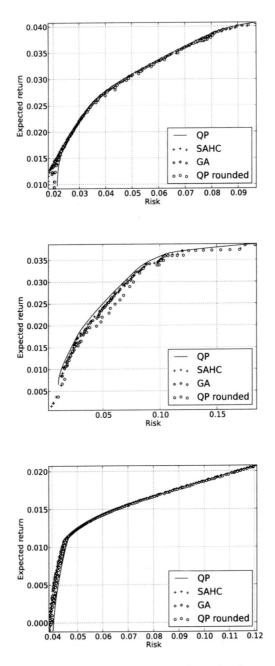

Fig. 2. Pareto front obtained by QP, SAHC and GA for the three problems considered: NASDAQ (top), FTSE (middle) and DAX (bottom)

Table 1. Performance comparison of the SAHC and the GA for the NASDAQ, FTSE and DAX problems. Non-dominated portfolios by self is the number of portfolios in the pareto frontier, more is better when compared with other methods.

		NASDAQ	FTSE	DAX
QP	Non-dominated by self	148	41	201
	Non-dominated by SAHC	44	13	82
	Non-dominated by GA	77	21	90
	Time	8	7	2
SAHC	Non-dominated by self	233	108	241
	Non-dominated by QP	226	107	240
	Non-dominated by GA	189	104	225
	Evaluations	3,010,989	3,009,437	2,957,817
	Time	343	341	236
GA	Non-dominated by self	233	64	246
	Non-dominated by QP	199	63	246
	Non-dominated by SAHC	127	37	159
	Evaluations	30,000,000	30,000,000	30,000,000
	Time	7,521	7,680	6,975

5   Conclusions

Throughout this paper we have presented a SAHC approach to solve the portfolio optimization problem with real-world constraints, such as portfolio value, cardinality constraints, rounded lots and maximum and minimum weights. The SAHC implementation is very simple, is fast and its performance is similar to a more complex approach, the GA. Because of its execution time lower than the GA and fewer evaluations needed to find good and even better solutions, the SAHC approach can be used as a method to quickly find practical solutions. The comparison was only with respect to a GA, as future work other algorithms can be considered such as simulated annealing and particle swarm optimization.

Acknowledgments. This work was supported in part by the Tecnológico de Monterrey Evolutionary Computation Research Chair CAT-044.

References

1. Aranha, C.C., Iba, H.: A tree-based ga representation for the portfolio optimization problem. In: Proceedings of the 10th Annual Conference on Genetic and Evolutionary Computation, GECCO 2008, pp. 873–880. ACM, New York (2008)
2. Aranha, C.d.C., Iba, H.: Using memetic algorithms to improve portfolio performance in static and dynamic trading scenarios. In: Proceedings of the 11th Annual Conference on Genetic and Evolutionary Computation, GECCO 2009, pp. 1427–1434. ACM, New York (2009)
3. Castillo Tapia, M., Coello, C.: Applications of multi-objective evolutionary algorithms in economics and finance: A survey. In: IEEE Congress on Evolutionary Computation, CEC 2007, pp. 532–539 (September 2007)

4. Chang, T.J., Meade, N., Beasley, J., Sharaiha, Y.: Heuristics for cardinality constrained portfolio optimisation. Computers and Operations Research 27, 1271–1302 (2000)
5. Chiam, S., Al Mamun, A., Low, Y.: A realistic approach to evolutionary multiobjective portfolio optimization. In: IEEE Congress on Evolutionary Computation, CEC 2007, pp. 204–211 (September 2007)
6. Gao, J., Chu, Z.: An improved particle swarm optimization for the constrained portfolio selection problem. In: International Conference on Computational Intelligence and Natural Computing, CINC 2009, vol. 1, pp. 518–522 (June 2009)
7. Hassan, G., Clack, C.D.: Multiobjective robustness for portfolio optimization in volatile environments. In: Proceedings of the 10th Annual Conference on Genetic and Evolutionary Computation, GECCO 2008, pp. 1507–1514. ACM, New York (2008)
8. Lu, J., Liechty, M.: An empirical comparison between nonlinear programming optimization and simulated annealing (sa) algorithm under a higher moments bayesian portfolio selection framework. In: Simulation Conference, Winter 2007, pp. 1021–1027 (December 2007)
9. Maringer, D.: Heuristic optimization for portfolio management [application notes]. IEEE Computational Intelligence Magazine 3(4), 31–34 (2008)
10. Maringer, D., Kellerer, H.: Optimization of cardinality constrained portfolios with a hybrid local search algorithm. OR Spectrum 25, 481–495 (2003), doi:10.1007/s00291-003-0139-1
11. Maringer, D., Parpas, P.: Global optimization of higher order moments in portfolio selection. Journal of Global Optimization 43, 219–230 (2009), doi:10.1007/s10898-007-9224-3
12. Maringer, D.G.: Distribution assumptions and risk constraints in portfolio optimization. Computational Management Science 2, 139–153 (2005), doi:10.1007/s10287-004-0031-8
13. Orito, Y., Yamamoto, H.: Index fund optimization using a genetic algorithm and a heuristic local search algorithm on scatter diagrams. In: IEEE Congress on Evolutionary Computation, CEC 2007, pp. 2562–2568 (September 2007)
14. Winker, P., Maringer, D.: The hidden risks of optimizing bond portfolios under var. Journal of Risk 9(4), 1–19 (2007)
15. Woodside-Oriakhi, M., Lucas, C., Beasley, J.: Heuristic algorithms for the cardinality constrained efficient frontier. European Journal of Operational Research 213(3), 538–550 (2011)
16. Xu, F., Chen, W.: Stochastic portfolio selection based on velocity limited particle swarm optimization. In: The Sixth World Congress on Intelligent Control and Automation, WCICA 2006, vol. 1, pp. 3599–3603 (2006)

A Neuro-evolutionary Approach to Intraday Financial Modeling

Antonia Azzini[1], Mauro Dragoni[2], and Andrea G.B. Tettamanzi[1]

[1] Università degli Studi di Milano
Dipartimento di Tecnologie dell'Informazione
via Bramante, 65 - 26013 Crema (CR) Italy
{antonia.azzini,andrea.tettamanzi}@unimi.it
[2] Fondazione Bruno Kessler (FBK-IRST)
Via Sommarive 18, Povo (Trento), Italy
dragoni@fbk.eu

Abstract. We investigate the correlations among the intraday prices of the major stocks of the Milan Stock Exchange by means of a neuro-evolutionary modeling method. In particular, the method used to approach such problem is to apply a very powerful natural computing analysis tool, namely evolutionary neural networks, based on the joint evolution of the topology and the connection weights together with a novel similarity-based crossover, to the analysis of a financial intraday time series expressing the stock quote variations of the FTSE MIB components. We show that it is possible to obtain extremely accurate models of the variations of the price of one stock based on the price variations of the other components of the stock list, which may be used for statistical arbitrage.

Keywords: Evolutionary Algorithms, Neural Networks, Intraday Trading, Statistical Arbitrage.

1 Introduction

Many successful applications of natural computing techniques to the modeling and prediction of financial time series have been reported in the literature [5,6,7].

To our knowledge, the vast majority of such approaches consider only the time series of daily closing prices only, or at most the opening and closing prices of each security, without considering what happens during the real-time market, i.e., the intraday prices.

However, recently, different authors [4,9,10] explored the direction of analyzing the behavior of the intraday stock prices in order to discover correlations between the behaviors of different stock prices.

In particular, Bi's work [4], which takes into account the Chinese stock market, highlights that it is possible to infer serial correlations between components,

C. Di Chio et al. (Eds.): EvoApplications 2012, LNCS 7248, pp. 155–164, 2012.
© Springer-Verlag Berlin Heidelberg 2012

while the work presented in [10] discusses the application of data mining techniques to the detection of stock price manipulations by incorporating the analysis of intraday trade prices, in addition to closing prices, for the investigation of intraday variations.

Another interesting direction early explored is related to the study of the impact of financial news on the intraday prices. In this sense, Mittermayer [14] describes a system implemented to predict stock price trends based on the publication of press releases. Its system mainly consists of three components: a crawler of relevant information from press releases, a press release categorizer, and a reasoner that is able to derive appropriate trading strategies. The results that he obtained is that, with an effective categorization of press releases and with an adequate trading strategy, additional information, useful to forecast stock price trends, can be easily provided. Two alternatives have been discussed also in [8] and [11]. In the former, the author focuses on the use of rough set theory for transforming the unstructured information into structured data; while in the latter, the author applies four different machine learning techniques in order to detect patterns in the textual data that could explain increased risk exposure of some stocks with respect to financial press releases.

The idea proposed in this work follows the first direction. We claim that the analysis of intraday variations could lead to the discovery of patterns that permit to improve the accuracy of the predictions of stock variations. In particular, in this work, we perform a preliminary investigation focusing on the components of the FTSE MIB index of the Milan Stock Exchange. We use an evolutionary algorithm to jointly evolve the structure and connection weights of feed-forward multilayer neural networks that predict the intraday price variations of one component of the index based on the price variations of the other components recorded at the same time. The resulting neural networks may be regarded as non-linear factor models of one security with respect to the others.

The rest of the paper is organized as follows: Section 2 presents the problem and the dataset, while a brief description of the evolutionary approach considered in this work is reported in Section 3. The experiments carried out by applying the evolutionary algorithm are presented in Section 4, together with a discussion of the results obtained. Finally, Section 5 provides some concluding remarks.

2   Problem Description

The object of our investigation may be formulated as a modeling problem, whereby we are seeking for a non-linear factor model that expresses the returns of one security as a function of the returns of a set of other securities at the same instant [12].

In particular, we use feed-forward multilayer neural networks to represent such non-linear factor models and we exploit a well-tested neuro-evolutionary algorithm [3] to optimize both the structure (number of hidden layers, number of neurons in each hidden layer) and the connection weights of the neural networks.

Factor models are used primarily for statistical arbitrage. A statistical arbitrageur builds a hedged portfolio consisting of one or more long positions and one or more short positions in various correlated instruments. When the price of one of the instruments diverges from the value predicted by the model, the arbitrageur puts on the arbitrage, by going long that instrument and short the others, if the price is lower than predicted, or short that instrument and long the others, if the price is higher. If the model is correct, the price will tend to revert to the value predicted by the model, and the arbitrageur will profit.

The simplest case of statistical arbitrage is pair trading, whereby the financial instruments considered for constructing the hedged portfolio are just two. The non-linear factor models that we obtain with our approach may be used for the more general case where the opportunity set is constituted by all the components of the index. However, the same models might also be used for pair trading of each individual stock against the index future, which is generally regarded as an accurate proxy of the index itself, at least at the time scales involved in intraday trading.

3   The Neuro Genetic Algorithm

The neuro-evolutionary algorithm that we use is based on the evolution of a population of individuals, each representing a feed-forward multilayer neural network, also known as a multilayer perceptron (MLP), through the joint optimization of their structures and weights, here briefly summarized; a more complete and detailed description can be found in the literature [3]. In this work, the algorithm uses the Scaled Conjugate Gradient method (SCG) [13] instead of the more traditional error back-propagation (BP) algorithm to decode a *genotype* into a *phenotype* NN, in order to speed up the convergence of such a conventional training algorithm. Accordingly, it is the genotype which undergoes the genetic operators and which reproduces itself, whereas the phenotype is used *only* for calculating the genotype's fitness. The rationale for this choice is that the alternative of applying SCG to the genotype as a kind of 'intelligent' mutation operator, would boost exploitation while impairing exploration, thus making the algorithm too prone to getting trapped in local optima.

The population is initialized with different hidden layer sizes and different numbers of neurons for each individual according to two exponential distributions, in order to maintain diversity among all of them in the new population. Such dimensions are not bounded in advance, even though the fitness function may penalize large networks. The number of neurons in each hidden layer is constrained to be greater than or equal to the number of network outputs, in order to avoid hourglass structures, whose performance tends to be poor. Indeed, a layer with fewer neurons than the outputs destroys information which later cannot be recovered.

3.1   Evolutionary Process

The initial population is randomly created and the genetic operators are then applied to each network until the termination conditions are not satisfied.

At each generation, the first half of the population corresponds to the best $\lfloor n/2 \rfloor$ individuals selected by truncation from a population of size n, while the second half of the population is replaced by the offsprings generated through the crossover operator. Crossover is then applied to two individuals selected from the best half of the population (parents), with a probability parameter p_{cross}, defined by the user together with all the other genetic parameters, and maintained unchanged during the entire evolutionary process.

It is worth noting that the p_{cross} parameter refers to a 'desired' crossover probability, set at the beginning of the evolutionary process. However, the 'actual' probability during a run will usually be lower, because the application of the crossover operator is subject to the condition of similarity between the parents.

Elitism allows the survival of the best individual unchanged into the next generation and the solutions to get better over time. Then, the algorithm mutates the weights and the topology of the offsprings, trains the resulting network, calculates fitness over the validation set, and finally saves the best individual and statistics about the entire evolutionary process.

The application of the genetic operators to each network is described by the following pseudo-code:

1. Select from the population (of size n) $\lfloor n/2 \rfloor$ individuals by truncation and create a new population of size n with copies of the selected individuals.
2. For all individuals in the population:
   (a) Randomly choose two individuals as possible parents.
   (b) Check their local similarity and apply crossover according to the crossover probability.
   (c) Mutate the weights and the topology of the offspring according to the mutation probabilities.
   (d) Train the resulting network using the training set.
   (e) Calculate the fitness f over the validation set.
   (f) Save the individual with lowest f as the best-so-far individual if the f of the previously saved best-so-far individual is higher (worse).
3. Save statistics.

The SimBa crossover starts by looking for a 'local similarity' between two individuals selected from the population. If such a condition is satisfied the layers involved in the crossover operator are defined. The contribution of each neuron of the layer selected for the crossover is computed, and the neurons of each layer are reordered according to their contribution. Then, each neuron of the layer in the first selected individual is associated with the most 'similar' neuron of the layer in the other individual, and the neurons of the layer of the second individual are re-ranked by considering the associations with the neurons of the first one. Finally a cut-point is randomly selected and the neurons above the cut-point are swapped by generating the offspring of the selected individuals.

Weights mutation perturbs the weights of the neurons before performing any structural mutation and applying SCG to train the network. All the weights and the corresponding biases are updated by using variance matrices and evolutionary strategies applied to the synapses of each NN, in order to allow a control parameter, like mutation variance, to self-adapt rather than changing their values by some deterministic algorithms. Finally, the topology mutation is implemented with four types of mutation by considering neurons and layer addition and elimination. The addition and the elimination of a layer and the insertion of a neuron are applied with three independent probabilities, indicated as p_{layer}^{+}, p_{layer}^{-} and p_{neuron}^{+}, while the elimination of a neuron is carried out only if the contribution of that neuron is negligible with respect to the overall network output.

For each generation of the population, all the information of the best individual is saved.

As previously considered [1,2], the evolutionary process adopts the convention that a lower fitness means a better NN, mapping the objective function into an error minimization problem. Therefore, the fitness used for evaluating each individual in the population is proportional to the mean square error (mse) and to the computational cost of the considered network. This latter term induces a selective pressure favoring individuals with reduced-dimension topologies.

The fitness function is calculated, after the training and the evaluation processes, by the Equation 1

$$f = \lambda kc + (1 - \lambda) * mse, \tag{1}$$

where λ corresponds to the desired tradeoff between network cost and accuracy, and it has been set experimentally to 0.2 to place more emphasis on accuracy, since the NN cost increase is also checked by the entire evolutionary algorithm. k is a scaling constant set experimentally to 10^{-6}, and c models the computational cost of a neural network, proportional to the number of hidden neurons and synapses of the neural network.

Following the commonly accepted practice of machine learning, the problem data is partitioned into training, validation and test sets, used, respectively for network training, to stop learning avoiding overfitting, and to test the generalization capabilities of a network. The fitness is calculated over the validation set.

4   Experiments and Results

In this section, we present the intraday dataset that has been created for performing our experiments and the discussion about the results that we have obtained.

4.1   Dataset and Experiment Set-Up

We created the dataset starting from raw data representing the intraday stock quotes of the FTSE MIB components in the period beginning on August the

1st, 2011 and ending on November the 20th, 2011, observed every 5 minutes. For each observation, we have computed the price variation (technically, a log-return) between the quote of the observation at instant t and the quote of the previous observation, at instant $t - 1$,

$$r(t) = \log \frac{x(t)}{x(t-1)}. \tag{2}$$

We discarded the observations for which t was the first observation of each day (i.e., the opening price). Therefore, knowing that the trading hours of the Milan Stock Exchange are from 9:00 am to 5.30 pm, we did not consider for our dataset any of the observations available whose time label is 9:00 am. The rationale behind this choice is that while markets are closed, perturbing events (economics, politics, etc.) that may strongly alter the opening quote of a stock may occur and, in that case, the variation between the first observation of the current day and the last observation of the day before could be orders of magnitude larger than the other stock quotes variations. As a matter of fact, that particular price variation is anyway of little or no interest to a day trader, who will typically refrain from maintaining open positions overnight, because of the risks doing that would involve.

Starting from the data thus obtained, we constructed a distinct dataset for each component stock of the list. In the case of FTSE MIB, we have 40 stocks, resulting in the construction of 40 datasets. In each dataset D_i, where i is the stock for which we want to build a model, for each instant t, we used all the log-returns $r_k(t)$, with $k \neq i$ as inputs, and we considered the log-return $r_i(t)$ as the desired output of the model. In other words, we are seeking for a model M_i such that, for all times t,

$$r_i(t) \sim M(r_1(t), \ldots, r_{i-1}(t), r_{i+1}(t), \ldots r_{40}(t)). \tag{3}$$

In compliance with the usual practice in machine learning, each dataset has been split into three subsets:

- training set: used for training the neural networks, it takes into account the stock quotes in the period between August 23rd, 2010 and October 31st, 2010;,
- validation set: used for selecting the best individuals returned by the neuro-evolutionary algorithm, it takes into account the stock quotes in the period between August 1st, 2010 and August 13th, 2010;
- test set: used for the final evaluation of the performance of the models obtained, it takes into account the stock quotes in the period between November 1st, 2010 and November 20th, 2010.

The experiments have been carried out by setting the parameters of the algorithm to the values obtained from a first round of experiments aimed at identifying the best parameter setting. These parameter values are reported in Table 1.

Table 1. Parameters of the Algorithm

Symbol	Meaning	Value
n	Population size	60
p^+_{layer}	Probability of inserting a hidden layer	0.05
p^-_{layer}	Probability of deleting a hidden layer	0.05
p^+_{neuron}	Probability of inserting a neuron in a hidden layer	0.05
p_{cross}	'Desired' probability of applying crossover	0.7
δ	Crossover similarity cut-off value	0.9
N_{in}	Number of network inputs	39
N_{out}	Number of network outputs	1
α	Cost of a neuron	2
β	Cost of a synapse	4
λ	Desired tradeoff between network cost and accuracy	0.2

For each modeling task, i.e., for each component of the FTSE MIB index, 40 runs were performed, with 40 generations and 60 individuals for each run. The number of maximum epochs used to train the neural network represented by each individual was 250.

4.2   Results and Discussion

In this subsection we report the results obtained by applying the neuro/evolutionary algorithm to the 40 FTSE MIB intraday datasets. Table 2 shows the average, over the 40 models returned by independent runs of the algorithm, of the mean square error for each of the 40 component of the FTSE MIB index, along with the standard deviation of the mean square errors. The average MSE are very small: in practice, the price variations are predicted by the model with an error between 2% and 3% of the log-return.

In addition, Table 2 reports, for each stock, the 5-minute volatility (computed as the square of the 5-minute log-returns over the entire period of observation) is reported, to allow for a comparison of the relative volatilites of the securities considered—of course, one would expect the most volatile stocks to be harder to model, and *vice versa*.

From the results, we can infer some patterns in the performance of the models discovered by the algorithm. The first aspect is that on stocks of the financial sector (for instance, BP.MI, BPE.MI, FSA.MI, ISP.MI, and UCG.MI) the approach obtains higher error values than on stocks related to the energy sector (ENEL.MI, ENI.MI, and SRG.MI). Related to the former set of stocks, the results suggest that the task of finding correlations between financial stocks and the other components of the index is harder; a possible reason is almost certainty related with the speculative nature that, especially in the last months,

Table 2. Summary of results obtained on the FTSE MIB intraday dataset

Ticker	Stock Name	5-min Volatility $(\times 10^{-5})$	Avg. MSE $(\times 10^{-4})$	St. Dev $(\times 10^{-5})$
A2A.MI	A2A	2.37	4.227	4.3
AGL.MI	AUTOGRILL	0.944	4.641	4.9
ATL.MI	ATLANTIA	0.788	3.783	5.1
AZM.MI	AZIMUT HOLDING	2.47	6.442	2.2
BMPS.MI	BANCA MPS	6.38	5.748	1.9
BP.MI	BANCO POPOLARE	3.14	6.367	6.4
BPE.MI	BCA POP. EMILIA R.	2.35	6.307	4.5
BZU.MI	BUZZI UNICEM	1.70	5.504	6.3
CPR.MI	CAMPARI	0.725	3.409	0.7
DIA.MI	DIASORIN	1.04	4.728	9.4
EGPW.MI	ENEL GREEN POWER	1.33	4.464	8.8
ENEL.MI	ENEL	0.913	3.841	6.6
ENI.MI	ENI	0.643	3.544	0.9
EXO.MI	EXOR	1.99	5.968	5.3
F.MI	FIAT	2.60	6.147	8.1
FI.MI	FIAT INDUSTRIAL	2.41	6.208	1.0
FNC.MI	FINMECCANICA	1.51	5.922	7.6
FSA.MI	FONDIARIA-SAI	3.64	7.322	4.7
G.MI	GENERALI	1.25	5.068	4.2
IPG.MI	IMPREGILO	2.03	5.764	3.1
ISP.MI	INTESA SANPAOLO	4.01	6.766	6.1
LTO.MI	LOTTOMATICA	1.54	5.665	2.6
LUX.MI	LUXOTTICA GROUP	0.723	3.647	0.3
MB.MI	MEDIOBANCA	1.52	4.964	7.9
MED.MI	MEDIOLANUM	2.51	6.544	4.4
MS.MI	MEDIASET	1.36	4.445	9.5
PC.MI	PIRELLI & C.	1.78	5.767	1.1
PLT.MI	PARMALAT	1.63	4.703	6.8
PMI.MI	BCA POP. MILANO	20.7	6.861	4.9
PRY.MI	PRYSMIAN	1.38	5.924	2.3
SPM.MI	SAIPEM	1.16	4.662	9.4
SRG.MI	SNAM RETE GAS	0.842	3.768	0.6
STM.MI	STMICROELECTRONICS	1.70	5.345	6.5
STS.MI	ANSALDO STS	1.95	5.464	7.1
TEN.MI	TENARIS	1.36	4.789	8.0
TIT.MI	TELECOM ITALIA	2.60	4.927	7.4
TOD.MI	TOD'S	1.45	5.968	9.3
TRN.MI	TERNA	1.01	3.907	0.5
UBI.MI	UBI BANCA	2.30	5.982	4.7
UCG.MI	UNICREDIT	4.93	7.083	5.5

characterizes these stocks categories. This is confirmed by the observation that the stocks of the financial sector are, on average, more volatile than stocks of the other sectors (4.6×10^{-5} vs. 1.2×10^{-5}). On the contrary, cyclic stocks like the ones of the energy sector, appear to be easier to model.

The second aspect is related to the set of manufacturing stocks (like F.MI, FI.MI, FNC.MI, and LUX.MI). By observing their MSE values, we can notice that the model performances vary widely. Our interpretation of these different behaviors is related to the very different contingent situations the companies are facing. Take, for instance, FIAT, who was facing financial tensions and a crisis of the automotive market during the period under investigation.

Anyway, Figure 1 shows that there is a clear correlation between the volatility of a security and the average MSE obtained by the neuro-evolutionary algorithm, as expected.

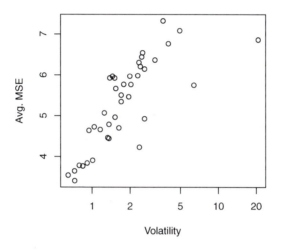

Fig. 1. Plot of the average MSE of evolved models against the 5-minute volatility of the security modeled

5   Conclusions

In this work we presented an approach whose aim is to discover correlations between stocks in the intraday market. We performed experiments by considering the components of the FTSE MIB index of the Milan Stock Exchange and our results suggest that it is possible to obtain a high prediction accuracy for a time scale of a few minutes. In general, we found that the accuracy of the evolved models is higher the less volatile is the stock under investigation, although even the accuracies of the models for the most volatile stocks are very good. Future work in this direction will aim at improving the approach by introducing a features selection module in order to refine input data by excluding those features that introduce noise into the learning process. Moreover, further experiments will be performed on several European and non-European stock markets in order to verify the existence of correlations both on the same or on different indexes.

References

1. Azzini, A., Dragoni, M., Tettamanzi, A.: A Novel Similarity-Based Crossover for Artificial Neural Network Evolution. In: Schaefer, R., Cotta, C., Kołodziej, J., Rudolph, G. (eds.) PPSN XI. LNCS, vol. 6238, pp. 344–353. Springer, Heidelberg (2010)
2. Azzini, A., Tettamanzi, A.: Evolving neural networks for static single-position automated trading. Journal of Artificial Evolution and Applications 2008 (Article ID 184286), 1–17 (2008)
3. Azzini, A., Tettamanzi, A.: A new genetic approach for neural network design. In: Engineering Evolutionary Intelligent Systems. SCI, vol. 82. Springer, Heidelberg (2008)
4. Bi, T., Zhang, B., Xu, R.: Dynamics of intraday serial correlation in china's stock market. Communications in Statistics - Simulation and Computation 40(10), 1637–1650 (2011)
5. Brabazon, A., O'Neill, M. (eds.): Natural Computing in Computational Finance. SCI, vol. 1. Springer, Heidelberg (2008)
6. Brabazon, A., O'Neill, M. (eds.): Natural Computing in Computational Finance. SCI, vol. 2. Springer, Heidelberg (2009)
7. Brabazon, A., O'Neill, M., Maringer, D. (eds.): Natural Computing in Computational Finance. SCI, vol. 3. Springer, Heidelberg (2010)
8. Cheng, S.H.: Forecasting the change of intraday stock price by using text mining news of stock. In: ICMLC, pp. 2605–2609. IEEE (2010)
9. Chicco, D., Resta, M.: An intraday trading model based on artificial immune systems. In: Apolloni, B., Bassis, S., Esposito, A., Morabito, F. (eds.) WIRN. Frontiers in Artificial Intelligence and Applications, vol. 226, pp. 62–68. IOS Press (2010)
10. Diaz, D., Theodoulidis, B., Sampaio, P.: Analysis of stock market manipulations using knowledge discovery techniques applied to intraday trade prices. Expert Syst. Appl. 38(10), 12757–12771 (2011)
11. Groth, S., Muntermann, J.: An intraday market risk management approach based on textual analysis. Decision Support Systems 50(4), 680–691 (2011)
12. Harris, L.: Trading and exchanges: market microstructure for practitioners. Oxford University Press (2003)
13. Hestenes, M., Stiefel, E.: Methods of conjugate gradients for solving linear systems. Journal of Research of the National Bureau of Standards 49(6) (1952)
14. Mittermayer, M.A.: Forecasting intraday stock price trends with text mining techniques. In: HICSS (2004)

A Comparative Study of Multi-objective Evolutionary Algorithms to Optimize the Selection of Investment Portfolios with Cardinality Constraints

Feijoo E. Colomine Duran[1], Carlos Cotta[2], and Antonio J. Fernández-Leiva[2]

[1] Universidad Nacional Experimental del Táchira (UNET)
Laboratorio de Computación de Alto Rendimiento (LCAR), San Cristóbal, Venezuela
fcolomin@unet.edu.ve
[2] Dept. Lenguajes y Ciencias de la Computación, ETSI Informática,
University of Málaga, Campus de Teatinos, 29071 - Málaga, Spain
{ccottap,afdez}@lcc.uma.es

Abstract. We consider the problem of selecting investment components according to two partially opposed measures: the portfolio performance and its risk. We approach this within Markowitz's model, considering the case of mutual funds market in Europe until July 2010. Comparisons were made on three multi-objective evolutionary algorithms, namely NSGA-II, SPEA2 and IBEA. Two well-known performance measures are considered for this purpose: hypervolume and R_2 indicator. The comparative analysis also includes an assessment of the financial efficiency of the investment portfolio selected according to Sharpe's index, which is a measure of performance/risk. The experimental results hint at the superiority of the indicator-based evolutionary algorithm.

1 Introduction

There are several theoretical studies related to risk-return interaction. The potential loss of performance or investment is not static, but it always depends on market developments. In the literature, we can find several proposals that model this scenario. For example, Markowitz's model [14] has become a key theoretical framework for the selection of investment portfolios. However, its application in practice has not been as extensive, mainly due to the mathematical complexity of the method.

Markowitz's model with multiple objectives is expressed as:

$$\min \sigma^2(R_p) = \sum_{i=1}^{n}\sum_{j=1}^{n} w_i w_j \sigma_{ij}, \quad \max E(R_p) = \sum_{i=1}^{n} w_i E(R_i) \qquad (1)$$

subject to:

$$\sum_{i=1}^{n} w_i = 1, \text{ and } w_i \geqslant 0 \, (i = 1, \cdots, n) \qquad (2)$$

C. Di Chio et al. (Eds.): EvoApplications 2012, LNCS 7248, pp. 165–173, 2012.
© Springer-Verlag Berlin Heidelberg 2012

where w_i is the investor's share of the budget for the financial asset i (to be found), $\sigma^2(R_p)$ is the variance of the portfolio p, and σ_{ij} is the covariance between the returns of the values i and j. $E(R_p)$ is the expected return or portfolio p, which are the set of proportions that minimize the risk of the portfolio and and its corresponding value. The set of pairs $[E(R_p), \sigma^2(R_p)]$ or combinations of all risk-return efficient portfolios is called the efficient frontier. Once known, the investor chooses according to his preferences the optimal portfolio. In this model some other restrictions can be added, such as cardinality (a maximum of K is non-zero weights) or limits on the percentage of an asset allocation. In this paper we focus on the constraints of the first type:

$$\sum_{w_i > 0} 1 = K \tag{3}$$

The research considers a comparative study between different multi-objective evolutionary algorithms. More specifically the comparative study considers three algorithms: SPEA2 [24], NSGA-II [5], and IBEA [23]. We also address the selection of a point in the Pareto front using Sharpe's index [19] whose expression is:

$$S_p = \frac{E(R_p) - R_0}{\sigma_p} \tag{4}$$

This index is a risk-return ratio. The numerator is the excess return defined by the difference between the yield on the portfolio $(E(R_p))$ and the risk free rate (R_0) in the same period of assessment. The portfolio risk is measured by the standard deviation of this (σ_p). That is, indicates the yield premium offered by a portfolio per unit of total risk of the same. It follows that the higher the risk-reward ratio, the greater the success of fund management.

The objective of this goal is to assess comparatively the performance of several multi-objective EAs on this problem scenario. For this purpose, we will firstly overview some related work in next section.

2   Related Work

There is a plethora of works in the literature dealing with the use of MOEAs in the area of investment portfolio optimization. Without being exhaustive, we can firstly cite the work by Diosan [6], who makes a comparison between PESA [4], NSGA II and SPEA2, and through empirical results suggests the adequateness of PESA. See also [1] for a comparison of these techniques in this context. Chang et al. [3] compare the use of different meta-heuristics (both evolutionary and local-search ones) for finding the efficient frontier by adding cardinality constraints. Perez et al. [15] make in turn a comparison between MOGA [9], NPGA [11] and NGGA [16] concluding the superiority of NGGA. On the other hand, Doerner et al. [7] make a comparison between Pareto Optimization

Using Ant Colony (PACO), simulated annealing and NSGA [21] in terms of both quality and computational cost, suggesting the superiority of PACO. Ehrgott et al. [8] propose an interesting variation of Markowitz's model adding additional objective functions to consider the individual preferences of the investor, and find genetic algorithms to perform better than local-search techniques. Skolpadungket et al. [20] perform a comparison between VEGA [17], MOGA, NSGA II and SPEA2 in the context of financial portfolio optimization, and find that SPEA2 provide the best performance.

3   Material and Methods

In the following we will address the data and algorithms used in the experimentation, as well as the performance measures considered for evaluating performance.

3.1   Data Analyzed

We consider data corresponding to mutual funds in Europe. More precisely, these data comprise the stock value of the funds, sampled on a monthly basis for five years. This period of time is long enough to span a full market cycle, with rises and falls, but not large enough to comprise profound changes in the reality of each of the fund shares, thus making past information little representative for foreseeing future performance. Based on this analysis we take a sample at the discretion of choosing funds that are no older than five years and are still available on the market. Table 1 shows the various funds used. The risk-free return R_0 is referenced is 0.04.

Table 1. Mutual funds considered [22]

Mutual Fund Europe
1 DFA United Kingdom Small Compan
2 Eastern European Equity A (VEEE)
3 Eastern European Equity C (VEEC)
4 Henderson European Focus A (HFE)
5 Henderson European Focus B (HFE)
6   ING Russia A (LETRX)
7   JPMorgan Russia A (JRUAX)
8   JPMorgan Russia Select (JRUSX)
9   Metzler Payden European Emergin
10   Mutual European A (TEMIX)
11   Mutual European B (TEUBX)
12   Mutual European C (TEURX)
13   Mutual European Z (MEURX)
14 Royce European Smaller Companie
15 Third Millennium Russia I (TMRI)

3.2   Algorithmic Methods

The optimization of investment portfolios according to a variety of performance-risk profiles lends itself very well to multi-objective optimization techniques in general, and multi-objective evolutionary algorithms (MOEAs). Aiming to compare the performance of three different MOEAs on the same problem setting and using the same experimental data, we have considered the following techniques:

1. Second-generation Pareto-based MOEAs: NSGA-II (Non-dominated Sorting Genetic Algorithm II) [5] and SPEA2 (Strength Pareto Evolutionary Algorithm 2) [24]. These MOEAs are based on the notion of Pareto-dominance, used for determining the solutions that will breed and/or the solutions that will be replaced. furthermore, as second-generation techniques they exploit elitism (an external archive of non-dominated solutions in the case of SPEA2, and a plus-replacement strategy in the case of NSGA-II). More precisely, NSGA-II sorts the population in non-domination levels (performing binary tournament on the so-obtained ranks for selection purposes), and uses crowding for performing replacement and spreading the Pareto front. As to SPEA2, it features an external archive of solutions which is used to calculate the "strength" of each individual i (the number of solutions dominated by or equal to i, divided by the population size plus one). This is used for selection purposes (aiming to minimize the strength of solutions which are non-dominated by tentative parents). SPEA2 also includes a nearest-neighbor density estimation technique to spread the front, and a sophisticated archive update strategy to preserve boundary conditions.
2. Indicator-based MOEAs: IBEA (Indicator-Based Evolutionary Algorithm) [23] attempts to incorporate practical decision-making and privileged information when searching for Pareto solutions. The question that arises is how to concentrate the search in regions of the Pareto front that are of interest to the person responsible for taking the decision. This is done in IBEA by maximizing (or minimizing) some performance indicator. By doing so, IBEA can be considered a collective approach, where selective pressure is exerted to maximize the performance of the whole population. In this work, we have considered an IBEA based on the ε-indicator [26].

As to the parameters considered, these are described in Table 2.

3.3   Performance Indicators

The evaluation of the performance of a MOEA is in itself a multi-objective problem, that can be approached in multiple ways. We have considered two well-known performance indicators: the hypervolume indicator [25] and the R_2 indicator [10]. The first one provides an indication of the region in the fitness space that is dominated by the front, and which must be maximized for better performance. As to the second indicator, it estimates the extent to which a certain front approximates another one (the Pareto-optimal front or a reference front if the former is unknown). We have considered the unary version of this

Table 2. Parameterization considered in the experimentation

Parameter	Value
representation	binary
number of genes	15
gene size	10 bits
size of chromosome	150
population size	300
generations	100
selection type	tournament/elitist
crossover operator	2-Point Crossover
crossover probability	0.8
mutation operator	bitflip
mutation probability	0.0666

indicator, taking the combined NSGA-II/SPEA2/IBEA Pareto front as a reference set. Being a measure of distance to the reference set, R_2 must be minimized for better performance

4   Experimental Results

Experimentation has been done before the three algorithms described, namely, NSGA-II, SPEA2 and IBEA, using the PISA library [2], the parameters described in Table 2, and the data described in Table 1. Four different values of the cardinality constraint K have been considered, namely $K \in \{2, 4, 7, 15\}$. For each algorithm and value of K, thirty runs have been done.

4.1   Analysis of the Pareto Front

The first part of the experimentation focused on the analysis of the obtained Pareto front. The three algorithms behave very similarly, with slight differences is the high-risk end of the fronts. To analyze in more detail these results we have applied the two performance indicators mentioned in Sect. 3.3, namely hypervolume and R_2. In the first case we have used as reference a point of maximum risk and minimum benefit. As to the latter, distance is measured against the best-known front (the combined front front the three algorithms). Figures 1–2 shows the distribution of values of the indicators for each algorithm for the two extreme cardinality values ($K = 2$ and $K = 15$).

Inspection of these figures indicates that SPEA2 provides better performance for a low value of K. In this constrained scenario, SPEA2 provides a broader dominance of the performance-risk fitness space, and is globally closer to the best-known Pareto front. However, in the other end of the spectrum (high value of K, and hence less constrained portfolios) IBEA stands out as a more effective algorithm. This is also the case for the intermediate values $K = 4$ and $K = 7$. In all cases, these performance differences can be shown to be statistically significant (at $\alpha = 0.05$) via the use of a non-parametric Wilcoxon test [13].

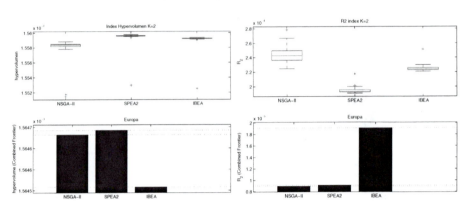

Fig. 1. Hypervolume and R_2 indicator for $K = 2$

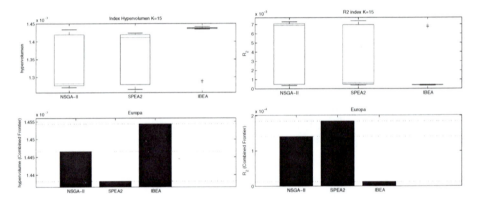

Fig. 2. Hypervolume and R_2 indicator for $K = 15$

4.2   Analysis through Sharpe's Ratio

Sharpe's index is used for decision-making once a front is obtained; it allows picking a single solution out of the whole efficient front, trying to maximize excess profit per risk unit. Geometrically, this can be interpreted as finding the straight line with highest slope that is tangent to the front and passes through the risk-free point $(0, R_0)$. This is obviously influence by the shape of the front; the analysis of the results through this index can thus be used how efficient are the MOEAs in terms of the performance-risk profile arising from this decision-making procedure. Figure 3 shows the distribution of Sharpe's index values for the different cardinality values.

Visual inspection of these results indicate a substantial advantage for IBEA. This advantage is again shown to be statistically significant via the use of a Wilcoxon ranksum test ($\alpha = 0.05$). We can also make an analysis of how different

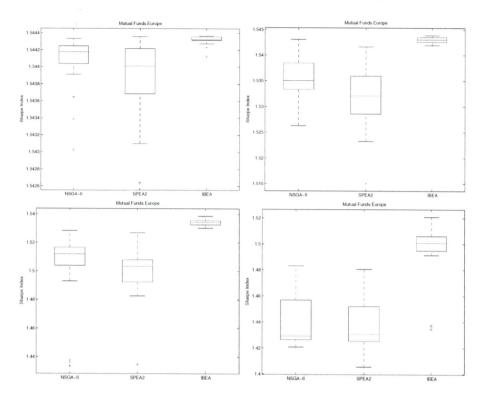

Fig. 3. Sharpe's index distribution for MOEAs under Markowitz's model. Top row: $K = 2$ (left) and $K = 4$ (right). Bottom row: $K = 7$ (left) and $K = 15$ (right).

algorithms behave when varying the risk-free return component (R_0). In this case, it can be observed that as the cardinality rises, IBEA returns slightly better results for increasing risk-free values.

5   Conclusions

The problem of portfolio optimization is a natural scenario for the use of multiobjective evolutionary algorithms, in which their power and flexibility can be readily exploited. In this sense this paper has analyzed three different state-of-the-art MOEAs, namely NSGA II, SPEA2 and IBEA under a common experimental framework centered in mutual funds in Europe. While the three algorithms provide a variety of solution profiles that can be considered optimal in a Pareto sense, a performance analysis conducted under two specific indicators (hypervolume and R_2) indicate that IBEA performs significantly better, in particular when the cardinality constraint K does not take an extremely low value. This can be interpreted in terms of the exploration capabilities of the multi-objective optimizer for the richer (less-constrained) fitness landscapes. As an additional means of comparison among the MOEAs, we have considered the outcome of a decision-making process based on the use of Sharpe's index. Again, IBEA stands out, indicating that it provides a better exploration capability in the area of fitness space around the knee of the front.

In future work we intend to explore other variants of the optimization problem by adding, e.g., maximum and minimum rates of investment. We also intend to study other model variations such as Jensen's alpha [12] under a market model such as CAPM (Capital Asset Pricing Model [18]).

Acknowledgements. This work is partially supported by Spanish MICINN under projects NEMESIS (TIN2008-05941) and ANYSELF (TIN2011-28627-C04-01), and by Junta de Andalucía under project TIC-6083.

References

1. Anagnostopoulos, K.P., Mamanis, G.: A portfolio optimization model with three objectives and discrete variables. Comput. Oper. Res. 37, 1285–1297 (2010)
2. Bleuler, S., Laumanns, M., Thiele, L., Zitzler, E.: PISA – A Platform and Programming Language Independent Interface for Search Algorithms. In: Fonseca, C.M., Fleming, P.J., Zitzler, E., Deb, K., Thiele, L. (eds.) EMO 2003. LNCS, vol. 2632, pp. 494–508. Springer, Heidelberg (2003)
3. Chang, T.-J., Meade, N., Beasley, J.E., Sharaiha, Y.M.: Heuristics for cardinality constrained portfolio optimisation. Computers and Operations Research 27, 1271–1302 (2000)
4. Corne, D., Knowles, J., Oates, M.: The Pareto Envelope Based Selection Algorithm for Multiobjective Optimization. In: Deb, K., Rudolph, G., Lutton, E., Merelo, J.J., Schoenauer, M., Schwefel, H.-P., Yao, X. (eds.) PPSN VI 2000. LNCS, vol. 1917, pp. 839–848. Springer, Heidelberg (2000)
5. Deb, K., Agrawal, S., Pratab, A., Meyarivan, T.: A Fast Elitist Non-Dominated Sorting Genetic Algorithm for Multi-Objective Optimization: NSGA-II. In: Deb, K., Rudolph, G., Lutton, E., Merelo, J.J., Schoenauer, M., Schwefel, H.-P., Yao, X. (eds.) PPSN VI 2000. LNCS, vol. 1917, pp. 849–858. Springer, Heidelberg (2000)
6. Diosan, L.: A multi-objective evolutionary approach to the portfolio optimization problem. In: Proceedings of the International Conference on Computational Intelligence for Modelling, Control and Automation and International Conference on Intelligent Agents, Web Technologies and Internet Commerce, CIMCA-IAWTIC 2006, vol. 2, pp. 183–187. IEEE, Washington, DC, USA (2005)
7. Doerner, K., Gutjahr, W., Hartl, R., Strauss, C., Stummer, C.: Pareto ant colony optimization: A metaheuristic approach to multiobjective portfolio selection. Annals of Operations Research 131, 79–99 (2004)
8. Ehrgott, M., Klamroth, K., Schwehm, C.: An MCDM approach to portfolio optimization. European Journal of Operational Research 155(3), 752–770 (2004)
9. Fonseca, C.M., Fleming, P.J.: Genetic algorithms for multiobjective optimization: Formulation, discussion and generalization. In: Forrest, S. (ed.) Proceedings of the Fifth International Conference Genetic Algorithms, pp. 416–423. University of Illinois at Urbana-Champaign, Morgan Kauffman Publishers, San Mateo, California (1993)
10. Hansen, M.P., Jaszkiewicz, A.: Evaluating the quality of approximations to the nondominated set. Technical Report IMM-REP-1998-7, Institute of Mathematical Modelling Technical University of Denmark (1998)

11. Horn, J., Nafpliotis, N., Goldberg, D.E.: A niched pareto genetic algorithm for multiobjective optimization. In: First IEEE Conference on Evolutionary Computation, pp. 82–87. IEEE (1994)
12. Jensen, M.C.: The performance of mutual funds in the period 1945 - 1964. Journal of Finance 23, 383–417 (1968)
13. Lehmann, E.L., D'Abrera, H.J.M.: Nonparametrics: Statistical Methods Based on Ranks. Prentice-Hall, Englewood Cliffs (1998)
14. Markowitz, H.M.: Portfolio selection. Journal of Finance 7, 77–91 (1952)
15. Pérez, M.E., del Olmo, R., Herrera, F.: The formation of efficient portfolios using multiobjective genetic algorithms. Technical Report DECSAI-00-01-26, Dept. CCIA, Universidad de Granada (2000)
16. Valenzuela Rendón, M., Uresti Charre, E.: A non-generational genetic algorithm for multiobjective optimization. In: Bäck, T. (ed.) Proceedings of the Seventh International Conference on Genetic Algorithms, pp. 658–665. Michigan State University, Morgan Kaufmann, San Mateo, California (1997)
17. David Schaffer, J.: Multiple objective optimization with vector evaluated geneticalgorithms. In: Grefenstette, J.J. (ed.) Proceedings of the First International Conference on Genetic Algorithms and their Applications, pp. 93–100. Lawrence Erlbaum, Hillsdale NJ (1985)
18. Sharpe, W.F.: Capital assets prices: A theory of market equilibrium under conditions of risk. Journal of Finance 19, 425–442 (1964)
19. Sharpe, W.F.: Mutual fund performance. Journal of Business 39, 119–138 (1966)
20. Skolpadungket, P., Keshav, D., Napat, H.: Portfolio optimization using multi objective genetic algorithms. In: IEEE Congress on Evolutionary Computation, pp. 516–523 (2007)
21. Srinivas, N., Deb, K.: Multiobjective optimization using nondominated sorting in genetic algorithms. Evolutionary Computation 2, 221–248 (1994)
22. Yahoo Finance. Mutual funds center (2011)
23. Zitzler, E., Künzli, S.: Indicator-Based Selection in Multiobjective Search. In: Yao, X., Burke, E.K., Lozano, J.A., Smith, J., Merelo-Guervós, J.J., Bullinaria, J.A., Rowe, J.E., Tiňo, P., Kabán, A., Schwefel, H.-P. (eds.) PPSN VIII 2004. LNCS, vol. 3242, pp. 832–842. Springer, Heidelberg (2004)
24. Zitzler, E., Laumanns, M., Thiele, L.: SPEA2: Improving the strength pareto evolutionary algorithm. In: Giannakoglou, K., et al. (eds.) EUROGEN 2001. Evolutionary Methods for Design, Optimization and Control with Applications to Industrial Problems, Athens, Greece, pp. 95–100 (2002)
25. Zitzler, E., Thiele, L.: Multiobjective Optimization Using Evolutionary Algorithms - A Comparative Case Study. In: Eiben, A.E., Bäck, T., Schoenauer, M., Schwefel, H.-P. (eds.) PPSN 1998. LNCS, vol. 1498, pp. 292–301. Springer, Heidelberg (1998)
26. Zitzler, E., Thiele, L., Laumanns, M., Fonseca, C.M., Grunert da Fonseca, V.: Performance Assessment of Multiobjective Optimizers: An Analysis and Review. IEEE Transactions on Evolutionary Computation 7(2), 117–132 (2003)

A GA Combining Technical and Fundamental Analysis for Trading the Stock Market

Iván Contreras[1], José Ignacio Hidalgo[1], and Laura Núñez-Letamendia[2]

[1] Facultad de Informática, Universidad Complutense de Madrid, Spain
ivancontrerasfd@gmail.com, hidalgo@dacya.ucm.es
[2] IE Business School, Madrid, Spain
Laura.Nunez@ie.edu

Abstract. Nowadays, there are two types of financial analysis oriented to design trading systems: fundamental and technical. Fundamental analysis consists in the study of all information (both financial and nonfinancial) available on the market, with the aim of carrying out efficient investments. By contrast, technical analysis works under the assumption that when we analyze the price action in a specific market, we are (indirectly) analyzing all the factors related to the market. In this paper we propose the use of an Evolutionary Algorithm to optimize the parameters of a trading system which combines Fundamental and Technical analysis (indicators). The algorithm takes advantage of a new operator called Filling Operator which avoids problems of premature convergence and reduce the number of evaluations needed. The experimental results are promising, since when the methodology is applied to values of 100 companies in a year, they show a possible return of 830% compared with a 180% of the Buy and Hold strategy.

1 Introduction

Investors and analysts have been using for the last few years several techniques to predict the value of the securities, indexes and market prices. Roughly speaking, there are two distinct (but not exclusive) streams, depending on the type of information they handle: Technical Analysis (TA) and Fundamental Analysis (FA). FA is based on the assumption that the value of a share is the discounted stream of future profits of the company. This analysis attempts to determine what those future benefits will be, and tries to know all possible information of the company (news and information that concern them, potential corporate moves, strategies, competitors, new products, etc..). The more information that can be obtained, the better. Any microeconomic information that is related to the company will have an impact on these future cash flows. Macroeconomic data is also important. Aspects to consider are the evolution of the overall business environment, regulatory and political environments, etc... FA attempts to transfer this information to the accounts of preliminary results, which can be deduced to find the present and actual value of the action. On the other hand, TA is based on the fact that the value of a share in the future is strongly related

C. Di Chio et al. (Eds.): EvoApplications 2012, LNCS 7248, pp. 174–183, 2012.

to the previous trajectory of this share. This analysis find patterns (in finance called up-trends and down-trends) and thresholds where the value of a share will bounce. Ultimately, TA uses mathematical models of different complexity that aim to predict the value of a share by studying historical patterns.

As mentioned above, the use of both types of analysis is not necessarily exclusive, i.e., an analyst can use fundamental and technical information together to make his operations. During the last years the investment systems have envolved into more complex models. The reasons for this change are mainly due to the large amount of information available as well as the use of sophisticated computer systems to make predictions and to set buy and sell orders. In this sense, Evolutionary Algorithms (EAs) provide a very useful tool to adjust all the parameters involved in the investment process. Thus, we can find different proposals on how to use EAs on this field. For example Allen and Karjalaein [1] proposed to use one of the first works on using Genetic Algorithms (GAs) to find technical trading Rules (this is precisely the title of the work). They applied a GA to obtain technical trading rules and compare the results with other models and the Buy and Hold (B&H) Strategy. In [14] Nuñez designed four GAs models incorporating different factors (e.g. risk, transaction costs, etc.) to obtain financial investment strategies. It is showed that all four GA models generate superior daily returns of long positions with lower risk than B&H strategy for 1987-1996 share price data from the Madrid Stock Exchange (Spain). Bodas et al. ([10] [4]) applied a MOEASI (Multi Objetive Evolutionary Algorithm with Super Individual), approach for obtaining the best parameters to apply MACD (Moving Average Convergence Divergence) and RSI (Relative Strength Index) technical indicators and show that this technique could work well on stock index trending comparing with a B&H strategy. In [11] the authors describe a trading system designed with GAs that uses different kind of rules with market and companies information. The system is applied to trade, on a daily base, to companies belonging to the S&P 500 index. They propose to apply the methodology to TA and FA separately and the authors claim that the main problem is the computational time required for training the trading system with daily data of stocks prices. This restriction is partially solved in [7] with the implementation on a parallel computer architecture to speed up the functioning of a GA-based trading system to invest in stocks: a GPU-CPU architecture. There are also other approaches that use Genetic Programming (GP) for obtaining a set of rules and signals for investing [12] [13].

In this paper we propose the use of EAs to optimize jointly the parameters of FA and TA. We have implemented two GA versions: a classical simple one (sGA) and a GA with a new operator, namely Filling Operator (GAwFO). This implementation helps to preserve the diversity of the population and solves the problem of premature convergence found when applying sGA. In addition GAwFO uses a particular structure of the population specifically designed to facilitate the implementation on GPUs and to reduce the number of evaluations. Experimental results indicate that the proposal is positive both in terms of quality of the solutions, and in terms of the convergence of the algorithm. The rest of the paper is organized as follows. Section 2 describes the implementation details of GAwFO, explains the GA encoding and other details of the GA. Section 3 presents the experimental results when applying the proposed methodology to a set of companies during 2004. We also analyse the evolution of the best individuals for GA and GAwFO. Finally, we conclude the paper and outline some future research lines in Section 4.

2   Genetic Algorithm and Trading System

In this paper we propose the use of an EA for optimizing the trading system. The objective is to obtain a set of trading signals indicating *buy* or *sell*. An individual represents a set of parameters and threshold values that are used by a trading system in order to obtain buy, maintain or sell signals. First we describe the details of the GAs, then we explain the encoding and the trading system.

2.1   GA Implementation

We have implemented two versions of a GA: a simple GA and an improved version called GA with Filling Operator (GAwFO). First we implemented a classic version of the genetic algorithm. Although the results are acceptable, in certain situations sGA suffers from problems of premature convergence, one of the first problems that we can found in a GA run. The problem is a fast lost in the diversity of the population when a single solution tends to transfer its genetic code to the entire population in a few generations. This would not be a problem if the solution was good enough. However, as the algorithm has not been left to evolve sufficiently, the solution obtained is usually a local optimum. Different techniques exist to prevent this. The most common is to increase the size of the population, but it has a high computational cost in our case. There are also strategies for regeneration, a mating strategy called incest prevention, to increase the value of the probability of mutation, etc .... In order to preserve the diversity of the population we propose a GA with a Filling operator (GAwFO). Basically, the GAwFO approach consists of a sGA with a modification of the selection and crossover operators, and as a consequence of this change it applies an operator driven to the generation of new random individuals.

Figure 1 shows the basic flow of GAwFO. We start from a population of n individuals obtained in the previous generation. As a first step, we retain the best individual which can not be changed with the implementation of other genetic operators but participate in the process of creating new individuals. We then select $s = (n/2) - 1$ individuals who participate in creating the rest of the population. This selection process is done by tournament among two individuals. With this s individuals we apply an uniform crossover operator, with a crossover probability (p_c), generating m new individuals. Since $p_c < 1$, in most generations we will get $m < (n/2)$. In order to preserve the size of the population (n individuals) we will need to generate $k = n - s - m$ individuals. This is done by the *Filling Operator*, which generates k random individuals in each generation (note that k is not a constant value for all generations, as it depends on the number of crossings made by the crossover operator). On this new population of $n = 1 + s + m + k$ individuals we apply a classical mutation operator with probability p_m only to the offspring of m individuals. The crossover operator follows a Uniform Point Crossover (UPX) strategy [16]. This election was made after testing other possibilities (results are not presented here due to the lack of space). We use a classic mutation operator which replaces the value of a random gene with another random value. For each gene the values of the bounds are represented in the *Range* column of Table 1.

As can be deduced from the previous explanation, this implementation does not use effectively the entire size of the population, since there are s individuals which do not vary for a generation. However, individuals do transmit their information to the next generation because they are precisely those involved in the crossover. Given the nature of the problem this can be useful to get several solutions, on the other hand it reduces the computation time, (it is not necessary to evaluate $n/2$ of the population in each generation) and it is also interesting for future implementations on GPU architectures that allow for faster execution.

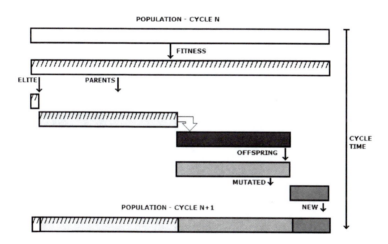

Fig. 1. General pattern of GAwFO operating in a cycle

2.2 Trading System

One solution is given by a set of values, which indicate weights and parameters or thresholds for a selected set of technical and fundamental indicators. The trading system works as follows:

1. The investor selects a set of Technical Indicators for TA
2. The investor selects a set of Fundamental Indicators for FA
3. Establish $Threshold_{buy}$ and $Threshold_{sell}$ ranges and $Weights$ ranges
4. For each company i

(a) Apply GAwFO over a period of years to obtain a solution
(b) Apply TA and FA using the parameters given by (a) to the target year
   i. For each indicator I_j generate the indicator signal $I_j s$ (Buy $= 1$, Sell $= -1$ or Neutral $= 0$) as follows:
      A. if $I_j >$ or $< Threshold_{buy}$ on a day Buy $I_j s = 1$
      B. elseif $I_j >$ or $< Threshold_{sell}$ on a day Sell $I_j s = -1$
      C. else Neutral $I_j s = 0$
   ii. Compute the Raw Trading System Signal by adding the indicators signals weighted by their weights $RTS_s = \sum_{j=1}^{n} I_j s \cdot W_j$
   iii. Compute the Net Trading System Signal choosing values for X and Y as follows:
      A. If $RTS_s =$ or $> X$ then $TS_s = 1$
      B. elseif $RTS_s =$ or $< -Y$ then $TS_s = -1$
      C. else $TS_s = 0$
   iv. Compute the profit given by the Trading System

TA is formulated by four Technical Indicators (TI) selected by the investor and four weights corresponding to each indicator (W_1, W_2, W_3 and W_4) . In this work the selected TIs are Moving Average(MA), market Volume (V), Relative Strength Index Divergences (RSI) and Support and Resistances (SR). Each one of these technical indicators gives us a signal of buy, sell or neutral. The values of W_1, W_2, W_3 and W_4 weigh the importance of each indicator in obtaining buy or sell signals. FA uses also four Fundamental Indicators (FIs) and four weights corresponding to each indicator (W_5, W_6, W_7 and W_8). We use Price Earning Ratio (PER), Price Book Value (PBV), Return On Assets (ROA) and, Sales Growth (SG) as FIs. Those indicators have been selected following the historical usefulness in the literature on investments and it's out of the scope of this paper to go into details about their interpretation and logic (we refer the interested reader to [2],[3],[5],[6],[9],[8] and [15]).

2.3   Genetic Encoding

The chromosome has a total of 23 positions or genes separated into two parts. The first 11 genes represent values that are interpreted by TA and the remainder (12 genes) affect FA indicators. Using the set of technical and fundamental indicators (previously selected by the investor) with the encoded values, we get a signal to buy, sell or remain inactive for each of the companies. With all the indicators (TA and FA) we obtain a value S_i for buying and selling (for example if we obtain 3 buying an 4 selling signals, the value will be $S_i = 3 - 4 = -1$. Once we have the sum of all trading signals, we use a determinate range for buy, sell or do nothing ($Threshold_{buy}$ and $Threshold_{sell}$). The higher the module of the thresholds are, the more conservative the trading system will be (less movements of trading). Then we compute the benefit of buying and selling following the signals indicated by the chromosome. Figure 2 represents an example of a member of the population. The meaning of the genes is explained on Table 1. It shows the indicator affected by each gene and the meaning (Long) of the acronyms

Table 1. Genetic Encoding. $*_1$ means Upper threshold and $*_2$ means Lower Threshold

Gene	Name		Range			Example	
#	Short	Long	Lower	Upper	Jumps	Coded	Decoded
1	MA$_S$	Short Moving Average	1	50	1	14	15
2	MA$_L$	Long Moving Average	51	150	1	49	100
3	RSI$_M$	Relative Strength Index	1	50	1	3	4
4	RSI$_D$	Relative Strength Index	1	50	1	21	22
5	V	Volumen	1	50	1	38	39
6	S	Supports	1	100	1	89	90
7	R	Resistances	1	100	1	75	76
8	W$_1$	Weight MA	0	4	1	1	1
9	W$_2$	Weight RSI	0	4	1	1	1
10	W$_3$	Weight V	0	4	1	2	2
11	W$_4$	Weight SR	0	4	1	0	0
12	PER$_U$	Price Earning Ratio$*_1$	9	18	0.5	18	18
13	PER$_L$	Price Earning Ratio$*_2$	20	29	0.5	10	25
14	PBV$_U$	Price Book Value$*_1$	1.5	3	0.1	15	3
15	PBV$_L$	Price Book Value$*_2$	3.25	4.75	0.1	0	3.25
16	ROA$_U$	Return on Assets$*_1$	6	10	0.1	40	10
17	ROA$_L$	Return on Assets$*_2$	1	5	0.1	11	2.1
18	SG$_U$	Sales Growth$*_1$	6	10	0.1	18	7.8
19	SG$_L$	Sales Growth$*_2$	1	5	0.1	6	1.6
20	W$_5$	Weight PER	0	4	1	4	4
21	W$_6$	Weight PBV	0	4	1	1	1
22	W$_7$	Weight ROA	0	4	1	1	1
23	W$_8$	Weight SG	0	4	1	0	0

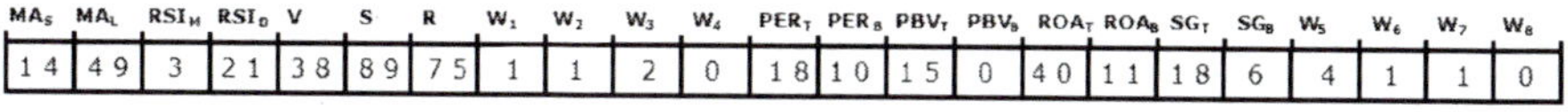

MA$_S$	MA$_L$	RSI$_M$	RSI$_D$	V	S	R	W$_1$	W$_2$	W$_3$	W$_4$	PER$_T$	PER$_B$	PBV$_T$	PBV$_B$	ROA$_T$	ROA$_B$	SG$_T$	SG$_B$	W$_5$	W$_6$	W$_7$	W$_8$
14	49	3	21	38	89	75	1	1	2	0	18	10	15	0	40	11	18	6	4	1	1	0

Fig. 2. Genetic Encoding

(Short). A gene can get a value on [Upper, Lower] with steps indicated on the Jump column. As we are using an integer encoding the encoded value differs from the actual (Decoded) value.

2.4 Fitness Function

The best way to evaluate an individual is calculating a the accumulated return obtained when applying the trading systems to the sample data computed as described by equations 1 and 2. AR_f is the accumulated return at the end of the trading period and DR_i is the daily return. P_i denotes the stock price at day "i", while $RFDR_i$ is the risk-free daily return given by the US Treasury

Bills, and TS stands for Trading System. We also compute the transaction costs ($Cost_{Trans}$) on each buying/selling operation as a commission fee of 0.1% in RFR strategy and a 0.5% in the B&H strategy.

$$AR_f = \prod_f^{i=1}(1 + DR_i) - Cost_{Trans} \qquad (1)$$

$$DR_i = \begin{cases} \frac{P_i - P_{i-1}}{P_{i-1}} & \text{if the TS gives a long signal} \\ \frac{-(P_i - P_{i-1})}{P_{i-1}} & \text{if the TS signal is short selling} \\ RFDR_i & \text{if the TS signal is neutral} \end{cases} \qquad (2)$$

3 Experimental Results

The experimental results has been obtained on a Intel i5 processor running at 2,67 GHz under Windows7 and with 4GB RAM. The data in the graphs represents the arithmetic average of 20 runs. sGA and GAwFO were run with a sample of 100 companies included in the S&P 500 Index Constituent List during the period January 1994 to December 2003. We apply FA and TA with the solutions during the year (2004). The data provides around 35000 observations for quarterly fundamental data and 41000 observations for monthly technical data. The source of the data is Compustat and CRSP databases. Table 2 shows the financial results for our trading system. We achieved a profit of the 830% using the GAwFO application, while the profit with Buy&Hold is 180%. Regarding the convergence of the GAs, as we have mentioned one of the objectives of GAwFO is to preserve the diversity of the population and improve the convergence of the GA. Figure 3 represents results for GA and GAwFO for 100 and 500 generations and for $10, 50, 100, 500, 1000, 5000$ and 10000 indivuals (n). The Y axis represents the average profit reached by our trading system, i.e. the quality of the best solution found by the algorithm. X axis represents the number of individuals in the population. As we can see, GAwFO obtains better results than GA for all the configurations. Here we can also observe a better convergence of GAwFO since the final value of the fitness is different for 100 generations and 500 generations, indicating that the algorithm is preserving the diversity along the generations.

Figure 4 represents the average execution time versus the population size (n), and shows that sGA needs more time than GAwFO (getting worse solutions). This fact is due to the reduction in the number of fitness evaluations. As we mentioned we are reducing to $n/2$ the evaluations on each generation, thus achieving an approximate speed-up of 2 when comparing GA and GAwFO.

Table 2. Experimental Results for 100 companies (ID #). Accumulated return: 180.62 for Buy and Hold and 830.09 if we apply TA and FA using the solution given by GAwFO (average of 20 runs).

ID #	B&H Return	GAwFO Return (Avg.)	ID #	B&H Return	GAwFO Return (Avg.)
1075	2.3991511	12.3095	3062	8.1966836	1.593483333
1078	18.7309602	9.471933333	3105	3.364995	-23.85003333
1161	3.7459	32.4946	3144	8.9026145	1.6268
1239	5.5369252	15.35606667	3170	8.8040119	1.6268
1279	4.0752351	42.82246667	3226	6.7905122	1.584183333
1300	-7.8202282	-6.080666667	3310	6.1251708	6.739133333
1318	-0.8569234	7.958883333	3336	3.2463518	18.5658
1356	2.5187882	-9.12545	3413	-2.1345836	-6.2236
1380	-13.3380317	27.41228333	3439	-0.6006628	6.3544
1408	0.2179087	1.316366667	3497	5.1964884	2.7564
1478	10.355741	4.6542	3505	0.2437237	14.1769
1567	0.6502365	5.502733333	3532	-30.3122337	-5.07805
1581	8.2909365	14.0832	3650	1.5798377	40.26825
1602	2.2006719	1.6268	3734	-7.3576148	9.36405
1632	-4.9004679	-22.8423	3813	2.5320106	30.9393
1651	-51.9001932	-30.6242	3851	18.4093679	5.22795
1661	-7.389481	4.327066667	3897	-0.9477856	6.16345
1678	-1.350981	13.3393	3964	3.6740088	54.7245
1690	-9.5208284	3.6859	4016	-0.6953763	1.6268
1704	6.1066691	1.529983333	4029	2.9502386	5.91695
1722	2.3449161	11.48945	4060	-2.8162409	9.8528
1794	-6.4084227	1.6268	4062	1.4673729	0.034616667
1878	11.5214366	73.56563333	4066	9.744019	4.134816667
1988	-6.712819	21.79263333	4087	12.5668186	0.19165
1995	-2.7094967	10.60435	4194	-12.0802362	11.7651
2044	-9.1449379	33.72938333	4242	-4.1682456	-0.90245
2085	5.912355	1.6268	4321	6.2214513	12.6339
2086	13.6199814	18.4906	4503	0.5317073	25.6679
2111	-5.8409623	8.095016667	4517	5.7802251	8.57505
2136	-6.8204425	1.8547	4560	10.042995	-21.96596667
2146	0.3304849	3.011516667	4598	3.9668814	46.3882
2154	7.5499425	14.4935	4611	4.7135776	3.424066667
2230	-16.0952894	1.6268	4843	-5.4197902	-37.82211667
2255	1.1289969	8.462733333	4988	2.9117564	-3.41645
2269	-1.8980754	-12.57478333	4990	16.607337	2.8008
2290	4.9147146	-3.539433333	5046	-4.6204849	2.36215
2312	8.2363498	1.6268	5071	8.9897884	1.585416667
2403	10.9443535	-6.4183	5074	5.2588111	3.007933333
2435	1.40653	2.075316667	5125	2.2202424	11.9096
2444	-8.0950327	2.27745	5134	10.7559239	11.42828333
2490	5.7200263	-5.585866667	5169	4.2730328	1.6268
2504	-4.5998717	3.1442	5256	6.1121823	34.32501667
2574	17.3010617	27.91153333	5439	-21.0643196	1.6268
2663	6.5206433	9.4525	5518	9.0290509	2.519233333
2710	2.1968028	2.8162	5597	9.1424048	39.5595
2783	4.0894619	1.899433333	5680	2.4276114	6.354983333
2817	12.4002508	19.67333333	5723	13.4362941	12.806
2884	26.8141074	29.3046	5860	-1.2318081	13.2949
2991	-3.7815371	1.6268	5878	5.2791739	12.44675
3054	-5.3061325	29.66486667	6008	9.286749	-3.271083333
Total			**Accum.**	180.62	830.09

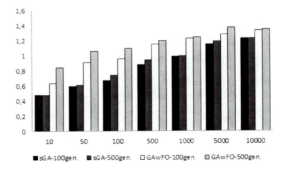

Fig. 3. Profit (Y axis is the return) for GA and GAwFO for 100 and 500 generations and for $10, 50, 100, 500, 1000, 5000$ and 10000 individuals

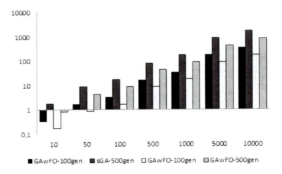

Fig. 4. Execution time (Y axis in sec.) analysis when running 100 and 500 generations of sGA and GAwFO for $n = 10, ...10000$ ind.

4   Conclusions

In this paper we have implemented a GA to obtain a trading system which combines Fundamental and Technical analysis. The algorithm (GAwFO) applies a new operator called Filling Operator which avoids problems of premature convergence and reduces the number of needed evaluations. Experimental results obtained when the methodology is applied to values of random 100 companies in the year 2004, show a possible return of 830% compared with a 180% of the Buy and Hold strategy, improving the outcome in a 65% of the companies. A more exhaustive analysis of the results should be done in order to improve the methodology and to include/exclude companies and other finantial information.

Acknowledgements. This work has been partially supported by Spanish Government grants Avanza Competitividad I+D+i TSI-020100-2010-962, INNPACTO IPT-2011-1198-430000, and TIN 2008-00508.

References

1. Allen, F., Karjalainen, R.: Using genetic algorithms to find technical trading rules. Journal of Financial Economics 51(2), 245–271 (1999)
2. Bali, T.G., Demirtas, O., Tehranian, H.: Aggregate earnings, firm-level earnings, and expected stock returns. JFQA 43(3), 657–684 (2008)
3. Basu, S.: The investment performance of common stocks in relation to their price-earnings ratios: A test of the efficient market hypothesis. Journal of Finance 32, 663–682 (1977)
4. Bodas-Sagi, D.J., Fernández, P., Hidalgo, J.I., Soltero, F.J., Risco-Martín, J.L.: Multiobjective optimization of technical market indicators. In: Proceedings of the GECCO 2009 Conference (Companion), pp. 1999–2004. ACM, New York (2009)
5. Campbell, Yogo: Efficient tests of stock return predictability. Journal of Financial Economics 81, 27–60 (2006)
6. Chan, L.K.C., Hamao, Y., Lakonishok, R.: Journal of finance. Fundamentals and Stock Returns in Japan, 1739–1764 (December 1991)
7. Contreras, I., Jiang, Y., Hidalgo, J., Núñez-Letamendia, L.: Using a gpu-cpu architecture to speed up a ga-based real-time system for trading the stock market. In: Soft Computing - A Fusion of Foundations, Methodologies and Applications, pp. 1–13 (2011)
8. Fama, E., French: The cross-section of expected stock returns. Journal of Finance 47(2), 427–465 (1992)
9. Fama, E.F., French, K.R.: Business conditions and expected returns on stocks and bonds. Journal of Financial Economics 25, 23–49 (1989)
10. Fernández, P., Bodas-Sagi, D.J., Soltero, F.J., Hidalgo, J.I.: Technical market indicators optimization using evolutionary algorithms. In: Proceedings of the GECCO 2008 Conference (Companion), pp. 1851–1858. ACM, New York (2008)
11. Jiang, Y., Núñez, L.: Efficient market hypothesis or adaptive market hypothesis? a test with the combination of technical and fundamental analysis. In: Proceedings of the 15th International Conference on Computing in Economics and Finance, University of Technology, Sydney, Australia, The Society for Computational Economics (July 2009)
12. Lohpetch, D., Corne, D.: Discovering effective technical trading rules with genetic programming: Towards robustly outperforming buy-and-hold. In: NaBIC, pp. 439–444. IEEE (2009)
13. Lohpetch, D., Corne, D.: Multiobjective algorithms for financial trading: Multiobjective out-trades single-objective. In: IEEE Congress on Evolutionary Computation, pp. 192–199. IEEE (2011)
14. Núñez, L.: Trading systems designed by genetic algorithms. Managerial Finance 28, 87–106 (2002)
15. Reinganum, M.: Selecting superior securities charlottesville. the tesearch foundation of the institute of chartered financial analysts. Technical report, The Research foundation of the institute of Chartered Financial Analysts (1988)
16. Sywerda, G.: Uniform crossover in genetic algorithms. In: Proceedings of the Third International Conference on Genetic Algorithms, pp. 2–9. Morgan Kaufmann Publishers Inc., San Francisco (1989)

Evolutionary Data Selection for Enhancing Models of Intraday Forex Time Series

Michael Mayo

University of Waikato, Hamilton, New Zealand
mmayo@waikato.ac.nz
http://www.cs.waikato.ac.nz/~mmayo/

Abstract. The hypothesis in this paper is that a significant amount of intraday market data is either noise or redundant, and that if it is eliminated, then predictive models built using the remaining intraday data will be more accurate. To test this hypothesis, we use an evolutionary method (called Evolutionary Data Selection, EDS) to selectively remove out portions of training data that is to be made available to an intraday market predictor. After performing experiments in which data-selected and non-data-selected versions of the same predictive models are compared, it is shown that EDS is effective and does indeed boost predictor accuracy. It is also shown in the paper that building multiple models using EDS and placing them into an ensemble further increases performance. The datasets for evaluation are large intraday forex time series, specifically series from the EUR/USD, the USD/JPY and the EUR/JPY markets, and predictive models for two primary tasks per market are built: intraday return prediction and intraday volatility prediction.

Keywords: intraday, forex, steady state, genetic algorithm, instance selection, data mining, return prediction, volatility prediction.

1 Introduction

This paper is concerned with using evolutionary algorithms to strengthen predictive models of market behaviour, in order to make them more robust to the noise typical in financial time series (especially intraday series). A new algorithm, Evolutionary Data Selection (EDS), is introduced, that uses a model building algorithm in conjunction with the available training data to find an optimal subset of that data. The search for the optimal subset is evolutionary, and the fitness measure used is the performance of the model built from the subset when tested against all the training data *not* included in the subset (which does not, of course, include the final testing data). At the conclusion of the search, the ultimate best subset is used to build the final predictive model. It is hypothesised that this method is helpful in eliminating noise and repetition from the training data that would otherwise confound the prediction models.

To test the algorithm, we have focussed on intraday (specifically, hourly) time series. The choice of intraday rather than daily series was made primarily because

C. Di Chio et al. (Eds.): EvoApplications 2012, LNCS 7248, pp. 184–193, 2012.

smaller time frames increase the number of samples with which to test predictions; therefore the statistics should have greater confidence, and also smaller predictive gains (from a trader's point of view) become much more significant intraday when there are multiple opportunities to trade compared to a daily time frame where there would be only a single opportunity to trade per day. It is also known that intraday series often exhibit non-zero autocorrelations (see, for example, Sewell [9] who gives an overview and Breedon [1] who has studied start and end of day effects in the forex markets). The aim of this research is to build models that attempt to capture these small but repeatedly measurable effects, as well as other more general intraday trends.

The evaluation of the approach is carried out against three different forex datasets. The markets used are the Eurodollar (EUR/USD), the Japenese Yen market (USD/JPY), and the Euro/Yen cross (EUR/JPY), with intraday price data ranging from 2002 to 2011. There are two prediction tasks per market: *return prediction*, which attempts to model the direction of the market over the near hour, and *volatility prediction*, which models the square of the return. Both tasks are important for understanding the markets.

It should be noted that the concept of subset selection from the training data is not new; in fact, there are several methods that use a mixture of nearest neighbour and evolutionary techniques to achieve this (and Cano [3] gives an overview in the context of general data mining). Most recently, in the financial modelling field, Larkin and Ryan [5] proposed a method that uses one evolutionary computation method (specifically, a genetic algorithm) for subset selection from the training data, and a second different method (genetic programming) for building the actual prediction model given the optimal subset. The work described here differs from that work in two main ways. Firstly, evolutionary methods are used here only for selecting the training data, wheras a machine learning techniques are utilized for making the predictions. This leverages the strength of disparate approaches with a single system and should increase robustness. Secondly, Larkin and Ryan's approach also extends the selection of cases to the test set; this results in a considerable variation in the number of trades for different strategies (from a lower limit of 30 to in the order of thousands of trades). In contrast, the models developed here are used to predict *every* data point in the test set and therefore we expect that whilst the overall average accuracy may be lower, the results should be more significant and comparable due to a much higher number of samples.

A third and final main point of difference between this work and others is that besides the basic EDS method described here, we also propose an ensemble method in which multiple runs of EDS are combined to a produce a single predictor. This is shown to further enhance accuracy.

2   Evolutionary Data Selection Algorithm

The EDS algorithm described here is derived from the well known steady-state genetic algorithm [6] in which a fixed population P of individuals is evolved. Each individual $p \in P$ has a fitness value $F(p)$, and the aim is to find the individual p^* that maximises F.

In the steady-state approach, each time step involves two "tournaments" in which T random individuals from P are selected. In each tournament, one winner is selected, the winner being the individual in the tournament with the highest fitness. Because two tournaments are performed per time step, there will be two winners, p_1 and p_2, which are then crossed-over to produce a new offspring o. This offspring may also be mutated with probability M.

After an offspring is generated, its fitness value is computed. If $F(o)$ exceeds either $F(p_1)$ or $F(p_2)$, then o replaces the parent with lowest fitness. In this way, evolution proceeds stochastically towards a population of higher fitness individuals. If no global improvements are made after N time steps, then the search finishes and the individual p^* with the highest fitness $F(p^*)$ is returned.

Clearly, the selection pressure can be adjusted by increasing or decreasing the T parameter: a smaller value of T will give weaker-fitness individuals a greater probability of producing offspring, whereas a higher T will result in only the strongest individuals being consistently selected. The choice of T needs to be made carefully, as too high a value may result in a greater chance of the GA converging on a local rather than a global or near-global maxima, and too low a value may unnecessarily prolong the search.

The specifics of how the individuals are represented, along with the fitness function, the crossover and the mutation operators, are now described.

Individual Representation. Each individual is represented as a dataset of *instances*, where an instance is defined as a vector of features that includes a single class value to predict. In this work, the datasets are subsets of the training data that is being used to build a predictive model. When the GA is initialised, the initial individuals are simply random subsets of the training data.

An additional parameter I is furthermore defined that governs the size of the initial subsets: each instance from the training data is included in an initial individual with probability I. Thus, the smaller the value of I, the smaller the size of the initial subsets/individuals; conversely the larger the value of I, the greater the size of the starting subsets/individuals.

Canonically speaking, this representation corresponds to a bit string of length equal to the number of instances in the data, where a "1" indicates the presence of an instance in the set, and a "0" the absence. (It was not efficient, however, to use this representation directly in the implementation.)

Section 3.1 discusses the exact datasets and instances that were generated in the experiments.

Fitness Function. Once an individual is defined, a model can then be built using only the instances in the subset as training data. The fitness of the individual is therefore defined as the model's accuracy, once the model is built, on those instances *not* included in the subset of selected instances. In other words, the unselected instances from the training data are used as proxy test data.

Crossover Operator. A very simple crossover operator is implemented. Since both parents are defined as sets of instances, we assign to the offspring each instance (excluding duplicates) from the parents with probability 50%. This ensures that, on average, offspring will be about the same size as the parents.

Mutation Operator. Mutation, if activated with probability M after an offspring is generated, removes a random 1% of the instances from the set. This method of mutation was designed because we were interested minimising the quantity of data used to train the model. Minimal data has several benefits, including simpler models and faster training times.

Other approaches allow subsets to grow as well as reduce in size, but include a term biasing for small subset sizes in the fitness function. This leads to a more complex fitness function with multiple components (i.e. accuracy as well as subset size) which must somehow then be combined, usually via the introduction of yet another parameter. In contrast, the opposite approach taken here starts with subsets as large as possible (defined by the I parameter) and gradually reduces their size via mutations. Thus, maximum subset size is strictly controlled.

Another reason for this approach is that it is important to keep the test sets large: if the test sets become too small (which can happen if an individual becomes too large), then random variations in the test instances will have greater impact on the accuracy and therefore the fitness, leading to weak individuals being scored unjustly highly.

Note that both crossover and mutation will never remove all the instances from an individual; subsets have a hard-coded minimum size of one.

3   Evaluation

In this section, an evaluation of the EDS method is described. We used data from three forex markets in this evaluation: EUR/USD, the USD/JPY and the EUR/JPY markets, and there were two basic prediction tasks: hourly return forecasts, and hourly volatility forecasts. In all the experiments, the EDS parameters are population size $P = 20$, no improvement limit $N = 200$, mutation rate M of 5%, tournament size $T = 2$ and initial subset size $I = 0.5$.

3.1   Data Preparation

The raw data for our evaluations are cleaned intraday forex time series for the three markets obtained from the Pi Trading Corporation [8]. Each series consists of open, high, low and closing prices for each *minute* that the given market was open where the price changed (i.e. there are no records where price did not change for the current minute), from 21 October 2002, 2am (EST), through to 18 March 2011, 5pm (EST). The total dataset sizes, therefore, are: 2,880,844 records for the EUR/USD market; 2,903,531 records for the USD/JPY market; and 3,009,328 records for the EUR/JPY market.

This minute time-scale data was then processed into hourly data series, by combining all records from the same hour into a single record giving the open, high, low and closing prices of a market for the entire hour. Again, hours where there was no change in price (and therefore no minute records) were excluded. This reduced the number of records by a factor of about 60, making the data much more wieldy for further investigation.

The hourly time series was then converted into a series of instances. One instance was produced for every hourly record in our processed series. The features used for every instance are defined as follows:

– the 24 previous hourly returns $rtn_{h-23}..rtn_h$
– the 24 previous hourly volatilities $vol_{h-23}..vol_h$
– the nominal day of week (a value from $0...6$)
– the nominal hour of the day, (a value from $0...23$)
– a class value to predict

The return and volatility values are defined by Equations 1 and 2. Return is normalized against the opening price for the hour in order to account for different price scales, and the scaling factor of 100.0 is applied to turn the values into percentages. Note that rtn_h is directional and can be positive or negative, whereas vol_h is never negative.

$$rtn_h = 100.0 \times \frac{close_h - open_h}{open_h} \tag{1}$$

$$vol_h = (rtn_h)^2 \tag{2}$$

The class value to predict was defined dependent on the prediction task. Recall that there are two prediction tasks: return prediction and volatility prediction. In the case of return prediction, that aim is to predict the *sign* of the return (positive or negative) of the market over the next hour; in the case of volatility prediction, the purpose is to predict whether volatility will increase or decrease over the next hour.

Let rtn_{h+1} and vol_{h+1} be the return and volatility of the next hour. Then the class variable for the prediction classes can be defined by Equations 3 and 4 respectively.

$$class_{h,rtn} = \begin{cases} 1 & \text{if } rtn_{h+1} \geq 0 \\ -1 & \text{otherwise} \end{cases} \tag{3}$$

$$class_{h,vol} = \begin{cases} 1 & \text{if } vol_{h+1} > vol_h \\ -1 & \text{otherwise} \end{cases} \tag{4}$$

There are no instances generated for any hour where the price does not change. That is, if the open, high, low and close price for an hour are all equal, then that hour is skipped, and the next instance will represent the next hour in which there is a price change.

After following this process, three datasets consisting of 52,092; 52,071; and 52,102 instances for the EUR/USD, USD/JPY and EUR/JPY markets respectively were generated. Each dataset comprised 50 features and a class variable,

Table 1. Predictive models used in the experiments

Model	Description
OR	Simple majority class classifier
RT-1	Single node decision tree, built using the REPTree algorithm [4] with depth limited to 1
RT	Full decision tree, built using the REPTree algorithm [4]
SMO	Sequential Minimal Optimization algorithm [7], a support vector machine learner with a linear kernel
RF	Random Forest algorithm [2] with an ensemble size of 10 random trees

and two versions were generated for each market, one for each of the prediction classes. This resulted in six different versions of the three datasets.

The next step in data preparation was to split the data into training and testing portions. Because financial time series are known to be non-stationary, it cannot be expected that the distribution of features and the relationship between the features and the class should be uniform over the entire period (from 2002 to 2011) covered by the data. To account for this, each dataset was divided into five 10,000 instance "segments", with the final 2,000 or so instances being discarded. The segments were maintained in chronological order, and it was decided that models should be evaluated in not one but four "sliding window" experiments. That is, rather than training a model on the first 50% of the data and testing on the following 50% of data, the better approach of training the model on the first segment (in chronological order) then testing on the second segment; training the model on the second segment and testing on the third; and so on, was utilized. This allowed four evaluations per prediction algorithm, and the resulting four accuracies could then be averaged to produce a more reliable overall estimate.

To summarise, six datasets were produced (two per market, for two different prediction tasks), and each dataset was divided into four train/test experiments. This meant that there would be $6 \times 4 = 24$ experiments per predictive model.

3.2 Base Classifiers

We considered a number of predictive models in our experiments, ranging from the very simple to the state-of-the-art. Two versions of each predictive model were considered: one version *without* the EDS algorithm (the control version); and another version *with* the EDS algorithm (the experimental condition).

The specific predictive models we chose are shown in Table 1. We used implementations of these models from WEKA version 3.7.3 [4], and all settings other than those specified in the Table are WEKA defaults.

The choices of the models can be explained as follows. The 0R model is simple and ignores the features in the datasets; instead, it simply predicts the majority class from the training data. This was chosen as a baseline with which to compare the other classifiers: any model should in theory perform better than 0R if the features are informative.

Table 2. Performance of models with and without EDS for the return prediction task, by dataset

	EUR/USD	USD/JPY	EUR/JPY	Average
0R	51.41	50.68	50.20	50.76
RT-1	51.57	50.58	50.50	50.88
EDS+RT-1	51.95	50.75	**50.72**	51.14
RT	51.61	50.41	50.29	50.77
EDS+RT	51.97	**51.06**	50.64	**51.22**
SMO	52.09	50.74	50.32	51.05
EDS+SMO	**52.15**	51.04	50.32	51.17
RF	51.40	50.79	50.20	50.80
EDS+RF	51.12	50.49	50.60	50.74

The RT-1 model is a single level decision tree, otherwise known as a stump, because it selects only a single feature with which to make a prediction. If RT-1 turns out to be the best model for any prediction task, then this finding implies that only one feature (e.g. hour of the day, or previous hour's return) is in fact relevant and the other features are noise. Thus, the performance of RT-1 can be thought of as a second, stronger, baseline for comparison. On the other hand, the RT model is a full decision tree: if RT performs better than RT-1, it implies that *more* than one feature is important, and that patterns in the data exist.

Decision tree models like RT and RT-1 are important because they are interpretable models, which means that humans can inspect them and determine the rules being used to make the predictions. This is important for traders and scientists interested in understanding markets.

The remaining algorithms in the Table, namely RF and SMO, however, are not interpretable: the former is an ensemble of decision trees, and the latter is a mathematical function based on the "support vectors" that it finds in the training data. These algorithms represent two of the state-of-the-art methods in machine learning and are included in our experiments because they should, empirically speaking, give the most accurate results.

3.3   Experiment 1: EDS Prediction

In this experiment, the performance of each prediction model described in Table 1 was compared with and without the EDS algorithm. Recall that each dataset was divided into five 10,000 hour segments, and each model was trained and tested four times. Thus, the results shown in Tables 2 and 3 represent an average over all four experiments.

Starting with results of the return prediction task (Table 2), a number of observations can be made.

Firstly, the EDS algorithm boosts the performance of every model in nearly every case. This is an encouraging result showing that the inituition behind the method is suitable for intraday financial data.

Table 3. Performance of models with and without EDS for the volatility prediction task, by dataset

	EUR/USD	USD/JPY	EUR/JPY	Average
0R	50.78	50.48	50.20	50.49
RT-1	70.30	71.86	70.55	70.90
EDS+RT-1	**70.32**	71.89	**70.69**	**70.97**
RT	70.21	70.47	69.90	70.19
EDS+RT	70.14	**71.94**	70.64	70.91
SMO	56.09	64.41	56.96	59.15
EDS+SMO	61.46	55.37	63.50	60.11
RF	68.26	68.49	67.81	68.18
EDS+RF	68.46	69.44	67.91	68.60

Secondly, return prediction on a "next hour" basis is very difficult; most of the models achieve only a very small excess over 50%, which would be expected by random guessing. However, they do *consistently* achieve more than 50%, which does imply that there is at least some kind of pattern to learn.

It should be noted at this point that it is very unlikely that higher probability patterns exist; if they did, they would have been exploited by traders already. Also note that a very small excess accuracy (e.g. 1%) can translate into a very significant "edge" if the number of samples is very high, as they are in these experiments. That is, across 40,000 tests, an accuracy of 51% implies that there were 400 more correct predictions than incorrect predictions.

In terms of which market predictability on an intraday basis, it appears that EUR/USD is the most predictable with accuracies achieving 52.15%. On the other hand, the EUR/JPY market is the least predictable, with no model accuracy exceeding 50.72%.

The performances of the different classifiers are also interesting. In both the EUR/USD and USD/JPY cases, classifiers other than 0R and RT-1 achieve the best performance. This implies that patterns encompassing more than a single feature exist in the data. In the case of EUR/USD, SMO performs the best, whilst for USD/JPY, the decision tree is the best performer. The decision tree RT also happens to be the best overall average performer.

For the volatility prediction task, the results shown in Table 3 are quite different. Although the frequency of the classes are balanced (0R achieves little over 50%), the typical classifier can achieve approximately 65-70% accuracy. However, unlike the previous case, the stronger classifiers perform no better than the weaker classifier RT-1 in two out of the three cases. This implies that only a single feature may be needed for volatility prediction, at least for those two markets.

Despite this, the Table shows that EDS again boosts every model's performance most of the time. In fact,EDS+RT-1 is the overall best model for volatility prediction in every case.

Table 4. Performance of ensembles of EDS-learned models for the return prediction task, by dataset

	EUR/USD	USD/JPY	EUR/JPY	Average
Ens(EDS+RT)	51.48	**51.23**	50.73	51.15
Ens(EDS+SMO)	**52.57**	**51.23**	50.58	**51.46**
Ens(EDS+RF)	52.01	51.06	**50.77**	51.28

Table 5. Performance of ensembles of EDS-learned models for the volatility prediction task, by dataset

	EUR/USD	USD/JPY	EUR/JPY	Average
Ens(EDS+RT)	71.48	72.31	71.03	71.61
Ens(EDS+SMO)	62.61	65.67	64.61	64.30
Ens(EDS+RF)	**71.89**	**72.79**	**71.85**	**72.18**

3.4 Experiment 2: Ensemble Prediction

One well-known drawback of randomized algorithms such as the steady state genetic algorithm is its dependence on the random initial conditions: on some runs, a highly accurate model may be built because the random initial solutions were good, but on other runs, only weaker models may be found because the initial population was largely poor. In other words, randomized algorithms have a performance variance dependent on the random number seed.

One method proposed here to reduce this variance is to build multiple models using EDS, and then combine them together into a single model by averaging their predictions. This should, in theory, produce more reliable estimates because the weaker and stronger classifiers will average out in the final results. However, this method does turn models that were originally interpretable (such as the decision tree model, RT) into black-box classifiers, because now each ensemble consists of multiple models whose predictions are now being combined in a new "second stage" of the prediction process.

To investigate this idea, EDS was applied to each base model (RT, SMO, and RF) not once but ten times, producing ten prediction models. These models were then combined into an averaging ensemble, and the ensemble was used to make the predictions. We applied this to both the return and volatility prediction tasks, with exactly the same experimental train/test setup as previously described. The results are shown in Tables 4 and 5.

The first point to note from these Tables (specifically comparing Table 2 to Table 4 and Table 3 to Table 5) is that uniformly, an ensemble of EDS-filtered classifiers outperforms a single equivalent EDS classifier by itself. For example, the accuracy of SMO on the EUR/USD return prediction task is increased by approximately 0.5% giving a new best accuracy. The best accuracies on the other markets are also increased, albeit not as much as in the case of EUR/USD. EUR/JPY still remains the most difficult market to predict with the ensemble only increasing the accuracy of the RF classifier to 50.77% from 50.72%.

An interesting point to note is that the best classifier for EUR/JPY is now RF, rather than the stump, which was the best performer in the previous experiment. This implies that the ensemble is able to detect a pattern involving more than a single attribute whereas the single EDS-tuned model by itself was unable to.

This observation is also true for the second volatility prediction experiment show in Table 5. In the previous experiment, no pattern beyond a single attribute was discovered for volatility prediction, and all the other models appeared to over-fit the data: hence, RT-1 uniformly performed with the highest accuracy. In the second experiment, however, the results are quite different: both the RT and RF models outperform RT-1, and the best accuracies are approximately 1% *at least* above the best results from that experiment.

4 Conclusion

To summarise, extensive experiments have shown that firstly, the EDS approach typically boosts the accuracy of the tested predictive models; and secondly, further accuracy increases are possible by using EDS to construct a diverse ensemble of models.

References

[1] Breedon, F., Ranaldo, A.: Intraday Patterns in FX Returns and Order Flow. Swiss National Bank Working Papers 2011-4 (2010)
[2] Breiman, L.: Random Forests. Machine Learning 45(1), 5–32 (2001)
[3] Cano, J., Herrera, F., Lozano, M.: Using Evolutionary Algorithms as Instance Selection for Data Reduction in KDD: An Experimental Study. IEEE Transactions on Evolutionary Computation 7(6), 561–575 (2003)
[4] Hall, M., Frank, E., Holmes, G., Pfahringer, B., Reutemann, P., Witten, I.H.: The WEKA data mining software: An update. SIGKDD Explorations 11(1) (2009)
[5] Larkin, F., Ryan, C.: Modesty Is the Best Policy: Automatic Discovery of Viable Forecasting Goals in Financial Data. In: Di Chio, C., Brabazon, A., Di Caro, G.A., Ebner, M., Farooq, M., Fink, A., Grahl, J., Greenfield, G., Machado, P., O'Neill, M., Tarantino, E., Urquhart, N. (eds.) EvoApplications 2010, Part II. LNCS, vol. 6025, pp. 202–211. Springer, Heidelberg (2010)
[6] Luke, S.: Essentials of Metaheuristics, Lulu (2009), http://cs.gmu.edu/~sean/book/metaheuristics/
[7] Platt, J.C.: Fast training of support vector machines using sequential minimal optimization. In: Schölkopf, B., Burges, C., Smola, A. (eds.) Advances in Kernel Methods – Support Vector Learning. MIT Press (1998)
[8] Pi Trading Corporation, http://pitrading.com/
[9] Sewell, M.: Characterization of Financial Time Series. Research Note RN/11/01, Dept. of Computer Science UCL (2011)

Initial Results from Co-operative Co-evolution for Automated Platformer Design

Michael Cook, Simon Colton, and Jeremy Gow

Computational Creativity Group, Imperial College, London
http://ccg.doc.ic.ac.uk

Abstract. We present initial results from ACCME, A Co-operative Co-evolutionary Metroidvania Engine, which uses co-operative co-evolution to automatically evolve simple platform games. We describe the system in detail and justify the use of co-operative co-evolution. We then address two fundamental questions about the use of this method in automated game design, both in terms of its ability to maximise fitness functions, and whether our choice of fitness function produces scores which correlate with player preference in the resulting games.

Keywords: automated game design, procedural generation, co-operative co-evolution.

1 Introduction

Procedural content generation (PCG) is a highly active area of research that offers effective methods for generating a wide variety of game content. PCG systems tend to work in isolation, often as a supplement to a human-designed system, designing aspects of the game's world [1,4,10,12]; generating items or abilities suited to the individual currently playing [6,7]; or generating quests or tasks for the player to undertake [2,8,11]. However, automating design as a whole – that is, the design of a game solely by a system and without direct human judgement – remains largely uninvestigated.

The problem of automated game design is an attractive one to address, because it not only provides us with a basis to build stronger, more capable procedural content generation systems, but also allows for more intelligent design systems, representing a move away from merely creating content and towards co-operating with a human designer on a shared creative task. We have shown in [5] that automated game design systems can help complete partially-specified designs – further work building systems such as this will drive the development of assistive design tools.

The remainder of this paper is organised as follows: in section 2, we introduce co-operative co-evolution, prior work in this area, and related work in automated game design. In section 3, we present ACCME, a system which employs co-operative co-evolution to automatically design 2D platform games. We give details of experiments conducted using ACCME to investigate the effectiveness of computational co-evolution as an automated design technique. In section 4, we present conclusions and look at future work in the area.

2 Background

A co-operative co-evolutionary (CCE) system solves a problem by decomposing it into several subtasks called *species*. These are represented as independent evolutionary

C. Di Chio et al. (Eds.): EvoApplications 2012, LNCS 7248, pp. 194–203, 2012.

processes that are evaluated by recomposing members from each population into solutions to the original problem, and evaluating the quality of the complete solution. A CCE system decomposes a problem, P, into n subtasks, $P_1, \ldots, P_n$, where a *solution* for P is a set $\{p_1, \ldots, p_n\}$ with $p_i \in P_i$.

A fitness function for such a system evaluates a solution to the larger supertask P, rather than evaluating members of a subtask's population. Therefore, to evaluate a member, p_x, of the subtask P_x's population, we extract the most fit member of every other subtask's population, compose this set to form a solution to P, and apply the fitness function to this hybrid solution. The notion of 'co-operative' evolution refers to the way in which the fitness of the solution is directly related to how well p_x *co-operates* with the other components of the solution. Since these other $n - 1$ components do not change during the evaluation of the population P_x, the fitness represents how well a member of the population of P_x contributes to the overall solution.

Co-operative co-evolution was proposed in [9] by Potter and De Jong, in the context of function optimisation problems. They state that "in order to evolve more complex structures, explicit notions of modularity need to be introduced in order to provide reasonable opportunities for complex solutions to evolve". We hold that for creative tasks such as game design, a similar level of modularity is desirable.

The ANGELINA System

In [5], we presented ANGELINA, a system that designs arcade games using co-operative co-evolution. We decomposed the task of designing such games into several species, each of which is responsible for a certain aspect of the design. The three species generate *maps*, two-dimensional arrays that describe passable and impassable areas in the game's level; *layouts*, which specify the arrangement of red, green and blue entities in the game world as well as the play character; and *rulesets*, which describe the way in which the non-player entities moved, and also define one or more rules that describe the effects of certain types of entity, player or obstacle colliding with one another. A combination of a map, layout and ruleset defined a game. We performed several experiments to support the claim that CCE is able to rediscover existing games in the target domain, such as PacMan, as well as games that were novel. In [5] we describe a demonstration of the software running independently to design games, and an assistive task where ANGELINA is given hand-designed maps and produces suitable rulesets and layouts.

Related Work

Although we are not aware of any other work which addresses the problem of automating whole game design, there are other related projects. In [11], the authors evolve rulesets for arcade games. The system uses a neural network to learn rulesets. The fitness of games is based on how hard or easy they were for the network to learn. This work inspired the design of our domain in [5]. In [8], Nelson and Mateas evolve simple 'minigames' by interpreting terms that describe actions or subjects. The work is interesting in terms of higher-level design tasks relating to the interpretation of themes and their relation to game mechanics, although the work does not specifically tackle interrelated or co-operating design tasks.

In [2], Browne and Maire present a system for automatically designing board games. The underlying task of designing a set of rules that govern ludic interactions is common to both projects. The work is primarily involved in both identifying 'indicators of game quality' and subsequently applying these as heuristics in an evolutionary process for generating games. The work culminated in major successes in the area, including the publishing of some computer-generated game designs as commercial board games.

3   ACCME

Metroidvania is a subgenre of 2D platform games. The term, a portmanteau of two games that popularised the genre, was coined by Sharkey [14]. Metroidvania games are "based... on exploration with areas that [can] only be reached after attaining items in other areas" [15]. Contemporary examples vary from casual to more challenging games [16,17]. The subgenre's core concepts lend themselves well to fitness functions.

ACCME is a system we have developed that designs such games using CCE, built on an evolutionary framework derived from [5]. ACCME is comprised of a Map species, a Layout species and a Powerset species. We first show how ACCME represents a game, then examine the species making up the CCE process and how playouts were implemented. We also detail some evaluative work which investigates the usefulness of CCE and the relationship between our definition of fitness and game quality.

3.1   Representation

A game is represented as a 3-tuple consisting of a Map, a Layout and a list of powerups called a *Powerset*. A map is a two-dimensional array of integers, where each integer in the array maps to an 8x8 pixel tile within the finished game. An zero value in the array describes an empty space, and any value greater than zero represents some tile texture (such as grass, or water). A *collision index*, i, is chosen such that any integer less than or equal to i is non-solid in the game world. This is used to define which integers describe solid platforms and walls, and which describe scenery. The collision index allows for tiles to change at runtime, allowing the representation of locked and unlockable doors.

A layout defines what we call *archetypes*, a description of a class of enemy. An archetype consists of one or more *actions* and a *movement behaviour*. Movement behaviours describe how the enemy moves through the game, selected from one of STATIC, where the enemy does not move, PATROLS, where the enemy moves horizontally until it meets an obstruction or there is no solid ground to walk on, and FLIES, which is similar to patrolling but does not require solid ground. Actions describe things that enemies can do during the game that provide a challenge to the player or somehow differentiate their behaviour from other enemies. An archetype has zero or more actions, selected from TURRET, which fires a projectile at the player whenever they are within a certain sight range, POUNCE, which causes the enemy to leap towards the player when they have an unbroken line of sight, and MISSILE, which fires a slower projectile that follows the player. A layout also contains a list of enemies, which are described by an (x, y) starting co-ordinate in the map, and an archetype number. The layout also describes the player's starting location and the location of the exit to the game.

A powerset is a list of powerups. A powerup is described by a co-ordinate representing its location in the map, as well as a *target variable* and a *target value*. When the player touches the powerup, the target variable is changed within the game code so

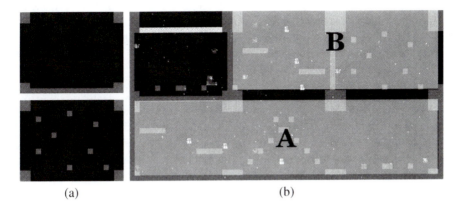

Fig. 1. Screenshots showing map templates (left) and regions (right)

that its value becomes that of the target value. There are three target variables available to ACCME - `jumpVelocity`, which describes the velocity applied to the player when the jump key is pressed, `globalAccelerationY`, which describes the effect of gravity applied to game objects, and `collisionIndex` which defines the integer value above which map tiles are considered solid. Target values are chosen from an integer range defined appropriately for each powerup.

3.2 CCE Species

Maps are constructed out of smaller two-dimensional arrays of fixed width and height called *Map Tiles*. The specific layout of a map tile is selected from one of 13 outer templates (defining the border around the tile). These outer templates define tile borders as blocked or unblocked - for instance, Figure 1(a) shows an outer template where the lower side of the tile is blocked. We provide all possible permutations, with the exception of the case where all sides are blocked. A map tile also selects one of 12 inner templates, which were hand-made. An example of such a hand-crafted template is provided in Figure 1a. Hand-designed templates were used to ensure some logical order to each tile, but with enough compositional variation that ACCME is responsible for the overall arrangement. A *Map*, therefore, is a two-dimensional array of map tiles, and the CCE process operates at no more detailed a level than map tiles when performing operations such as crossover.

The fitness function scores highly those maps which do not allow the player to leave the map bounds. It also heavily relies on playouts, assigning a higher fitness to those maps which have initially small reachable fractions, but whose maximal reachable fraction (having collected all relevant powerups) is high. We discount the fitness for contributions made to reachability early on in the game and towards the end. The intention here is to encourage steady progress throughout the game, where powerups make an increasing contribution to the player's abilities, and then after the game's midpoint, the player makes progressively smaller advances towards the exit. To describe the fitness function, first consider the game as a list of 'stages', beginning with the player start (p_s), culminating with the exit (p_e), and with intermediate stages representing the collection of a powerup. We represent the list of stages as: $[\, p_s, pup^1, \ldots, pup^n, p_e \,]$.

Let $rch(x)$ be a function that returns the percentage of the map that is reachable at stage x of the game (but not reachable in the previous state). Then this list of stages contributes to the overall fitness proportionally, using fractional variables d_i as discounting factors to reduce the contributions made by each stage:

$$d_1 \times rch(p_s) + d_2 \times rch(pup^1) + \ldots d_x \times rch(pup^m) + \ldots d_2 \times rch(pup^n) + d_1 \times rch(p_e)$$

where $\forall x \forall y \ x < y \implies d_x < d_y$. The fitness function also assigns higher fitness to maps with longer paths between the start and the exit. In many games, it is considered poor practice to arbitrarily extend the player's path; however, due to the large state spaces inherent in ACCME, it is important to keep the maps small, while utilising as much of the space available as possible. Our intention was to generate games in which the optimal path through the map passes through as many map tiles as possible, to maximise the utilisation of the space available.

Crossover of two maps produces child amps that inherit either alternating rows or columns of the two parent maps. Mutation of a map replaces a number of randomly selected map tiles with newly generated ones, with a maximum of four replacements per map per mutation.

Powersets. The fitness function for a powerset assigns fitness proportional to the amount of increase in reachability each powerup provides. Playout data is used to calculate this, and to calculate which powerups are reachable (and whether powerups can be collected in multiple orders, or whether there is a linear progression through the game). We employ the notion of a *trace object* which describes all possible routes through the game, recording the order in which powerups are collected, as well as if the exit is reachable. This is expressed as an ordering on the set of powerups, P. We are interested in traces where this ordering is partial, rather than total, as player choice is a desirable factor in the design of Metroidvania games. We define a trace T as a list of powerups, $\{p_1, \ldots, p_n\} \subseteq P$, where P is the set of all powerups in the game. We denote that the predicate $term(T)$ holds if, after executing the trace T, the player is able to reach the exit. We increase the fitness of a powerset relative to the number of *legitimate traces* it has in its trace object, where T is legitimate $\iff \forall T' \in (\mathcal{P}(T) \setminus \{T\}) \ . \ \neg term(T')$.

Note that $\mathcal{P}(T)$ is the power set of the set of powerups T. The above states that any sequence of powerups smaller than T would not permit the player to reach the exit. Preliminary experimentation showed this to be a useful balancing factor which encourages multiple traces through a game, but penalises designs in which the player is able to bypass a section of the game and ignore some powerups entirely. We also add value to a powerset's fitness relative to the average distance between each powerup. We calculate distance between objects by performing an A* search on the reachability map.

Powersets are crossed over by creating child powersets that randomly select powerups from the two parents, with a small chance to generate an entirely new powerup instead of inheriting from either. Mutation of a powerup randomises the magnitude of the change the powerup makes to its target variable.

Layouts. The layout species is concerned with designing the enemy types present in the game and placing them within the map along with the player's starting location and the level's exit location. The task of enemy design is similar to the design of entities in the experiment described in [5]. We initially give very low fitnesses to any illegal or invalid placements. For ACCME, this involves penalising for enemies, player character or exit locations that are placed in walls. We penalise heavily for layouts that do not allow the player to reach the exit. To evaluate this, we use the same reachability calculations as present in the powerset evaluation described above.

Figure 1(b) shows a subsection of a game design. The player begins in section A, in which there is a powerup that allows access to section B. We identify these sections by calculating the player's reachability potential after picking up powerups in the game. We then reward layouts that introduce archetypes gradually, so that sections that are explored later in the game are more likely to have the full selection of archetypes, whereas sections explored early in the game may only have a subset.

Layouts are crossed over by exchanging locations of enemy archetypes, exit and player locations, and designs for archetypes themselves. Crossover can also switch the enemies of map tiles, in much the same way that map crossover exchanges map tiles, that is either by row, column or single tile. Mutation is applied to make small changes to the location of enemies, player start location and game exit. Mutation can also randomly change features in an enemy archetype, altering the movement type or adding and removing behaviours.

3.3   Playouts and Reachability

ACCME performs playouts in order to take a game state and establish which regions of its map are currently reachable. The system can apply powerups to change variables that affect reachability. Calculation of the reachable area is computationally expensive given the number of games assessed in a run of the system (although an individual reachability check merely tests each reachable tile for nearby reachable tiles, a sample run described in section 3.5 evaluates over 240,000 games). ACCME maintains an *open list* of map locations that are known to be reachable, initialised with the starting location. Upon removing a new location from the open list, it checks three possible scenarios:

Jumping. If the player is standing on solid ground, they are capable of jumping. The formula used to calculate the potential height of the jump is $V_{start}^2/2g$, where g is gravity, expressed in pixels per second per second, and V_{start} is the upward force applied by the jump operation, also expressed in pixels per second.

The formula for jump height, combined with the knowledge that horizontal force can be applied regardless of the player's position, allows us to calculate the space in which the player has a positive vertical velocity (the *rising area*) by simply applying the maximum horizontal force in both directions for the duration of the jump. We then traverse the map locations in this area, and for each location we test to see if there are obstructions between the starting location and the target. If there is not, the area is reachable, and is added to the open list. We found that using line-of-sight as a check for accessibility is a cheap but effective method for deciding whether or not an area was reachable.

Walking. If the player is on solid ground, we perform a Walking check. If a contiguous area of solid ground extends left or right of the current location, then the locations above this solid ground are also considered reachable. This helps cover some map areas that would otherwise take a longer time to detect using only jumping and falling.

Falling. If the player is not on solid ground, then they are falling. This may be because they have walked off the edge of a platform, or are jumping. In this case, we calculate the horizontal extent of a jump to simulate the player's descent and label areas that are reachable during the fall. Because the player can apply horizontal velocity during a fall, this is different to a real-world physics simulation.

Quality and Accuracy of Reachability Estimations. The above cases provide an estimate of reachability, allowing ACCME to assess levels and infer where the player can and cannot reach without having to simulate a full game playout. By avoiding such extensive simulation, we are able to greatly reduce the complexity of evaluating a game without much loss in reachability data; however, extensions to ACCME's domain that allow for other kinds of obstacles (such as enemies which cannot be destroyed) would, we think, require a full simulation in order to fully assess runtime reachability.

In deciding how best to estimate reachability, we opted for a system which, at worst, *underestimates* the amount of reachable map space. Overestimating in this case would produce games that were potentially unsolvable, but by underestimating we merely allow for the fact that through application of skill the player may be able to bypass certain sections of the game level (by reaching areas which ACCME had flagged as unreachable). Such situations are not uncommon for games, and give rise to *speed runs*, where players use such design flaws to complete a game in the fastest possible time[18].

A pilot study outlined in section 3.5 showed that, since reachability was not bidirectional, ACCME was unable to differentiate between areas that were reachable, and areas that could be reached and then returned from. This caused ACCME to design games with one-way jumps and inescapable pits. We modified the software to use a single additional check per reachable map tile to detect if it can be exited as well as entered. We proportionally reduce the fitness of maps that contain dead ends – this still allows for situations where a player is able to escape a dead end by obtaining a powerup.

3.4   Evolutionary Setup

A typical execution of the software is composed of 400 generations, undertaken with each species maintaining a population of 200 solutions. We utilise a steady-state, elitist selection method, with the fittest 10% comprising the parents of the next generation. The parents are also included for another generation of evolution; this is to allow trends to emerge more readily from the co-operating species, as existing progress towards co-operation is not lost between generations. We experimented with the application of some other selection techniques such as roulette-based approaches, but found them to be considerably less reliable. We plan to explore other such techniques in future work.

3.5   Evaluation

Effectiveness of CCE. To compare the results of CCE with selecting from a comparable population of randomly-generated games, we generated 240,000 game designs at random and evaluated them using the same fitness functions that ACCME uses. The fitness of the highest-scoring game is shown as a dashed line in Figure 2. On the same graph, the line shows the fitness change over 400 generations for ACCME running as a co-operative co-evolutionary process with three species, each with a population of 200 members. Each species evaluates against only the fittest members of the other species, hence ACCME evaluated 600 games per generation. The graph shows a clear improvement over random generation after only a handful of generations, and also highlights the

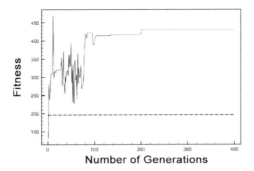

Fig. 2. A standard run of ACCME against a comparable random search

fast convergence of ACCME. The strength of the convergence may point to a weakness in the CCE system, a matter which we discuss in section 4.

The early spike seen in figure 2 highlights a game with higher fitness than those of the final generations. We select our output from the final generation run, because we posit that these games exhibit the strongest co-operative traits, so a spike of this nature is, we believe, caused by one of the three species presenting a very high-fitness solution that counterbalances lower-fitness solutions in other species. CCE processes fluctuate wildly during the early stages of generation. At these stages, the species are far from co-operative, which means that the change between generations is often very large as they try to compensate for the lack of co-operation. We plan to examine such spikes in further work, as it may indicate that our fitness functions are not balanced in their evaluation of co-operative fitness, or that our method of synthesising fitness evaluations from the co-operating processes is not the best way of evaluating overall fitness.

Pilot Study. We performed a pilot study to assess player response to the games produced, and to highlight issues in player evaluation of the games. 180 players played the same game and rated it between 1 and 5. Qualitative feedback was also sought to gain insight into player reactions. From the responses, we made several improvements to ACCME, including developing its understanding of reachability to include dead end detection (described in section 3.3). We noted that players' responses frequently highlighted areas of the game that were not ACCME's responsibility – such as control schemes or art direction. We discuss this in section 4 in the context of future work.

Comparative Study. Following the pilot study, we performed a second smaller study in which 35 participants, responding to a call sent out to the 180 pilot study participants, were asked to play three games designed by ACCME. We chose three games with fitness valuations of 436, 310 and 183, labelled high, medium and low fitness respectively, to represent a range of game fitnesses. The unlabelled games were presented to each participant in a randomly-selected order, and the participants were asked to rank the games in terms of perceived quality after playing all three. Our hypothesis was that higher-fitness games should be preferable to players than low-fitness games. For our study data, we found a greater proportion of high fitness games were ranked highest: 49% compared to 25% for low/medium fitness games. However, the effect was not significant (chi-squared, p=0.15). We found a very weak but insignificant rank correlation

	Best Rank	Middle Rank	Worst Rank
HighFitnessGame	19	9	11
MedFitnessGame	9	15	15
LowFitnessGame	11	15	13

Fig. 3. Data showing frequencies of ranks for the comparative study

between fitness and player preference, (Kendall's $\tau = 0.11$, p = 0.17). In both tests, we were unable to reject the null hypothesis. Although these results are inconclusive, the data suggests to us that there may be some effect of fitness on preference, but further study is required to investigate the relationship.

One key piece of written feedback we received was that some players felt the games were too similar. This is partly down to commonality in features not designed by AC-CME. However, the repeated nature of the goals and the restrictive set of powerups from which ACCME chooses from also contributed to this. In a creative task such as game design, the ability to create novelty is crucial, and ACCME does not appear capable of this in its current state. The brevity of the three games was also mentioned in some written feedback as a negative quality. The games in the study were restricted to 3x3 map tiles; given that a key element in Metroidvania games is gradual exploration, it is conceivable that the games did not last long enough for the players to experience this sense of gradual exploration. Hence, larger game sizes might improve future studies.

4 Conclusions and Further Work

We have introduced co-operative co-evolution in the context of automated game design, and presented ACCME, a system for designing simple Metroidvania platform games using CCE, reachability analysis and a flexible powerup system. We have shown CCE to be effective at developing high-fitness solutions, but a comparative study shows a gap between ACCME's concept of fitness and player preference.

The pilot study highlighted many difficulties in the evaluation of automatically-designed games by humans. The task of distinguishing decisions made by the automated designer, and decisions made by the authors in constructing the framework, is difficult. We received comments on aspects of the design that ACCME was responsible for, as well as things inherent in the template game supplied to ACCME. This shows a need for more forethought in presenting automatically designed games to players in future.

Our later comparative study highlighted the similarities in games produced by AC-CME, even when the system considers there to be large differences in fitness. This leads us to two areas of further work. Firstly, we plan to reconsider the fitness valuations used, in order to strengthen some areas of evaluation, and add in new valuations to emphasise some areas that playtesters perceived as lacking, such as difficulty. One approach might be to simulate simplified combat, reducing the game to a turn-based simulation, thereby discretising and simplifying the evaluation. Secondly, it may be necessary to focus our surveys in future to avoid general concepts such as preference or fun. Designing experiments to evaluate specific parts of a design, such as level layout or powerup design, may provide a better way of estimating the impact of a system such as ACCME.

We noted earlier that ACCME converges on a solution very quickly. CCE has unique problems associated with it that affects its ability to locate global optima, which we are yet to investigate in ACCME. Such problems are discussed in [13] and some solutions proposed in [3]. We aim to apply these ideas to ACCME in the hope that it will improve the process. We also wish to investigate alternative selection methods for the evolution. All of the games listed in this paper can be played online at http://bit.ly/gbangelina.

Acknowledgements. The authors would like to thank Zack Johnson, Kevin Simmons and Riff Conner for their insight into Metroidvania design. We also thank the anonymous reviewers for their helpful comments and suggestions which have helped improve the paper.

References

1. Ashlock, D., Lee, C., McGuinness, C.: Search based procedural generation of maze-like levels. IEEE Transactions on Computational Intelligence and AI in Games 3(3), 260–273 (2011)
2. Browne, C., Maire, F.: Evolutionary game design. IEEE Transactions on Computational Intelligence in AI and Games 2(1) (2010)
3. Bucci, A., Pollack, J.B.: On identifying global optima in cooperative coevolution. In: Proc. of the 2005 Conf. on Genetic and Evolutionary Computation (2005)
4. Cardamone, L., Yannakakis, G.N., Togelius, J., Lanzi, P.L.: Evolving Interesting Maps for a First Person Shooter. In: Di Chio, C. (ed.) EvoApplications 2011, Part I. LNCS, vol. 6624, pp. 63–72. Springer, Heidelberg (2011)
5. Cook, M., Colton, S.: Multi-faceted evolution of simple arcade games. In: Proc. of 2011 IEEE Conference on Computational Intelligence and Games (2011)
6. Hastings, E.J., Guha, R.K., Stanley, K.O.: Evolving content in the galactic arms race video game. In: Proc. of 2009 IEEE Conf. on Computational Intelligence and Games (2009)
7. Liapis, A., Yannakakis, G., Togelius, J.: Neuroevolutionary constrained optimization for content creation. In: Proc. of 2011 IEEE Conf. on Computational Intelligence and Games (2011)
8. Nelson, M., Mateas, M.: Towards automated game design. In: Artificial Intelligence and Human-Oriented Computing, pp. 626–637 (2007)
9. Potter, M., de Jong, K.: A Cooperative Coevolutionary approach to Function Optimization. In: Davidor, Y., Männer, R., Schwefel, H.-P. (eds.) PPSN III 1994. LNCS, vol. 866, pp. 249–257. Springer, Heidelberg (1994)
10. Smith, G., Treanor, M., Whitehead, J., Mateas, M.: Rhythm-based level generation for 2D platformers. In: Proc. of the 4th International Conf. on Foundations of Digital Games (2009)
11. Togelius, J., Schmidhuber, J.: An experiment in automatic game design. In: Proceedings of 2008 IEEE Conference on Computational Intelligence and Games (2008)
12. Togelius, J., Preuss, M., Beume, N., Wessing, S., Hagelbäck, J., Yannakakis, G.: Multiobjective exploration of the Starcraft map space. In: Proc. of 2010 IEEE Conf. on Computational Intelligence and Games (2010)
13. Wiegand, R.: An analysis of cooperative coevolutionary algorithms. Ph.D. dissertation, George Mason University, USA (2004)
14. Sharkey, S., Parish, J.: Debunking Metroidvania, http://www.bit.ly/wiredmv/
15. Metroidvania, Gaming Wikia, http://gaming.wikia.com/wiki/Metroidvania
16. Knytt Stories, Nifflas Games (2007), http://nifflas.ni2.se/
17. Spelunky, Mossmouth Games (2009), http://www.spelunkyworld.com
18. Portal Done Pro - Speedrun, DemonStrate (2010), http://www.j.mp/ygoJLh

Evolving Third-Person Shooter Enemies
to Optimize Player Satisfaction in Real-Time

José M. Font

Departamento de Inteligencia Artificial, Universidad Politécnica de Madrid, Campus
de Montegancedo, 28660, Boadilla del Monte, Spain
jm.font@upm.es

Abstract. A grammar-guided genetic program is presented to automatically build and evolve populations of AI controlled enemies in a 2D
third-person shooter called Genes of War. This evolutionary system constantly adapts enemy behaviour, encoded by a multi-layered fuzzy control
system, while the game is being played. Thus the enemy behaviour fits a
target challenge level for the purpose of maximizing player satisfaction.
Two different methods to calculate this challenge level are presented:
"hardwired" that allows the desired difficulty level to be programed at every stage of the gameplay, and "adaptive" that automatically determines
difficulty by analyzing several features extracted from the player's gameplay. Results show that the genetic program successfully adapts armies
of ten enemies to different kinds of players and difficulty distributions.

Keywords: Evolutionary computation, fuzzy rule based system,
grammar-guided genetic programming, player satisfaction.

1   Introduction

Game development is a complex and multidisciplinary task that involves character design, environment modeling, level planning, story writing, music composition and, finally, programming [15]. Procedural content generation (PCG) is
a research field that studies the application of algorithmic methods from many
areas, such as computational intelligence, computer graphics, modeling and discrete mathematics, to the automatic generation of game content. Having access
to these tools helps game developers to reduce the design costs when creating
games that include huge amounts of content [4].

Search-based PCG is a research area in which evolutionary computation techniques are used to automatically generate game content [16]. Some examples in
this area include the on-line generation of weapons for the space shooter Galactic Arms Race based on player preferences [8], the off-line creation of tracks for
racing games focusing on diversity and driving experience [9] or personalization
to player driving style [14], and the level and game mechanics customization for
platform games [11].

Evolution of AI controlled characters in games enhances the creation of intelligent behaviors that raise player interest during gameplay [10]. For example,
the NERO video game involves players training non-player characters to learn
to target tactical directives by using a neuro-evolution approach [13]. Genetic

C. Di Chio et al. (Eds.): EvoApplications 2012, LNCS 7248, pp. 204–213, 2012.

programming has been applied to the generation of challenging opponents in several competitive games [1,2,12].

Nowadays, shooters are one of the most popular as well as worldwide best-selling game genres[1]. A grammar-guided genetic programming (GGGP) system is proposed here to automatically generate and evolve enemy behavior in a 2D shooter called Genes of War. GGGP has been successfully applied to the generation of both symbolic and sub-symbolic self-adapting intelligent systems [7,6,5]. GGGP enhances search space exploration because of the usage of the grammatical crossover operator [3], which does not generate invalid individuals during the evolutionary process.

Enemies in Genes of War are controlled by a multi-layered fuzzy ruled-based system. A population of these systems is constantly evolved in real-time in order to maximize player satisfaction at each stage of the gameplay. Satisfaction is measured from an implicit perspective [17] by matching the challenge preferred by the player with the one offered by the game. Two methods are proposed to measure this challenge level: one "adaptive" that fits the enemy population to player skills, and one "hardwired" to match it with programmers preferences.

Results show that the proposed GGGP system adapts the enemy population to two kinds of players, one beginner and one experienced, as well as to two fixed challenge distributions. Modularity and flexibility of the proposed system make it suitable to be exported to other competitive game genres.

2   The Genes of War Game

Genes of War is a 2D scrolling shooter with top down perspective (Figure 1) specifically developed for this research as a test-bed application. In Genes of War, a human player takes the role of a soldier that must defeat an unlimited supply of AI controlled enemies. The player controls the soldier with a simple point and click system, which allows the soldier to move along a fixed size map. The soldier can aim in 360 degrees around itself and right clicking fires the soldier's weapon, even when the soldier is moving. The player can also make the soldier crouch by pushing the left control button in the keypad. Crouching allows the soldier to dodge enemy attacks, but prevents it from moving.

Enemies are AI controlled robots that can move along the map, crouch, and shoot in 360 degrees as the soldier does. Unlike the soldier, enemies are not allowed to shoot while they are moving. The player kills enemies by shooting them until their health points are depleted. Analogously, enemies defeat the player by shooting the soldier until it has no health points left. Friendly fire between enemies does not affect health points. When the soldier or an enemy is killed, it is respawned in a fixed map location with full health points and the default weapon.

Both soldier and enemies start the game equipped with their basic weapons: a machine gun and a cannon respectively. They can upgrade them to more pow-

---

[1] Walton, B.: Video game chartz (Last accessed in November, 2011), http://www.vgchartz.com

Fig. 1. A snapshot from Genes of War

erful weapons by picking up power-ups that are spawned over the game map. These powerful weapons are classified into short-range (shotgun, laser gun and flamethrower) and long-range weapons (rifle and missile launcher). Each has its own identifying power-up icon and unique values for the following characteristics: power, range, rate of fire, recoil, and ammunition. When the soldier runs out of ammunition, its weapon is automatically downgraded to the basic machine gun that has infinite ammunition. Unlike the soldier, enemies have unlimited ammunition for all weapons. There is a sixth power-up that heals a character (soldier or enemy) by increasing its health points. Power-ups can also be destroyed by firing at them, which causes an explosion that damages any character located close to it.

2.1   Multi-layered Enemy Control System

Enemies are fully controlled by the computer AI. Each enemy has a multi-layered control system composed by two modules. The upper module, called the strategy layer, holds a fuzzy rule-based system (FRBS) that processes a set of environmental variables whose values are extracted from the gameplay. As a result of its inference process, the FRBS determines the strategy that the enemy must follow. A strategy is composed by a destination, the location in the game map where the enemy is heading for, and a target, the location where the enemy aims at.

The lower module, called the action layer, holds a finite state machine composed by five states that represent the five basic actions that an enemy can do during gameplay. These states names are: "move", "fire", "cfire", "stop" and "die". The action layer takes both destination and target variables as input, performing the set of basic actions needed to follow strategy given by the upper layer.

The strategy layer is explained in detail in Figure 2a. The input of the FRBS is a set of seven environmental variables: $\delta_{soldier}$, $\delta_{shortRange}$, $\delta_{longRange}$, δ_{health}, life, weapon and soldier weapon. The first four of them make reference to the distance (in pixels) from the enemy to the soldier, to a short range weapon power-up, to a long range weapon power-up, and to a health power-up respectively. The linguistic labels close, med, and far have been defined for these four input variables, making reference to the three fuzzy sets in which they can take values.

a)

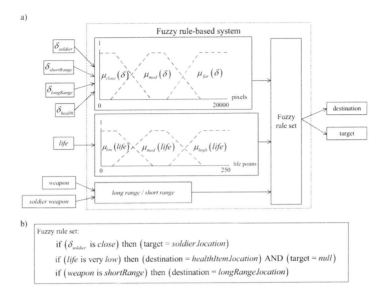

b)

Fuzzy rule set:

   if $(\delta_{soldier}$ is *close*) then (target = *soldier location*)

   if (*life* is very *low*) then (destination = *healthItem location*) AND (target = *null*)

   if (*weapon* is *shortRange*) then (destination = *longRange location*)

Fig. 2. a) Description of the strategy layer in the multi-layered enemy control system. b) Sample fuzzy rule set.

The membership of each variable to the fuzzy sets represented by these labels is defined by the membership functions $\mu_{close}(\delta)$, $\mu_{med}(\delta)$, and $\mu_{far}(\delta)$.

Variable *life* contains the enemy's health points. This variable can take values in three fuzzy sets represented by the linguistic labels *low*, *med* and *high*. The membership functions related to these fuzzy sets are $\mu_{low}(life)$, $\mu_{med}(life)$, and $\mu_{high}(life)$. Variables *weapon* and *soldier weapon* describe the kind of weapon that the enemy and the soldier hold, respectively. These are non-fuzzy variables, so the only values they can take are *long range* and *short range*.

The set of input variables together with a fuzzy rule set are processed by the FRBS inference system, producing values for the output variables destination and target. These values are valid game map locations expressed in Cartesian coordinates, such as the location of the soldier or the location of a power-up. A sample fuzzy rule set is displayed in Figure 2b. Fuzzy rules may contain any input variable as an antecedent as well as any output variable as a consequent. Two different clauses can be linked in the consequent by means of the AND operator.

Figure 3a represents the finite state machine (FSM) in the action layer as a graph. During gameplay, a FSM can be placed only in one state (node) at a time. If a fixed transition condition (arc) is verified, the control system jumps from its actual state to another one that is connected to it. The basic action that an enemy performs is given by the state in which its control system is placed: "move" towards the destination specified by the strategy layer, "fire" to the target specified by the strategy layer, "stop", and "die", the final state only reached when an enemy has no health points. The "cfire" behaves like the "fire" state but makes the enemy shoot in a crouched position.

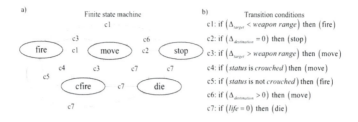

Fig. 3. a) Description of the finite state machine in the action layer. b) The set of transition conditions between states.

Figure 3b shows the set of transitions between the states of the FSM written in conditional rule form, in which the antecedent is the condition and the consequent is the arrival state. Every condition matches at least one arc in the graph. The input variables for the FSM are: the distance between the enemy and its destination ($\Delta_{destination}$), the distance between the enemy and its target (Δ_{target}), the *status* of the target (*crouched* or *stand up*) and the enemy's *life* measured in health points.

3   The Evolutionary System

The action layer is common to every enemy control system. Unlike this, every strategy layer has its own unique fuzzy rule set, making every enemy control system, and therefore, every enemy in the game behave differently under the same environmental conditions. Since fuzzy logic uses a human-like knowledge representation, hand coding different fuzzy rule sets for every enemy in the game seems a feasible task. Nevertheless, the automatic generation of fuzzy rule sets from scratch seems to be a more interesting goal to achieve.

For this purpose, a grammar-guided genetic program has been included in Genes of War. A grammar-guided genetic program is a system that is able to find solutions to any problem whose syntactic restrictions can be formally defined by a context-free grammar. The context-free grammar G_{FRBS} has been designed to generate the language composed by all the valid fuzzy rule sets that match the features required by Genes of War enemy control system. Using G_{FRBS}, a grammar-guided genetic program is able to automatically generate a population of fuzzy rule sets, and thus, a population of enemy control systems, being able to evolve them during the gameplay to create a set of enemies that autonomously adapt themselves to maximize player satisfaction. For more information about encoding FRBS in context-free grammars, please refer to [7].

At the beginning of the gameplay, an army of ten enemies is created. During the initialization step of the genetic program, a population composed by ten individuals is randomly generated. Each of them is assigned to only one enemy in the army. Each individual's genotype is a derivation tree belonging to G_{FRBS}, which codifies a fuzzy rule set. Every enemy decodes its assigned fuzzy rule set and stores it in the FRBS of its strategy layer, creating a control system that is different from any other else in the army.

3.1 Fitness Evaluation

A generation is defined as the time period that starts when the soldier is spawned in the game and ends when the soldier has died l times, that is, when the player has lost l lives. When a generation ends, every enemy in the army is assigned a score that measures the performance of its control system. This score is based on several values gathered from the gameplay during that generation. Given an enemy E_i, this score is calculated as $enemyScore(E_i) = e_1 \cdot Kills_i + e_2 \cdot Damage_i + e_3 \cdot Destroy_i + lifepoints_i$, where $Kills_i$ is the number of times that the soldier was beaten by E_i, $Damage_i$ is the damage dealt by E_i, $Destroy_i$ is the number of items destroyed by E_i, $lifepoints_i$ are the remaining health points of E_i, and e_1, e_2 and e_3 are adjustment constants.

An implicit approach to player satisfaction has been implemented, so it is assumed that satisfaction is achieved by matching the challenge preferred by the player with the one offered by the game. Due to this, the fitness of E_i is calculated as $Fitness(E_i) = |\sigma_{challenge} - enemyScore(E_i)|$, where $\sigma_{challenge}$ is a parameter whose value reflects a target challenge level. The grammar-guided genetic program in Genes of War offers two different methods to set the value of this parameter.

The first, called the "adaptive method", is based on the score obtained by the player during the generation. This score is calculated analogously to an enemy's score: $playerScore = s_1 \cdot Kills + s_2 \cdot Damage + s_3 \cdot Time + s_4 \cdot Life$, where $Kills$ is the number of enemies beaten by the soldier, $Damage$ is the damage dealt by the soldier measured in health points, $Time$ is the duration of the generation in milliseconds, $Life$ is the number of health points recovered by the soldier by picking up health power-ups, and s_1, s_2, s_3 and s_4 are adjustment constants. The value of $\sigma_{challenge}$ is equal to the average player's score obtained in the last five generations.

When using the "adaptive method", individuals in the population of the genetic program (and thus, the enemies in the army) are evolved to minimize the distance between their scores and the score obtained by the player at each generation.

The second method is called the "hardwired method" and allows the shape of the learning curve, that the player must face during gameplay, to be directly programed. This curve is defined by a function $C : \mathbb{N} \to \mathbb{Q}$, defined in such a way that, given the number of the actual generation g, $C(g)$ returns the value for $\sigma_{challenge}$ in that generation.

By using this method, enemies evolve to minimize the distance between their scores and the target score programmed in $C(g)$ for each generation.

3.2 Crossover and Replacement

After fitness evaluation, the ten enemies are sorted by fitness. The enemies ranked second to ninth are stored in a mating pool. Each member of the mating pool is then submitted to a crossover operation with the fittest enemy.

During a crossover operation, the genotype of both enemies is combined by means of the grammar-based crossover operator [3], creating one new genotype that encodes a new fuzzy rule set. The genotype of the non-fittest enemy is replaced by this new one.

The genotype of the tenth (the last) fittest enemy in the army is replaced by a randomly generated genotype in order to increase the exploration capability of the genetic program. The genotype of the fittest individual remains unchanged.

After crossover and replacement operations, the soldier is respawned with renewed health points and the next generation starts. All values used to calculate player's and enemies' scores are reset, as well as all scores and fitness measures.

4 Experimental Results

Two experiments have been run using Genes of War. In the first, the "adaptive method" is used to calculate the target challenge level. Two players, one experienced and one beginner, have been asked to play the game during 30 generations (g) with $l = 1$, that is, one generation lasts one player's life. Based on previous experimental analyses, the following values for the adjustment constants have been chosen: $e_1 = 100$, $e_2 = 8$, $e_3 = 30$, $s_1 = 100$, $s_2 = 0.125$, $s_3 = 0.001$, and $s_4 = 0.5$.

Figure 4a shows the results obtained from this first experiment by the experienced player. The bold line displays the values taken by $\sigma_{challenge}$ and the dotted line shows the evolution of the score obtained by the fittest enemy in the army during 30 generations. This score evolves accordingly to $\sigma_{challenge}$ in a way similar to a predator-prey system, in which the fittest enemy chases the player generation by generation.

The value of $\sigma_{challenge}$ increases until generation 12 because the experienced player can easily handle their task (beating enemies). Then, the genetic program evolves the population to produce smarter enemies that obtain higher scores. In generation twelve the enemies are very smart, and thus harder to beat, so the player's score decreases until generation 26. This causes the genetic program to remove smart enemies from the population and produce easier ones. After generation 26 the player is skilled enough to overtake the enemies. Consequently, $\sigma_{challenge}$ increases again and the genetic program starts producing smart enemies that raise the difficulty level.

Figure 4b shows the results obtained by the beginner player. Here, $\sigma_{challenge}$ takes low values because of player's lack of experience. Initial enemies are too smart, so the genetic program evolves the population to obtain easy-to-beat enemies that better fit player's capability. After generation 9 the player is skilled enough to overtake the enemies. This leads to an increase in $\sigma_{challenge}$ that is

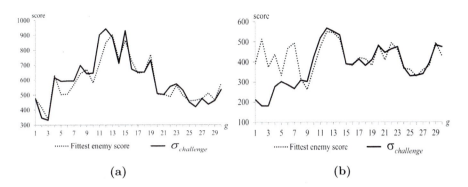

Fig. 4. Evolution of the fittest enemy using the "adaptive method" with a) an experienced player and b) a beginner player

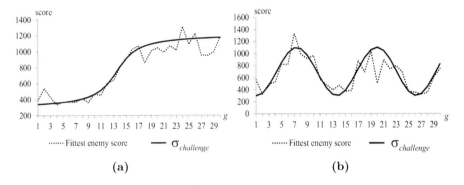

Fig. 5. Evolution of the fittest enemy using the hardwired method a) with function $C_1(g)$ b) and $C_2(g)$

followed by a predator-prey behavior like the one shown in Figure 4a, but with a smaller amplitude in this case.

In the second experiment, the hardwired method has been used to calculate the challenge level. Two different functions, $C_1(g) = 300 \cdot arctan(g/2,5 - 5) + 750$ and $C_2(g) = 400 \cdot cos(g/2 + 600) + 700$ have been programmed to determine the values taken by $\sigma_{challenge}$ given the generation number (g).

Figure 5a shows the results obtained from the application of $C_1(g)$. As it is shown by the bold line, this function shapes an ideal learning curve in which the challenge level increases slowly during the first stages of the gameplay, then it raises quickly by the middle game to finally return to a slow growth by the end of the game. The dotted line shows how the score of the fittest enemy closely evolves to the values assigned to $\sigma_{challenge}$ by this function.

Figure 5b shows the results obtained from the application of function $C_2(g)$, that represents an oscillatory behavior. This experiment shows how the genetic program is able to adapt the army of enemies to a variable environment, in which the target challenge level in constantly oscillating between two scores: 300 and 1100.

5   Conclusions and Future Work

The grammar-guided genetic program in Genes of War generates and evolves populations of enemies that match different target challenge levels to optimize player satisfaction. The difficulty in Genes of War can be adapted in real time to fit player skills, when using the "adaptive method", or to match programmers preferences, when using the "hardwired method".

By using this genetic program, no artificial behavior has to be implemented. If using the "hardwired method", programmers only have to set the target difficulty level desired at every stage of the gameplay, and armies of enemies are automatically generated to fit them. Using the "adaptive method" is even easier since the game is capable of finding the challenge level that better fits to every player at every moment. This is a key advantage because it lets programmers focus on any other aspects of game development.

In both cases, changes in game difficulty are only achieved by getting more or less intelligent enemies. Enemies "physical" attributes, like health, speed, endurance or strength, are never modified for this purpose. Smarter enemies develop intelligent behaviors that make them harder to beat. These behaviors include equipping different weapons depending on the situation, looking for health power-ups when an enemy is running low on health points, and running away from the soldier when it is too close.

Insofar as evolution is implemented as a continuous and unending task, the genetic program can operate during the whole gameplay without requiring to stop or restart the game.

The grammar-guided genetic program is a very flexible tool that can be easily modified to include any change made in later development phases. Changing the used context-free grammar makes the system capable of generating more complex fuzzy rule-based systems that deal with more environmental variables, fuzzy sets or membership functions.

The layered design of the presented control system provides it with modularity, granting that any changes performed at the strategy level will not prevent the whole system to work properly.

Due to its flexibility and modularity, it seems feasible to export the presented evolutionary system to other competitive game genres, like 3D third-person and first-person shooters, fighting and racing games.

References

1. Azaria, Y., Sipper, M.: Gp-gammon: Using genetic programming to evolve backgammon players. Genetic Programming, pp. 143–143 (2005)
2. Benbassat, A., Sipper, M.: Evolving board-game players with genetic programming. In: Proceedings of the 13th Annual Conference Companion on Genetic and Evolutionary Computation, pp. 739–742. ACM (2011)

3. Couchet, J., Manrique, D., Ríos, J., Rodríguez-Patón, A.: Crossover and mutation operators for grammar-guided genetic programming. Soft Computing: A Fusion of Foundations, Methodologies and Applications 11(10), 943–955 (2007)
4. Doull, A.: The death of the level designer,
   http://roguelikedeveloper.blogspot.com/2008/01/death-of-level
   -designer-procedural.html (last accessed November 2011)
5. Font, J.M., Manrique, D., Pascua, E.: Grammar-Guided Evolutionary Construction of Bayesian Networks. In: Ferrández, J.M., Álvarez Sánchez, J.R., de la Paz, F., Toledo, F.J. (eds.) IWINAC 2011, Part I. LNCS, vol. 6686, pp. 60–69. Springer, Heidelberg (2011)
6. Font, J.M., Manrique, D.: Grammar-guided evolutionary automatic system for autonomously building biological oscillators. In: 2010 IEEE Congress on Evolutionary Computation, pp. 1–7 (July 2010)
7. Font, J.M., Manrique, D., Ríos, J.: Evolutionary construction and adaptation of intelligent systems. Expert Systems with Applications 37, 7711–7720 (2010)
8. Hastings, E., Guha, R., Stanley, K.: Evolving content in the galactic arms race video game. In: IEEE Symposium on Computational Intelligence and Games, CIG 2009, pp. 241–248. IEEE (2009)
9. Loiacono, D., Cardamone, L., Lanzi, P.: Automatic track generation for high-end racing games using evolutionary computation. IEEE Transactions on Computational Intelligence and AI in Games 3(3), 245–259 (2011)
10. Lucas, S.: Computational intelligence and games: Challenges and opportunities. International Journal of Automation and Computing 5(1), 45–57 (2008)
11. Pedersen, C., Togelius, J., Yannakakis, G.: Modeling player experience in super mario bros. In: IEEE Symposium on Computational Intelligence and Games, CIG 2009, pp. 132–139. IEEE (2009)
12. Shichel, Y., Ziserman, E., Sipper, M.: Gp-robocode: Using genetic programming to evolve robocode players. Genetic Programming, pp. 143–143 (2005)
13. Stanley, K., Bryant, B., Miikkulainen, R.: Real-time neuroevolution in the nero video game. IEEE Transactions on Evolutionary Computation 9(6), 653–668 (2005)
14. Togelius, J., De Nardi, R., Lucas, S.: Towards automatic personalised content creation for racing games. In: IEEE Symposium on Computational Intelligence and Games, CIG 2007, pp. 252–259. IEEE (2007)
15. Togelius, J., Whitehead, J., Bidarra, R.: Guest editorial: Procedural content generation in games. IEEE Transactions on Computational Intelligence and AI in Games 3, 169–171 (2011)
16. Togelius, J., Yannakakis, G., Stanley, K., Browne, C.: Search-Based Procedural Content Generation. In: Di Chio, C., Cagnoni, S., Cotta, C., Ebner, M., Ekárt, A., Esparcia-Alcazar, A.I., Goh, C.-K., Merelo, J.J., Neri, F., Preuß, M., Togelius, J., Yannakakis, G.N. (eds.) EvoApplicatons 2010. LNCS, vol. 6024, pp. 141–150. Springer, Heidelberg (2010)
17. Yannakakis, G., Hallam, J.: Real-time game adaptation for optimizing player satisfaction. IEEE Transactions on Computational Intelligence and AI in Games 1(2), 121–133 (2009)

Why Simulate? Hybrid Biological-Digital Games

Maarten H. Lamers[1] and Wim van Eck[1,2]

[1] Media Technology Research Group, Leiden Institute of Advanced Computer Science,
Leiden University, The Netherlands
[2] AR Lab, Royal Academy of Art, The Hague, The Netherlands
lamers@liacs.nl, mail@wimeck.com

Abstract. Biologically inspired algorithms (neural networks, evolutionary computation, swarm intelligence, etcetera) are commonly applied in development of digital games. We argue that there are opportunities and possibilities for integrating real biological organisms inside computer games, with potential added value to the game's player, developer and integrated organism. In this approach, live organisms are an integral part of digital gaming technology or player experience.

To spark further thought and research into the concept of hybrid biological-digital games, we present an overview of its opportunities for creating computer games. Opportunities are categorized by their mainly affected stakeholder: game player, game designer, and bio-digital integrated organism. We clarify the categorization via numerous examples of existing hybrid bio-digital games. Based on our review work we present conclusions about the current state and future outlook for hybrid bio-digital games.

Keywords: Hybrid, bio-digital, computer games, animals, biological organisms, biotic games, bio-art.

1 Introduction

We have become used to applying biologically inspired algorithms in the development and design of computer games. Neural and evolutionary computation, swarm systems and other forms of self-organization, and various other biologically inspired techniques are now fairly commonplace in both science and computer games.

In a way, such techniques are simulations of biological processes, either abstracted, altered or enhanced to fit a particular purpose within the process of game design and development. We argue, however, that there are opportunities and possibilities for integrating *real* biological systems inside computer games, with potential added value to the game player, game developer and organism. With real biological systems we refer to plants, microorganisms, animals, and even complete ecosystems.

Using biological systems and components as part of (digital) technological solutions is applied in practice already. Particularly within the realms of robotics (e.g. [4], [13], [14], [15]) and artistic- and entertainment computing (e.g. [5], [9], [12]), but also more general engineering (e.g. [1], [19], [20], [23]) examples exist of biological systems that are intricately entwined within mainly technological systems. Perhaps the

C. Di Chio et al. (Eds.): EvoApplications 2012, LNCS 7248, pp. 214–223, 2012.

most famous example of such interactions is the Second World War project by famous behaviorist B.F. Skinner [26], in which he successfully trained real pigeons to guide missiles to strategic targets, for lack of an equally reliable technological solution for this task. Although this famous endeavor does not relate to computer games, it demonstrates how integration of a biological system solves a problem to which no technological solution is yet available.

More recently animal-computer interaction has appeared on the scientific agenda, a field analogous to that of human-computer interaction. It places the animal perspective at the heart technological development aimed at animals [17].

Naturally, very real disadvantages exist for integrating biological components into digital computer games. Practical issues arise in game maintenance, shipping, sales, and more. Dealing with microorganisms and cells may require a well-balanced biochemical environment. Games requiring exotic or trained animals are naturally commercially unattractive, and possibly illegal due to endangered animal protection laws.

Perhaps most importantly, animal welfare must be respected at all times. Ethical treatment of animals is a major concern within the bio-digital approach to systems design. Several commercial arcade crane-games exist[1] that challenge players to catch a live lobster from an aquarium with the aid of an electrically operated claw. Such games were criticized for causing harm to animals [22].

Legal restrictions have been implemented in many countries to fight maltreatment of animals for purposes of science, commerce, entertainment and otherwise. We strongly support and respect such restrictions. Moreover, we point out that within the emerging and dynamic realm of bio-digital systems particular concern is in place regarding the ethical treatment of animals and other organisms.

Despite obvious disadvantages, let us view the opportunities of this approach. To spark further investigation and creativity around the topic of hybrid bio-digital computer games, we discuss these opportunities for the three major stakeholders of this approach to computer game design:

1. the player,
2. the developer,
3. the organism integrated into the computer game system.

As the developer, we mean anyone with an interest in designing and developing games for any purpose (e.g. commercial gain or scientific data gathering). Interests of player and developer may overlap, since what is good for the player should be at least appealing to developers. Similarly, interests of organisms and players may overlap.

We clarify the proposed opportunities with examples of existing integrations of real biological systems within computer games. For the sake of brevity we do not discuss examples from the realm of artistic computing, unless they take the form of a computer game. Bio-art, the artistic discipline that works with real biological systems,

---

[1] E.g. "The Lobster Zone" (Lobster Zone Inc., USA), "Love Maine Lobster Claw" (Marine Ecological Habitats Inc., USA), and "Sub Marine Catcher" (unknown manufacturer, Japan).

is currently receiving much attention[2]. Overlap exists between bio-art and hybrid bio-digital computer games, and concerns regarding ethics of working with organisms are shared between disciplines.

Also, games exist that integrate real plants into their digital systems (e.g. [29], [32]). Although these are highly interesting and involve biological components, they are not discussed here. We choose to focus on more active biological entities, such as animals, microorganisms, and neurons.

Moreover, our attention is directed towards the integration of organisms within *digital* games only. Outside this scope falls for example "BioPong" [25], a Pong-style arcade cabinet game in which the pixelated ball is in fact a cardboard square attached to a real cockroach. Players must prevent the cockroach from passing their physical Pong-paddle. Although BioPong emulates a classic computer game, it is in fact not a digital game itself.

For a more widely scoped overview of organisms integrated in digital artistic and entertainment systems we refer to another publication by the authors [10]. This study focuses on opportunities for hybrid bio-digital computer games.

2   Opportunities for Players

2.1   Interspecies Awareness

For the benefit of the player, an important distinction that can be made is whether or not one is aware of the biological organism's role within the game. In particular with real-time integration of animal behavior in the game, the realization that one plays against (or with) an animal may change the player's view. In a sense this distinction is comparable to that between playing against a simulated or a live opponent, in that it may affect aspects of competition, willingness to collaborate, and even endurance.

Stephen Wilson's "Protozoa Games" [30] is a series of digital games that let humans compete against a variety of protozoa, single-celled parasitic organisms. Human players are tracked with a camera; the protozoa are recorded inside a petri dish via a microscope. In several games humans and protozoa can compete in agility. With lights and audio picked up via microphones, players can attempt to influence the protozoa as part of the gameplay. Similarly, in Wilson's "IntroSpection" installation [31], players can play games with projections of cells taken directly from their own mouths and placed under a microscope.

2.2   Enabling Care

The care-relationship that exists between a pet and its owner was extensively exploited in computer games. Tamagotchi (Bandai Co. Ltd., Japan, 1996), Nintendogs (Nintendo Co. Ltd., Japan, 2005), and many similar games offer the enjoyment of

---

2   For a list and discussions of various bio-artworks, we recommend the following webpage: http://www.we-make-money-not-art.com/archives/bioart/

caring for a virtual pet. Similarly, digital games can mediate between caregiver and real pet. In situations where physically interacting with a real pet is not possible, digital interaction may be a useful alternative. Moreover, such interaction makes multiple caregivers and a more transient pet-caregiver relationship possible.

In analogy of the once popular Tamagotchi digital pet game, the "Tardigotchi" artwork [16] houses a living tardigrade organism inside a brass sphere. While the organism can be seen through a viewing hole, its digital caricature is visualized on an LED screen. Buttons feed both the digital and real creature, and players can activate a heating lamp for the tardigrade by sending the digital creature an e-mail message.

The "Cat Cat Revolution" [21] interspecies computer game (cat-human) attempts to include pets into their owner's digital gaming experience. The cat owner controls a virtual mouse via his/her smartphone. The virtual mouse appears on a tablet-computer, where it can be chased by the cat. The game detects when the cat's paw hits the virtual mouse. Users expressed positive feelings towards the game's role in pet-owner daily activities and the pet's well-being and freedom to play.

2.3   Education

Traditionally, games (both digital and non-digital) have been widely applied as educational tools. By extension, dealing with real biological systems via digital games can educate about biology, offering realizations and interactions that would not be available through simulation.

The only projects that mention educating the player as a *specific* purpose are Stephen Wilson's aforementioned "Protozoa Games" [30] and "IntroSpection" [31] installations. However, it holds true for many other existing projects that through awareness and curiosity about the biological component, informal learning may be expected to occur.

2.4   Behavioral Variability

Even when a player is unaware of the interaction with biological organisms, its effects may nonetheless be relevant. For example, the behavior of virtual characters may be steered by real-time behavior of organisms. In this way behavioral models can be surpassed, potentially leading to more natural and unpredictable behavior.

In our earlier study entitled "Animal Controlled Computer Games: Playing Pac-Man against Real Crickets" [11] we researched the possible use of live animals for real-time behavior generation in computer games. In a Pac-Man style video game, the behavior of virtual opponent characters (ghosts) was derived from that of live crickets inside a real maze (Figure 1). The location of the player-controlled Pac-Man character within the virtual game was translated to the real maze via vibrations. Use of real crickets led to unexpected and interesting behavior of the virtual game characters.

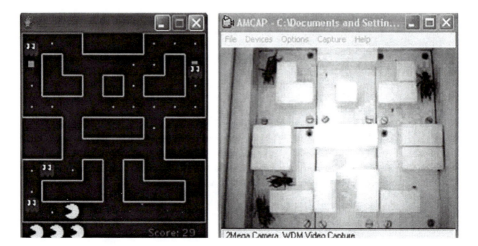

Fig. 1. Screenshot of Pac-Man style game (*left*) with four "ghost" opponent characters, and camera capture of the physical facsimile maze (*right*) containing four live crickets (Images from [11]). The crickets control the ghosts in real-time. The position of the player-controlled "Pac-Man" character is translated into vibrations within the physical maze.

3   Opportunities for Developers

3.1   Perception and Processing

As demonstrated in B.F. Skinner's "Project Pigeon" [26], complex control issues can be solved by using natural sensing and processing capabilities of organisms. The same concept was demonstrated in a digital environment by Garnet Hertz's "Cockroach Controlled Mobile Robot" artistic project [13]. Similarly, designing complex (behavioral) control models in games could be avoided by stimulation and sensing of real organisms in a suitable environment.

Using a grid of 60 electrodes, DeMarse and Dockendorf [8] stimulated and recorded the activity from a network of 25.000 rat neurons. This network was connected to a consumer flight simulator, and successfully trained to act as an autopilot. By stimulating the neurons with information about deviation from level flight, they slowly learned to control the flight simulator's pitch and roll, until they were able to maintain straight and level flight.

3.2   Crowdsourcing and Gamification

Given the ongoing scientific interest in biological systems, it is not surprising that interest has been expressed to apply hybrid bio-digital games for scientific data gathering. Moreover gaming can offer a platform and community for crowdsourcing projects. Applying games for crowdsourcing is closely linked to the strongly emerging paradigm of *gamification*, the use of gaming elements to engage potential audiences in various tasks [18], such as problem solving, data collection, and learning.

Riedel-Kruse *et al.* [24] incorporated biological processes of real microorganisms (not simulations) into several variations of classic game titles as Pac-Man, Pong and pinball. To achieve their game-tasks, players control in real-time microscopically observed paramecium organisms by way of electrical fields or chemicals released from a micro-needle. As such, biological processes are an integral part of the gameplay – a concept they term "biotic games".

Based on their successful implementations, the authors propose to realize complex bio-engineering tasks by applying crowdsourcing mechanics to biotic games [24, p.19]. Essentially, by playing the biotic games players would perform experiments on actual living biological matter, and thus contribute to solving scientific problems. Unfortunately, no concrete bio-digital crowdsourcing was yet realized.

3.3   Organic Design

Perhaps slightly more experimental are efforts to employ organisms for level-design of games. Through the process of self-organization, groups of microorganisms and cells (and even larger organisms) can collaboratively create structured spatial patterns [7]. Think of honeycomb patterns in a beehive, and zebra skin stripes generated by collaborating pigment cells. Such evolving structures and patterns could be used to create or dynamically grow intricate landscapes, structures and levels for digital games.

In a recently started project, Wim van Eck aims to derive virtual worlds in real-time from microscopically observed living materials, such as growing cellular cultures or fungi. Visitors of these virtual worlds are confronted with their constantly changing and transient nature, reflecting the process of organic growth. By concurrently changing the conditions to which the organism is subjected, its growth dynamics can be affected, and consequently the virtual landscape. The virtual worlds and interactions designed in this project will be applied to gaming applications, yielding organic and transient level design.[3]

4   Opportunities for Animals

4.1   Welfare and New Forms of Care

As mentioned earlier, the animal-caregiver relation can be focus of bio-integrated computer games. Although animal welfare should come naturally for those providing care, computer games can aid in several ways. Firstly, mutual games that are typically shared between pet and owner can be played over greater distance via tele-operating methods. This makes it possible to maintain levels of care in physical absence. Secondly, via hosted games multiple players can act as caregiver for a single animal.

---

[3]   For information about this project, contact author Wim van Eck.

Remote pet-owner interaction is exactly what the "Metazoa Ludens" project [27] [28] pursues. The position of a hamster or other small pet inside a closed environment is tracked, and represented as an avatar inside the player's computer game. Simultaneously, the pet is tempted to chase moving bait inside its enclosure – the bait's movement representing the movement of the player's avatar inside the game. Even the terrain inside the pet's enclosure is manipulated by actuators to mimic the virtual terrain of the computer game.

According to the creators, their setup enables human-pet interaction "on an equal level in the virtual world (which is impossible in the physical world)" [28, p.308]. From studies using hamsters, it was furthermore reported that regular play in Metazoa Ludens increased overall body fitness in the hamsters and that over the study period they increasingly chose to play, which indicates a positive desire to play the game.

4.2   Fighting Stereotypy

A special case for animal welfare deserves attention here, since interspecies computer games are a very real solution to the problem of animal *stereotypy*: repetitive behaviors in captive animals caused by inadequate mental stimulation [6]. Computer games that let human players interact with captive animals have been proposed, and are currently researched, as a method to enhance mental stimulation for the animal. Naturally, actions directed towards the animal must be bound along multiple parameters, such as type of interaction, intensity and time.

In the Netherlands captive pigs have the legal right to be provided with a toy. However, in practice the toys provided are insufficient for adequate mental stimulation. In April 2011, the Dutch Cultural Media Fund awarded a grant to the "Playing with Pigs" project [2]. This research project plans to develop a tablet-computer based game entitled "Pig Chase" [3] that lets users play with captive pigs in an effort to provide pigs with more fitting mental stimulation (Figure 2). Further goals are to design new forms of human-pig interaction, to study opportunities for new human-pig relations, and to let both species experience the cognitive capabilities of each other.

5   Discussion and Future Outlook

We have presented an overview of opportunities of the hybrid bio-digital approach to creating computer games. In this approach, live organisms are an integral part of digital gaming technology or player experience. The overview categorizes opportunities by their mainly affected stakeholder: game player, game designer, and bio-integrated animal. Overlap in these categories exists, and perhaps further opportunities were overlooked. Also, project examples that were mentioned to illustrate a particular opportunity, may have well applied to other opportunities also. Nonetheless we view this endeavor as a successful attempt to organize what hybrid bio-digital work is relevant to the domain of computer games.

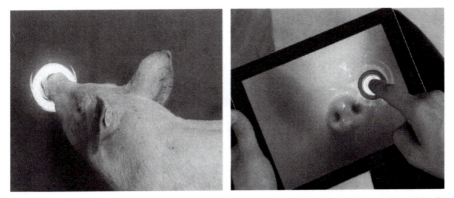

Fig. 2. Interaction modes within the inter-species game "Pig Chase": pig interacting with virtual object on a large touch sensitive display (*left*), and human interacting with tablet-computer (*right*) to control the virtual object and view the pig interaction (Images from [3]). The game objective for the human is to guide the pig's snout to an on-screen target, whereas successful actions by the pig trigger colorful visual projections on the large display.

Overseeing the reviewed projects, several observations arise. Firstly, without exception, all projects can be classified as experimental, exploratory and often artful endeavors. No truly commercial applications in computer game design were realized, nor were indications given for any commercial use.

Secondly, the applications of biological components in digital games have thus far been very diverse with respect to their aims. From controlling part of a flight simulator to growing virtual gaming environments, organic components have been applied for a plethora of purposes. No single "killer app", or highly desirable use of biological systems in digital gaming, stands out from the overview.

Thirdly, no paradigms have been put forward that offer handlebars to hybrid bio-digital games design. The number of working methods and working hypotheses in the reviewed projects seems to rival the number of projects itself. None of the proposed working methods was developed to the extent that it can be named a possible paradigm.

Fourthly, it is yet to be discovered which of the proposed bio-digital solutions (be it neurons to solve a multivariate control task, or growing cell cultures to design game worlds) will outperform their fully technological counterparts. In other words, when do we need organisms, and when does technology suffice? With regards to this question, we present the reviewed projects with an open mind, but acknowledge that modern technology is not only more practical, but often sufficiently potent.

Fifthly, given the above observation, it appears that most interesting opportunities benefit players and animals, and to a lesser extent game engineering. Therefore, we expect future advances in hybrid bio-digital game development to be stronger on the game-conceptual front, than in terms of engineering success.

All this is not to say that the bio-digital approach to computer gaming has a limited future outlook. Perhaps some proposed ideas will not advance to become wider applied, while others may open doors to gaming concepts that are yet to be envisioned.

Perhaps behavioral variability of in-game characters cannot be simulated to the extent desired by players, necessitating use of real organisms. Perhaps inter-species gaming will develop to become popular in its own right. One approach that in our view stands out in terms of future applicability and potential benefit to both animal welfare and innovative gameplay, is exemplified in the project "Playing with Pigs" [3].

Finally, we trust that our contribution, in the form of this combined position paper and short review, sparks further thought and research into the concept of hybrid biological-digital games.

References

1. Adamatzky, A.: Physarum Machines: Computers from Slime Mould. World Scientific Books (2010)
2. Alfrink, K., van Peer, I., Rengen, W.J.: Spelen met Varkens. Dutch Cultural Media Fund E-Culture grant #83178 (2011),
   http://www.mediafonds.nl/toekenning/83178/
3. Alfrink, K., van Peer, I., Lagerweij, H., Driessen, C., Bracke, M.: Playing with Pigs (2012), http://www.playingwithpigs.nl
4. Bakkum, D.J., Shkolnik, A.C., Ben-Ary, G., Gamblen, P., DeMarse, T.B., Potter, S.M.: Removing Some 'A' from AI: Embodied Cultured Networks. In: Iida, F., Pfeifer, R., Steels, L., Kuniyoshi, Y. (eds.) Embodied Artificial Intelligence. LNCS (LNAI), vol. 3139, pp. 130–145. Springer, Heidelberg (2004)
5. Bakkum, D.J., Gamblen, P.M., Ben-Ary, G., Chao, Z.C., Potter, S.M.: MEART: The Semi-living Artist. Frontiers in NeuroRobotics 1(5), 1–10 (2007)
6. Bolhuis, J.J., Giraldeau, L.-A.: The Behavior of Animals: Mechanisms, Function, and Evolution. Wiley-Blackwell (2005)
7. Camazine, S., Deneubourg, J.-L., Franks, N.R., Sneyd, J., Theraulaz, G., Bonabeau, E.: Self-Organization in Biological Systems. Princeton University Press (2003)
8. DeMarse, T., Dockendorf, K.P.: Adaptive Flight Control with Living Neuronal Networks on Microelectrode Arrays. In: Proceedings of the International Joint Conference on Neural Networks, pp. 1548–1551 (2005)
9. Easterly, D.: Bio-Fi: Inverse Biotelemetry Projects. In: 12th ACM International Conference on Multimedia, pp. 182–183 (2004)
10. van Eck, W., Lamers, M.H.: Hybrid Biological-Digital Systems in Artistic and Entertainment Computing. To appear in Leonardo, vol. 45. MIT Press (2012)
11. van Eck, W., Lamers, M.H.: Animal Controlled Computer Games: Playing Pac-Man Against Real Crickets. In: Harper, R., Rauterberg, M., Combetto, M. (eds.) ICEC 2006. LNCS, vol. 4161, pp. 31–36. Springer, Heidelberg (2006)
12. Studio EDHV: Debug (2009), http://www.edhv.nl
13. Hertz, G.: Control and Communication in the Animal and the Machine. Master's Thesis, University of California Irvine (2004)
14. Holzer, R., Shimoyama, I.: Locomotion Control of a Bio-robotic System via Electric Stimulation. In: IEEE/RSJ International Conference on Intelligent Robots and Systems (IROS 1997), pp. 1514–1519 (1997)
15. Jones, J., Tsuda, S., Adamatzky, A.: Towards *Physarum* Robots. In: Meng, Y., Jin, Y. (eds.) Bio-Inspired Self-Organizing Robotic Systems. SCI, vol. 355, pp. 215–251. Springer, Heidelberg (2011)

16. Kenyon, M., Easterly, D., Rorke, T.: Tardigotchi (2009),
http://www.tardigotchi.com
17. Mancini, C.: Animal-Computer Interaction: a Manifesto. Interactions 18(4), 69–73 (2011)
18. McGonigal, J.: Reality Is Broken: Why Games Make Us Better and How They Can Change the World. The Penguin Press, New York (2011)
19. Moar, P., Guthrie, P.: Biocomponents – Bringing Life to Engineering. Ingenia 27, 24–30 (2006)
20. Nakagaki, T., Yamada, H., Tóth, Á.: Maze-solving by an Amoeboid Organism. Nature 407, 470 (2000)
21. Noz, F., An, J.: Cat Cat Revolution: An Interspecies Gaming Experience. In: proceedings of ACM Conference on Computer Human Interaction, Vancouver, pp. 2661–2664 (2011)
22. PETA: Lobster Zone 'Games' Cause Pain and Suffering (2010),
http://www.peta.org/features/Lobster-Zone.aspx
23. Pickering, A.: Beyond Design: Cybernetics, Biological Computers and Hylozoism. Synthese 168, 469–491 (2009)
24. Riedel-Kruse, I.H., Chung, A.M., Dura, B., Hamilton, A.L., Lee, B.C.: Design, Engineering and Utility of Biotic Games. Lab on a Chip 11, 14–22 (2011)
25. Savičić, G.: BioPong (2005),
http://www.yugo.at/processing/?what=biopong
26. Skinner, B.F.: Pigeons in a Pelican. American Psychologist 15, 28–37 (1960)
27. Tan, R.K.C., Cheok, A.D., James, K.S.: The: Mixed Reality Environment for Playing Computer Games with Pets. International Journal of Virtual Reality 5(3), 53–58 (2006)
28. Tan, R.K.C., et al.: MetazoaLudens: Mixed Reality Interactions and Play for Small Pets and Humans. Leonardo 41(3), 308–309 (2008)
29. Vermeulen, A.: Biomodd (2007), http://www.biomodd.net
30. Wilson, S.: Protozoa Games (2003),
http://userwww.sfsu.edu/~wilson/art/protozoagames/protogames10.html
31. Wilson, S.: IntroSpection (2005),
http://userwww.sfsu.edu/~swilson/art/guests/guests.html
32. Young, D.: Lumberjacked (2005),
http://www.newdigitalart.co.uk/lumberjacked

Spicing Up Map Generation

Tobias Mahlmann, Julian Togelius, and Georgios N. Yannakakis

IT University of Copenhagen, Rued Langaards Vej 7, 2300 Copenhagen, Denmark
{tmah,juto,yannakakis}@itu.dk

Abstract. We describe a search-based map generator for the classic real-time strategy game *Dune 2*. The generator is capable of creating playable maps in seconds, which can be used with a partial recreation of Dune 2 that has been implemented using the Strategy Game Description Language. Map genotypes are represented as low-resolution matrices, which are then converted to higher-resolution maps through a stochastic process involving cellular automata. Map phenotypes are evaluated using a set of heuristics based on the gameplay requirements of Dune 2.

1 Introduction

Procedural Content Generation (PCG) for Games is a field of growing interest among game developers and academic game researchers alike. It addresses the algorithmic creation of new game content. Game content normally refers to weapons, textures, levels or stories etc. and may — to distinguish PCG from other fields of research — exclude any aspect connected to agent behaviour, although generating behavioural policies might be considered PCG in some contexts. One particular approach to PCG which has gained traction in recent years is the *search-based* paradigm, where evolutionary algorithms or other stochastic optimisation algorithms are used to search spaces of game content for content artefacts that satisfy gameplay criteria [14]. In search-based PCG, two of the main concerns are how this content is represented and how it is evaluated (the fitness function). The key to effective content generation is largely to find a combination of representation and evaluation such that the search mechanism quickly zooms in on regions of interesting, suitable and diverse content.

We are addressing the problem of map generation, in particular the generation of maps for a strategy game. A "map" is here taken to mean a two-dimensional spatial structure (though maps for some other types of games might be three-dimensional) on which objects or features of some kind (e.g. trees, tanks, mountains, oil wells, bases) are placed and on which gameplay takes place. While the generation of terrains without particular reference to gameplay properties is a fairly well-studied problem [13,5,6,3,2], a smaller body of work has addressed the problem of generating maps such that the maps support the game mechanics of a particular game or game genre.

One example of the latter is the cave generator by Johnson et al. [8], which generates smooth two-dimensional cave layouts, that support the particular design needs of a two-dimensional abusive endless dungeon crawler game. This basic

C. Di Chio et al. (Eds.): EvoApplications 2012, LNCS 7248, pp. 224–233, 2012.
© Springer-Verlag Berlin Heidelberg 2012

principle of that generator is to randomly sprinkle "rock" and "ground" on an open arena, and then use cellular automata (CA) to "smelt the rock together" in several steps, after which another heuristic ensures that rooms are connected to each other. While the resulting generator is fast enough for on-the-fly generation and generates natural-looking and adequately functional structures, the CA-based method lacks controllability and could not easily be adapted to generate maps that satisfy other functional constraints (e.g. reachability).

Another example is the search-based map generator for the real-time strategy game *StarCraft* by Togelius et al. [15]. Recognising that devising a single good evaluation function for something as complex as a strategy game map is anything but easy, the authors defined a handful of functions, mostly based on distance and path calculations, and used multi-objective evolutionary algorithms to study the interplay and partial conflict between these evaluation dimensions. While providing insight into the complex design choices for such maps, it resulted in a computationally expensive map generation process and problems with finding maps that are "good enough" in all relevant dimensions. The map representation is a combination of direct (positions of bases and resources) and indirect (a turtle-graphics-like representation for rock formations), with mixed results in terms of evolvability.

We propose a new search-based method for generating maps that draws heavily on the two very different approaches described above. Like in the StarCraft example, we use an evolutionary algorithm to search for maps and a collection of heuristics derived from an analysis of the game's mechanics to evaluate them. The embryogeny is borrowed from the cave generator. The transformation from genotype (which is evolved) to phenotype (which is evaluated) is happening through a process of sprinkling and smelting trough cellular automata. These steps will be described in some detail below. Our results show that this process effectively generates maps that look good and satisfy the specifications. The target game in this paper is *Dune 2*, which has the advantage of being in several respects simpler than StarCraft, which makes it easier to craft heuristic evaluation functions based on its mechanics, and also makes it easier to re-implement it in our own strategy game modelling framework for validating the results.

This paper is an integral part of the Strategy Games Description Language (SGDL) project at IT University of Copenhagen. SGDL is an initiative to model game mechanics of strategy games. Our previous work consisted of evolving heterogeneous unit sets [10], different approximations of game play quality [9], and general purpose agents for strategy games [12]. The re-creation of Dune 2 as a turn-based strategy game is a continuation of this research. An example map, created by the generator described in this paper, loaded into the SGDL game engine can be seen in Figure 2.

2   Background

Dune 2 (Westwood 1992) is one of the earliest examples of real-time strategy games, and came to strongly influence this nascent genre. The game is loosely based on Frank Herbert's Dune [7] but introduces new plots and acting parties. The player takes the role of a commander of one of three dynasties competing in the production of "spice", a substance that can only be gathered on the desert planet "Arrakis", also known as "Dune". In the dune universe, spice is required for inter-stellar travel, making it one of the most valuable substance in the universe. Dune 2 simplifies this relation slightly, treating spice as a resource which can be used to build new units and buildings. The only way to gain spice is sending harvester units to the sand parts of the map, where the spice is located. Apart from opposing parties that try to harvest the same fields, the sand parts are also habited by the native animals of the planets: the sandworms. Menace and important resource alike, these non-controllable units are involved in the generation of new spice on the map, but also occasionally swallowing units of the player - or his enemies if he uses the sand as a tactical element.

The main objective of the player on each map is to harvest spice and use the gathered resources to build new buildings and produce military units to ultimately destroying one or two enemies' bases. As mentioned, compared to modern real-time strategy games the game is rather simple: there is only one resource, two terrain types and no goals beside eliminating the enemies' forces. The two terrain types are "rocky terrain" and "sand", and both can be passed by all units. Two game mechanics involve the terrain types: buildings can only be constructed on rocky terrain, and spice and sandworms can only exist on sand. For completeness it should be mentioned that the game also contains cliffs that are only passable by infantry, but those have negligible effect on gameplay. Although the game does not contain any mechanics to model research, buildings and units are ordered in tiers. As the single player campaign progresses, the game simply unlocks additional tiers as the story progresses. This removes the necessity to model additional mechanics. An exemplary screenshot of the original game can be seen in Figure 1.

3   Map Generator

The map generator consists of two parts: the genotype-to-phenotype mapping and the search-based framework that evolves the maps. The genotypes are vectors of real numbers, which serve as inputs for a process that converts them to phenotypes, i.e. complete maps, before they are evaluated. The genotype-to-phenotype mapping can also be seen as, and used as, a (constructive) map generator in its own right. (The relationship between content generators at different levels, where one content generator can be used as a component of another, is discussed further in [14].)

The genotype-to-phenotype mapping is a constructive algorithm that takes an input as described in the following and produces an output matrix o. Based on tile types of the original Dune 2, the elements of o can assume the value $0 = $ SAND, $1 = $ ROCK, and $2 = $ SPICE. The matrix o is then later interpreted by a game engine into an actual game map. Our current implementation contains only an SGDL backend, but using an open source remake of the game and its

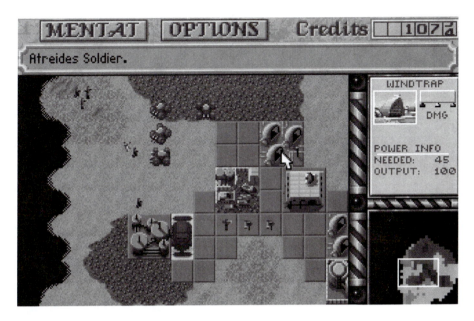

Fig. 1. Screenshot from the original Dune 2 showing the player's base with several buildings, units, and two spice fields in direct proximity

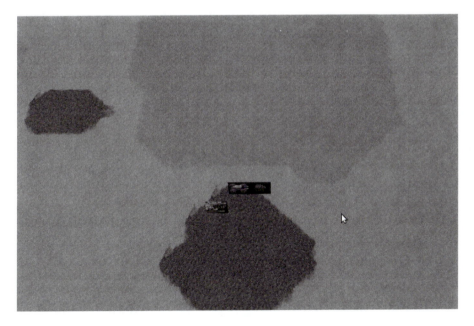

Fig. 2. A Dune 2 map loaded into the SGDL Game Engine. Terrain and unit textures are taken from the original asset set, but actors are, due to the lack of 3D models, placed as billboards into the game world.

tools (e.g. Dune II The Maker [1]) should make creating maps for the original Dune 2 easy.

The input vector is structured as followed ($mapSize$ refers to the map's edge length):

- n the size of the Moore-neighbourhood $[1, \frac{mapSize}{2}]$
- n_t the Moore-neighbourhood threshold $[2, mapSize]$
- i the number of iterations for the CA $[1, 5]$
- $w_{00}..w_{99}$ members the weight matrix w for the initial noise map $[0, 1]$
- s the number of spice blooms to be placed on the map $[1, 10]$

The generator starts with creating the initial map based on the values w. The $10x10$ matrix is scaled to the actual map size and used as an overlay to determine the probability of a map tile starting as rock or sand. For each iteration i_n a CA is invoked for each map tile to determine its new type. If the number of rock tiles in the n-Moore-Neighbourhood is greater or equal than n_t the tile is set to *Rock* in the next iteration.

The next step is the determination of the start zones, where the players' first building will be placed. We always use the largest rock area available as the starting zones. The selection is done by invoking a 2D variant of Kadane's algorithm [4] on o to find the largest sub-matrix containing ones. To prevent players from spawning too close to each other, we invoke Kadane's algorithm on a sub-matrix of o that only represents the i top rows of o for one player, and only the i bottom rows for the other player. We let i run from 8 to 2 until suitable positions for both players are found. This operation ensures that one player starts in the upper half of the map and one in the lower. It also restricts us to maps that are played vertically, but this could be changed very easily. At this step we don't assert that the start positions are valid in terms of gameplay. Broken maps are eliminated through the fitness functions and the selection mechanism.

The last step is the placement of the spice blooms and filling their surrounding areas. Since Kadane's algorithm finds sub-matrices of ones, we simply clone o and negate its elements with $o_{nm} = 1 - o_{nm}$; whereas o_{nm} is the m-th member of the n-th row of o. We use the computed coordinates to fill the corresponding elements in o with spice. In order to make the fields look a bit more organic, we use a simple quadratic falloff function: a tile is marked as spice if its distance d from the center of the spice field (the bloom) fulfils the condition $\frac{1}{d^2} \geq t$. Where t is the width of the spice field multiplied by 0.001. We created a simple frontend application to test the generator. A screenshot with a basic description can be seen in Figure 3.

The genetic algorithm optimises a genome in the shape of a vector of real-numbers, using a fitness function we created. Since a desert is very flat, there exists almost no impassable terrain, hence choke points (as introduced in [15]) is not a useful fitness measure. The challenge of choke points was instead replaced by the assumption that passing sand terrain can be rather dangerous due to sandworms. Furthermore, it should be ensured that both players have an equally sized starting (rock) zone and the distance to the nearest spice bloom should be

Options

Width x Height:
64   64

Iterations/Evolutions:
2

Population:
200

Moore neighborhood:
15

Neighborhood threshhold:
130

Spice blooms:
7

Iterate

Reset

Evolve

Find start points

The spice must flow!

Weight Matrix

0,67	0,69	0,52	0,41	0,08	0,39	0,86	0,9	0,85	0,69
0,59	0,07	0,53	0,64	0,05	0,38	0,68	0,34	0,29	0,94
0,51	0,17	0,21	1	0,54	0,24	0,13	0,83	0,14	0,75
0,21	0,59	0,34	0,47	0,06	0,56	0,33	0,48	0,57	0,69
0,34	0,81	0,75	0,56	0,47	0,34	0,07	0,38	0,17	0,77
0,33	0,63	0,56	0,32	0,46	0,71	0,55	0,04	0,41	0,71
0,69	0,22	0,06	0,3	0,39	0,42	0,07	0,17	0,71	0,18
0,64	0,19	0,03	0,51	0,51	0,69	0,66	0,7	0,49	0,18
0,94	0,14	0,94	0,02	0,96	0,2	0,07	0,51	0,96	0,96
0,66	0,09	0,31	0,55	0,45	0,6	0,3	0,18	0,88	0,8

Nullify   Oneify   Randomize

Fig. 3. Screenshot of the generator application. The right pane lets the user input a seed matrix directly, or observe the result of the evolution. The middle pane can be used to either invoke the generator directly ("Iterate") or start the non-interactive evolution ("Evolve"). The other buttons allow the user to go through the map generation step-by-step. The left pane shows a preview of the last map generated: yellow = sand, gray = rock, red = spice. The blue and green dot symbolise the start positions.

equal. All values were normalised to $[0, 1]$. To summarise, the following features were part of the fitness function:

– the overall percentage of sand in the map s
– the euclidean distance between the two starting points d_{AB}
– the difference of the starting zones' sizes Δ_{AB} (to minimise)
– the difference of the distance from each starting position to the nearest spice bloom Δd_s (to minimise)

Apart from these criteria a map was rejected with a fitness of 0 if one of the following conditions was met:

– There was a direct path (using A^*) between both starting positions, only traversing rock tiles. (Condition c_1)
– One or both start positions' size was smaller than a neighbourhood of eight. (Condition c_2)

The resulting fitness function was:

$$f_{map} = \begin{cases} 0 & \text{if } c_0 \vee c_1, \\ \frac{s + d_{AB} + (1 - \Delta_{AB}) + (1 - \Delta d_x)}{3} & \text{else} \end{cases}$$

In other words: the average of the components if the map passed the criteria, 0 otherwise.

We ran the genetic algorithm over 150 generations with a population size of 200. Each generation took between three and ten seconds on a modern 3.2GHz desktop PC to compute. The genetic algorithm was an off-the-shelf implementation (using the JGAP library) [11] using a uniform random distribution for the genome creation and fitness driven selection probability (40% of the top scoring genomes preserved each generation).

4   Results

We present the result of an example run of the GA in Figure 4. The graph shows the average fitness value for each component and the overall fitness. The increasing rock coverage slightly influences the start zone size differences, as there is less rocky terrain in the map and therefore chances are higher that it is unequally distributed. There is a steady increase of the distance between the two starting zones, but this doesn't seem to have an impact on distance to the nearest spice bloom. The development of the overall fitness shows that the excluding case (where the fitness is set to zero if the map fails one or two conditions) has a high impact on the average overall score in the first 80 generations. In the same interval, the average component scores seem steady, although eliminated maps are not removed from the average component score calculations. This suggests that these maps might be enjoyable to play despite having a continuous path between starting zones.

Instead, we ran into an interesting problem with setting the elitism threshold too low (thus preserving too many genomes unaltered every generation): on rare occasions each genome in the start population would score as zero. The GA then converged quickly towards two pathological cases, which can be seen in Figure 5(a) and 5(b). The first one only consist of sand and one spice field, and the second map only consists of rock. While the second one might not be very interesting to play, it is actually playable, given that both players start with a sufficient amount of money to build units. The sand-only map on the other hand makes it impossible to win the game, since there is no space to build any buildings.

Table 1. Aggregated results of 30 runs: the minimal maximum fitness, the maximal maximum fitness, the average maximum fitness, and the standard deviation of the maximum fitness in each the first and last generation

Generation	minMax	maxMax	avgMax	stdMax
first	0	0.82	0.38	0.39
last	0.86	0.92	0.89	0.02

5   Discussion

With appropriate parameters, we were able to generate several maps that resembled the style of the original Dune 2 maps. The GA was able to adapt to our fitness function and produced "good" maps on every single run (see Table 1).

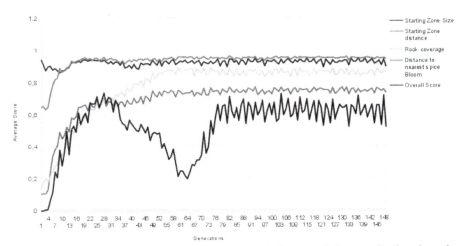

(a) The development of the component scores and the overall fitness, displayed as the population average per generation.

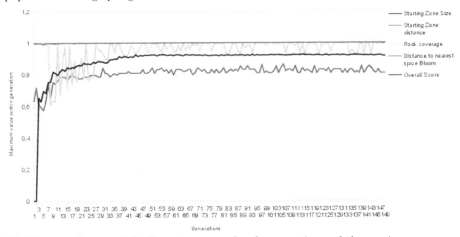

(b) The overall score of the fittest genome of each generation and the maximum component value encountered in each generation. The component values are tracked individually and might come from a different individual than the fittest genome.

Fig. 4. Results from an exemplary run of the genetic algorithm

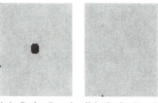

(a) Only Sand   (b) Only Rock

Fig. 5. Two pathological, non-functional, generated maps

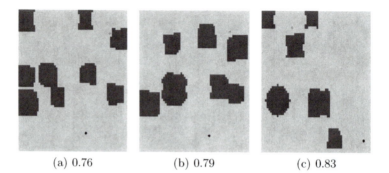

(a) 0.76   (b) 0.79   (c) 0.83

Fig. 6. The evolution of a map over three generations with slightly improving overall fitness

Our fitness was based on heuristic created from expert knowledge. If this actually resembles players' preferences is clearly something that requires further examination. From an aesthetic point of view, the maps look interesting enough to not bore the player and remind them of the original Dune 2 maps, while still presenting fresh challenges.

We are currently working on modelling the complete mechanics of the original Dune 2 game in SGDL, so that both humans and AIs can play full games. We will then load the maps generated through methods described in this paper into the game and gather gameplay information and player preference reports in order to test the validity of our fitness function.

6   Conclusion

We have presented a fast search-based map generator that reliably generates playable and good-looking maps for Dune 2. By using a cellular automata-based genotype-to-phenotype mapping we have avoided some problems associated with other map phenotype representations, and by using a search-based mechanism with direct evaluation functions built on game mechanics we have retained controllability. We believe this method, with minor modifications, can be used to generate maps for a large variety of games.

References

1. Dune II: The Maker, http://d2tm.duneii.com/
2. Ashlock, D.: Automatic generation of game elements via evolution. In: 2010 IEEE Symposium on Computational Intelligence and Games (CIG), pp. 289–296 (August 2010)
3. Ashlock, D., Gent, S., Bryden, K.: Embryogenesis of artificial landscapes. In: Hingston, P.F., Barone, L.C., Michalewicz, Z. (eds.) Design by Evolution. Natural Computing Series, pp. 203–221. Springer, Heidelberg (2008)
4. Bentley, J.: Programming pearls: algorithm design techniques. Commun. ACM 27, 865–873 (1984)
5. Doran, J., Parberry, I.: Controlled procedural terrain generation using software agents. IEEE Transactions on Computational Intelligence and AI in Games 2(2), 111–119 (2010)
6. Frade, M., de Vega, F., Cotta, C.: Evolution of Artificial Terrains for Video Games Based on Accessibility. In: Di Chio, C., Cagnoni, S., Cotta, C., Ebner, M., Ekárt, A., Esparcia-Alcazar, A.I., Goh, C.-K., Merelo, J.J., Neri, F., Preuß, M., Togelius, J., Yannakakis, G.N. (eds.) EvoApplicatons 2010. LNCS, vol. 6024, pp. 90–99. Springer, Heidelberg (2010)
7. Herbert, F.: Dune. New English Library (1966)
8. Johnson, L., Yannakakis, G.N., Togelius, J.: Cellular automata for real-time generation of infinite cave levels. In: Proceedings of the 2010 Workshop on Procedural Content Generation in Games, PCGames 2010, pp. 10:1–10:4. ACM, New York (2010)
9. Mahlmann, T., Togelius, J., Yannakakis, G.: Modelling and evaluation of complex scenarios with the strategy game description language. In: Proceedings of the Conference for Computational Intelligence, CIG 2011, KR 2011, Seoul (2011)
10. Mahlmann, T., Togelius, J., Yannakakis, G.: Towards Procedural Strategy Game Generation: Evolving Complementary Unit Types. In: Di Chio, C., Cagnoni, S., Cotta, C., Ebner, M., Ekárt, A., Esparcia-Alcázar, A.I., Merelo, J.J., Neri, F., Preuss, M., Richter, H., Togelius, J., Yannakakis, G.N. (eds.) EvoApplications 2011, Part I. LNCS, vol. 6624, pp. 93–102. Springer, Heidelberg (2011)
11. Meffert, K., Rotstan, N., Knowles, C., Sangiorgi, U.: Jgap-java genetic algorithms and genetic programming package (2008), http://jgap.sf.net
12. Nielsen, J.L., Jensen, B.F.: Artificial Agents for the Strategy Game Description Language. Master's thesis, ITU Copenhagen (2011)
13. Smelik, R.M., Kraker, K.J.D., Groenewegen, S.A., Tutenel, T., Bidarra, R.: A survey of procedural methods for terrain modelling. In: Proc. of the CASA Workshop on 3D Advanced Media In Gaming And Simulation (3AMIGAS) (2009)
14. Togelius, J., Yannakakis, G., Stanley, K., Browne, C.: Search-based procedural content generation: A taxonomy and survey. IEEE Transactions on Computational Intelligence and AI in Games 3(3), 172–186 (2011)
15. Togelius, J., Preuss, M., Beume, N., Wessing, S., Hagelbäck, J., Yannakakis, G.: Multiobjective exploration of the starcraft map space. In: 2010 IEEE Conference on Computational Intelligence and Games, CIG (2010)

Dealing with Noisy Fitness in the Design of a RTS Game Bot*

Antonio M. Mora, Antonio Fernández-Ares,
Juan-Julián Merelo-Guervós, and Pablo García-Sánchez

Departamento de Arquitectura y Tecnología de Computadores,
Universidad de Granada, Spain
{amorag,antares,jmerelo,pgarcia}@geneura.ugr.es

Abstract. This work describes an evolutionary algorithm (EA) for
evolving the constants, weights and probabilities of a rule-based decision
engine of a bot designed to play the Planet Wars game. The evaluation
of the individuals is based on the result of some non-deterministic combats, whose outcome depends on random draws as well as the enemy
action, and is thus noisy. This noisy fitness is addressed in the EA and
then, its effects are deeply analysed in the experimental section. The
conclusions shows that reducing randomness via repeated combats and
re-evaluations reduces the effect of the noisy fitness, making then the EA
an effective approach for solving the problem.

1 Introduction

Bots are autonomous agents that interact with a human user within a computer-
based framework. In the games environment they run automated tasks for competing or cooperating with the human player in order to increase the challenge
of the game, thus making their *intelligence* one of the fundamental parameters
in the video game design. In this paper we will deal with real-time strategy
(RTS) games, which are a sub-genre of strategy-based video games in which the
contenders control a set of units and structures distributed in a playing area.
A proper control of these units is essential for winning the game, after a *battle*.
Command and Conquer™, Starcraft™, Warcraft™ and Age of Empires™ are
some examples of these type of games.

RTS games often employ two levels of AI: the first one, makes decisions on the
set of units (workers, soldiers, machines, vehicles or even buildings); the second
level is devoted to every one of these small units. These two level of actions,
which can be considered *strategical* and *tactical*, make them inherently difficult;
but they are made even more so due to their real-time nature (usually addressed
by constraining the time that can be used to make a decision) and also for the
huge search space (plenty of possible behaviours) that is implicit in its action.
Such difficulties are probably one of the reasons why Google chose this kind of

---

* This work has been supported in part by project P07-TIC-03044, awarded by the
Andalusian Regional Government.

C. Di Chio et al. (Eds.): EvoApplications 2012, LNCS 7248, pp. 234–244, 2012.
© Springer-Verlag Berlin Heidelberg 2012

games for their AI Challenge 2010. In this contest, real time is sliced in one second *turns*, with players receiving the chance to play sequentially. However, *actions* happen at the *simulated* same time.

This paper describes an evolutionary approach for generating the decision engine of a bot that plays *Planet Wars*, the RTS game that was chosen for the commented competition. The decision engine was implemented in two steps: first, a set of parametrised rules that models the behaviour of the bot was defined by means of human players testing; the second step of the process applied a Genetic Algorithm (GA) for evolving these parameters offline (i.e., not during the match, but prior to the game battles).

The evaluation of the quality (fitness) of each set of rules in the population is made by playing the bot against predefined opponents, being a pseudo-stochastic or *noisy* function, since the results for the same individual evaluation may change from time to time, yielding good or bad values depending on the battle events and on the opponent's actions.

In the experiments, we will show that the set of rules evolve towards better bots, and finally an efficient player is returned by the GA. In addition, several experiments have been conducted to analyse the issue of the cited *noisy fitness* in this problem. The experiments show its presence, but also the good behaviour of the chosen fitness function to deal with it and yield good individuals even in these conditions.

2   State of the Art

Video games have become one of the biggest sectors in the entertainment industry; after the previous phase of searching for the graphical quality perfection, the players now request opponents exhibiting intelligent behaviour, or just human-like behaviours [1].

Most of the researches have been done on relatively simple games such as Super Mario [2], Pac-Man [3] or Car Racing Games [4], being many bots competitions involving them.

RTS games show an emergent component [5] as a consequence of the cited two level AI, since the units behave in many different (and sometimes unpredictable) ways. This feature can make a RTS game more entertaining for a player, and maybe more interesting for a researcher. There are many research problems with regard to the AI for RTSs, including planning in an uncertain world with incomplete information; learning; opponent modelling and spatial and temporal reasoning [6].

However, the reality in the industry is that in most of the RTS games, the bot is basically controlled by a fixed script that has been previously programmed (following a finite state machines or a decision tree, for instance). Once the user has learnt how such a game will react, the game becomes less interesting to play. In order to improve the users' gaming experience, some authors such as Falke et al. [7] proposed a learning classifier system that can be used to equip the computer with dynamically-changing strategies that respond to the user's strategies, thus greatly extending the games playability.

In addition, in many RTS games, traditional artificial intelligence techniques fail to play at a human level because of the vast search spaces that they entail [8]. In this sense, Ontano et at. [9] proposed to extract behavioural knowledge from expert demonstrations in form of individual cases. This knowledge could be reused via a case-based behaviour generator that proposed advanced behaviours to achieve specific goals.

Evolutionary algorithms have been widely used in this field, but they involve considerable computational cost and thus are not frequently used in on-line games. In fact, the most successful proposals for using EAs in games correspond to off-line applications [10], that is, the EA works (for instance, to improve the operational rules that guide the bot's actions) while the game is not being played, and the results or improvements can be used later during the game. Through offline evolutionary learning, the quality of bots' intelligence in commercial games can be improved, and this has been proven to be more effective than opponent-based scripts.

This way, in the present work, an offline GA is applied to a parametrised tactic (set of behaviour model rules) inside the Planet Wars game (a basic RTS), in order to build the decision engine of a bot for that game, which will be considered later in the online matches.

3   The Planet Wars Game

It is a simplified version of the game Galcon, aimed at performing bot's fights which was used as base for the Google AI Challenge 2010 (GAIC)[1].

A Planet Wars match takes place on a map (see Figure 1) that contains several planets (neutral or owned), each one of them with a number assigned to it that represents the quantity of starships that the planet is currently hosting. The objective of the game is to defeat all the starships in the opponent's planets. Although Planet Wars is a RTS game, this implementation has transformed it into a turn-based game, in which each player has a maximum number of turns to accomplish the objective. At the end of the match (after 200 actions, in Google's Challenge), the winner is the player owning more starships.

There are two strong constraints (set by the competition rules) which determine the possible methods to apply to design a bot: a simulated turn takes *just one second*, and the bot is *not allowed to store any kind of information* about its former actions, about the opponent's actions or about the state of the game (i.e., the game's map). Therefore, the goal in this paper is to design a function that, according to the state of the map in each simulated turn (input) returns a set of actions to perform in order to fight the enemy, conquer its resources, and, ultimately, win the game.

For more details, the reader is invited to revise the cited webs and our previous work [11].

---

[1] http://ai-contest.com

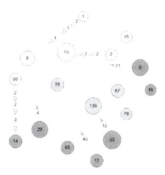

Fig. 1. Simulated screen shot of an early stage of a run in Planet Wars. White planets belong to the player (blue colour in the game), dark grey belong to the opponent (red in the game), and light grey planets belong to no player. The triangles are fleets, and the numbers (in planets and triangles) represent the starships. The planet size means growth rate of the amount of starships in it (the bigger, the higher).

4   Genetic Approach for the Planet Wars Game

The competition restrictions strongly limit the design and implementation possibilities for a bot, since many algorithms are based on a memory of solutions or on the assignment of payoffs to previous actions in order to improve future behaviour. Moreover most of them are quite expensive in running time. Due to these reasons, there was defined a set of rules which models the on-line (during the game) bot's AI. The rules have been formulated through exhaustive experimentation, and are strongly dependent on some parameters, which ultimately determine the behaviour of the bot.

Anyway, there is only one type of action: move starships from one planet to another; but the nature of this movement will be different depending on whether the target planet belongs to oneself or the enemy. As the action itself is very simple, the difficulty lies in choosing which planet creates a fleet to send forth, how many starships will be included in it and what will the target be.

Three type of bots are going to be tested in this paper, all of them previously introduced in our previous work [11][2].

The first one is **GoogleBot**, the basic bot provided by Google for testing our own. It is quite simple, since it has been designed for working well independently of the map configuration, so it may be able to defeat bots that are optimised for a particular kind of map. It just choose a planet as a base (the one with most of its starships) and a target chosen by calculating the ratio between the growth rate and the number of ships for all enemy and neutral planets. It wastes the rest of time until the attack has finished.

The second bot is known as **AresBot**, and it was defined as the first approach for solving the problem. It models a new hand-coded strategy better than the

---

[2] The source code of all these bots can be found at:
   forja.rediris.es/svn/geneura/Google-Ai2010

one scripted in the GoogleBot. So, several rules were created based on the experience of a human player. As a summary, this bot tries to firstly find a *base planet* depending on a score function, the rest of its planets are considered as *colonies*. Then, it determines which *target planet* to attack or to reinforce, if it already belongs to it in the next turns (since it can take some turns to get to that planet). The base planet is also reinforced with starships coming from colonies; this action is called *tithe*. Furthermore, colonies that are closer to the target than to the base also send fleets to attack the target instead of reinforcing the base. The internal flow of AresBot's behaviour with these states is shown in Figure 2. It can be seen in that figure a set of seven parameters (weights, probabilities and amounts to add or subtract) which has been included in the rules that model the bot's behaviour. These parameters have been adjusted by hand, and they totally determine the behaviour of the bot. Their values and meaning can be consulted in the previous work [11]. As stated in that paper, their values are applied in expressions used by the bot to take decisions. For instance, the function considered to select the *target planet*.

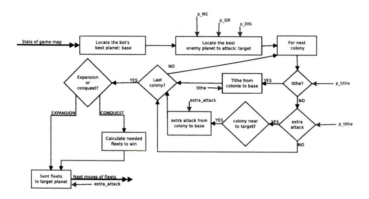

Fig. 2. Diagram of states governing the behaviour of AresBot and GeneBot. The parameters that will be evolved are highlighted.

This bot already had a behaviour more complex than GoogleBot, and was able to beat it in 99 out of 100 maps; however, it needed lots of turns to do that; this means that faster bots or those that developed a strategy quite fast would be able to beat it quite easily. That is why we decided to perform a systematic exploration of the values for the parameters shown above, in order to find a bot that is able to compete successfully (to a certain point) in the google AI challenge.

The third bot is an evolutionary approach, called **GeneBot**, and it performs an offline AresBot's parameter set optimisation (by means of a GA). The objective is to find the parameter values that maximise the efficiency of the bot's behaviour. The proposed GA uses a floating point array to codify the parameters, and follows a *generational* scheme with *elitism* (the best solution always survives). The genetic operators include a *BLX-α* crossover [12], very common in this kind of chromosome codification to maintain the diversity, and a *gene* *mutator* which mutates the value of a random gene by adding or subtracting a random quantity in the $[0, 1]$ interval. The *selection mechanism* implements a *2-tournament*. Some other mechanisms were considered (such as roulette wheel), but eventually the best results were obtained for this one, which represents the lowest selective pressure. The elitism has been implemented by replacing a random individual in the next population with the global best at the moment. The worst is not replaced in order to preserve diversity in the population.

The evaluation of one individual is performed by setting the correspondent values in the chromosome as the parameters for GeneBot's behaviour, and placing the bot inside *five different maps* to fight against a GoogleBot. These maps were chosen for its significance and represent a wide range of situations: bases in the middle and planets close to them, few and spread planets, planets in the corners, bases in the corners, both planets and bases in the corners. The aim is to explore several possibilities in the optimisation process so, if the bot is able to beat GoogleBot in all of them, it would have a high probability of succeeding in the majority of 'real' battles. The bots then fight five matches (one in each map). The result of every match is non-deterministic, since it depends on the opponent's actions and the map configuration, conforming a *noisy fitness* function, so the main objective of using these different maps is dealing with it, i.e. we try to test the bot in several situations, searching for a good behaviour in all of them, but including the possibility of yielding bad results in any map (by chance). In addition, there is a reevaluation of all the individuals every generation, including those who remain from the previous one, i.e. the elite. These are mechanisms implemented in order to avoid in part the noisy nature of the fitness function, trying to obtain a real (or reliable) evaluation of every individual.

The performance of the bot is reflected in two values: *the number of turns* that the bot has needed to win in each arena, and the second is *the number of games that the bot has lost.* In every generation the bots are ranked considering this last value; in case of coincidence, then the number of turns value is also considered, so the best bot is the one that has won every single game or the one that needs less turns to win. Thus the *fitness* associated to an individual (or bot in this case) could be considered as the minimum aggregated number of turns needed for winning the five battles.

5   Experiments and Results

In order to test the GA proposed in previous section, several experiments and studies have been conducted. These are different from those performed in the previous work [11], being more complete in the first step (parameter optimisation), and analysing the pseudo-stochastic fitness function, and the value of the dealing mechanisms for avoiding it.

First of all, the (heuristically found) parameter values used in the algorithm can be seen in Table 1. 15 runs have been performed in the optimisation of the AresBot's behaviour parameter, in order to calculate average results with a certain statistical confidence. Due to the high computational cost of the evaluation of one individual (around 40 seconds each battle), a single run of the GA takes

Table 1. Parameter setting considered in the Genetic Algorithm

Num. Generations	Num. Individuals	Crossover prob.	α	Mutation prob.	Replacement policy
100	200	0.6	0.5	0.02	2-elitism

around two days with this configuration. The previously commented evaluation is performed by playing in 5 representative maps, but besides, Google provides 100 example/test maps to check the bots, so they will be used to evaluate the value of the bots once they (their parameters) have been evolved. The following sections describe each one of the studies developed for demonstrating the value of the presented method and also the correct performance of the noisy fitness function.

5.1   Parameter Optimisation

In the first experiment, the parameters which determine the bot's behaviour have been evolved (or improved) by means of a GA, obtaining the so-called GeneBot. The algorithm yields the evolved values shown in Table 2.

Table 2. Initial behaviour parameter values of the original bot (AresBot), and the optimised values (evolved by a GA) for the best bot and the average obtained using the evolutionary algorithm (GeneBot)

	$tithe_{perc}$	$tithe_{prob}$	w_{NS-DIS}	w_{GR}	$pool_{perc}$	$support_{perc}$	$support_{prob}$
AresBot	0.1	0.5	1	1	0.25	0.5	0.9
GeneBot (Best)	0.018	0.008	0.509	0.233	0.733	0.589	0.974
GeneBot (Average)	0.174	0.097	0.472	0.364	0.657	0.524	0.599
	±0.168	±0.079	±0.218	±0.177	±0.179	±0.258	±0.178

Looking at Table 2 the evolution of the parameters can be seen. If we analyse the new values for the best bot of all the 15 executions, yielded in run number 8, it can be seen that the best results are obtained by strategies where colonies have a low probability of sending tithe, $tithe_{prob}$, to the base planet (only 0.008 or 0.09 in average value). In addition, those tithes send ($tithe_{perc}$) just a few of the hosted starships, which probably implies that colonies should be left on its own to defend themselves, instead of supplying the base planet. On the other hand, the probability for a planet to send starships to attack another planet, $support_{prob}$, is quite high (0.97 or 0.59 in average), and the proportion of units sent, $support_{perc}$, is also elevated, showing that it is more important to attack with all the available starships than wait for reinforcements. Related to this property is the fact that, when attacking a target planet, the base also sends ($pool_{perc}$) a large number of extra starships (73.3% or 65.7% in average of the hosted ships). Finally, to define the target planet to attack, the number of starships hosted in the planet, w_{NS-DIS}, is much more important than the growth range w_{GR}, but also considering the distance as an important value to take into account.

In order to analyse the value of these bots, a massive battle against AresBot has been conducted. The best bot in every run is confronted against it in 100 battles (one in each of the example maps provided by Google in the competition pack). Table 3 shows the percentage of battles won by each bot. It can be seen that the best individual is the one of execution 8, i.e. the one considered as the best in the previous experiment (meaning a robust result). In addition, the improvement of the best bots with regard to AresBot can be noticed, since all of them win at least 82 out of 100 matches.

Table 3. Winning percentage of the best individuals against AresBot in 100 battles

	E1	E2	E3	E4	E5	E6	E7	E8	E9	E10	E11	E12	E13	E14	E15
Winning Percentage	90	82	87	83	98	85	98	**99**	98	94	91	89	90	98	84

One fact to take into account in this work is that even as this solution looks like a simple GA (since it just evolves seven parameters) for a simple problem, it becomes more complicated due to the noisiness and complex fitness landscape; that is, small variations in parameter values may imply completely different behaviours, and thus, big changes in the battle outcome. This fitness nature is studied in the next section.

5.2   Noisy Fitness Study

A good design of the fitness function is a key factor in any EA for getting success. It has to model the value of every individual. In a pseudo-stochastic environment (the victory or defeat depends on the opponent's actions) as is this, it is important to test the stability of the evaluation function, i.e. check if the fitness value is representative of the individual quality, or if it has been yielded by chance. In order to avoid this random factor a re-evaluation of the fittest individuals has been implemented, even if they survive for the next generation, testing continuously them in combat. In addition (and as previously commented), the fitness function performs 5 matches in 5 representative maps for calculating an aggregated number of turns, which ensures (in part) strongly penalising an individual if it gets a bad results.

The first study in this line is the evolution of fitness along the generations. Since the algorithm is a GA, it would be expected that the fitness is improved (on average) in every generation. It is shown in Figure 3, where it can be seen that as the evolution progresses (number of generations increases), the aggregate number of turns needed to win on five maps decreases (on average) on the three cases; however, since the result of every combat, and thus the fitness is pseudo-stochastic, it can increase from one generation to the next (it oscillates more than usual in GAs).

The second study tries to show the fitness tendency or stability, that is, if a bot is considered as a good one (low aggregated number of turns), it would be desirable that its associated fitness remains being good in almost all the battles, and the other way round if the bot is considered as a bad one (high aggregated number of turns). We are interested in knowing whether the fitness we are considering actually reflects the ability of the bot in beating other bots. It could be considered as a measure to determine if the algorithm is robust.

Figure 4 shows the fitness associated to two different GeneBots when fighting against the GoogleBot 100 times (battles) in the 5 representative maps. Both of

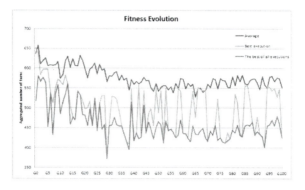

Fig. 3. The graphs show the complete execution of the best bot (best execution), the distribution of the best individuals in every run (best of all executions), and the average of the best bots in 15 runs

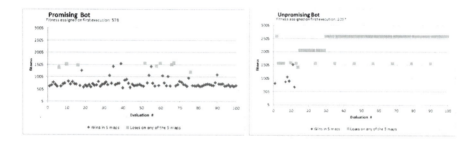

Fig. 4. Fitness tendency of two different and random individuals (bots) in 100 different battles (evaluations), everyone composed by 5 matches in the representative maps, against the GoogleBot

them have been chosen randomly among all the bots in the 15 runs, selecting one with a good fitness value (578 turns), called *Promising Bot*, and another bot with a bad fitness value (2057 turns), called *Unpromising Bot*.

As it can be seen, both bots maintain their level of fitness in almost every battle, winning most of them in the first case, and losing the majority in the second case. In addition, both of them win and lose battles in the expected frequency, appearing some outlier results due to the pseudo-stochastic nature of the fights.

5.3   GeneBots Fighting

Finally, a study concerning the behaviour of several GeneBots (the best in every execution) has been conducted to establish the validity of the fitness choice (better fitness means better bot). To do this, battles of 5 matches (in the 5 representative maps) have been performed. The winner in each battle is the bot who wins 3 out of 5 combats. Figure 5 shows the results, along with the fitness value for each bot.

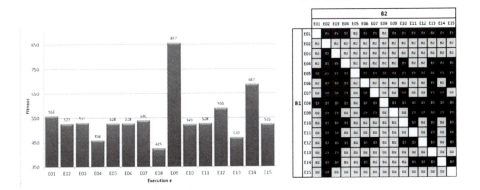

Fig. 5. Battles results between each pair of bots (the best in every execution). The winner in 3 out of 5 matches is marked (black for B1 victory and grey for B2 victory). The associated fitness to each bot (in its corresponding execution) is shown on the left graph.

The results demonstrate that individuals with lower fitness can hardly win to the fittest ones and the other way round, as it is desirable. However, it also proves the *noisy* nature of fitness, with a non-zero chance of the worst bot beating the best one.

6   Conclusions and Future Work

This paper shows how Genetic Algorithms (GAs) can be applied to the design of one autonomous player (bot) for playing Planet Wars game, which held the Google AI Challenge 2010. It have been proved that Genetic Algorithms (GeneBot) can improve the efficiency of a hand-coded bot (AresBot), winning more battles in a lower number of turns.

Besides, from looking at the parameters that have been evolved, we can draw some conclusions to improve overall strategy of hand-designed bots; results show that it is important to attack planets with almost all available starships, instead of keeping them for future attacks, or that the number of ships in a planet and its distance to it, are two criteria to decide the next target planet, much more important than the growing rate.

In addition, the presence of noisy fitness (the evaluation of one individual may strongly vary from one generation to the next) has been addressed by performing several battles in each evaluation, in addition to a re-evaluation of all the individuals in each generation. This subject has been studied in several experiments, concluding that the proposed algorithm yields results which have a good deal with it, being quite robust.

As future work, we intend apply some other techniques (such as Genetic Programming or Learning Classifier Systems) for defining the initial set of rules which limit the improving range of the bot by means of GAs. In the evolutionary algorithm front, several improvements might be attempted. For the time being, the bot is optimised against a single opponent; instead, several opponents might be tried, or even other individuals from the same population, in a co-evolutionary approach. Another option will be to change the bot from a single optimised strategy to a set of strategies and rules that can be chosen also using an evolutionary algorithm. Finally, a multi-objective EA will be able to explore the search space more efficiently, although in fact the most important factor is the overall number of turns needed to win.

References

1. Lidén, L.: Artificial stupidity: The art of intentional mistakes. In: AI Game Programming Wisdom 2, pp. 41–48. Charles River Media, Inc. (2004)
2. Togelius, J., Karakovskiy, S., Koutnik, J., Schmidhuber, J.: Super mario evolution. In: Proceedings of the 5th IEEE Symposium on Computational Intelligence and Games (CIG 2009). IEEE Press, Piscataway (2009)
3. Martín, E., Martínez, M., Recio, G., Saez, Y.: Pac-mant: Optimization based on ant colonies applied to developing an agent for ms. pac-man. In: Yannakakis, G.N., Togelius, J. (eds.) IEEE Symposium on Computational Intelligence and Games, CIG 2010, pp. 458–464 (2010)
4. Onieva, E., Pelta, D.A., Alonso, J., Milanés, V., Pérez, J.: A modular parametric architecture for the torcs racing engine. In: Proceedings of the 5th IEEE Symposium on Computational Intelligence and Games (CIG 2009), pp. 256–262. IEEE Press, Piscataway (2009)
5. Sweetser, P.: Emergence in Games. Game Development. Charles River Media, Boston (2008)
6. Hong, J.H., Cho, S.B.: Evolving reactive NPCs for the real-time simulation game. In: Proceedings of the 2005 IEEE Symposium on Computational Intelligence and Games, CIG 2005 (2005)
7. Falke-II, W., Ross, P.: Dynamic Strategies in a Real-Time Strategy Game. In: Cantú-Paz, E., Foster, J.A., Deb, K., Davis, L., Roy, R., O'Reilly, U.-M., Beyer, H.-G., Kendall, G., Wilson, S.W., Harman, M., Wegener, J., Dasgupta, D., Potter, M.A., Schultz, A., Dowsland, K.A., Jonoska, N., Miller, J., Standish, R.K. (eds.) GECCO 2003. LNCS, vol. 2724, pp. 1920–1921. Springer, Heidelberg (2003)
8. Aha, D.W., Molineaux, M., Ponsen, M.: Learning to Win: Case-Based Plan Selection in a Real-Time Strategy Game. In: Muñoz-Ávila, H., Ricci, F. (eds.) ICCBR 2005. LNCS (LNAI), vol. 3620, pp. 5–20. Springer, Heidelberg (2005)
9. Ontanon, S., Mishra, K., Sugandh, N., Ram, A.: Case-Based Planning and Execution for Real-Time Strategy Games. In: Weber, R.O., Richter, M.M. (eds.) ICCBR 2007. LNCS (LNAI), vol. 4626, pp. 164–178. Springer, Heidelberg (2007)
10. Spronck, P., Sprinkhuizen-Kuyper, I., Postma, E.: Improving opponent intelligence through offline evolutionary learning. International Journal of Intelligent Games & Simulation 2(1), 20–27 (2003)
11. Fernández-Ares, A., Mora, A.M., Merelo, J.J., García-Sánchez, P., Fernandes, C.: Optimizing player behavior in a real-time strategy game using evolutionary algorithms. In: IEEE Congress on Evolutionary Computation, CEC 2011 (2011) (accepted for publication)
12. Herrera, F., Lozano, M., Sánchez, A.M.: A taxonomy for the crossover operator for real-coded genetic algorithms: An experimental study. International Journal of Intelligent Systems 18, 309–338 (2003)

On Modeling, Evaluating and Increasing Players' Satisfaction Quantitatively: Steps towards a Taxonomy

Mariela Nogueira[1], Carlos Cotta[2], and Antonio J. Fernández-Leiva[2]

[1] University of Informatics Sciences, La Habana, Cuba
mnogueira@uci.cu
[2] University of Málaga, Málaga, Spain
{ccottap,afdez}@lcc.uma.es

Abstract. This paper shows the results of a review about modeling, evaluating and increasing players' satisfaction in computer games. The paper starts discussing the main stages of development of *quantitative* solutions, and then it tries to propose a taxonomy that represents the most common trends. In the first part of this paper we take as base some approaches that were already described in the literature for quantitatively capturing and increasing the real-time entertainment value in computer games. In a second part we analyze the stage in which the game's environment is adapted in response to player needs, and the main trends on this theme are discussed.

Keywords: player satisfaction, player modeling, adaptive game, taxonomy.

1 Introduction

Most of the games' genres assume as an important goal the entertainment of the players, which can be different for distinct player (e.g., players may not enjoy the same challenges). If the preferences of the player could be modeled, we might be able to adapt the gameplay to each player [1] and try to increase players' satisfaction during the play. The IEEE Task Force on Player Satisfaction Modeling [2] was created with the primary focus on the use of Computational Intelligence for modeling and optimizing the player's perceived satisfaction during gameplay, and grouped many of the most relevant events and results on this topic.

In [1], a new taxonomy is defined about the player modeling, in which models are distinguished according to their purpose: satisfaction, knowledge, position and strategy. Some of the most common models' applications can be: the classification of players according to their skills or preferences; the training of bots to simulate human's behavior [3]; the analysis of physical and emotional states of the player, and the prediction of behaviors, among others. For the specific topic of modeling focused on measuring the level of player satisfaction two main trends were categorized in [4]. One of them approaches the subject from a *qualitative*

C. Di Chio et al. (Eds.): EvoApplications 2012, LNCS 7248, pp. 245–254, 2012.
© Springer-Verlag Berlin Heidelberg 2012

point of view, closer to psychology, whereas the another proposes alternatives to measure fun *quantitatively*.

With respect to *qualitative* approaches, we can mention a number of works that can be considered pioneer; for instance, the theory of the intrinsic motivation of Thomas W. Malone [5] or the theory of Flow defined by Czikszentmihalyi [6]. Also, a very influential work is the adaptation of this latter theory to the game's field (made by Penelope Sweetser and Peta Wyeth in [7]), and also the contributions in the understanding of the entertainment in games proposed by Lazzaro [8] and Calleja [9]. The research on qualitative approaches is often useful in conceptualization of a modeling process, because some of them allow the classification of different types of players, their preferences, and trends in behavior [10]. In this aspect, two interesting studies were addressed in [11] and [12], both works focused on identifying behaviors that distinguish the human players from the bots, in the game of *Pong* and in a strategy game respectively.

All these works based on the qualitative approach have limitations that decrease the robustness of the result, since most of the studies are based on empirical observations or linear correlations established between the provided information in the player's profile and reported emotions [13].

On the other hand, quantitative contributions are focused on the attempt to formally model the behavior of the player based on her preferences, skills, emotions, and other elements that influence the decision-making process. These models are then used in conjunction with the online information that is being received from the user, to define a measure of the level of fun that the player is obtaining in the game.

The work presented here focuses on the quantitative approach and tries to identify the main stages of development of quantitative solutions with the aim of easing the definition of a taxonomy that represents the most common trends used in each of them.

This paper is organized as follows. Section 2 shows a taxonomy which includes the main trends in the process of modeling and quantifying player's satisfaction. In the third section we analyze the stage in which the game's environment is adapted in response to player's needs, and we discuss a taxonomy for this theme. Finally, Section 4 provides some conclusions and gives some indications for future work.

2   Players' Satisfaction Approaches

It is not an easy task to determine the satisfaction an activity causes to a person, since the mechanisms to manage the human emotional states are complex. Many factors influence a change in mood, and seeking for a generality is not simple because each person has her own characteristics as well as particular preferences. In the following we discuss different attempts to formally model the fun that a player obtains during the game; this analysis allows to identify the fundamental stages of this process and distinguishes taxonomies between most used trends. Each of them is explained below.

2.1 Selection of Relevant Information

This task represents a basic process that should be done as initial stage; its goal is to identify the elements that will influence the amount of players' fun; to do so, researchers usually base their analysis on the qualitative studies mentioned in the introduction. It is thus necessary to have a broad knowledge of the game functions in order to establish a direct projection of the psychological elements in real variables (that are assumed to be measurable) to describe the behavior of the user.

The information obtained in this process can be classified according to its nature under different points of views: for example, *offline* and *online* [14], *observational* and *in-game* [15], and *subjective, objective* and *gameplay-based,* [16]. In general all of them can be summarized in three categories with respect to the nature of the information: *reported, in-game,* and *sensorial.*

Reported Information. It represents the information that is requested directly from the user, for example, when the player has to create a game-profile, or answer a questionnaire designed to know her predilections (for instance, [16] proposed to adapt the game not only to the skills of players, but also to their preferences. To do so, a model of the player experience can be created from the answers provided by the player, after a gameplay session, to specifically designed preference questionnaires). The main goal is to identify the player via her preferences. The reliability of the information collected is completely dependent on the consistency of the responses provided by the players. This information is usually employed to validate players' models. Also [17] proposed the use of questionnaires that should be filled by the players to measure their satisfaction.

In-game Information. It comprises the data that are generated and processed within the game engine (and during the game); this task usually involves the gathering of numerical data describing players' performance. For example, in a combat game, we may consider kill counts, death counts, and use of sophisticated weapons.

Sensorial Information. Here, physical sensing of the player during play is obtained from one or more specialized devices; it is representative of emotional reactions in players. Sensors measure players' attributes including: galvanic skin response, facial reactions, heart-rate, and temperature, among others. The objective here is to increase the amount of information that can be obtained during a game session and that can be complementary to that obtained as in-game information (in the sense explained above). By doing so the game designer can have more arguments to manage fun in the game with more assurance, and might try the adaptation of the play to the player, with the goal for instance to improve her 'immersion' in the game [9].

In fact, the design of game interfaces is nowadays one of the most interesting topics in game development and there is a growing tendency to use multi-sensory (e.g., visual, auditory and haptic) interfaces to broaden the game experience (i.e., sensation) of the player. Precisely [18] analyzed if *by displaying different information to different senses, it is possible to increase the amount of information available to players and so assist their performance*; in general, the conclusions obtained in this analysis shown that players had improved not only 'immersion' but also 'confidence' and 'satisfaction' when additional sensory cues were included.

2.2   Capture Players' Fun

In this stage the aim is to determine how the value of fun can be defined. Two main approaches, explained by Yannakakis and Hallam in [19], are usually considered., and in this section we try to refine their classifications. The first one proposes to find an scalar value of fun, and the second focuses in the creation of a model which defines the relation between variables and entertainment's level (i.e., a model of players' fun).

2.2.1   Scalar Value of Fun

This approach proposes the empirical definition of a mathematical formula to quantify players' fun, according to their behavior. This way allows a fast path to know the player's status during the game, and further to employ this information for assisting her with the aim of increasing her entertainment. An example of this approach is described in [19] where a quantitative metric of the *interestingness* of opponent behaviors is designed on the basis of qualitative considerations of what is enjoyable in predator/prey games. A mathematical formulation of those considerations, based upon observable data that are taken into account during game sesions, is derived. This metric is validated successfully when it is compared with the human notion of entertainment in the context of the well-known *Pac-Man* computer game [19].

2.2.2   Model of Players' Fun

Here it is necessary to quantify the variables that influence the fun in order to have notion of its evolution in every moment of the game; these values will be use as inputs to the model construction process. The main difference with the previous approach is that the relationships between variables and the level of entertainment will be defined through machine learning techniques. We can mention here the two main approaches that have been proposed in the literature and that are discussed below:

Empirical evaluation
In this case the model is obtained from any metaheuristics algorithm (or soft computing technique in general), and the objective function defined to guide the optimization process is derived from the author's appreciation. An example of this approach was presented in [20] where authors consider that some change in the rules of the game *Commons Game* would make it much more exciting, in this way, game players are modeled with Artificial Neural Networks (ANNs). The weights of the neural network based model are evolved by a multi-Objective evolutionary algorithm [21]. In order to evaluate each individual they defined

two objective functions: the variance of the total number of each card chosen in each game run, and the efficiency of played cards, respectively.

Relative evaluation

This variant has been the most widely used in the literature. Here, the metric that guides the process of models' optimization is directly based on the results that the learning mechanism shows, and this represents precisely the primary distinction with the approach previously discussed where the metric is defined by authors. The basic process is carried out by a training of models that is followed by a supervised approach, so that one can identify when a generated model is correct. Then, the function is defined on the basis of analyzing the balance between correct and incorrect models, which depends on the effectiveness of the learning's mechanism. For example, in [22] an artificial neural network (ANN) representing the user's preference model is constructed using a preference learning approach in which a fully-connected ANN of fixed topology is evolved by a generational genetic algorithm which uses a fitness function that measures the difference between the preferences of entertainment (treported by a group of children) and the output value of fun returned by the model. Another instance than can be catalogued in this category was presented in [23]; here the authors do not use neuronal techniques but a different linear model obtained with Linear Discriminant Analysis; this model follows a supervised approach in search of a correlation between physiological features and the reported subject enjoyment. Also, [17] proposed a combination of ANNs with the technique of *preference learning* to assist in the prediction of player preferences; here players are requested to explicitly report their preferences on variants of the game via questionnaires, and computational models are built on the preference data.

3   Game's Adjustment

This will be the final stage of an attempt to optimize the players's satisfaction. After having obtained the models that identify the player, and having a measure of her entertainment, it is the moment to use that information and adapt or adjust the game to the characteristics of the user with the aim of providing a personalized match according to her preferences, resulting in an entertaining experience that at the same time meets her expectations.

The processes of modeling and satisfaction evaluation are closely related to the implemented adjustment mechanism. The indicators that were considered for the evaluation of satisfaction must match up with the adjustable elements of the game, in a way that manipulating them will influence the level of satisfaction. Some of these elements could be: aesthetic aspects, auxiliary contents that can serve as a guide to the player, the drama, the level of difficulty of the terrain and opponents, among others; but selection of these elements is not a trivial task; this is precisely the goal of Procedural Content Generation for games [24] that represents one of the most exciting lines of research inside the community of computational intelligence applied to videogames. Moreover, it is also true that it is not clear the impact of game difficulty and player performance on game enjoyment. This was precisely the analysis conducted in [25] although the authors could not give concrete conclusions.

From the conceptualization of the game, the script and the design should be developed with a generic approach that allows the flexibility in each game be adaptable to the wide range of preferences imposed by any group of users. The previous issue is also important to reduce the probability that the new game variants might be not well accepted. For example: causing a dramatic change in the rules might frustrate the player, or conducting the game towards unknown status, which is indeed possible when machine learning techniques are used.

With regard to the scope of the game settings we can categorize two approaches that comprise many works described in the literature and that are discussed in the following.

3.1   Circumstantial Adjustment

Let's call the first one *circumstantial adjustment* which embraces only the action of changing the specific game elements according to the needs of the player; for example, the difficulty of the opponents - i.e., the game artificial intelligence (AI) - is often decreased because we have previously identified that the level of challenge goes beyond the users' skills. This approach focuses on managing the elements that will directly influence the level of satisfaction of the player. The change it will cause to the game is something particular to that play, which do not lead to a persistent change in the player's model, or in the decision making rules, because online learning don't occurred.

A successful application of this approach can be found in the experiment described in [26] and [22] where the aim was to increase, in real time, the satisfaction of the player in a game with physiological devices. Here, the authors, starting from collected data from several studies conducted with children, constructed a model of the user preferences using ANNs which proved to have a high precision. They implemented a mechanism of adaptation which allows to customize the game to the individual needs of each user. The logic of the used game was based on well-established rules, which allowed the authors to identify the specific parameters that handled the level of challenge and curiosity of the player, and to obtain an adaptive version of the game turned out to be preferred by the majority of users in the validation tests.

3.2   Constructive Adjustment

This approach refers to the *constructive adjustment*, and the difference with the approach previously mentioned is that here not only the elements that determine the level of entertainment vary but also a transformation (or reconstruction) in the operation of the AI mechanism is carried out as a result of the online learning; an example of this transformation could be to adapt the game strategy that rules the decision-making of the *non-player characters* (NPCs); another example might be to vary the model that identifies the player taking into account the information is being received online (i.e., during the game session).

In [14], this latter issue is called *dynamic modeling* and has a corrective nature because the player's skill (as the game progresses) tends to improve and thus the player progressively polished her technique as part of her own adaptation, and these changes have a direct impact on her preferences. This line of research represents a very interesting field that promises to get a more reliable representation of the human player preferences.

The *constructive* approach offers advantages over the circumstantial one as regards the customization of models with the use of machine learning techniques but it is also more complex to implement; as a consequence we cannot affirm that one is better or worse than the another. In the following we discuss some examples where good results were obtained with this approach.

In a recent proposal made in [27] an evolutionary algorithm to adapt the AI strategy governing the opponent army not controlled by the player (in a strategy game in real time) to the ability of each player is developed; the objective was to catch the interest of the player in every game with the hope of increasing, as a result, her satisfaction. The idea is developed in two processes: the first one takes place during the game execution and consists of extracting a formal model to imitate the behavior of the player (i.e., the way that the player plays and the decisions that she takes during the game); in the second step authors try to generate automatically, through an evolutionary algorithm, an optimized AI adequate to the player's level (i.e., player's skill) in correspondence with the model previously obtained. These two processes are repeated indefinitely during the game, the first one is conducted on-line during the game whereas the second one is executed in-between games. The interesting fact of this proposal is that the AI level depends specifically on the player and is adapted to her with the aim of increasing player's satisfaction by engaging the player to play the game again.

Another example of the application of this paradigm is debated in [14], where authors described a framework for dealing with this issue and providing more adaptable games, and in particular approaches for dealing with two particularly current issues: that of monitoring the effectiveness of adaptation through affective and statistical computing approaches, and the dynamic remodeling of players based on ideas from *concept drift*. This article also discussed the use of ANN with supervised and non-supervised learning which are feasible to implement similar applications.

3.3   Who Makes the Adjustment?

In the majority of the works that have been focused in the topic of the adaptive games, the adjustment is started by the own game, as part of the software's adaptation, without the player noticing that this is happening, as we have seen in the examples previously analyzed. In this case we can name *auto-adaptation* to this approach. Nevertheless, the attempt of personalizing the game can be seen from another perspective where the player is the protagonist of managing the adjustable elements of the game. This seems to be evident, but it marks a difference from the design of the game. It is thus a question of giving the player at all time the control so that she can plan her own way towards the satisfaction. Let us say they are games with *controllable adaptation.*

An illustrative example of *controllable adaptation* in games is proposed by Jenova Chen in [28] as an implementation of the Theory of Flow from Czikszentmihalyi. Here, the author uses the concept of Dynamic Adjustment of the Difficulty. His aim was to design an adaptive game that would show the user the way to her zone of flow. One of the games implemented for this design was *Flow* and proved to have a great acceptance. In *Flow* the players use the cursor to sail, simulating an organism inside a virtual biosphere, where they can eat other organisms, evolve, and advance. Twenty levels were designed; every level introduces new creatures that symbolize new challenges. Unlike the traditional games in which the player finishes a level and advances progressively towards upper levels, *Flow* offers to the user the total control of the progress in the game. In every moment of the game the player is continuously being informed about the possible organisms that she can eat, and according to her choice she will be able to advance towards top levels or to return to a lower one. The fact of offering the total control on the difficulty of the game, allows the own managing of the balance among the challenges and the skills which at the same time control the immersion in the zone of flow. Doing so, Czikszentmihalyi makes possible that a very simple game become *adaptive* to every player, without getting into the intrinsic complications that have the processes of modeling and auto adjustment previously analyzed.

4   Conclusions

Nowadays, increasing player's satisfaction in (video)games is an exciting (and sometimes a very hard to achieve) challenge. This paper has discussed a number of different approaches that try to intensify the diversion of the player from a quantitative point of view, and can be considered a first (and preliminary) attempt to extract a taxonomy of this issue.

Most of the proposed approaches whose primary objective leads to increment quantitatively player's satisfaction can be catalogued in two main categories: one that tries to quantify user's entertainment in a game, and another which focuses in adapting the game in response to player's needs. Several proffers have been proposed in both themes, and some of them have been validated and shown interesting results.

However, as the investigation continues, there are several open research questions, and further attempts will be developed for obtaining more accurately models that represent player's preferences, more complete metrics of entertainment, and more powerful adaptation mechanisms to personalize the games. For these reasons, any proposal of taxonomy, will be temporal, and should be extended in a near future. Future work will be focused on enriching this initial taxonomy, refining its classifications and embracing other aspects of the *modeling and increasing player satisfaction* issue.

Acknowledgements. This work is partially supported by Spanish MICINN under project ANYSELF (TIN2011-28627-C04-01), and by Junta de Andalucía under project TIC-6083.

References

1. Machado, M.C., Fantini, E.P.C., Chaimowicz, L.: Player modeling: Towards a common taxonomy. In: 16th International Conference on Computer Games (CGAMES), pp. 50–57 (July 2011)
2. Official wiki of IEEE Task Force on Player Satisfaction Modeling (2011), http://gameai.itu.dk/psm
3. Fernández-Leiva, A.J., Barragán, J.L.O.: Decision Tree-Based Algorithms for Implementing Bot AI in UT2004. In: Ferrández, J.M., Álvarez Sánchez, J.R., de la Paz, F., Toledo, F.J. (eds.) IWINAC 2011, Part I. LNCS, vol. 6686, pp. 383–392. Springer, Heidelberg (2011)
4. Yannakakis, G.: How to model and augment player satisfaction: A review. In: Proceedings of the 1st Workshop on Child, Computer and Interaction, WOCCI 2008. ACM Press (2008)
5. Malone, T.: What makes things fun to learn? Heuristics for designing instructional computer games. In: Proceedings of the 3rd ACM SIGSMALL Symposium and the First SIGPC Symposium on Small Systems, vol. 162, pp. 162–169. ACM (1980)
6. Czikszentmihalyi, M.: Flow: The psychology of optimal experience. Harper & Row, New York (1990)
7. Sweetser, P., Wyeth, P.: Gameflow: a model for evaluating player enjoyment in games. Computers in Entertainment (3), 3 (2005)
8. Lazzaro, N.: Why we play games: Four keys to more emotion without story. Technical report, XEODesign, Inc. (2005)
9. Calleja, G.: Revising immersion: A conceptual model for the analysis of digital game involvement. In: Akira, B. (ed.) Situated Play: Proceedings of the 2007 Digital Games Research Association Conference, pp. 83–90. The University of Tokyo, Tokyo (2007)
10. Aarseth, E.: Playing research: Methodological approaches to game analysis. In: Digital Games Research Conference 2003. University of Utrecht, The Netherlands (November 2003)
11. Livingstone, D.: Turing's test and believable ai in games. Computers in Entertainment 4(1) (2006)
12. Hagelbäck, J., Johansson, S.J.: A study on human like characteristics in real time strategy games. In: [29], pp. 139–145
13. Yannakakis, G., Togelius, J.: Tutorial on measuring and optimizing player satisfaction. In: IEEE Symposium on Computational Intelligence and Games, CIG 2008, pp. xiv –xvi. IEEE Press (2008)
14. Charles, D., Black, M.: Dynamic player modeling: A framework for player-centered digital games. In: Mehdi, Q., Gough, N., Natkin, S., Al-Dabass, D. (eds.) Proc. of 5th Game-on International Conference on Computer Games: Artificial Intelligence, Design and Education, CGAIDE 2004, pp. 29–35. University of Wolverhampton School of Computing (2004)
15. Cowley, B., Charles, D., Black, M., Hickey, R.: Using decision theory for player analysis in pacman. In: Proceedings of SAB 2006 Workshop on Adaptive Approaches for Optimizing Player Satisfaction in Computer and Physical Games, Rome, Italy, pp. 41–50 (2006)

16. Yannakakis, G.N., Togelius, J.: Experience-driven procedural content generation. T. Affective Computing 2(3), 147–161 (2011)
17. Martínez, H.P., Hullett, K., Yannakakis, G.N.: Extending neuro-evolutionary preference learning through player modeling. In: [29], 313–320
18. Nesbitt, K.V., Hoskens, I.: Multi-sensory game interface improves player satisfaction but not performance. In: Proceedings of the Ninth Conference on Australasian User Interface, AUIC 2008, pp. 13–18. Australian Computer Society, Inc., Darlinghurst (2008)
19. Yannakakis, G.N., Hallam, J.: Capturing player enjoyment in computer games. In: Baba, N., Jain, L.C., Handa, H. (eds.) Advanced Intelligent Paradigms in Computer Games. SCI, vol. 71, pp. 175–201. Springer, Heidelberg (2007)
20. Baba, N., Handa, H., Kusaka, M., Takeda, M., Yoshihara, Y., Kogawa, K.: Utilization of Evolutionary Algorithms for Making COMMONS GAME Much More Exciting. In: Setchi, R., Jordanov, I., Howlett, R.J., Jain, L.C. (eds.) KES 2010. LNCS, vol. 6278, pp. 555–561. Springer, Heidelberg (2010)
21. Coello, C.A.C.: Evolutionary multi-objective optimization: Current state and future challenges. In: Nedjah, N., de Macedo Mourelle, L., Abraham, A., Köppen, M. (eds.) 5th International Conference on Hybrid Intelligent Systems (HIS 2005), p. 5. IEEE Computer Society, Rio de Janeiro (2005)
22. Yannakakis, G.N., Hallam, J.: Real-time game adaptation for optimizing player satisfaction. IEEE Trans. Comput. Intellig. and AI in Games 1(2), 121–133 (2009)
23. Tognetti, S., Garbarino, M., Bonarini, A., Matteucci, M.: Modeling enjoyment preference from physiological responses in a car racing game. In: [29], 321–328
24. Togelius, J., Yannakakis, G.N., Stanley, K.O., Browne, C.: Search-based procedural content generation: A taxonomy and survey. IEEE Trans. Comput. Intellig. and AI in Games 3(3), 172–186 (2011)
25. Klimmt, C., Blake, C., Hefner, D., Vorderer, P., Roth, C.: Player Performance, Satisfaction, and Video Game Enjoyment. In: Natkin, S., Dupire, J. (eds.) ICEC 2009. LNCS, vol. 5709, pp. 1–12. Springer, Heidelberg (2009)
26. Yannakakis, G.N., Hallam, J.: Real-time adaptation of augmented-reality games for optimizing player satisfaction. In: Hingston, P., Barone, L. (eds.) 2008 IEEE Symposium on Computational Intelligence and Games (CIG 2008), pp. 103–110. IEEE, Perth (2008)
27. García, J.A., Cotta, C., Fernández-Leiva, A.J.: Design of Emergent and Adaptive Virtual Players in a War RTS Game. In: Ferrández, J.M., Álvarez Sánchez, J.R., de la Paz, F., Toledo, F.J. (eds.) IWINAC 2011, Part I. LNCS, vol. 6686, pp. 372–382. Springer, Heidelberg (2011)
28. Chen, J.: Flow in games (and everything else). Commun. ACM 50(4), 31–34 (2007)
29. Yannakakis, G.N., Togelius, J. (eds.): Proceedings of the 2010 IEEE Conference on Computational Intelligence and Games, CIG 2010, Copenhagen, Denmark, August 18-21. IEEE (2010)

Monte-Carlo Tree Search
for the Physical Travelling Salesman Problem

Diego Perez, Philipp Rohlfshagen, and Simon M. Lucas

School of Computer Science and Electronic Engineering,
University of Essex, Colchester CO4 3SQ, United Kingdom
{dperez,prohlf,sml}@essex.ac.uk

Abstract. The significant success of MCTS in recent years, particularly in the game Go, has led to the application of MCTS to numerous other domains. In an ongoing effort to better understand the performance of MCTS in open-ended real-time video games, we apply MCTS to the Physical Travelling Salesman Problem (PTSP). We discuss different approaches to tailor MCTS to this particular problem domain and subsequently identify and attempt to overcome some of the apparent shortcomings. Results show that suitable heuristics can boost the performance of MCTS significantly in this domain. However, visualisations of the search indicate that MCTS is currently seeking solutions in a rather greedy manner, and coercing it to balance short term and long term constraints for the PTSP remains an open problem.

1   Introduction

Games such as Chess have always been a popular testbed in the field of Artificial Intelligence to prototype, evaluate and compare novel techniques. The majority of games considered in the literature are two-player turn-taking zero-sum games of perfect information, though in recent years the study of AI for video game agents has seen a sharp increase. The standard approach to the former type of game is *minimax* with $\alpha\beta$ pruning which consistently chooses moves that maximise minimum gain by assuming the best possible opponent. This technique is optimal given a complete game tree, but in practice needs to be approximated given time and memory constraints: a value function may be used to evaluate nodes at depth d. The overall performance of $\alpha\beta$ depends strictly on the quality of the value function used. This poses problems in games such as Go where a reliable state value function has been impossible to derive to date. It is possible to approximate the minimax tree, without the need for a heuristic, using Monte Carlo Tree Search (MCTS), though in practice MCTS still benefits significantly from good heuristics in most games.

MCTS is a best-first tree search algorithm that incrementally builds an asymmetric tree by adding a single node at a time, estimating its game-theoretic value by using self-play from the state of the node to the end of the game: each iteration starts from the root and descends the tree using a *tree policy* until a leaf node has been reached. The simulated game is then continued along a previously unvisited state, which is subsequently added to the tree, using the *default policy* until the end of the game. The actual outcome of the game is then back-propagated and used by the tree policy in subsequent roll-outs.

C. Di Chio et al. (Eds.): EvoApplications 2012, LNCS 7248, pp. 255–264, 2012.
© Springer-Verlag Berlin Heidelberg 2012

The most popular tree policy π_T is UCB1, based on upper confidence bounds for bandit problems [7]:

$$\pi_T(s_i) = \arg\max_{a_i \in A(s_t)} \left\{ Q(s_t,\ a_i) + K \sqrt{\frac{\log N(s_t)}{N(s_t, a_i)}} \right\} \qquad (1)$$

where $N(s_t)$ is the number of times node s_t has been visited, $N(s_t,\ a_i)$ the number of times child i of node s_t has been visited and $Q(s_t,\ a_i)$ is the expected reward of that state. K is a constant that balances between exploitation (left-hand term of the equation) and exploration (right-hand term). MCTS using UCB1 is generally known as UCT. In the simplest case, the default policy π_D is uniformly random: $\pi_D(s_t) = rand(A(s_t))$.

In this paper we study the performance of MCTS on the single-player real-time Physical Travelling Salesman Problem (PTSP). This study is part of an ongoing effort that explores the applicability of MCTS in the domain of real-time video games: the PTSP, where one has to visit n cities as quickly as possible by driving a simple point-mass, provides an excellent case study to better understand the strengths and weaknesses of MCTS in open-ended[1] and time-constrained domains. The PTSP also has the feature that the best possible score for a map is usually unknown. Hence, even assigning a value to a roll-out that does terminate (cause the salesman to visit all cities) raises interesting issues. The experimental studies presented in this paper offer new insights into the behaviour of MCTS in light of these attributes and may be used to create stronger AI players for more complex video games in the future, or perhaps more importantly, create intelligent players with very little game-specific programming required.

2   Literature Review

MCTS was first proposed in 2006 (see [3,4]) and rapidly became popular due to its significant success in computer Go [6], where traditional approaches had been failing to outplay experienced human players. MCTS has since been applied to numerous other games, including games of uncertain information and general game playing. In this section we review applications of MCTS to domains most closely related to the PTSP, including optimisation problems, single-player games (puzzles) and real-time strategy games.

MCTS and other Monte Carlo (MC) methods have been applied to numerous combinatorial optimisation problems, including variations of the classical Travelling Salesman Problem (TSP). For instance, Rimmel et al. [9] used a nested MC algorithm to solve the TSP with time windows, reaching state of the art solutions in problems with no more than 29 cities. Bnaya et al. [2] obtained near-optimal results using UCT to solve the Canadian Traveller Problem, a variation of the TSP where some edges of the graph might be blocked with some probability.

Related to optimisation problems are single player games, also known as puzzles. The Single-Player MCTS (SP-MCTS) was introduced by Shadd et al. [13], where a modification of UCT was proposed in order to include the effect of not having an opponent to play against. The authors found that restarting the seed of the random simulations periodically, while saving the best solution found so far, increases the performance of the

---

[1] Open-ended in this context means that many lines of play will never terminate.

algorithm for single player games. Another puzzle, SameGame, has also been addressed by many researchers, including Schadd et al. [13]. They used SP-MCTS with modified back propagation, parameter tuning and a meta-search extension, to obtain the highest score ever obtained by any AI player so far. Matsumoto et al. [8] incorporated domain knowledge to guide the MC roll-outs, obtaining better results with a little more computational effort. Similarly, Björnsson and Finnsson [1] proposed a modification of standard UCT in order to include the best results of the simulations (in addition to average results) to drive the search towards the most promising regions. They also stored good lines of play found during the search which may be used effectively in single-player games.

MCTS has also been applied widely to *real-time* games. The game Tron has been a benchmark for MCTS in several studies. Samothrakis et al. [11] apply a standard implementation of MCTS, including knowledge to avoid self-entrapment in the MC roll-outs. The authors found that although MCTS works well a significant number of random roll-outs produce meaningless outcomes due to ineffective play. Den Teuling [5] applied UCT to Tron with some modifications, such as progressive bias, simultaneous moves, game-specific simulation policies and heuristics to predict the score of the game without running a complete simulation. The enhancements proposed produce better results only in certain situations, depending on the board layout. Another real-time game that has frequently been considered by researchers is Ms. Pac-Man. Robles et al. [10] expand a tree with the possible moves that Ms. Pac-Man can perform, evaluating the best moves with hand-coded heuristics, and a flat MC approach for the end game prediction. Finally, Samothrakis et al. [12] used MCTS with a 5-player max^n game tree, where each ghost is treated as an individual player. The authors show how domain knowledge produced smaller trees and more accurate predictions during the simulations.

3   The Physical Travelling Salesman Problem

The Physical Travelling Salesman Problem (PTSP) is an extension of the Travelling Salesman Problem (TSP). The TSP is a very well known combinatorial optimisation problem in which a salesperson has to visit n cities exactly once using the shortest route possible, returning to the starting point at the end. The PTSP converts the TSP into a single-player game and was first introduced as a competition at the Genetic and Evolutionary Computation Conference (GECCO) in 2005. In the PTSP, the player always starts in the centre of the map and cities are usually distributed uniformly at random within some rectangular area; the map itself is unbounded.

Although the original PTSP challenge was not time constrained, the goal of the current PTSP is to find the best solution in real-time. At each game tick the agent selects one of five force vectors to be applied to accelerate a point mass around the map with the aim of visiting all cities. The optimality of the route is the time taken to traverse it which differs from its distance as the point-mass may travel at different speeds. At any moment in time, a total of 5 actions may be taken: forward, backward, left, right and neutral. At each time step, the position and velocity of the point-mass is updated using Newton's equations for movement: $v = v_i + a\Delta t$ and $s = s_i + v_i\Delta t + \frac{1}{2}a(\Delta t)^2$ with $\Delta t = \sqrt{0.1}$.

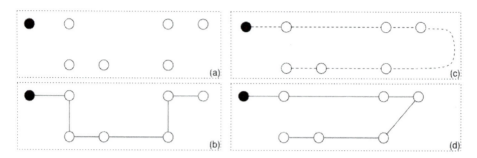

Fig. 1. Example of a 6 city problem where the optimal TSP route differs from the optimal PTSP route: (a) the six cities and starting point (black circle); (b) the optimal TSP solution to this problem without returning to the start; (c) optimal PTSP route and (d) equivalent TSP route which is worse than the route shown in (b)

There are at least two high-level approaches to confront this problem: one possibility is to address the order of cities and the navigation (steering) of the point mass independently. However, it is important to keep in mind that the physics of the game make the PTSP quite different from the TSP. In particular, the optimal order of cities for a given map which solves the TSP does not usually correspond to the optimal set of forces that can be followed by an agent in the PTSP. This is illustrated in Figure 1. Another possible approach to tackle the PTSP is thus to attempt to determine the optimal set of forces and order of cities simultaneously.

The PTSP can be seen as an abstract representation of video games characterised by two game elements: order selection, and steering. Examples of such games include CrystalQuest, XQuest and Crazy Taxi. In particular, the PTSP has numerous interesting attributes that are commonly found in these games: players are required to act quickly as the game progresses at every time step. Furthermore, the game is open-ended as the point-mass may travel across an unbounded map – it is thus highly unlikely that MCTS with a uniform random default policy would be able to reach a terminal state. This requires the algorithm to (a) limit the number of actions to take on each roll-out (depth); and (b) implement a value function that scores each state. Finally, it is important to note that there is no win/lose outcome which affects the value of K for the UCB1 policy (see Equation 1).

4   Preliminary Experimental Study

The application of MCTS to the PTSP requires a well-defined set of states, a set of actions to take for each of those states and a value function that indicates the quality of each state. Each state is uniquely described by the position of the point-mass, its velocity, and the minimum distance ever obtained to all cities. The actions to take are identical across all states (forward, backward, left, right and neutral). Finally, the value (fitness) function used is the summation of the values v_i, based on the minimum distances ever obtained to all cities, plus some penalty due to travelling outside the boundaries of the map. The number of steps is also considered. The value v_i is calculated for each city as follows:

$$v_i = \begin{cases} 0 & \text{if } d_i < c_r \\ f_m - \frac{f_m}{d_i - c_r + 2} & \text{otherwise} \end{cases} \tag{2}$$

where d_i represents the distance between the point-mass and the city i, c_r is the radius of each city and f_m is the maximum value estimated for the fitness. This equation forces the algorithm to place more emphasis on the positions in the map that are very close to the cities. The score associated with each state is normalised: the maximum fitness is equivalent to the number of cities multiplied by f_m, plus the penalties for travelling outside the boundaries of the map and the number of steps performed so far. The normalisation is very important to identify useful values of K which has been set to 0.25 in this study following systematic trial and error.

Two default policies have been tested. The first uses uniform random action selection while the second, DRIVEHEURISTIC, includes some domain knowledge to bias move selection: it penalises actions that do not take the agent closer to any of the unvisited cities. This implies that actions which minimise the fitness value are more likely to be selected. Four algorithms are considered in this preliminary experiment: the simplest is 1-ply MC Search which uses uniform random action selection for each of the actions available from the current state, selecting the best one greedily. A slight modification of this is Heuristic MC which biases the simulations using the DRIVEHEURISTIC. The first MCTS variant is using UCB1 as tree policy and uniform random rollouts. The heuristic MCTS implementation also uses UCB1 as its tree policy, but biases the rollouts using the DRIVEHEURISTIC.

To compare the different configurations, the following experiments have been carried out on 30 different maps that have been constructed uniformly at random. A minimum distance between cities prevents overlap. The same set of 30 maps has been used for all experiments and configurations were tested over a total of 300 runs (10 runs per map). Finally, the time to make a decision at each stage of the problem has been set to 10 milliseconds.

The results are shown in Table 1.[2] It is evident that MCTS outperforms, with respect to the number of best solutions found, both 1-ply MC Search and MCTS with a heuristic in the roll-outs. The differences in the average scores are, however, insignificant. The observation that 1-ply MC Search is achieving similar results to MCTS suggests that the information obtained by the roll-outs is either not utilised efficiently or is simply not informative enough. If this is the case, the action selection and/or tree selection mechanism cannot be effective.

Figures 2 and 3 depict a visualisation of the MCTS tree at a particular iteration: each position, represented in the figure as a pixel on the map, is drawn using a grey scale that represents how often this position has been occupied by the point-mass during the MC simulations. Lighter colours indicate more presence of those positions in the simulations, while darker ones represent positions less utilised. Figure 3 is of a special interest: the simulations are taking the point-mass to positions where no cities can be found. This happens because no other portion of the map (where cities are located) is explored and thus MCTS is unable to steer the point-mass towards regions of higher fitness.

---

[2] The experiments were executed on an Intel Core i5 PC, with 2.90GHz and 4GB of memory.

Table 1. Results for 10 city maps compared in terms of the number of time steps required to solve the problem. The *Best count* indicates how often the algorithm produced the overall best solution, while *Not solved* shows the number of trials where the algorithm was unable to find a solution. Finally, *Average simulations* indicates how many simulations were performed, on average, in each execution step.

Algorithm	Average	Standard Error	Best count	Not solved	Average simulations
1-ply Monte Carlo Search	539.65	3.295	0	1	816
Heuristic MC	532.63	3.302	3	0	726
MCTS UCB1	531.25	3.522	10	2	878
Heuristic MCTS UCB1	528.77	3.672	2	0	652

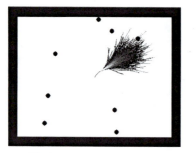

Fig. 2. Tree exploration at the start

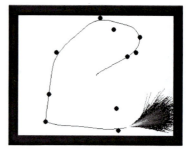

Fig. 3. No cities close to tree exploration

5   Extended Experimental Study

The objective of the extended experimental study is to analyse the impact of additional domain knowledge on the performances of the algorithms. In particular, we exploit the concept of a *centroid*, calculated as the centre of all unvisited cities (and the centre of the map). The idea is to prevent the selection of actions that take the point-mass away from that point. However, this penalty cannot be used all the time, because otherwise the point-mass would always be drawn towards the centroid, not visiting any cities. In order to decide when this heuristic is to be enabled, we define a circle with a certain radius r, centred on the centroid, to be the centroid's influence.

The maps depicted in Figures 4 and 5 show a set of cities, the centroid (located near the center of the map) and the centroid's area of influence.[3] The value of r used is the distance to the farthest city from the centroid, multiplied by a factor e; the value of e can be used to modulate how far the point-mass is allowed to go from the centroid. In this study the value is set to 1.05.

We define the following new algorithms using the CENTROIDHEURISTIC: the *Centroid Heuristic MC* is identical to the *Heuristic MC* but the heuristic used to guide the MC simulations ignores actions that do not take the point-mass towards the centroid

---

[3] Videos of the tree and the *CentroidHeuristic* may be found at
www.youtube.com/user/MonteCarloTreeSearch.

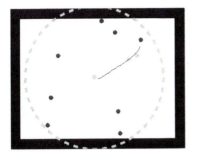

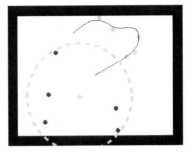

Fig. 4. Centroid and influence   **Fig. 5.** Centroid and influence update

Table 2. 10 city result comparison, 10ms limit

Algorithm	Average	Standard Error	Best count	Not solved	Average simulations
1-ply Monte Carlo Search	539.65	3.295	1	1	816
Heuristic MC	532.63	3.302	1	0	726
MCTS UCB1	531.25	3.522	2	2	878
Heuristic MCTS UCB1	528.77	3.672	1	0	652
Centroid Heuristic MC	552.87	4.158	2	0	915
Centroid Heuristic MCTS	524.13	3.443	2	0	854
Centroid MCTS only UCT	599.38	10.680	6	76	1009
Centroid MCTS & UCT	**481.85**	6.524	**12**	0	659

(if within the centroid influence). Likewise, the *Centroid Heuristic MCTS* is similar to the *Heuristic MCTS* but uses the *CentroidHeuristic* during the default policy. The *Centroid MCTS only UCT* is identical to the standard *MCTS* algorithm but uses the *CentroidHeuristic* in the tree policy, by not allowing the selection of those actions that do not take the point-mass towards the centroid (if within the centroid influence). Finally, the *Centroid MCTS & UCT* is similar to the *Centroid Heuristic MCTS only UCT* using the *CentroidHeuristic* also in the default policy.

5.1   Random Maps of 10 Cities

The results for random maps of 10 cities are shown in Table 2. It is evident that solution quality was improved by the CENTROIDHEURISTIC. The average solution quality, using the centroid heuristic for both the tree selection and MC roll-outs, is 481.85, with a low standard error and a very good count of best solutions found: both *Centroid MCTS only UCT* and *Centroid MCTS & UCT*, with $K = 0.05$, achieve more than the 50% of the best scores. The Kolmogorov-Smirnov test confirms that these results are significant. The test provides a *p-value* of 1.98×10^{-22} when comparing *1-ply MC Search* and *Centroid MCTS only UCT*, and 1.98×10^{-22} for *1-ply Monte Carlo Search* against *Centroid MCTS & UCT*.

Similar results have been obtained for time steps of 50 milliseconds. In this case, the *1-ply Monte Carlo Search* algorithm achieves an average time of 514.31, while *MCTS UCB1*, *Centroid MCTS only UCT* and *Centroid MCTS & UCT* obtain 511.87, 556.53 and 469.72 respectively. Several things are worth noting from these results: first, the algorithms perform better when the time for simulations is increased. Second, it is interesting to see how the different MCTS configurations (specially *MCTS UCB1* and *Centroid MCTS only UCT*) improve more than the MC techniques when going from 10 to 50ms. The third MCTS configuration, which obtains the best results for both time limits, does not improve its solution quality as much as the other algorithms when increasing the simulation time. However, it is important to note that the results obtained by this configuration given 10ms are better than the best solution found by any other algorithm given 50ms. It is highly significant that the solutions obtained by this algorithm are the best ones found for this problem, showing an impressive performance even when the available time is very limited. This makes the approach very suitable for time-constrained real-time games.

5.2   Random Maps of 30 Cities

To check if the results of the previous section are consistent, some experiments were performed with 30 cities. Table 3 shows the results of these algorithms for a time limit of 10ms. The results are similar to the ones recorded in the 10 cities experiments, although in this case *Centroid MCTS only UCT* with $K = 0.05$ is the algorithm that solves the problem in the least number of time steps. Figure 6 shows the performance of some of the configurations tested for the different time limits considered. Comparing *1-ply Monte Carlo Search* with the *Centroid MCTS only UCT* using the Kolmogorov-Smirnov test gives the following p-values for 10ms, 20ms and 50ms respectively: 0.349, 0.0005 and 1.831×10^{-7}. This confirms significance in the case of 20ms and 50ms, but not in the case of 10ms.

Table 3. 30 city result comparison, 10ms limit

Algorithm	Average	Standard Error	Best count	Not solved	Average simulations
1-ply Monte Carlo Search	1057.01	6.449	2	0	562
Heuristic MC	1133.10	6.581	0	11	319
MCTS UCB1	1049.16	6.246	6	0	501
Heuristic MCTS UCB1	1105.46	5.727	2	3	302
Centroid Heuristic MC	1119.93	6.862	0	1	441
Centroid Heuristic MCTS	1078.51	5.976	1	0	428
Centroid MCTS only UCT	1032.94	6.365	7	14	481
Centroid MCTS & UCT	1070.86	6.684	7	0	418

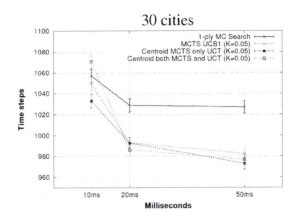

Fig. 6. Performance of the algorithms when time limit changes (30 cities)

6   Conclusions

This paper analyses the performance of Monte Carlo Tree Search (MCTS) on the Physical Travelling Salesman Problem (PTSP), a real-time single player game. The two experimental studies outlined in this paper focus on the impact of domain knowledge on the performance of the algorithms investigated and highlight how a good heuristic can significantly impact the success rate of an algorithm when the time to select a move is very limited.

The results show that the *CentroidHeuristic* helps the algorithm to find better solutions, especially when the time allowed is very small (10ms). As shown in the results, when the time limit is 10ms, some approaches (like MCTS without domain knowledge) are not able to provide significantly better results than 1-ply Monte Carlo Search. They are, however, able to produce superior results when the time limit is increased. The main contribution of this research is evidence to show that it is possible to effectively utilise simple domain knowledge to produce acceptable solutions, even when the time to compute the next move is heavily constrained.

The off-line version of the problem has also been solved with evolutionary algorithms, notably in the GECCO 2005 PTSP Competition. In fact, the winner of that competition utilised a genetic algorithm, using a string with the five available forces as a genome for the individuals (results and algorithms employed can be found at cswww.essex.ac.uk/staff/sml/gecco/ptsp/Results.html).   This PTSP solution format, a string of forces, is a suitable representation for evolutionary algorithms that may be applied to this problem. A thorough comparison of evolution versus MCTS for this problem would be interesting future work.

Other ongoing work includes the use of more interesting maps (by introducing obstacles, for instance), modified game-physics to steer a vehicle rather than a point-mass, and the inclusion of more players that compete for the cities. Competitions based on these variations are already in preparation. The results of these should provide further insight into how best to apply MCTS to the PTSP, as well as its strengths and weaknesses compared to evolutionary and other optimisation methods.

Finally, a particular challenge for MCTS applied to the PTSP is how to persuade it to make more meaningful simulations that consider the long-term plan of the order in which to visit the cities, together with the short term plan of how best to steer to the next city or two.

Acknowledgements. This work was supported by EPSRC grant EP / H048588 / 1.

References

1. Björnsson, Y., Finnsson, H.: CadiaPlayer: A Simulation-Based General Game Player. IEEE Trans. on Computational Intelligence and AI in Games 1(1), 4–15 (2009)
2. Bnaya, Z., Felner, A., Shimony, S.E., Fried, D., Maksin, O.: Repeated-task Canadian traveler problem. In: Proceedings of the International Symposium on Combinatorial Search, pp. 24–30 (2011)
3. Chaslot, G.M.J.-B., Bakkes, S., Szita, I., Spronck, P.: Monte-Carlo Tree Search: A New Framework for Game AI. In: Proc. of the Artificial Intelligence for Interactive Digital Entertainment Conference, pp. 216–217 (2006)
4. Coulom, R.: Efficient Selectivity and Backup Operators in Monte-Carlo Tree Search. In: van den Herik, H.J., Ciancarini, P., Donkers, H.H.L.M(J.) (eds.) CG 2006. LNCS, vol. 4630, pp. 72–83. Springer, Heidelberg (2007)
5. Den Teuling, N.G.P.: Monte-Carlo Tree Search for the Simultaneous Move Game Tron. Univ. Maastricht, Tech. Rep. (2011)
6. Gelly, S., Silver, D.: Monte-Carlo tree search and rapid action value estimation in computer Go. Artificial Intelligence 175(11), 1856–1875 (2011)
7. Kocsis, L., Szepesvári, C.: Bandit Based Monte-Carlo Planning. In: Fürnkranz, J., Scheffer, T., Spiliopoulou, M. (eds.) ECML 2006. LNCS (LNAI), vol. 4212, pp. 282–293. Springer, Heidelberg (2006)
8. Matsumoto, S., Hirosue, N., Itonaga, K., Yokoo, K., Futahashi, H.: Evaluation of Simulation Strategy on Single-Player Monte-Carlo Tree Search and its Discussion for a Practical Scheduling Problem. In: Proc. of the International Multi Conference of Engineers and Computer Scientists, vol. 3, pp. 2086–2091 (2010)
9. Rimmel, A., Teytaud, F., Cazenave, T.: Optimization of the Nested Monte-Carlo Algorithm on the Traveling Salesman Problem with Time Windows. In: Di Chio, C., Brabazon, A., Di Caro, G.A., Drechsler, R., Farooq, M., Grahl, J., Greenfield, G., Prins, C., Romero, J., Squillero, G., Tarantino, E., Tettamanzi, A.G.B., Urquhart, N., Uyar, A.Ş. (eds.) EvoApplications 2011, Part II. LNCS, vol. 6625, pp. 501–510. Springer, Heidelberg (2011)
10. Robles, D., Lucas, S.M.: A Simple Tree Search Method for Playing Ms. Pac-Man. In: Proc. of the IEEE Conference on Computational Intelligence and Games, pp. 249–255 (2009)
11. Samothrakis, S., Robles, D., Lucas, S.M.: A UCT Agent for Tron: Initial Investigations. In: Proc. of IEEE Conference on Computational Intelligence and Games, pp. 365–371 (2010)
12. Samothrakis, S., Robles, D., Lucas, S.M.: Fast Approximate Max-n Monte-Carlo Tree Search for Ms Pac-Man. IEEE Trans. on Computational Intelligence and AI in Games 3(2), 142–154 (2011)
13. Schadd, M.P.D., Winands, M.H.M., van den Herik, H.J., Chaslot, G.M.J.-B., Uiterwijk, J.W.H.M.: Single-Player Monte-Carlo Tree Search. In: van den Herik, H.J., Xu, X., Ma, Z., Winands, M.H.M. (eds.) CG 2008. LNCS, vol. 5131, pp. 1–12. Springer, Heidelberg (2008)

Diversified Virtual Camera Composition

Mike Preuss, Paolo Burelli, and Georgios N. Yannakakis

Computational Intelligence Group, Dept. of Computer Science,
Technische Universität Dortmund, Germany and
Center for Computer Games Research, IT University of Copenhagen, Denmark
mike.preuss@tu-dortmund.de, {pabu,yannakakis}@itu.dk

Abstract. The expressive use of virtual cameras and the automatic generation of cinematics within 3D environments shows potential to extend the communicative power of films into games and virtual worlds. In this paper we present a novel solution to the problem of virtual camera composition based on niching and restart evolutionary algorithms that addresses the problem of diversity in shot generation by simultaneously identifying multiple valid camera camera configurations. We asses the performance of the proposed solution against a set of state-of-the-art algorithms in virtual camera optimisation.

1 Introduction

In computer games, as well as in most 3D applications, effective camera placement is fundamental for the user to understand the virtual environment and be able to interact. Camera settings for games are usually directly controlled by the player or statically predefined by designers. Direct control of the camera by the player increases the complexity of the interaction and reduces the designer's ability to control game storytelling (e.g. the player might manually look at an object revealing an unwanted information). Statically defined cameras, on the other hand, release the player from the burden of controlling the point of view, but often fail to correctly frame the game actions. Moreover, when the game content is procedurally generated, the designer might not have the necessary information to define, a priori, the camera positions and movements.

Automatic camera control aims to define an abstraction layer that permits the designers to instruct the camera with high-level and environment-independent rules. The camera controller should dynamically and effectively translate these rules into camera movements. Most researchers model this problem as an optimisation problem [8] in which the search space is the space of all the possible camera configurations and high level properties are modelled as an objective function to be optimised.

Although the space of possible camera configurations is relatively low dimensional (at least 5 dimensions to define position and orientation), automatic camera control is a complex optimisation problem for two reasons: the evaluation functions corresponding to frame properties often generate landscapes that are very rough for a search algorithm to explore [7] and the evaluation of such properties is computationally expensive with respect to the time available for computation (16ms for real time applications), significantly reducing the number of evaluations available for the search process. In general, these problems seem to be highly multimodal, but the degree of ruggedness and the number of basins may vary a lot across different instances [7].

C. Di Chio et al. (Eds.): EvoApplications 2012, LNCS 7248, pp. 265–274, 2012.

To the authors knowledge, all the research carried out to solve this optimisation problem focuses on providing more accurate, robust and efficient algorithms to find the best possible camera configuration given the objective function defined by the designer's requirements. However, as pointed out by Thawonmas et al. [22], one single best solution is often unsatisfactory. When filming a scene with little movement, such as a dialogue, selecting always the same solution will lead to a repetitive direction. While this might be the explicit will of the designer, it is often an issue for media such as films and comics. Thawonmas et al. address this problem by randomizing the shot definition; such a solution, however, acts on the design of the shot rather than on the implementation, potentially disrupting the intended message. We consider the problem of providing multiple alternative good solutions as largely unsolved, and it naturally calls for application of niching methods because they are designed for providing more than one solution. However, as the *black box optimization benchmark* (BBOB) competitions[1] at GECCO 2009 and 2010 conferences have shown (see [2] for data and [13] for a comprehensive analysis and summary), the CMA-ES also copes well with multimodal functions due to its clever restart mechanisms and naturally, each restart delivers an approximation for a local optimum. Consequently, we intend to pursue the following tasks in this work:

a) Assess if modern evolutionary algorithm approaches as niching and restart based variants of the CMA-ES [10] are capable of reliably providing multiple diverse good solutions to the problem quickly;
b) investigate the trade-off between diversity and quality (in solutions) by setting up specific performance criteria and comparing our suggested methods with different state-of-the-art ones;
c) collect some (experimentally based) knowledge about the landscape structure, following the idea of *exploratory landscape analysis* (ELA) [14], in order to allow for even faster future algorithm implementations.

Our approaches exploit the multi-modal nature of the camera optimisation problem and identify multiple alternative solutions basins, thereby also revealing much information about the fitness landscape of the problem. Each basin contains potentially optimal camera configurations that have comparable fitness, but different visual aspect; such configurations can be used to diversify the shots while maintaining the designers requirements. However, in order to correctly estimate the suitability of the different algorithms, we make several simplifications that have to be rethought when applying them under real-time conditions:

a) We relax the runtime limit by allowing longer runs than would be possible in 16ms. This follows the *make it run first, then make it run fast* principle. Once good methods are found, they can be further adjusted to the problem to increase performance.
b) For now, we ignore the multi-objective nature of the problem as this will most likely make it even harder. This must be considered later on when already challenging single-objective formulation is solved sufficiently.

---

[1] http://coco.gforge.inria.fr/doku.php

In the remaining of the paper we describe the current state-of-the-art in virtual camera composition, we present our algorithmic approaches and showcase their capabilities and performance in a set of test environments.

2   Related Work

Since the introduction of virtual reality, virtual camera control attracted the attention of a large number of researchers [8]. Early studies on virtual camera [23] investigated manual camera control metaphors for exploration of virtual environments and manipulation of virtual objects. However, direct control of the several degrees of freedom of the camera showed often to be problematic for the user [9] leading researchers to investigate for the automation of camera control.

In 1988, Blinn [4] showcased one of the first examples of an automatic camera control system. Blinn designed a system to automatically generate views of planets in a NASA space simulator. Although limited in its expressiveness and flexibility, Blinn's work inspired many other researchers trying to investigate efficient solutions and more flexible mathematical models able to handle more complex aspects such as camera motion and frame composition [1].

More generic approaches model camera control as an optimisation problem by requiring the designer to define a set of targeted frame properties which are then put into an objective function. These properties describe how the frame should look like in terms of object size, visibility and positioning. Olivier et al. [15] first formalised the camera control problem as an optimisation problem and introduced detailed definition of these properties. Since then, numerous search strategies have been applied to solve the problem, including population based algorithms, local search algorithms and combinations of the two [8]. These approaches offer different performances with respect to computational cost, robustness and accuracy; however, none of them regards diversity of solutions as a key characteristic.

Thawonmas et al. [22] identify variety of shots as a major problem in automatic generation of cinematics and they introduce a roulette-wheel selection mechanism to force variety in shot descriptions. However, by altering the shot properties, this approach does not only vary the shot visual aspect but potentially changes the shot meaning.

We propose the application of niching and restart evolutionary algorithms based on the real-valued blackbox optimization method CMA-ES to the virtual camera composition problem, to find multiple alternative solutions during the optimisation process and we showcase its performance with respect of a selection of state-of-the-art algorithms.

3   Virtual Camera Composition

An optimal camera configuration is defined as the combination of camera settings which maximises the satisfaction of the requirements imposed on the camera, known as camera profile. A camera profile describes the characteristics of the image that the camera should generate in terms of composition properties. Based on the author's previous work on automatic camera control [6], the properties that can be imposed are: *Object Visibility*, *Object Projection Size*, *Object View Angle* and *Object Frame Position*.

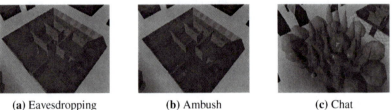

(**a**) Eavesdropping   (**b**) Ambush   (**c**) Chat

Fig. 1. Test problems' virtual environments

The first property defines whether an object (or a part of it) should be visible in the frame, the second defines the size an object should have in the frame, the third one defines the angle from which the camera should frame the object and the fourth one defines the position that the projected image of the object should have in the frame.

Each composition property corresponds to an objective function which describes the satisfaction of such property. The complete virtual camera composition objective function F is a linear combination of the objective functions corresponding to each property included in the camera profile.

3.1 Test Problems

In order to assess the performance of the proposed solutions we compare their convergence behaviour with a set of state-of-the-art algorithms across three test problems. Each test problem requires the camera to frame a common game situation (e.g. a dialogue between virtual characters) according to a set of standard cinematographic visual properties. The problems are set in a virtual 3D environment including a large variety of geometrical features of modern computer games such as closed rooms, walls or trees. The set of properties of the desired camera configuration and the virtual environments are designed to include alls the typical optimisation challenges of the virtual camera composition problem such as lack of gradient or multi-modality.

In the first problem (Fig. 1a) the environment includes three characters, with two of them facing each other and ideally chatting, while the third one eavesdropping. The properties for this problem include full visibility of all characters and a projection size equal to one third of the screen for all characters. In the second problem (Fig. 1b) the environment includes two characters on two sides of a wall. The properties for this problem include full visibility of all characters and a projection size equal to half of the screen for all characters and an horizontal angle of 90 degrees to the right of each character. The last problem is based on the chat scene by Thawonmas et al. [22] and it includes three characters with one ideally chatting to the other two. The visibility and projection size properties are equal to the ones in the first problem but the camera is also expected to be on the back of the listening characters.

The first and the second problems are set in an indoor environment with closed spaces separated by solid walls. As described in [7], walls act as large occluders inducing large areas of the objective function landscape to have little or no gradient. Figures 2a and 2b display the aforementioned characteristic which are smoothed by the presence of other properties besides visibility in the problem description. The third problem is set

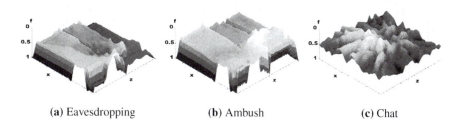

(a) Eavesdropping   **(b)** Ambush   **(c)** Chat

Fig. 2. Maximum value of the problems' objective function sampled across the X and Z axis of the virtual test environments

in an outdoor environment composed by a cluster of trees. As displayed in Fig 2c, such environment influences the objective function landscape by increasing the modality.

4   Niching and Restart CMA-ES Variants under Test

Niching in evolutionary optimization dates back at least to the 1970s with the suggestion of Sharing and Crowding. Its general idea is that by organizing the search process and keeping several populations/local searches separate from each other we can obtain multiple good solutions at once, which is not that far from the scheme of modern real-valued memetic seach algorithms. In the biological protoype (Earth), niching works well because the surface on which most lifeforms move around is only 2 dimensional. However, in optimization, we usually have a larger number of dimensions, so that human intuition can get very wrong about distances, relative positions and volume sizes and the principles of geometry get less and less applicable. The test case we have here is interesting, as its 5 dimensions place it somewhere between 'well suited' (2D) and 'not applicable' ($> 20D$) with respect to niching algorithms. The number of available niching algorithms is quite large, recent suggestions include e.g. [19], [20], [16], and we by no means claim that we are able to select the most appropriate niching EA (this would hardly be possible without knowing much more about the problem properties).

We therefore resolve to an algorithm that is a further development of [16] which is to date the only niching method with documented results on the BBOB test set. We call the original version (also labelled as NBC-CMA) *niching evolutionary algorithm 1* (NEA1) here to differentiate it from the newer version we term NEA2. NEA1 highly relies on the CMA-ES as local searcher, but uses a much larger starting population ($40 \times D$) on which the nearest-better clustering method is run to separate it into populations representing different basins of attraction [18]. This topological clustering method connects every search point in the population to the nearest one that is better and cuts the connections that are longer than $2\times$ the average connection. The remaining connections determine the found clusters by computing the weakly connected components. This works very well for a reasonably large population in two dimensions, but increasingly fails if the number of dimensions increases. Therefore, in NEA2, a second additional cutting rule has been implemented: For all search points that have at least 3 incoming connections (it is the nearest better point for at least 3 others), we divide the length of its own nearest-better connection (in case it has none it is the best point and has surely been treated by the old rule) by the median of its incoming connections. If this is larger than a precomputed correction factor, the outgoing connection is cut (and we have one additional cluster). The correction factor cf has been experimentally derived and depends on D and the population size $\#elems$. This works astonishingly well for up to around 20D and not too complex landscapes.

$$cf = -4.69 * 10^{-4} * D^2 + 0.0263 * D + 3.66/D - 0.457 * log10(\#elems)$$
$$+7.51e - 4 * D^2 - 0.0421 * D - 2.26/D + 1.83 \qquad (1)$$

As both cutting rules are heuristics that work well in many cases but come without guarantee, the number of resulting clusters had to be limited in NEA1, as it processes all clusters as separate CMA-ES populations in parallel. This can result in very long runtimes in cases where the clustering was not very accurate. NEA2 overcomes this problem by switching from a BFS-like to a DFS-like search in which the clusters are treated sequentially sorted according to their best members (best first, see [17] for details). Should the problem be less multimodal then detected, (e.g. unimodal), NEA2 would perform very similar to the CMA-ES as every start point leads to the same optimum. Although these niching methods are still much simpler than many other ones suggested in literature, they are arguably still much more complex than a restart CMA-ES.

However, there is a much simpler way to cope with organizing the search, and that is by just randomly chosing a new starting position as soon as stagnation is detected. Of course, this does not require to compare positions in search space and should work especially well in higher dimensions, when the geometry-based niching must fail. The CMA-ES does just that and is currently one of the leading algorithms in real-valued black-box optimization. We therefore add it to the algorithm test set, as a reference and reliable default solution. As the problem is highly time-critical and thus only very few evaluations are allowed, the CMA-ES is run without heuristic population enlargement as e.g. proposed with the IPOP- [3] and BIPOP-CMA variants. All parameters are left at their default values with the exception of the `TolFun` stopping criterion which is highly connected to the desired accuracy [11]. This is set to a value of 10^{-3} which is still below the needed accuracy. The effect of this setting is that fruitless searches in local optima are stopped earlier, thus more restarts can be done. As the NEA2 internally also heavily relies on the CMA and its stopping criteria, it is also affected by this change.

5   Experimental Analysis

5.1   Measures

Next to the raw performance (best obtained objective value over time), we measure the diversity target by first defining the properties of one/multiple suitable solutions. A solution is considered good enough if its fitness value is ≤ 0.05 (fitness values range from 0 to 1). This is an ad-hoc definition, but the first test runs told us that this quality can be achieved for all 3 problem instances. The motivation for 0.05 is that for a human, it will be hard to discriminate these solutions from the one with 0 values, thus they can be considered good enough for the practical application.

It is somewhat harder to determine when several good solutions are useful (this would not be the case if they are too similar). For discriminating useful alternatives, we demand a minimal Euclidean distance of at least 1 in the three spatial coordinates, regardless of the camera angles. The expected time to reach the desired quality is computed over several repeated runs after the *expected runtime* (ERT)[2] definition suggested in [3], with $\#fevals$ being the sum of all evaluations that were spend before reaching the target value $f_{target} = 0.05$, and $\#succ$ standing for the number of successful runs:

$$ERT = \frac{\#fevals > f_{target}}{\#succ} \tag{2}$$

As we desire several good solutions, we denote the ERT for the first one by ERT1, and the running times for the next ones (that have to fulfill the distance criterion concerning all the already detected ones) as ERT2 and ERT3, respectively. The diversity of the attained solutions is also of interest and it is measured by taking the average over the distance sums from the first solution to every other solution per run.

5.2 Experiment

With the following experiment we want to find out which of the suggested algorithms, CMA-ES, NEA2, Particle Swarm Optimisation (PSO) [12], Differential Evolution (DE) [21], or Sliding Octree (SO) [5] is capable of reliably delivering multiple, diverse and good solutions quickly, and to pursue the goals named in the introduction. NEA1 is only added for a performance comparison to NEA2. Standard variants of DE, PSO and SO methods have been included as representatives of previous approaches to the problem.

Pre-experimental planning. During the first test runs, we found that a run length of 5000 evaluations is usually enough for the algorithms to converge to (best) solutions of around 0.05 or better.

Setup. We run each algorithm on each of the three problem instances (scenarios) 20 times for 5000 evaluations. Performance is measured as given in sec. 5.1. All parameters are kept at default values, except for the `TolFun` stopping criterion (applying to CMA-ES, NEA2 and NEA1) which is set to 10^{-3}. Default values for NEA2 resemble the ones of NEA1, given in [16]. The start population is determined randomly for the CMA-ES based methods and the stepsize start value is set to 0.15 in the normalized parameter space $[0, 1]$.

Task. We do not dare to declare a clearly winning algorithm, instead we demand that the methods find at least 2 sufficiently good and diverse solutions reliably and call these algorithms 'suitable' to the problem, to be considered for further work. However, we take out Wilcoxon rank-sum tests between the time needed to the first optimum as measured in each run (together resembling ERT1), between the different algorithms.

Results/Visualization. Figure 3 shows the table of the diversity measures and depicts the median best solutions over the number of spent evaluations for all three instances.

---

[2] The term may be misleading as it is defined in evaluations, for absolute times it has to be multiplied with 16 ms.

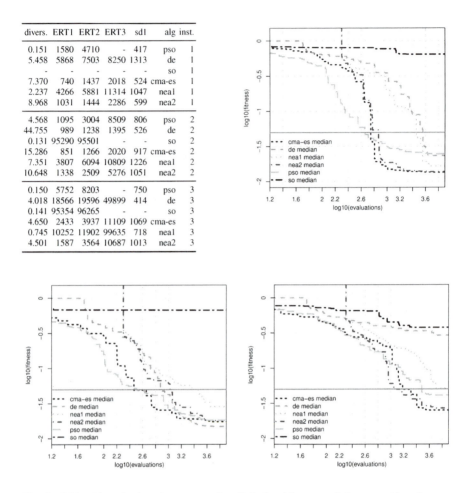

divers.	ERT1	ERT2	ERT3	sd1	alg	inst.
0.151	1580	4710	-	417	pso	1
5.458	5868	7503	8250	1313	de	1
-	-	-	-	-	so	1
7.370	740	1437	2018	524	cma-es	1
2.237	4266	5881	11314	1047	nea1	1
8.968	1031	1444	2286	599	nea2	1
4.568	1095	3004	8509	806	pso	2
44.755	989	1238	1395	526	de	2
0.131	95290	95501	-	-	so	2
15.286	851	1266	2020	917	cma-es	2
7.351	3807	6094	10809	1226	nea1	2
10.648	1338	2509	5276	1051	nea2	2
0.150	5752	8203	-	750	pso	3
4.018	18566	19596	49899	414	de	3
0.141	95354	96265	-	-	so	3
4.650	2433	3937	11109	1069	cma-es	3
0.745	10252	11902	99635	718	nea1	3
4.501	1587	3564	10687	1013	nea2	3

Fig. 3. Table: Diversity based measures for all 6 algorithms on all 3 test problem instances, sd1 resembles the standard deviation over the successful detections of the first solution. Figures: Empirical attainment surface plots of the best obtained solutions over time (only 50% attainment surface), for the three problem instances (first in the upper row). The red line marks the required quality for an applicable solution. Logarithmic scaling on both axes.

Observations. As the variances in the ERT values are quite high (see e.g. the sd1 value), it is dangerous to read too much out of the obtained result. However, from the pictured median performance values, we can clearly see that the third scenario is the hardest, followed by the first one, and the second scenario is the easiest. Concerning the different algorithms, SO does not solve any test case, DE does not solve instance 3 and is very slow on instance 1, and NEA1 is not much better. It is noteworthy that DE is the fastest method to obtain at least 2 or 3 diverse solutions for instance 2. PSO mostly converges quickly to the first solution but needs a lot of time to provide the second one. CMA-ES and NEA2 are both reliable in detecting several solutions, where CMA-ES looks clearly favourable for the simple and the medium instance, and NEA2 a bit better on the hard one. With the noteable exception of DE on instance 2, the diversity values obtained by the best algorithms are comparable. We review the results of our speed based statistical tests only for the leading algorithms: in scenario 1, PSO is significantly worse than CMA-ES and NEA2, but CMA-ES and NEA2 cannot be differentiated. In scenario 2, the leading three (DE, PSO, CMA-ES) are not distinguishable, only between CMA-ES and NEA2 we get significance (at the 5%-level). For scenario 3, the difference of CMA-ES and NEA2 is just significant, while the CMA-ES itself is significantly faster than all others.

Discussion. Why DE fails to solve medium or hard instances cannot be easily seen, possibly this is due to premature convergence to a bad local optimum. PSO clearly needs a better restart mechanism as the convergence is often fast but no second best solution can be obtained. However, we would not recommend to use both algorithms for these kind of problems in their current form. More instances would be needed to collect better evidence on the relationship between problem hardness and algorithm performance, but it seems that as a default method, one should employ a CMA-ES unless it is known that the problem is very hard, then niching methods as NEA2 can pay off. At least in the case of given quality and distance requirements, it seems that concentrating on the diversity instead of convergence speeds does not change much, the good algorithms are still the same. This may of course change if no concrete quality and distance tasks are provided.

6 Summary and Conclusions

This paper proposed the application of niching and restart evolutionary algorithms to the problem of diversity of shot generation in virtual camera composition. The suggested algorithms are compared against state-of-the-art algorithms for optimisation of virtual camera composition and have been evaluated in their ability to find up to three different valid solutions on three different problems with varying complexity. Both NEA2 and CMA-ES show at least comparable performance to the standard optimisation algorithms in terms of number of evaluations required to find the first solution; however, for the second and third solution, these two algorithms demonstrated a clear advantage compared to all others included in the experiment.

The actual analysis has been performed using the Euclidean distance as a diversity measure between the solutions. Even though it is effective, this solution does not evaluate accurately how different are the shots generated by two solutions. In the future it is advisable to investigate different objectives next to visibility and also new diversity measurements such as the Euclidean distance in the multi-objective (target) space.

References

1. Arijon, D.: Grammar of the Film Language. Silman-James Press, LA (1991)
2. Auger, A., Finck, S., Hansen, N., Ros, R.: BBOB 2010: Comparison Tables of All Algorithms on All Noiseless Functions. Technical Report RT-388, INRIA (September 2010)

3. Auger, A., Hansen, N.: A restart cma evolution strategy with increasing population size. In: Proceedings of the IEEE Congress on Evolutionary Computation, CEC 2005, Edinburgh, UK, September 2-4, pp. 1769–1776. IEEE Press (2005)
4. Blinn, J.: Where Am I? What Am I Looking At? IEEE Computer Graphics and Applications 8(4), 76–81 (1988)
5. Bourne, O., Sattar, A., Goodwin, S.: A Constraint-Based Autonomous 3D Camera System. Journal of Constraints 13(1-2), 180–205 (2008)
6. Burelli, P., Yannakakis, G.N.: Combining Local and Global Optimisation for Virtual Camera Control. In: IEEE Conference on Computational Intelligence and Games, p. 403 (2010)
7. Burelli, P., Yannakakis, G.N.: Global Search for Occlusion Minimisation in Virtual Camera Control. In: IEEE Congress on Evolutionary Computation, pp. 1–8. IEEE, Barcelona (2010)
8. Christie, M., Olivier, P., Normand, J.M.: Camera Control in Computer Graphics. Computer Graphics Forum 27, 2197–2218 (2008)
9. Drucker, S.M., Zeltzer, D.: Intelligent camera control in a virtual environment. In: Graphics Interface, pp. 190–199 (1994)
10. Hansen, N., Ostermeier, A.: Completely derandomized self-adaptation in evolution strategies. Evolutionary Computation 9(2), 159–195 (2001)
11. Hansen, N.: The cma evolution strategy: A tutorial, http://www.lri.fr/~hansen/cmatutorial.pdf (version of June 28, 2011)
12. Kennedy, J., Eberhart, R.C.: Particle swarm optimization. In: IEEE Conference on Neural Networks, pp. 1942–1948 (1995)
13. Mersmann, O., Preuss, M., Trautmann, H., Bischl, B., Weihs, C.: Analyzing the bbob results by means of benchmarking concepts. Evolutionary Computation (accepted, 2012)
14. Mersmann, O., Bischl, B., Trautmann, H., Preuss, M., Weihs, C., Rudolph, G.: Exploratory landscape analysis. In: Proceedings of the 13th Annual Conference on Genetic and Evolutionary Computation, GECCO 2011, pp. 829–836. ACM, New York (2011)
15. Olivier, P., Halper, N., Pickering, J., Luna, P.: Visual Composition as Optimisation. In: Artificial Intelligence and Simulation of Behaviour (1999)
16. Preuss, M.: Niching the cma-es via nearest-better clustering. In: Proceedings of the 12th Annual Conference Companion on Genetic and Evolutionary Computation, GECCO 2010, pp. 1711–1718. ACM (2010)
17. Preuss, M.: Improved Topological Niching for Real-Valued Global Optimization. In: Di Chio, C., et al. (eds.) EvoApplications 2012. LNCS, vol. 7248, pp. 386–395. Springer, Heidelberg (2012)
18. Preuss, M., Schönemann, L., Emmerich, M.: Counteracting genetic drift and disruptive recombination in $(\mu + /, \lambda)$-EA on multimodal fitness landscapes. In: Proc. Genetic and Evolutionary Computation Conf. (GECCO 2005), vol. 1, pp. 865–872. ACM Press (2005)
19. Shir, O.M., Emmerich, M., Bäck, T.: Adaptive niche radii and niche shapes approaches for niching with the cma-es. Evolutionary Computation 18(1), 97–126 (2010)
20. Stoean, C., Preuss, M., Stoean, R., Dumitrescu, D.: Multimodal optimization by means of a topological species conservation algorithm. IEEE Transactions on Evolutionary Computation 14(6), 842–864 (2010)
21. Storn, R., Price, K.: Differential Evolution A Simple and Efficient Heuristic for global Optimization over Continuous Spaces. Journal of Global Optimization 11(4), 341–359 (1997)
22. Thawonmas, R., Oda, K., Shuda, T.: Rule-Based Camerawork Controller for Automatic Comic Generation from Game Log. In: Yang, H.S., Malaka, R., Hoshino, J., Han, J.H. (eds.) ICEC 2010. LNCS, vol. 6243, pp. 326–333. Springer, Heidelberg (2010)
23. Ware, C., Osborne, S.: Exploration and virtual camera control in virtual three dimensional environments. ACM SIGGRAPH 24(2), 175–183 (1990)

Digging Deeper into Platform Game Level Design: Session Size and Sequential Features

Noor Shaker, Georgios N. Yannakakis, and Julian Togelius

IT University of Copenhagen, Rued Langaards Vej 7, 2300 Copenhagen, Denmark
{nosh,yannakakis,juto}@itu.dk

Abstract. A recent trend within computational intelligence and games research is to investigate how to affect video game players' in-game experience by designing and/or modifying aspects of game content. Analysing the relationship between game content, player behaviour and self-reported affective states constitutes an important step towards understanding game experience and constructing effective game adaptation mechanisms. This papers reports on further refinement of a method to understand this relationship by analysing data collected from players, building models that predict player experience and analysing what features of game and player data predict player affect best. We analyse data from players playing 780 pairs of short game sessions of the platform game Super Mario Bros, investigate the impact of the session size and what part of the level that has the major affect on player experience. Several types of features are explored, including item frequencies and patterns extracted through frequent sequence mining.

1   Introduction

What makes a good computer game? What features should be presented in the game, where in the game should they be presented, how often and in which order? What features should be manipulated to alter specific player experience? And what is the minimum length of time a player need to play in order to elicit a particular affective state? We describe a method that we believe can be used to help answer these questions, and exemplify it with an investigation based on data from hundreds of players playing Super Mario Bros. In the process, we arrive at tentative partial answers to these questions in the context of Super Mario Bros levels.

Many analyses of computer games can be found in the literature, both in terms of game mechanics and from a player perspective based on how the player can interact with the game. For example, some researchers have analysed game content into its constituent parts, or "design patterns" [1], [5], [17]; others have tried to state general facts about what makes games enjoyable [9], [8]. Most of this research, however, tackles this problem from a top-down perspective, creating theories of player experience based on qualitative methods. Some attempts have been made to construct computational models form qualitative theories [23], [26].

Another direction that is related to this work is the procedural generation of game content (PCG) with no or limited human designer input. PCG has recently received increasing attention with the use of artificial and computational intelligence methods to generate different aspects of game content such as maps [3], [21], levels [18], [11]

C. Di Chio et al. (Eds.): EvoApplications 2012, LNCS 7248, pp. 275–284, 2012.

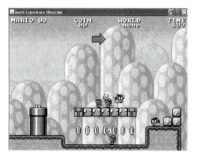

Fig. 1. Snapshot from Infinite Mario Bros, showing Mario standing on horizontally placed boxes surrounded by different types of enemies

and racing tracks [4], [20]. One interesting direction in PCG is the online generation of personalized game content [6], [7], [16]. One approach towards achieving this goal is first to model the relationship between player experience and game content. This requires the construction of data-driven models based on data collected about the game, the player behaviour and correlating this data with data annotated with player experience tags [25] .

This paper continues our previous work on player experience modeling in a version of Super Mario Bros [10], [12], [14]. The main focus of those experiments was on modelling the relationship between direct features of game content, players' playing styles and reported players experience by using features extracted from the full game sessions. The dataset used for those studies constitutes of 480 game pairs constructed using four different controllable features. In a recent paper [15] we reported preliminary explorations of predicting player experience of engagement based only on level features and introduced the use of sequence mining to extract simple patterns from game content, using an incomplete version of a larger and more detailed dataset. We also explored constructing models from parts of levels in order to find the minimum segment length which would allow us to perform meaningful adaptation.

In this paper, we explore the full dataset of 780 game pairs played by hundreds of players. We draw upon the approach proposed in [15] and we extend it in through (1) investigating the three emotional states; engagement frustration and challenge; (2) constructing player experience models based on game content, player behaviour and reported player experience; (3) investigating the impact of the size of game session on the accuracy of predicting players' reported emotion; (4) analysing the importance of the features for each emotional state with respect to their relative placement within the game and (5) exploring direct and sequential feature representations.

The testbed platform game we are using for this study is a modified version of Markus "Notch" Persson's *Infinite Mario Bros*. The game is well known and the benchmark software has been used relatively extensively as a testbed for research [22], [12], [13], [2], [14] and for the Mario AI Championship [16]. Please refer to [22] for details about the game and the gameplay experience it provides.

The paper is organized as follows. Section 2 explains the process followed to collect data from players. Section 3 presents the two forms that have been used to represent the collected data. A method that has been used to mine sequential data is presented in

Section 4. Section 5 describes player experience modeling via preference learning. The process of segmenting the levels into smaller chucks and constructing models based on the segments is discussed in Section 6, while Section 7 presents the experiments conducted and the analysis of the results. Finally, Section 8 presents our conclusions.

2 Experiment Design

The following section describes the level generation process, the survey that has been designed to collect the data and the types of data that were extracted. The level generator of the game has been modified to generate levels according to the following six *controllable features*

- The number of gaps in the level, G and the average with of gaps, $\bar{G}_w$.
- The number of enemies, E.
- Enemies placement, E_p. The way enemies are placed around the level is determined by three probabilities which sum to one; on or under a set of horizontal blocks, P_x; within a close distance to the edge of a gap, P_g and randomly placed on a flat space on the ground, P_r.
- The number of powerups, N_w.
- The number of boxes, B. We define one variable to specify the number of the different types of boxes that exist; *blocks* and *rocks*. Blocks (which look like squares with question marks) contain hidden elements such as coins or powerups. Rocks (which look like squares of bricks) may hide a coin, a powerup or simply be empty. The generator randomly select one of these two types for each box generated.

The selection of these particular controllable features was done with the intent to cover the features that have the most impact on the investigated affective states. The placement of gaps, powerups and boxes is approximately uniformly random. Two states (low and high) are set for each of the controllable parameters above except for enemies placement which has been assigned three different states allowing more control over the difficulty and diversity of the generated levels. For example, setting P_g to 80% results in a level with a majority of enemies placed around gaps, increasing the level difficulty. Other features of the level have been assigned fixed values. For example, the number of gaps in the level can be either two or six, while the number of free coins is fixed to seven in all generated levels. The level generator constructs level by exploring the total number of pairwise combinations of these states (96). This number can be reduced to 40 by analysing the dependencies between these features and eliminating the combinations that contain independent variables.

2.1 Data Collection

The game survey study [15] has been designed to collect subjective affective reports expressed as pairwise preferences of subjects playing the different levels of the game by following the experimental protocol proposed in [28]. The game sessions have been constructed using a level width of 100 Super Mario Bros units (blocks) based on all

Table 1. Gameplay features extracted from data recorded during gameplay

Category	Feature	Description
Time	t_{comp}	Completion time
	$t_{lastLift}$	Playing duration of last life over total time spent on the level
	t_{duck}	Time spent ducking (%)
	t_{jump}	Time spent jumping (%)
	t_{left}	Time spent moving left (%)
	t_{right}	Time spent moving right (%)
	t_{run}	Time spent running (%)
	t_{small}	Time spent in Small Mario mode (%)
	t_{big}	Time spent in Big Mario mode (%)
Interaction with items	n_{coins}	Free coins collected (%)
	$n_{coinBlocks}$	Coin blocks pressed or coin rocks destroyed (%)
	$n_{powerups}$	Powerups pressed (%)
	n_{boxes}	Sum of all blocks and rocks pressed or destroyed (%)
Interaction with enemies	$k_{cannonFlower}$	Times the player kills a cannonball or a flower (%)
	$k_{goombaKoopa}$	Times the player kills a goomba or a koopa (%)
	k_{stomp}	Opponents died from stomping (%)
	$k_{unleash}$	Opponents died from unleashing a turtle shell (%)
Death	d_{total}	Total number of deaths
	d_{cause}	Cause of the last death
Miscellaneous	n_{mode}	Number of times the player shifted the mode (Small, Big, Fire)
	n_{jump}	Number of times the jump button was pressed
	n_{gJump}	Difference between the number of gaps and the number of jumps
	n_{duck}	Number of times the duck button was pressed
	n_{state}	Number of times the player changed the state between: standing still, run, jump, moving left, and moving right

combinations of the controllable features. 780 pairs of games (exhausting the space of controllable features) were played by hundreds of players. Participants' age covers a range between 16 and 64 years from different origins. Complete games were logged enabling complete replays. The following types of data were extracted.

- Gameplay Data: All player actions and interactions with game items and their corresponding time-stamps have been recorded with the full trajectory of Mario.
- Reported Player Experience: A 4-alternative forced choice questionnaire is presented to the players after playing each pair asking them to report their emotional preferences across three user states: *engagement*, *challenge* and *frustration*. The selection of these states is based on earlier game survey studies [12] and our intention to capture both affective and cognitive components of gameplay experience [25].

3   Data Representation

3.1   Direct Features

Several features have been directly extracted from the data recorded during gameplay. Most of these features appear in our previous studies [10], [14] and the choice of them is made in order to be able to represent the difference between a large variety of Super Mario Bros playing styles. These features are presented in Table 1.

3.2   Sequential Features

We investigate another form of data representation that allows including features based on ordering in space or time by means of sequences. Sequences of game content and players' behaviour yields patterns that might be directly linked to player experience. These patterns provide a mean for an in-depth analyses of the relationship between the player and the game. There are several possible approaches to generate sequences from interaction logs. In this paper we concentrate on two types of sequences:

- Content corresponding to gameplay events: Game content at the specific player position is recorded whenever the player performs an action or interacts with game items. Different content events are used: increase/decrease in platform height, $P^\uparrow/P^\downarrow$; existence of an enemy, P_e; existence of a coin, block or rock, P_d; existence of a coin, block or rock with an enemy, P_{ed}; and the beginning/ending of a gap P_{gs}/P_{ge}.
- Sequential Gameplay Features: Sequences representing different players' behaviour have been generated by recording key pressed/released events (*action event*). The list of events that have been considered includes: pressing an arrow key to move right, left, or duck (▶, ◀, ▼); pressing the jump key, ⇑; pressing the jump key in combination with right or left key (⇑▶, ⇑◀); pressing the run key in combination with right or left key ($R^\blacktriangleright$, $R^\blacktriangleleft$); pressing the run and jump keys in combination with right/left ($R^{\blacktriangleright\Uparrow}$, $R^{\blacktriangleleft\Uparrow}$); and not pressing any key, S.

4   Sequence Mining

Sequence mining techniques have been applied to extract useful information from the different types of the sequences generated. Generalised Sequential Pattern (GSP) algorithm [19] has been used to mine the sequences and find frequent patterns within the dataset of sequences. The GSP algorithm has been chosen because of its advantages over other apriori-based sequence mining algorithms. Using GSP, we can discover patterns with a predefined minimum support, min_{sup} (the minimum number of times a pattern has to occur in the data-sequences to be considered frequent), and specify a time constraints within which adjacent events can be considered elements of the same pattern, max_{gap}. Different min_{sup} values have been explored to obtain a reasonable trade off between considering patterns that are generalised over all players and more specific patterns. For the experiments presented in this paper, we use a min_{sup} of 500 which forces a sequence pattern to occur in at least 31.8% of the samples to be considered frequent. The max_{gap} has a great impact on the number of frequent patterns that can be extracted. By assigning a large value to this parameter, we allow more generalised patterns to be taken into account. The experiments conducted for tuning the value of this parameter showed that a max_{gap} of 1 second provides a good trade off between the number of patterns extracted and their expressiveness value. Different sequence length values have been explored, the experiments showed that the number of extracted subsequences is quite large for sequences containing information about players' behaviour. In order to lower the feature space dimensionality and the computational cost of searching for relevant features we chose to use only frequent sequences of length three.

5   Preference Learning for Modelling Playing Experience

Neuroevolutionary preference learning has been used in order to construct models that approximate the function between gameplay features, controllable features, and reported affective preferences. We start the models' constructing procedure by selecting the relevant subset of features for predicting each emotional state, this is done by using Sequential Forward Selection (SFS) to generate the input vector of single-layer perceptrons (SLPs) [27]. The feature subset derived from SFS using SLP is then used as the input to small multi-layer perceptron (MLP) models containing one hidden layer of two neurons. The quality of a feature subset and the performance of each MLP is obtained through the average classification accuracy in three independent runs using 3-fold cross validation across five runs. Parameter tuning tests have been conducted to set up the parameters' values for neuroevolutionary user preference learning that yield the highest accuracy and minimise computational effort. A population of 100 individuals is used, and evolution run for 20 generations. A probabilistic rank-based selection scheme is used, with higher ranked individuals having higher probability of being chosen as parents. Finally, reproduction was performed via uniform crossover, followed by Gaussian mutation of 1% probability.

6   Level Segmentation

The purpose of segmenting the level is to draw a picture of the importance of the features with respect to player experience; different features correlated with player experience for each emotional state could be extracted from each segment of the game pointing out to positions in the games where these features play a role in triggering particular affective state. By segmenting the levels we can also identify the size of the level segment that generates the best prediction accuracy of the three emotional states. That segment size can then potentially be used to set the frequency of a real-time adaptation mechanism for maximising particular player experience (as in [24], [14]).

We follow the same process presented in [15] for segmenting the level into half and one-third width segments but instead of using the same feature set and calculating each feature value across all segments, we run SFS to select the most relevant feature subset from all direct and sequential features for each segment across all emotional states.

7   Experiments and Analysis

We ran a number of experiments to select features from the full levels and from different segments into which the levels have been divided. Player experience models were constructed based on the different subset of features selected from direct and sequential features for each segment across the three emotional states. Table 2 presents the selected features and the models' accuracies.

The networks found vary in the number of selected features and performance. The most accurate model is the one for predicting challenge (91.23%) with a large subset of 13 features selected from the full levels. Engagement comes next with a best model accuracy of 86.43% obtained from features extracted from the first segment out of two followed by frustration which can be predicted with an accuracy up to 85.88% from a subset of ten features extracted from the full level.

Segmenting the sessions resulted in a performance increase for the models of predicting engagement while a performance decrease has been observed for the experience

Table 2. The features selected from the set of direct and sequential features for predicting engagement, frustration and challenge using sequential feature selection with SLP and simple MLP models and the corresponding models' performance

	Full level	1st seg/2	2nd seg/2	1st seg/3	2nd seg/3	3rd seg/3
			Engagement			
Selected features	t_{comp} n_{coins} d_{cause} t_{small} t_{jump} E $n_{coinBlocks}$ t_{big} t_{run} n_{jump} $P^{\downarrow}P_{gs}P_{ge}$ ⮕⇑⮕ S	n_{state} t_{big} d_{cause} t_{right} E S ⇑⇑⇑⮕ ⮕⮕◄ ⇑⇑⮕ ⇑ ⇑⮕ ⮕◄	B d_{cause} N_w E n_{boxes} ⮕S⮕ ⮕R⮕S ⮕⇑ S	n_{gJump} t_{left} n_{boxes} B G_w d_{cause} E_p n_{coins} ⮕R⮕S S ◄S	t_{right} E n_{gJump} d_{cause} t_{duck} G	B G t_{jump} d_{cause} $k_{unleash}$
MLP_{perf}	75.21%	86.43%	72.03%	72.19%	71.69%	72.86%
			Frustration			
Selected features	t_{right} d_{total} d_{cause} $t_{lastLift}$ G_w G n_{jump} $P_dP_dP^{\uparrow}$ ⇑⮕ ⮕⮕ ⇑⮕ S⮕	n_{gJump} G n_{coins} E t_{left} R⮕ ⮕⮕ R⮕R⮕⇑⮕	G n_{gJump} G_w d_{cause} k_{stomp}	n_{jump} t_{small} t_{left} S ⇑⮕ S	G B t_{small}	E_p $k_{goombaKoopa}$ B G n_{gJump} t_{left} t_{right} SS⮕ R⮕R⮕⇑S
MLP_{perf}	85.88%	81.73%	79.85%	78.72%	78.15%	73.45%
			Challenge			
Selected features	$t_{lastLift}$ n_{jump} d_{total} n_{coins} t_{right} G_w E_p t_{left} k_{stomp} $P_dP_dP_d$ $P^{\downarrow}P_{gs}P_{ge}$ ⇑⮕S $P^{\uparrow}P^{\downarrow}P^{\downarrow}$	t_{right} n_{gJump} n_{state} G t_{small} B ⮕⇑ S	G B d_{cause} n_{gJump}	n_{gJump} n_{boxes} $n_{powerups}$ t_{right} n_{mode} t_{left} t_{big}	G E $k_{cannonFlower}$ n_{jump} $k_{unleash}$ ⇑ S ⮕	E_p k_{stomp} B E $k_{unleash}$ ⮕SS
MLP_{Perf}	91.23%	75.6%	77.19%	72.41%	73.52%	69.38%

models of frustration and challenge. The model constructed based on features selected from the first half of the session for predicting engagement significantly (significant effect is determined by $p < 0.01$ over 10 runs in this paper) outperforms all other models constructed on full and other partial information. A significant performance decrease was found for predicting frustration when constructing the models based on features extracted from segments with one third of the full size. Using features from the full sessions, we were able to predict challenge with high accuracy that is significantly better than all other models constructed on partial information.

The results suggest that different sizes of game session are needed to elicit different affective states. While challenge can be predicted with high accuracy from the full sessions, smaller session size somewhat count-intuitively appears to give better results for predicting engagement, and frustration can be predicted with high accuracy from full and half size sessions.

The different subsets of features selected from each segment draw a picture of the importance of the positioning of the features within the game and the different impacts this has on the different emotional states under investigation. Some content features have been selected from the full sessions and also appear in the subset of features selected from the parts, such as the number of enemies (E) and the number of gaps (G) for predicting engagement and frustration respectively. This suggests the importance of these features for eliciting a particular emotional state regardless of their specific positioning within the game. Other features like the number of powerups (N_w) appears to have an impact on engagement when presented in the second half of the game. This can be explained by the fact that powerups are more important to the players towards the end of the game since this increases their chance of winning, the selection of *cause of death* feature in all segments also supports this assumption. It's worth noting that only one controllable feature has been selected for the best model for predicting engagement and the rest of the features relate to the particular playing style for each player. Most aspects of level design appear to have a large impact on challenge since five content features (direct and sequential) have been selected for the best model of predicting challenge.

8   Discussion and Conclusion

In this paper, we reported on the creation of data-driven computational models that predict three players' reported affective states based on level design and gameplay features. We investigated direct and sequential features of both game content and gameplay, as well as the impact of session sizes. We were able to predict players' reported levels of engagement, frustration and challenge with high accuracy.

Our previous study [15] concluded that segmenting the levels yields a performance decrease when constructing models based only on game content features. The results presented in this paper show that different session sizes should be considered for investigating the different emotional states when gameplay data is also considered as input to the models. Challenge can be best predicted with longer sessions' size than the ones needed for predicting frustration or engagement. The results indicate that the models performance in general significantly decreases when segmenting the session into more than two segments. This suggests that segmenting the data into more than two segments causes information loss. Another possible explanation is that the session size should be longer than a particular length to elicit a specific emotional state, and it appears that one third of the level size is too small to consider the reported player experience valid while the gameplay experience and the reported affects can still be considered valid for one half of the session size for engagement and frustration.

The different subsets of features selected for predicting affective states suggest differing relative importance of design elements for different aspects of player experience. This has the potential to partly decouple dissimilar aspects of player adaptation.

The results presented here will feed into our ongoing research on modelling player affect and preferences in Super Mario Bros, with the ultimate goal of producing an effectively player-adaptive version of the game, but could also inform separate studies.

Acknowledgments. The research was supported in part by the Danish Research Agency, Ministry of Science, Technology and Innovation; project "AGameComIn" (274-09-0083).

References

1. Björk, S., Holopainen, J.: Patterns in game design. Cengage Learning (2005)
2. Bojarski, S., Congdon, C.: Realm: A rule-based evolutionary computation agent that learns to play mario. In: 2010 IEEE Symposium on Computational Intelligence and Games (CIG), pp. 83–90 (2010)
3. Cardamone, L., Yannakakis, G.N., Togelius, J., Lanzi, P.L.: Evolving Interesting Maps for a First Person Shooter. In: Di Chio, C., Cagnoni, S., Cotta, C., Ebner, M., Ekárt, A., Esparcia-Alcázar, A.I., Merelo, J.J., Neri, F., Preuss, M., Richter, H., Togelius, J., Yannakakis, G.N. (eds.) EvoApplications 2011, Part I. LNCS, vol. 6624, pp. 63–72. Springer, Heidelberg (2011)
4. Cardamone, L., Loiacono, D., Lanzi, P.L.: Interactive evolution for the procedural generation of tracks in a high-end racing game. Interface, 395–402 (2011)
5. Hullett, K., Whitehead, J.: Design patterns in fps levels. In: FDG 2010: Proceedings of the Fifth International Conference on the Foundations of Digital Games, pp. 78–85. ACM, New York (2010)
6. Jennings-Teats, M., Smith, G., Wardrip-Fruin, N.: Polymorph: dynamic difficulty adjustment through level generation. In: Proceedings of the 2010 Workshop on Procedural Content Generation in Games, PCGames 2010, pp. pp. 11:1–11:4. ACM, New York (2010)
7. Kazmi, S., Palmer, I.: Action recognition for support of adaptive gameplay: A case study of a first person shooter. International Journal of Computer Games Technology 1 (2010)
8. Koster, R.: A theory of fun for game design. Paraglyph Press (2004)
9. Malone, T.: What makes computer games fun (abstract only). In: Proceedings of the Joint Conference on Easier and More Productive use of Computer Systems (Part - II): Human Interface and the user Interface, CHI 1981, vol. 1981, p. 143. ACM, New York (1981)
10. Pedersen, C., Togelius, J., Yannakakis, G.N.: Modeling player experience in super mario bros. In: CIG 2009: Proceedings of the 5th International Conference on Computational Intelligence and Games, pp. 132–139. IEEE Press, Piscataway (2009)
11. Pedersen, C., Togelius, J., Yannakakis, G.N.: Modeling player experience for content creation. IEEE Transactions on Computational Intelligence and AI in Games 2(1), 54–67 (2010)
12. Pedersen, C., Togelius, J., Yannakakis, G.N.: Modeling player experience for content creation. IEEE Transactions on Computational Intelligence and AI in Games 2(1), 54–67 (2010)
13. Perez, D., Nicolau, M., O'Neill, M., Brabazon, A.: Evolving Behaviour Trees for the Mario AI Competition Using Grammatical Evolution. In: Di Chio, C., Cagnoni, S., Cotta, C., Ebner, M., Ekárt, A., Esparcia-Alcázar, A.I., Merelo, J.J., Neri, F., Preuss, M., Richter, H., Togelius, J., Yannakakis, G.N. (eds.) EvoApplications 2011, Part I. LNCS, vol. 6624, pp. 123–132. Springer, Heidelberg (2011)
14. Shaker, N., Yannakakis, G.N., Togelius, J.: Towards Automatic Personalized Content Generation for Platform Games. In: Proceedings of the AAAI Conference on Artificial Intelligence and Interactive Digital Entertainment (AIIDE). AAAI Press (2010)
15. Shaker, N., Yannakakis, G.N., Togelius, J.: Feature Analysis for Modeling Game Content Quality. IEEE Transactions on Computational Intelligence and AI in Games, CIG (2011)
16. Shaker, N., Togelius, J., Yannakakis, G.N., Weber, B., Shimizu, T., Hashiyama, T., Sorenson, N., Pasquier, P., Mawhorter, P., Takahashi, G., Smith, G., Baumgarten, R.: The 2010 Mario AI championship: Level generation track. IEEE Transactions on Computational Intelligence and Games (2011)

17. Smith, G., Cha, M., Whitehead, J.: A framework for analysis of 2d platformer levels. In: Sandbox 2008: Proceedings of the 2008 ACM SIGGRAPH Symposium on Video Games, pp. 75–80. ACM, New York (2008)
18. Smith, G., Whitehead, J., Mateas, M.: Tanagra: A mixed-initiative level design tool. In: Proceedings of the International Conference on the Foundations of Digital Games (2010)
19. Srikant, R., Agrawal, R.: Mining Sequential Patterns: Generalizations and Performance Improvements. In: Apers, P.M.G., Bouzeghoub, M., Gardarin, G. (eds.) EDBT 1996. LNCS, vol. 1057, pp. 1–17. Springer, Heidelberg (1996)
20. Togelius, J., De Nardi, R., Lucas, S.: Towards automatic personalised content creation for racing games. In: IEEE Symposium on Computational Intelligence and Games, CIC 2007, pp. 252–259. IEEE (2007)
21. Togelius, J., Preuss, M., Yannakakis, G.: Towards multiobjective procedural map generation. In: Proceedings of the 2010 Workshop on Procedural Content Generation in Games, p. 3. ACM (2010)
22. Togelius, J., Karakovskiy, S., Koutník, J., Schmidhuber, J.: Super mario evolution. In: Proceedings of the 5th International Conference on Computational Intelligence and Games, CIG 2009, pp. 156–161. IEEE Press, Piscataway (2009),
    http://dl.acm.org/citation.cfm?id=1719293.1719326
23. Tognetti, S., Garbarino, M., Bonanno, A., Matteucci, M.: Modeling enjoyment preference from physiological responses in a car racing game. IEEE Transactions on Computational Intelligence and AI in Games (2010)
24. Yannakakis, G.N., Hallam, J.: Real-time Game Adaptation for Optimizing Player Satisfaction. IEEE Transactions on Computational Intelligence and AI in Games 1(2), 121–133 (2009)
25. Yannakakis, G.N., Togelius, J.: Experience-Driven Procedural Content Generation. IEEE Transactions on Affective Computing (2011)
26. Yannakakis, G., Hallam, J.: A generic approach for generating interesting interactive pac-man opponents. In: Proceedings of the IEEE Symposium on Computational Intelligence and Games, pp. 94–101 (2005)
27. Yannakakis, G.N., Hallam, J.: Entertainment modeling through physiology in physical play. Int. J. Hum.-Comput. Stud. 66, 741–755 (2008),
    http://dl.acm.org/citation.cfm?id=1410473.1410682
28. Yannakakis, G.N., Maragoudakis, M., Hallam, J.: Preference learning for cognitive modeling: a case study on entertainment preferences. Trans. Sys. Man Cyber. Part A 39, 1165–1175 (2009), http://dx.doi.org/10.1109/TSMCA.2009.2028152

Robot Base Disturbance Optimization with Compact Differential Evolution Light*

Giovanni Iacca, Fabio Caraffini, Ferrante Neri, and Ernesto Mininno

Department of Mathematical Information Technology,
P.O. Box 35 (Agora), 40014 University of Jyväskylä, Finland
{giovanni.iacca,fabio.caraffini,ferrante.neri,ernesto.mininno}@jyu.fi

Abstract. Despite the constant growth of the computational power in consumer electronics, very simple hardware is still used in space applications. In order to obtain the highest possible reliability, in space systems limited-power but fully tested and certified hardware is used, thus reducing fault risks. Some space applications require the solution of an optimization problem, often plagued by real-time and memory constraints. In this paper, the disturbance to the base of a robotic arm mounted on a spacecraft is modeled, and it is used as a cost function for an on-line trajectory optimization process. In order to tackle this problem in a computationally efficient manner, addressing not only the memory saving necessities but also real-time requirements, we propose a novel compact algorithm, namely compact Differential Evolution light (cDElight). cDElight belongs to the class of Estimation of Distribution Algorithms (EDAs), which mimic the behavior of population-based algorithms by means of a probabilistic model of the population of candidate solutions. This model has a more limited memory footprint than the actual population. Compared to a selected set of memory-saving algorithms, cDElight is able to obtain the best results, despite a lower computational overhead.

1 Introduction

Some real-world problems, due to real-time, space, and cost requirements, impose the solution of an optimization problem, sometimes even complex, on a device with limited memory and computational resources. This situation is typical, for instance, in mobile robots and real-time control systems, where all the computation is performed on board of an embedded system. Some examples of this class of problems can be found in home automation, mobile TLC devices, smart sensors and biomedical devices. Among these applications, space control systems represent an interesting exception. Despite the constant growth of the computational power in consumer electronics, very simple and dated hardware is indeed still used on spacecrafts. It must be remarked that the computational devices on board of a spaceship should reliably work without any kind of rebooting for

---

* This research is supported by the Academy of Finland, Akatemiatutkija 130600, Algorithmic Design Issues in Memetic Computing and Tutkijatohtori 140487, Algorithmic Design and Software Implementation: a Novel Optimization Platform.

C. Di Chio et al. (Eds.): EvoApplications 2012, LNCS 7248, pp. 285–294, 2012.

months or, in some cases, even for years. In this sense, the use of tremendously simple but fully tested and certified hardware allows a high reliability of the computational cores, thus reducing fault risks. For example, since over twenty years, National Aeronautics and Space Administration (NASA) employs, within the space shuttles, IBM AP-101S computers [12]. These computers constitute an embedded system for performing the control operations. The memory of computational devices is only 1 Mb, i.e. much less capacious than any modern device. Thus, the necessity of having an efficient control notwithstanding the hardware limitations arises.

In these cases, an optimization algorithm should perform the task without limited requirements of memory and computational resources. Unfortunately, high performance algorithms are usually fairly complex structures which make use of a population of candidate solutions (requiring a high memory footprint) and other components such as learning systems or classifiers, see e.g. [13,18].

In order to address problems characterized by a limited hardware, compact Evolutionary Algorithms (cEAs) have been designed. A cEA is an EA belonging to the class of Estimation of Distribution Algorithms (EDAs) [7]. Compact algorithms mimic the behavior of population-based algorithms but, instead of storing and processing an entire population, make use of a probabilistic representation of the population. In this way, the algorithm saves at least a part of the advantages of population-based algorithms but requires a much smaller memory with respect to their corresponding population-based versions. Thus, a run of these algorithms requires a small amount of memory compared to their correspondent standard EAs. Recently, a compact algorithm based on Differential Evolution, called cDE, has been proposed in [9]. This algorithm encodes the population within a probabilistic distribution and employs the standard DE logic for generating new trial solutions and selecting the most promising search directions. The cDE algorithm showed a performance superior to other compact algorithms for a large set of test problems. Some work has been performed to enhance the performance of the cDE algorithm, for example embedding it in a memetic framework [10,11], or combining multiple cDE core [5,6].

In this work we propose a modified cDE scheme for minimizing the disturbance to the base of a robotic arm working in a non-gravitational environment on board of a spacecraft. The proposed algorithm, namely compact Differential Evolution light (cDElight), includes a *light mutation*, which allows a unique solution sampling instead of the multiple sampling typical of cDE frameworks, and a *light crossover*, which allows to save on the random number generation. Simulation results show that the proposed algorithm has the same performance as the original cDE but with a lower computational overhead, thus leading to important advantages for real-time online optimization.

The remainder of this paper is organized as follows. Section 2 describes the mechanical model of the robotic arm and the fitness function. Section 3 describes the proposed cDElight. Section 4 shows the simulation results. Section 5 gives the conclusion of this work.

2   Base Disturbance Optimization in Space Robotic Arms

Space robots are crucially important in current space operations as they can prevent humans from having to perform extremely risky operations, e.g. extra-vehicular activities such as reparations outside the spacecraft. Due to the enormous distances, the robot cannot be fully remotely controlled manually from the Earth because the communication delay between the command and the execution of the robot operation can likely be unacceptable in several cases. For this reason an efficient real-time control system is essential. The absence of gravity plays an important role in the dynamics of the robot and must be taken into account when the control system is designed. In this case of study, a robotic arm connected to a base, e.g. a spacecraft or a satellite, is considered. In a nutshell, the control system aims to perform the robot movements in order to minimize the disturbances, i.e. inertial movements, on the base. More specifically, each new trajectory step is optimized online using the look-ahead optimized algorithm for trajectory planning proposed in [14].

Space robots are highly nonlinear, coupled multi-body systems with nonlinear constraints. Moreover, the dynamic coupling between the manipulator (robotic arm) and the base usually affects the performance of the manipulator. The dynamic coupling is important to understand the relationship between the robot joint motion and the resultant base motion, and it is useful in minimizing fuel consumption for base attitude control. The measure of dynamic coupling has been formulated in [17].

Let us consider a manipulator composed of n_b links (bodies) interconnected by joints and connected by means of an external joint to the base. With reference to Fig. 1, let V_i and Ω_i be linear and angular velocities of the ith body of the manipulator arm with respect to the inertial reference system Σ_I, and let v_i and ω_i be linear and angular velocities of the ith body of the manipulator arm with respect to the base Σ_B. Thus, we can obtain that the velocities of the ith body are:

$$V_i = v_i + V_0 + \Omega_0 \times r_i$$
$$\Omega_i = \omega_i + \Omega_0 \tag{1}$$

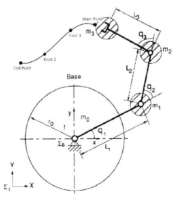

Fig. 1. Space robotic arm scheme

where the operator $\times$ represents the outer product of $\mathbb{R}^3$ vectors. V_0 and Ω_0 are, respectively, linear and angular velocities of the centroid of the base with respect to Σ_I. The variable r_i represents the position vector related to the ith body pointing towards the center of the base with reference to Σ_B, see [17] for details. The velocities in the base of the coordinates of the reference system Σ_B can be calculated as:

$$\begin{bmatrix} v_i \\ w_i \end{bmatrix} = J_i(q)\dot{q} \tag{2}$$

where q is the vector of the angular positions of each joint of the manipulator arm (see q_1, q_2, and q_3 in Fig. 1) and $J_i(q)$ is thus the Jacobian of the ith body of manipulator arm. The Jacobian can be then decomposed into two sub-matrices related to its translational and rotational movements [17]:

$$J_i(q) = \begin{bmatrix} J_{Ti}(q) \\ J_{Ri}(q) \end{bmatrix} \tag{3}$$

The total linear (P) and angular (L) momenta of the entire robotic arm can be expressed as:

$$P = m_0 V_0 + \sum_{i=1}^{n_b} (m_i v_i) \tag{4}$$

$$L = I_0 \Omega_0 + m_0 R_B \times V_0 + \sum_{i=1}^{n_b} (I_i w_i + m_i r_i \times v_i) \tag{5}$$

where I_i and m_i are the inertia momentum and mass of each body composing the robot manipulator, and R_B is a positioning vector pointing towards the centroid of the base with reference to Σ_I. Equations (4) and (5) can then be combined:

$$\begin{bmatrix} P \\ L \end{bmatrix} = H_B \begin{bmatrix} V_0 \\ \Omega_0 \end{bmatrix} + H_m \dot{q}. \tag{6}$$

The details about the structures of the matrices H_B and H_m are given in [4]. In a free-floating situation (due to the fact that both robot and base are in outer space), there are no external forces or momenta. If we consider the gravitational force to be negligible, linear and angular momenta are conserved. We assume that the initial state of the system is stationary, so that the total linear and angular momenta are zero. Hence, from Eq. (6), the mapping relationship between the manipulator joint motion and the base motion is thus given by:

$$\begin{bmatrix} V_0 \\ \Omega_0 \end{bmatrix} = -H_B^{-1} H_m \dot{q}. \tag{7}$$

For a given trajectory that must be followed by the robot, the optimization problem under investigation consists of detecting the angular positions q_i, angular velocities $\dot{q}_i$, and angular accelerations $\ddot{q}_i$ of each joint i in each knot k identifying the trajectory such that the disturbance on the base is minimized. The fitness to be minimized is, in our study, the integral over time of the norm of the acceleration vector on the base. The acceleration values can be derived by Eq. (7). Since the trajectory must be continuous, the function describing the position of the joints over time must also be continuous. In order to satisfy this condition, we modeled each function $q_i(t)$ (where t is the time) as a set of 5th polynomial splines and imposed the continuity of $q_i(t)$, $\dot{q}_i(t)$, and $\ddot{q}_i(t)$. Each spline is a polynomial of the 5th order because six conditions are imposed by the physics of the phenomenon (continuity of the function, its first and second derivative in the knots). With reference to Fig. 1, considering that the robot manipulator contains three joints, the trajectory is marked by two knots, and for each joint it is necessary to control angular position, velocity and acceleration, our fitness function depends on $3 \cdot 2 \cdot 3 = 18$ variables.

3   Compact Differential Evolution Light

In order to solve the online trajectory optimization problem described in the previous section, we considered a compact algorithm which proved to be successful in a wide range of landscapes, namely the compact Differential Evolution (cDE) [9]. More specifically, we introduced two minor modifications into the original cDE framework, in order to make the algorithm computationally less expensive without compromising its efficiency. Most importantly, these modifications allow to use cDE in real-time environments.

The resulting algorithm, named compact Differential Evolution light (cDElight), see Algorithm 1, consists of the following steps. Without loss of generality, let us assume that the decision variables are normalized so that each search interval is $[-1, 1]$. At the beginning of the optimization, a $2 \times n$ matrix, where n is the problem dimension, is generated. This matrix $PV = [\mu, \sigma]$ is called probability vector. The initial values of μ and σ are set equal to 0 and 10, respectively. The value of σ is empirically set large enough to simulate a uniform distribution.

A solution x_e, called *elite*, is then sampled from PV. More specifically, the sampling mechanism of a design variable $x_r[i]$ associated to a generic candidate solution x from PV consists of the following steps. For each i-th design variable indexed, $i = 1, \cdots, n$, a truncated Gaussian Probability Distribution Function (PDF) characterized by a mean value $\mu[i]$ and a standard deviation $\sigma[i]$ is associated. The formula of the PDF is:

$$PDF\left(truncNorm\left(x\right)\right) = \frac{e^{-\frac{(x-\mu[i])^2}{2\sigma[i]^2}}\sqrt{\frac{2}{\pi}}}{\sigma[i]\left(\mathrm{erf}\left(\frac{\mu[i]+1}{\sqrt{2}\sigma[i]}\right) - \mathrm{erf}\left(\frac{\mu[i]-1}{\sqrt{2}\sigma[i]}\right)\right)} \tag{8}$$

where erf is the error function [3]. From the PDF, the corresponding Cumulative Distribution Function (CDF) is constructed by means of Chebyshev polynomials according to the procedure described in [2]. It must be observed that the codomain of CDF is $[0, 1]$. In order to sample the normalized design variable $x[i]$ from PV, a random number $r = rand(0, 1)$ is sampled from a uniform distribution. Finally, the inverse function of CDF, representing $x[i]$ in correspondence of r, is then calculated and scaled in the original decision space.

```
counter t = 0
{** PV initialization **}
initialize μ = 0̄ and σ = 1̄ · 10
sample elite from PV
while budget condition do
    {** Mutation Light**}
    generate x'_off from a modified PV: PV' = [μ, √((1 + 2F²)σ²)]
    {** Crossover Light**}
    x_off = elite
    generate i_start = round (n · rand (0, 1))
    x_off[i_start] = x'_off[i_start] {** Deterministic Copy**}
    xoverL = round ( log(rand(0,1)) / log(Cr) )
    i = i_start + 1
    j = 1
    while i ≠ i_start AND j ≤ xoverL + 1 do
        x_off[i] = x'_off[i]
        i = i + 1
        j = j + 1
        if i == n then
            i = 1
        end if
    end while
    {** Elite Selection **}
    [winner, loser] = compete (x_off, elite)
    if x_off == winner then
        elite = x_off
    end if
    {** PV Update **}
    update μ and σ according to Eq. 12
    t = t + 1
end while
```

Algorithm 1. cDElight pseudo-code

At each step t, cDElight generates an offspring by means of two operators called *mutation light* and *crossover light*, which represent two modifications of the original cDE framework. Mutation light consists of the following steps. According to a DE/rand/1 mutation scheme, three individuals x_r, x_s, and x_t are needed to generate a provisional offspring x'_{off} computing $x'_{off} = x_t + F(x_r - x_s)$. Instead of sampling three separate individuals, like the original cDE algorithm does, the light mutation generates x'_{off} by performing only one sampling. In order to do that, we intentionally confuse Gaussian and truncated Gaussian PDF, and we apply the properties of the algebraic sum of normally distributed variables. In this way, given that $\{x_r, x_s, x_t\} \sim \mathcal{N}(\mu, \sigma^2)$, and considering the DE/rand/1 scheme, it is possible to sample directly the x'_{off} from the following modified Gaussian distribution:

$$x'_{off} \sim \mathcal{N}(\mu, \sigma^2) + F(\mathcal{N}(\mu, \sigma^2) - \mathcal{N}(\mu, \sigma^2)) =$$
$$= \mathcal{N}(\mu + F(\mu - \mu), \sigma^2 + F^2(\sigma^2 + \sigma^2)) = \mathcal{N}(\mu, (1 + 2F^2)\sigma^2) \quad (9)$$

It must be remarked that this operation is not fully equivalent to sampling three solutions from a truncated Gaussian distribution but it is an approximation which neglects the tails of the distribution outside the interval $[-1, 1]$.

Similarly to mutation light, crossover light is a computationally light version of the exponential crossover used in the cDE schemes, described in [10]. Instead of generating a set of random numbers until the condition $rand\,(0,1) \le Cr$ is no longer satisfied, only the length of the section of solutions to be inherited, i.e. how many genes should be swapped, is randomly generated. Obviously, since only one random number is generated, this light exponential crossover has a lighter computational overhead with respect to a traditional exponential crossover. When crossover is applied, the offspring x_{off} is initialized to the current *elite*, and one randomly selected decision variable is deterministically copied from x'_{off} to the corresponding position in x_{off}. In order to simplify the notation let us consider the deterministic copy of the first gene and the probabilistic copy of the other genes as independent events. The probability that m genes, on the top of the first one, are copied from x'_{off} to x_{off} is Cr^m. More formally the discrete probability Pr that the crossover length $xoverL$ is equal to m is known as geometric distribution and is given by:

$$Pr\,(xoverL = m) = Cr^m \tag{10}$$

where $m = 1, 2, \ldots, n - 1$. In order to extract the number of genes m to be copied, it is enough to apply the inverse formulas and obtain:

$$xoverL \sim round\,(\log_{Cr}\,(rand\,(0,1))) = round\left(\frac{\log\,(rand\,(0,1))}{\log\,(Cr)}\right) \tag{11}$$

where the last equality is due to the change of base of a logarithm. In other words, crossover light consists of performing the deterministic copy and subsequently the copy of $xoverL$ genes, where $xoverL$ is determined by formula (11). To further control the crossover effect, cDElight makes use of parameter, namely the proportion of genes undergoing exponential crossover α_m, so that, for a chosen α_m, the crossover rate can be set as $Cr = \frac{1}{\sqrt[n\alpha_m]{2}}$, as proposed in [10]. In this way, it is possible to estimate the proportion of genes to be inherited from the provisional offspring independently on problem dimension.

Finally, when x_{off} is generated, its fitness is computed and compared with the fitness of the *elite* (see the function *compete()* in Alg. 1). On the basis of this comparison, a *winner* solution (solution displaying the best fitness) and a *loser* solution (solution displaying the worst fitness) are detected. The winner solution biases the virtual population by affecting the PV values, according to the following update rules:

$$\mu^{t+1} = \mu^t + \frac{1}{N_p}\,(winner - loser)$$
$$\sigma^{t+1} = \sqrt{(\sigma^t)^2 + (\mu^t)^2 - (\mu^{t+1})^2 + \frac{1}{N_p}\,(winner^2 - loser^2)} \tag{12}$$

where N_p is a parameter, namely virtual population size. Details for constructing Eg. 12 are given in [8]. In addition to the PV values, also the *elite* is updated, according to a persistent elitism scheme [1].

4   Simulation Results

In order to minimize the fitness function described in Section 2, the following memory saving algorithms have been implemented and compared:

- Simplified Intelligence Single Particle Optimization: ISPO [19], with acceleration $A = 1$, acceleration power factor $P = 10$, learning coefficient $B = 2$, learning factor reduction ratio $S = 4$, minimum threshold on learning factor $E = 1e - 5$, and particle learning steps $PartLoop = 30$;
- Non uniform Simulated Annealing (nuSA) [16] with mutation exponent $B = 5$, temperature reduction ratio $\alpha = 0.9$, temperature reduction period $L_k = 3$, and $initialSolutions = 10$;
- compact Differential Evolution (cDE) [9] with rand/1 mutation, exponential crossover and persistent elitism. The cDE algorithm has been run with virtual population size $N_p = 300$, scale factor $F = 0.5$, and $\alpha_m = 0.25$;
- compact Differential Evolution light (cDElight), as described above, with $N_p = 300$, scale factor $F = 0.5$, and $\alpha_m = 0.25$.

We decided to compare the proposed cDElight to ISPO and nuSA because these two methods employ a completely different logic. On the other hand, cDE has been included in the experiments because it represents the original version of cDElight. For each competing algorithm, the parameter setting suggested in the original paper was used. It must be remarked that, in order to perform a fair comparison, cDE and cDElight employ the same parameter setting. All the algorithms can be considered memory saving heuristics, as they require a fixed amount of memory slots which does not depend on the problem dimension. In other words, if one of these algorithms is used to tackle a large scale problem, although the slot length is proportional to the dimension of the problem, these slots do not increase in number. More specifically, ISPO and nuSA are typical single solution algorithms, requiring only two memory slots, one for the current best solution and the other for a trial candidate solution. On the other hand, the cDE schemes are memory-wise slightly more expensive as they require, on the top of the two slots for single solution algorithms, two extra slots for the virtual population PV. This compromise is made in order to have the advantages of a population-based search and a still low memory usage.

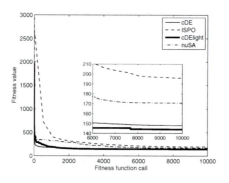

Fig. 2. Average performance trend of the algorithms for the space robot application. The inset figure shows a close-up of the fitness trend after 6000 fitness evaluations.

For each algorithm, 30 independent runs have been performed. The budget of each single run has been fixed equal to 10000 fitness evaluations. In Fig. 2, the average performance trend of the algorithms considered in this study is represented. Table 1 shows the obtained numerical results. Average final fitness values are computed for each algorithm over the 30 runs available. The best result is highlighted in bold face. In order to strengthen the statistical significance of the results, the Wilcoxon Rank-Sum test has also been applied according to the description given in [15], where the confidence level has been fixed to 0.95. With respect to Table 1, the results of the Wilcoxon test for cDElight against the other algorithms are displayed. The symbol "+" indicates that cDElight statistically outperforms both ISPO and nuSA, while the symbol "=" indicates that, on the basis of the Wilcoxon Rank-Sum test, the null hypothesis is accepted, i.e. cDE and cDElight have similar performance. However, as we have seen in the previous section, the computational overhead of cDElight is definitely lower.

Table 1. Compared results on the space robot application

cDE	W	ISPO	W	nuSA	W	cDElight
1.480e+02 ± 5.55e+00	=	1.959e+02 ± 1.08e+02	+	1.709e+02 ± 7.19e+00	+	**1.440e+02 ± 1.08e+00**

5 Conclusion

In this paper we modeled the disturbance to the base of a robotic arm mounted on a spacecraft, and we used a measure of the acceleration at the base as cost function to be minimized. In order to avoid waste of fuel and disturbances to the spacecraft orbit, an online optimization of the robot trajectory is proposed. A specifically designed compact algorithm, namely compact Differential Evolution light (cDElight), was used to tackle this problem in a computationally efficient manner. Using cDElight, it was possible to address not only the memory saving necessities of the hardware typically used in space applications, but also the real-time requirements of the specific application. Compared to the other memory saving meta-heuristics considered in this study, cDElight proved to obtain the best results, still having a limited computational overhead and a reduced memory footprint. In our view, the proposed cDElight algorithm represents an interesting optimization solution when a limited hardware (memory and CPU) is available and a fast response of the system is required.

References

1. Ahn, C.W., Ramakrishna, R.S.: Elitism based compact genetic algorithms. IEEE Transactions on Evolutionary Computation 7(4), 367–385 (2003)
2. Cody, W.J.: Rational chebyshev approximations for the error function 23(107), 631–637 (1969)
3. Gautschi, W.: Error function and fresnel integrals. In: Abramowitz, M., Stegun, I.A. (eds.) Handbook of Mathematical Functions with Formulas, Graphs, and Mathematical Tables, ch.7, pp. 297–309 (1972)

4. Huang, P., Chen, K., Xu, S.: Optimal path planning for minimizing disturbance of space robot. In: Proceedings of the IEEE International Conference on on Control, Automation, Robotics, and Vision (2006)

5. Iacca, G., Mallipeddi, R., Mininno, E., Neri, F., Suganthan, P.: Global supervision for compact differential evolution. In: Proceedings IEEE Symposium on Differential Evolution, pp. 25–32 (2011)

6. Iacca, G., Mininno, E., Neri, F.: Composed compact differential evolution. Evolutionary Intelligence 4(1), 17–29 (2011)

7. Larrañaga, P., Lozano, J.A.: Estimation of Distribution Algorithms: A New Tool for Evolutionary Computation. Kluwer (2001)

8. Mininno, E., Cupertino, F., Naso, D.: Real-valued compact genetic algorithms for embedded microcontroller optimization. IEEE Transactions on Evolutionary Computation 12(2), 203–219 (2008)

9. Mininno, E., Neri, F., Cupertino, F., Naso, D.: Compact differential evolution. IEEE Transactions on Evolutionary Computation 15(1), 32–54 (2011)

10. Neri, F., Iacca, G., Mininno, E.: Disturbed exploitation compact differential evolution for limited memory optimization problems. Information Sciences 181(12), 2469–2487 (2011)

11. Neri, F., Mininno, E.: Memetic compact differential evolution for cartesian robot control. IEEE Computational Intelligence Magazine 5(2), 54–65 (2010)

12. Norman, P.G.: The new AP101S general-purpose computer (gpc) for the space shuttle. IEEE Proceedings 75, 308–319 (1987)

13. Qin, A.K., Huang, V.L., Suganthan, P.: Differential evolution algorithm with strategy adaptation for global numerical optimization. IEEE Transactions on Evolutionary Computation 13, 398–417 (2009)

14. Ren, K., Fu, J.Z., Chen, Z.C.: A new linear interpolation method with lookahead for high speed machining. In: Technology and Innovation Conference, pp. 1056–1059 (2006)

15. Wilcoxon, F.: Individual comparisons by ranking methods. Biometrics Bulletin 1(6), 80–83 (1945)

16. Xinchao, Z.: Simulated annealing algorithm with adaptive neighborhood. Applied Soft Computing 11(2), 1827–1836 (2011)

17. Xu, Y.: The measure of dynamic coupling of space robot system. In: Proceedings of the IEEE Conference on Robotics and Automation, pp. 615–620 (1993)

18. Zhang, J., Sanderson, A.C.: Jade: Adaptive differential evolution with optional external archive. IEEE Transactions on Evolutionary Computation 13(5), 945–958 (2009)

19. Zhou, J., Ji, Z., Shen, L.: Simplified intelligence single particle optimization based neural network for digit recognition. In: Proceedings of the Chinese Conference on Pattern Recognition (2008)

Electrocardiographic Signal Classification with Evolutionary Artificial Neural Networks

Antonia Azzini[1], Mauro Dragoni[2], and Andrea G.B. Tettamanzi[1]

[1] Università degli Studi di Milano
Dipartimento di Tecnologie dell'Informazione
via Bramante, 65 - 26013 Crema (CR) Italy
{antonia.azzini,andrea.tettamanzi}@unimi.it
[2] Fondazione Bruno Kessler (FBK-IRST)
Via Sommarive 18, Povo (Trento), Italy
dragoni@fbk.eu

Abstract. This work presents an evolutionary ANN classifier system as an heart beat classification algorithm suitable for implementation on the PhysioNet/Computing in Cardiology Challenge 2011 [14], whose aim is to develop an efficient algorithm able to run within a mobile phone, that can provide useful feedback in the process of acquiring a diagnostically useful 12-lead Electrocardiography (ECG) recording.

The method used in such a problem is to apply a very powerful natural computing analysis tool, namely evolutionary neural networks, based on the joint evolution of the topology and the connection weights together with a novel similarity-based crossover.

The work focuses on discerning between usable and unusable electrocardiograms tele-medically acquired from mobile embedded devices. A prepropcessing algorithm based on the Discrete Fourier Trasform has been applied before the evolutionary approach in order to extract the ECG feature dataset in the frequency domain. Finally, a series of tests has been carried out in order to evaluate the performance and the accuracy of the classifier system for such a challenge.

Keywords: Signal Processing, Heartbeat Classification, Evolutionary Algorithms, Neural Networks.

1 Introduction

In the last decades, cardiovascular diseases have represented one of the most important causes of death in the world [11] and the necessity of a trustworthy heart state evaluation is increasing. Electrocardiography (ECG) is one of the most useful and well-known methods for heart state evaluation. Indeed, ECG analysis is still one of the most common and robust solutions in the heart diseases diagnostic domain, also due to the fact that it is one of the simplest non-invasive diagnostic methods for various heart diseases [10].

In such a research field, one of the most important critical aspects regards the quality of such heart state evaluations, since, often, the lack of medically trained

C. Di Chio et al. (Eds.): EvoApplications 2012, LNCS 7248, pp. 295–304, 2012.

experts, working from the acquisition process to the discernment between usable and unusable medical information, increases the need of easy and efficient measuring devices, which can send measured data to a specialist. Furthermore, the volume of the data that have to be recorded is huge and, very often, the ECG records are non-stationary signals, and critical information may occur at random in the time scale. In this situation, the disease symptoms may not come across all the time, but would show up at certain irregular intervals during the day.

In this sense, the Physionet Challenge [14], on which this work focuses, aims at reducing, if not eliminating, all the fallacies that currently plague usable medical information tele-medically provided, by obtaining efficient measuring systems through smart phones.

In this challenge, several approaches were explored; in particular, in order to inform inexperienced user about the quality of measured ECGs, artificial-intelligence-based (AI-based) systems have been considered, by reducing the quantum of worst quality ECGs sent to a specialist, this contributing to a more effective use of her time.

Moody and colleagues [16] reported that some of the top competitors in this challenge employed a variety of techniques, using a wide range of features including entropy, higher order moments, intra-lead information, etc, while the classification methods also included Decision Trees, Support Vector Machines (SVMs), Fuzzy Logic, and heuristic rules.

An example of SVM-based approach is reported in [11], where the authors developed a decision support system based on an algorithm that combines simple rules in order to discard recordings of obviously low quality and a more sophisticated classification technique for improving quality of AI-based decision system for mobile phones, showing the fine tuning of sensitivity and specificity of detection. Another example has been also given in [17], where a rule-based classification method that mimics the SVM has been implemented, by using a modified version of a real time QRS-Complex algorithm and a T-Wave detection approach.

Anyway, according to [16], Artificial Neural Networks (ANNs) have been extensively employed in computer aided diagnosis because of their remarkable qualities: capacity of adapting to various problems, training from examples, and generalization capabilities with reduced noise effects. Also Jiang and colleague confirmed the usefulness of ANNs as heartbeat classifiers, emphasizing in particular evolvable ANNs, due to their ability to change the network structure and internal configurations as well as the parameters to cope with dynamic operating environments. In particular, the authors developed an evolutionary approach for the structure and weights optimization of block-based neural network (BbNN) models [8] for a personalized ECG heartbeat pattern classification.

We approach the heartbeat classification problems with another evolutionary algorithm for the joint structure and weights optimization of ANNs [3], which exploits an improved version of a novel similarity-based crossover operator [4], based on the conjunction of topology and connection weight optimization.

This paper is organized as follows: Section 2 briefly presents the problem, while a summary description of the evolutionary approach considered in this work is reported in Section 3. The results obtained from the experiments carried out are presented in Section 4, together with a discussion of the performances obtained. Finally, Section 5 provides some concluding remarks.

2 Problem Description

As previously reported, the ECG is a bio-electric signal that records the electrical activities of the heart. It provides helpful information about the functional aspects of the heart and cardiovascular system, and the state of cardiac health is generally reflected in the shape of ECG waveform, that is a critical information. For this reason, computer-based analysis and classification and automatic interpretation of the ECG signals can be very helpful to assure a continuous surveillance of the patients and to prepare the work of the cardiologist in the analysis of long recordings.

Moreover, as indicated by the main documentation of Physionet, according to the World Health Organization, cardiovascular diseases (CVD) are the number one cause of death worldwide. Of these deaths, 82% take place in low- and middle-income countries. Given their computing power and pervasiveness, the most important question is to check the possibility, for mobile phones, to aid in delivery of quality health care, particularly to rural populations distant from physicians with the expertise needed to diagnose CVD.

Advances in mobile phone technology have resulted in global availability of portable computing devices capable of performing many of the functions traditionally requiring desktop and larger computers. In addition to their technological features, mobile phones have a large cultural impact. They are user-friendly and are among the most efficient and most widely used means of communication. With the recent progress of mobile-platforms, and the increasing number of mobile phones, a solution to the problem can be the recording of ECGs by untrained professionals, and subsequently transmitting them to a human specialist.

The aim of the PhysioNet/Computing in Cardiology Challenge 2011 [15] is to develop an efficient algorithm able to run in near real-time within a mobile phone, that can provide useful feedback to a layperson in the process of acquiring a diagnostically useful ECG recording. In addition to the approaches already cited in Section 1, referring to such a challenge, Table 2 reports other solutions already presented in the literature, capable of quantifying the quality of the ECG looking at leads individually and combined, which can be implemented on a mobile-platform. As reported later, all such approaches are used to compare their results with those obtained in this work.

3 The Neuro-evolutionary Algorithm

The overall algorithm is based on the evolution of a population of individuals, represented by Multilayer Perceptrons neural networks (MLPs), through a joint optimization of their structures and weights, here briefly summarized; a more complete and detailed description can be found in the literature [3].

In this work the algorithm uses the Scaled Conjugate Gradient method (SCG) [7] instead of the more traditional error back-propagation (BP) algorithm to decode a *genotype* into a *phenotype* NN, in order to speed up the convergence of such a conventional training algorithm. Accordingly, it is the genotype which undergoes the genetic operators and which reproduces itself, whereas the phenotype is used *only* for calculating the genotype's fitness. The rationale for this choice is that the alternative of applying SCG to the genotype as a kind of 'intelligent' mutation operator, would boost exploitation while impairing exploration, thus making the algorithm too prone to being trapped in local optima.

The population is initialized with different hidden layer sizes and different numbers of neurons for each individual according to two exponential distributions, in order to maintain diversity among all of them in the new population. Such dimensions are not bounded in advance, even though the fitness function may penalize large networks. The number of neurons in each hidden layer is constrained to be greater than or equal to the number of network outputs, in order to avoid hourglass structures, whose performance tends to be poor. Indeed, a layer with fewer neurons than the outputs destroys information which later cannot be recovered.

3.1 Evolutionary Process

The initial population is randomly created and the genetic operators are then applied to each network until the termination conditions are not satisfied.

At each generation, the first half of the population corresponds to the best $\lfloor n/2 \rfloor$ individuals selected by truncation from a population of size n, while the second half of the population is replaced by the offsprings generated through the crossover operator. Crossover is then applied to two individuals selected from the best half of the population (parents), with a probability parameter p_{cross}, defined by the user together with all the other genetic parameters, and maintained unchanged during the entire evolutionary process.

It is worth noting that the p_{cross} parameter refers to a 'desired' crossover probability, set at the beginning of the evolutionary process. However, the 'actual' probability during a run will usually be lower, because the application of the crossover operator is subject to the condition of similarity between the parents.

Elitism allows the survival of the best individual unchanged into the next generation and the solutions to get better over time. Following the commonly accepted practice of machine learning, the problem data is partitioned into training, validation and test sets, used, respectively for network training, to stop learning avoiding overfitting, and to test the generalization capabilities of a network. Then, the algorithm mutates the weights and the topology of the offsprings, trains the resulting network, calculates fitness on the validation set, and finally saves the best individual and statistics about the entire evolutionary process.

The application of the genetic operators to each network is described by the following pseudo-code:

1. Select from the population (of size n) $\lfloor n/2 \rfloor$ individuals by truncation and create a new population of size n with copies of the selected individuals.
2. For all individuals in the population:
   (a) Randomly choose two individuals as possible parents.
   (b) Check their local similarity and apply crossover according to the crossover probability.
   (c) Mutate the weights and the topology of the offspring according to the mutation probabilities.
   (d) Train the resulting network using the training set.
   (e) Calculate the fitness f on the validation set.
   (f) Save the individual with lowest f as the best-so-far individual if the f of the previously saved best-so-far individual is higher (worse).
3. Save statistics.

The SimBa crossover starts by looking for a 'local similarity' between two individuals selected from the population. If such a condition is satisfied the layers involved in the crossover operator are defined. The contribution of each neuron of the layer selected for the crossover is computed, and the neurons of each layer are reordered according to their contribution. Then, each neuron of the layer in the first selected individual is associated with the most 'similar' neuron of the layer in the other individual, and the neurons of the layer of the second individual are re-ranked by considering the associations with the neurons of the first one. Finally a cut-point is randomly selected and the neurons above the cut-point are swapped by generating the offspring of the selected individuals.

Weights mutation perturbs the weights of the neurons before performing any structural mutation and applying SCG to train the network. All the weights and the corresponding biases are updated by using variance matrices and evolutionary strategies applied to the synapses of each NN, in order to allow a control parameter, like mutation variance, to self-adapt rather than changing their values by some deterministic algorithms. Finally, the topology mutation is implemented with four types of mutation by considering neurons and layer addition and elimination. The addition and the elimination of a layer and the insertion of a neuron are applied with three independent probabilities, indicated as p_{layer}^{+}, p_{layer}^{-} and p_{neuron}^{+}, while the elimination of a neuron is carried out only if the contribution of that neuron is negligible with respect to the overall network output.

For each generation of the population, all the information of the best individual is saved.

As previously considered [2,1], the evolutionary process adopts the convention that a lower fitness means a better NN, mapping the objective function into an error minimization problem. Therefore, the fitness used for evaluating each individual in the population is proportional to the mean square error (mse) and to the computational cost of the considered network. This latter term induces a selective pressure favoring individuals with reduced-dimension topologies.

The fitness function is calculated, after the training and the evaluation processes, by the Equation 1 and it is defined as a function of the confusion matrix M obtained by that individual:

$$f_{multiclass}(M) = N_{\text{outputs}} - \text{Trace}(M), \tag{1}$$

where N_{outputs} is the number of output neurons and $\text{Trace}(M)$ is the sum of the diagonal elements of the row-wise normalized confusion matrix, which represent the conditional probabilities of the predicted outputs given the actual ones.

4   Experiments and Results

The data used for the PhysioNet/CINC 2011 Challenge consist of 2,000 twelve-lead ECGs (I, II, III, aVR, aVF, aVL, V1, V2, V3, V4, V5, and V6), each 10 second long, with a standard diagnostic bandwidth defined in the range (0.05–100 Hz). The twelve leads are simultaneously recorded for a minimum of 10 seconds; each lead is sampled at 500 Hz with 16-bit resolution (i.e., 16 bits per sample).

The proposed approach has been evaluated by using the dataset provided by the challenge organizers. This dataset, described above in Section 2, is public and has been distributed in two different parts:

- Set A: this dataset has to be used to train the approach. It is composed of 998 instances provided with reference quality assessments;
- Set B: this dataset has to be used for testing the approach. It is composed of 500 instances and the reference quality assessments are not distributed to the participants. The reports generated by the approach have to be sent to the submission system in order to directly receive the results from the system used for the challenge.

We split the Set A in two parts: a training set composed of the 75% of the instances contained in the Set A, and a validation set, used to stop the training algorithm, composed of the remaining 25%. While the Set B is used as test set for the final evaluation of the approach.

Each instance of the dataset represents an ECG signal composed of 12 series (one for each lead) of 5,000 values representing the number of recordings performed for each lead. These data have been preprocessed in order to extract the features that we used to create the datasets given in input to the algorithm. We have applied to each lead the fast Fourier transform function in order to transform each lead to the frequency domain. After the transformation, we summed the 5,000 values by groups of 500 in order to obtain 10 features for each lead. Finally, The input attributes of all datasets have been rescaled, before being fed as inputs to the population of ANNs, through a Gaussian distribution with zero mean and standard deviation equal to 1.

The experiments have been carried out by setting the parameters of the algorithm to the values obtained from a first round of experiments aimed at identifying the best parameter setting. These parameter values are reported in Table 1. We performed 40 runs, with 40 generations and 60 individuals for each run, while the number of epochs used to train the neural network implemented in each individual has been set to 250.

Table 1. Parameters of the Algorithm

Symbol	Meaning	Default Value
n	Population size	60
p_{layer}^{+}	Probability of inserting a hidden layer	0.05
p_{layer}^{-}	Probability of deleting a hidden layer	0.05
p_{neuron}^{+}	Probability of inserting a neuron in a hidden layer	0.05
p_{cross}	'Desired' probability of applying crossover	0.7
δ	Crossover similarity cut-off value	0.9
N_{in}	Number of network inputs	120
N_{out}	Number of network outputs	1
α	Cost of a neuron	2
β	Cost of a synapsis	4
λ	Desired tradeoff between network cost and accuracy	0.2
k	Constant for scaling cost and MSE in the same range	10^{-6}

The challenge has been organized in two different events: a closed event and an open one. While in the close event it is possible to develop the classification algorithm in any language, in the open event it is mandatory to develop the algorithm by using the Java language. For this reason, by considering that the proposed approach has been developed in Java too, we compared the obtained results with the results obtained by the other systems that participated to the challenge in the open event. It is important to highlight that we do not claim to obtain the best performance. Our goal was to show that, even if our system is trained with a training set that exploits very little information, the performance obtained by our approach does not lag too much behind the one obtained by the best state-of-the-art systems.

Table 2 shows the results obtained by the other participants compared with the results obtained by the proposed approach. We have inserted both the best and the average performance obtained by the proposed approach. It is possible to observe that, if we consider the best performance, we obtained the second best accuracy; while the average accuracy, computed over the 40 runs, obtained the fourth performance. The robustness of the approach is also proved by observing the low value of the standard deviation that, in the performed experiments, was 0.011.

Table 2. Results of the open event challenge

Participant	Score
Xiaopeng Zhao [18]	0.914
Proposed Approach (Best)	**0.902**
Benjamin Moody [13]	0.896
Proposed Approach (Average)	**0.892**
Lars Johannesen [9]	0.880
Philip Langley [12]	0.868
Dieter Hayn [6]	0.834
Vclav Chudcek [5]	0.833

Besides the evaluation on the test set, we performed also a ten-fold cross validation on the training set. We split the training set in ten fold F_i and we performed ten different set of 10 runs in order to observe which is the behavior of the algorithm when training, validation, and test data change. Table 3 shows the results of the ten-fold cross validation. By observing the results we can observe the robustness of the algorithm. In fact, the accuracies obtained by changing the folds used for training, validation, and test are very close; moreover, the standard deviation of the results is very low.

Table 3. Results of the ten-fold cross validation

Training Set	Validation Set	Test Set	Average Accuracy	Standard Deviation
F1...F7	F8, F9	F10	0.8984	0.0035
F2...F8	F9, F10	F1	0.8988	0.0067
F3...F9	F10, F1	F2	0.9002	0.0075
F4...F10	F1, F2	F3	0.9022	0.0107
F5...F10, F1	F2, F3	F4	0.9040	0.0071
F6...F10, F1, F2	F3, F4	F5	0.9002	0.0029
F7...F10, F1...F3	F4, F5	F6	0.9002	0.0018
F8...F10, F1...F4	F5, F6	F7	0.8976	0.0054
F9, F10, F1...F5	F6, F7	F8	0.9032	0.0090
F10, F1...F6	F7, F8	F9	0.8986	0.0047

5 Conclusions

In this study, we have proposed an ECG classification scheme based on a neuro-evolutionary approach, based on the joint evolution of the topology and the connection weights together with a novel similarity-based crossover, to aid classification of ECG recordings. The signals were first preprocessed into the frequence domain by using a Fast Fourier Trasform algorithm, and then they were normalized through a gaussian distribution with 0 mean and standard deviation equal to 1. The present system was validated on real ECG records taken from the PhysioNet/Computing in Cardiology Challenge 2011.

A series of tests has been carried out in order to evaluate the capability of the neuro-evolutionary approach to discern between usable and unusable electrocardiograms tele-medically acquired from mobile embedded devices. The obtained results show an overall satisfactory accuracy and performances in comparison with other approaches carried out in this challenge and presented in the literature.

It is important to stress the fact that the proposed method was able to achieve top-ranking classification accuracy despite the use of a quite standard preprocessing step and a very small number of input features. No attempt was made yet to fine tune the signal pre-processing and the feature selection steps. On the other hand, it is well known that these two steps are often critical for the success of a signal classification methods. For this reason, we believe that the proposed neuro-evolutionary approach has a tremendous improvement potential.

Future works will involve the adoption of more sophisticated preprocessing techniques, by working, for example, on a multi-scales basis, where each scale represents a particular feature of the signal under study. Other ideas could regard the study and the implementation of feature selection algorithms in order to provide an optimized selection of the signal given as inputs to the neural networks.

References

1. Azzini, A., Dragoni, M., Tettamanzi, A.: A Novel Similarity-Based Crossover for Artificial Neural Network Evolution. In: Schaefer, R., Cotta, C., Kołodziej, J., Rudolph, G. (eds.) PPSN XI. LNCS, vol. 6238, pp. 344–353. Springer, Heidelberg (2010)
2. Azzini, A., Tettamanzi, A.: Evolving neural networks for static single-position automated trading. Journal of Artificial Evolution and Applications 2008 (Article ID 184286), 1–17 (2008)
3. Azzini, A., Tettamanzi, A.: A new genetic approach for neural network design. In: Engineering Evolutionary Intelligent Systems. SCI, vol. 82. Springer, Heidelberg (2008)
4. Azzini, A., Tettamanzi, A., Dragoni, M.: SimBa-2: Improving a Novel Similarity-Based Crossover for the Evolution of Artificial Neural Networks. In: 11th International Conference on Intelligent Systems Design and Applications (ISDA 2011), pp. 374–379. IEEE (2011)
5. Chudáček, V., Zach, L., Kužílek, J., Spilka, J., Lhotská, L.: Simple Scoring System for ECG Signal Quality Assessment on Android Platform. Contribution sent to the 38th Physionet Cardiology Challenge (2011)
6. Hayn, D., Jammerbund, B., Schreier, G.: Real-time Visualization of Signal Quality during Mobile ECG Recording. Contribution sent to the 38th Physionet Cardiology Challenge (2011)
7. Hestenes, M., Stiefel, E.: Methods of conjugate gradients for solving linear systems. Journal of Research of the National Bureau of Standards 49(6) (1952)
8. Jiang, W., Kong, S.G.: Block-Based Neural Networks for Personalized ECG Signal Classification. IEEE Transactions on Neural Networks 18(6) (2007)
9. Johannesen, L.: Assessment of ECG Quality on an Android Platform. Contribution sent to the 38th Physionet Cardiology Challenge (2011)
10. Jokic, S., Krco, S., Delic, V., Sakac, D., Jokic, I., Lukic, Z.: An Efficient ECG Modeling for Heartbeat Classification. In: IEEE 10th Symposium on Neural Network Applications on Electrical Engineering, NEUREL 2010, Belgrade, Serbia, September 23-25 (2010)
11. Kuzilek, J., Huptych, M., Chudacek, V., Spilka, J., Lhotska, L.: Data Driven Approach to ECG Signal Quality Assessment using Multistep SVM Classification. Contribution sent to the 38th Physionet Cardiology Challenge (2011)
12. Langley, P., Di Marco, L., King, S., Di Maria, C., Duan, W., Bojarnejad, M., Wang, K., Zheng, D., Allen, J., Murray, A.: An Algorithm for Assessment of ECG Quality Acquired Via Mobile Telephone. Contribution sent to the 38th Physionet Cardiology Challenge (2011)
13. Moody, B.E.: A Rule-Based Method for ECG Quality Control. Contribution sent to the 38th Physionet Cardiology Challenge (2011)

14. Moody, G.B.: Improving the quality of ECGs collected using mobile phones: The 12th Annual Physionet/Computing in Cardiology Challenge. Computing in Cardiology Challenge 38 (2011)
15. PhysioNet: Research Resource for Complex Physiologic Signals, http://www.physionet.org
16. Silva, K., Moody, G.B., Celi, L.: Improving the Quality of ECGs Collected Using Mobile Phones: The PhysioNet/Computing in Cardiology Challenge 2011. Contribution sent to the 38th Physionet Cardiology Challenge (2011)
17. Tat, T.H.C., Chen Xiang, C., Thiam, L.E.: Physionet Challenge 2011: Improving the Quality of Electrocardiography Data Collected Using Real Time QRS-Complex and T-Wave Detection. Contribution sent to the 38th Physionet Cardiology Challenge (2011)
18. Xia, H., McBride, J., Sullivan, A., De Bock, T., Bains, J., Wortham, D., Zhao, X.: A Multistage Computer Test Algorithm for Improving the Quality of ECGs. Contribution sent to the 38th Physionet Cardiology Challenge (2011)

A Genetic Fuzzy Rules Learning Approach for Unseeded Segmentation in Echography

Leonardo Bocchi and Francesco Rogai

Università degli Studi di Firenze, Dipartimento di Elettronica e Telecomunicazioni

Abstract. Clinical practice in echotomography often requires effective and time-efficient procedures for segmenting anatomical structures to take medical decisions for therapy and diagnosis. In this work we present a methodology for image segmentation in echography with the aim to assist the clinician in these delicate tasks. A generic segmentation algorithm, based on region evaluation by means of a fuzzy rules based inference system (FRBS), is refined in a fully unseeded segmentation algorithm. Rules composing knowledge base are learned with a genetic algorithm, by comparing computed segmentation with human expert segmentation. Generalization capabilities of the approach are assessed with a larger test set and over different applications: breast lesions, ovarian follicles and anesthetic detection during brachial anesthesia.

1 Introduction

Segmentation in echo-tomography imaging is an important task for the clinician to speed up his work and improving phases of measurement and diagnosis. Several algorithms have been proposed for this purpose, tailored for applications spanning, among others, from neoplastic lesions assessment in oncology, to ventricle volume determination and myocardial efficiency in cardiology and to therapy planning in ovarian follicles measurements for assisted fecundation. Different applications and the intrinsic artifacts of ultrasound images make difficult the definition of a general approach to the segmentation problem. Low resolution, speckle noise, shadowing/enhancement artifacts make the contour of anatomic structures difficult to discriminate [9].

The following considerations where taken into account to describe the framework on which the proposed segmentation system is designed. Assuming an image $\mathscr{I}$ as a map defined on an ordered grid of coordinates (x, y) in a planar domain $\mathscr{D}$ of the image:

$$\mathscr{I}(x, y), (x, y) \in \mathscr{D} = M \times N \tag{1}$$

It is possible to consider a generic partitioning of $\mathscr{D}$ as a mapping that generates different Boolean mask for each value of a choice parameter θ:

$$\mathscr{P}_\theta \{\mathscr{D}\} = \{(x, y) \in \mathscr{D} \mid m_\theta(x, y) = \text{TRUE}\} \tag{2}$$

C. Di Chio et al. (Eds.): EvoApplications 2012, LNCS 7248, pp. 305–314, 2012.

We define as a generic or totipotent segmentation function - borrowing the term from cell biology - a partitioning function which is capable to generate all possible partitions for an image $\mathscr{I}$ varying the choice parameter θ in a finite set. The cardinality of the set of all possible partitions of an image of size $X \times Y$ increases exponentially with image size. However, common segmentation algorithms work on a very restrictive subclass of the partition set. Thus they can be referred as quasi-generic or multipotent segmentation functions and their use is licit assuming that the correct segmentation is part of the set of possible regions generated by the method (which has a very lower cardinality respect that of a true generic segmentation function). In this work, we will evolve a specialized, or unipotent, algorithm - tailored for a particular purpose - by simple introduction of a selection procedure which selects the correct connected components among all partitions produced by a quasi-generic segmentation function. A soft computing approach is proposed to define a function to operate this choice.

2   Unseeded Segmentation

Proposed system is composed of a three-stage procedure: preprocessing, primary segmentation and fine segmentation. The preprocessing step enhances image quality, by removing speckle noise, for improving the subsequent segmentation. Several techniques have been proposed for attenuation of speckle noise, and one of the most recent ones is wavelet denoising [12]. In this work it is used a cellular automata approach proposed for image enhancement in [7] and shown to be feasible for the particular purpose in [2].

The proposed approach is based on a primary segmentation algorithm which can be implemented using any algorithm satisfying the only requirement of being capable to generate a set of image partitions containing the correct segmentation. The correctness is evaluated by minimizing a function of discrepancy of each partition against the segmentation performed by a human radiology expert, representing the segmentation error. The choice of the primary segmentation algorithm has been guided considering the cardinality of the image partitions set feasible by the candidate algorithm - through which the best solution has to be searched - which can be up to $2^{XY} - 2$ for an image of size $X \times Y$ pixel. An additional constraint on the algorithm requires the partitions chosen depend on a partition choice parameter, possibly with a smooth behavior (small variation in the parameter implies small variation in the partitions). Several different quasi-generic segmentation methods were experimented (such as watershed and k-means clustering), and we selected an algorithm based on a sort of filtered thresholding for the efficiency and smoothness of dependence of its choice parameter (threshold level) versus the generated partitions and for its computational efficiency.

The preprocessed image is filtered with a combined Gaussian and Laplacian of Gaussian kernel to regularize the image and enhance the contrast on blob-like regions. Filtered image is subject to thresholding, at a generic level θ, and a connected-component labeling algorithm determines each hypothetical region of

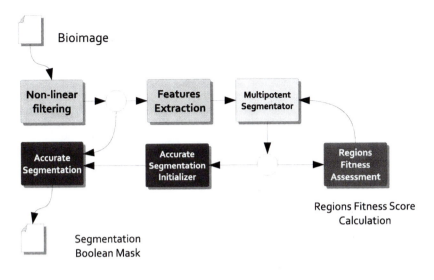

Fig. 1. Overall representation of the segmentation system

interest. Searching space is constituted by all regions individuated by all values of θ. A regions fitness score function (following referred as ρ) evaluates each region and assigns a score value. The function ρ is computed by means of a fuzzy inference system (FIS) whose inference rules are defined by a genetic approach against an objective function constituted by the mean error on a set of manually segmented images.

Coarse defined regions from the previous step can be regularized with morphological operations and utilized as proper initialization for the use in an instance of GrowCut algorithm [13,10] as described in [2].

3   Method

The core of the proposed method consists of a fuzzy inference system which selects the optimal region, among all the regions generated by the segmentation algorithm. Each region is characterized by a set of features, which are chosen as a trade off between representation (features should be similar to region properties used in standard medical practice) and parsimony (too many features can make the algorithm too slow or unstable). The selected feature set is composed of $N = 7$ features, including: coordinates of the center of mass, mean and standard deviation of gray level, solidity, eccentricity and orientation.

Solidity is defined as the ratio between the number of pixels in the region and the number of pixel in its convex hull which can also be defined as the region with lesser area in the set of convex regions containing the area, or, equivalently, the intersection between all convex sets containing the region (the intersection of convex sets is a convex set). Eccentricity is defined as the ratio between major and minor axis of an ellipse with the same second order moment of surface

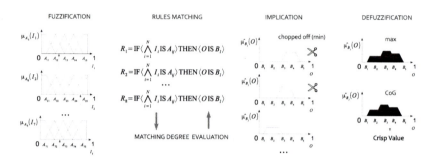

Fig. 2. Principal steps in the fuzzy inference system used

(polar barycentric). Orientation is the angle in radians of the major axis of the ellipse with respect to the horizontal axis.

Fuzzy inference systems where introduced by considerations of Zadeh in [14] and used in the seminal paper [11]. Nowadays their use is spread in many fields and they are employed in many different applications spanning from industrial control[6] to advanced image processing [15]. Fuzzy inference systems are characterized by the ability of handling the uncertainty and the inference based on approximate reasoning is more robust than in classical rule systems. In particular, Mamdani type FRBSs provide a highly flexible means to formulate knowledge, while at the same time remain interpretable.

The fuzzy inference system methodology used for this work is based on the so called Mamdani inference. Although this approach does not scale well as the dimensionality and complexity of the input-output relations increases, it has the major advantage of being well interpretable [5] and this requisite is important for acceptability in a medical context.

Accordingly to the Mamdami approach, the proposed implementation consists of an algorithm based on the following steps:

1. *Fuzzyfication:* Every input range is partitioned on $M = 5$ equally spaced membership functions, which are triangular shaped (triangular degenerated at borders) as shown in 2. Every crisp input (region feature) I_p, $p = 1..N$, is represented by means of the corresponding value of membership function (MF) of each fuzzy partitions the input range is subdivided in. The output of this stage is an array of membership values for each input, i.e. N arrays of M elements (each input is partitioned with the same number of MF).

2. *Rules matching:* Every rule in the rule knowledge base which refers to a partition whose current membership value for the current inputs is not null contributes to the matching degree of the rule itself which is calculated assuming the conventions of fuzzy logic algebra. Fuzzy rule set is an aggregate of Q rules expressed as follows:

$$R_q = \text{IF} \left\langle \bigwedge_{i=1}^{N} I_i \text{ IS } A_{ij} \right\rangle \text{ THEN } \langle O \text{ IS } B_l \rangle \tag{3}$$

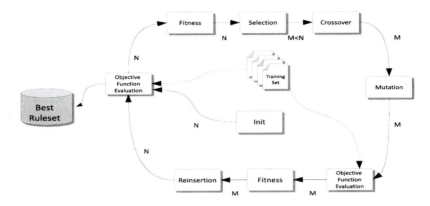

Fig. 3. Genetic algorithm for fuzzy rules learning

3. *Rules implication:* The matching degree of the antecedent is passed to the consequent. Every partition of the output variable is subject to the effect of the min t-norm implication operator, corresponding to the chopping off of each MF at the corresponding matching degree.
4. *Aggregation:* The output partitions are aggregated using the max operator.
5. *Defuzzification:* The center of gravity is calculate on the aggregation to achieve a crisp value for the output.

The fuzzy inference system assesses the region fitness score ρ_p for each region p determined by the multipotent segmentation method. All candidate regions are sorted accordingly to their score ρ_p from min to max, and the cumulative score $R = \sum \rho_p$ is evaluated. The algorithm selects the k regions with higher score, where k is the largest integer satisfying $\sum_{p=1}^{k} \rho_p < 0.05R$. This apparently simple assumption is sufficient to discern regions of medical interest from others by means of the capabilities of scoring algorithm itself wich determines a frequency distribution of region score that is always notch-shaped and practically bimodal with the interest regions laying in the second modal group (higher score) which generally covers not more than 5% of the entire cumulative score and well separated from the other group.

Each selected region is used to evaluate a single region accuracy (SRA):

$$SRA_p = \frac{\sum \hat{c}_p \wedge \hat{h}_p}{\sum \hat{c}_p \oplus \hat{h}_p} \tag{4}$$

where $\hat{c}_p$ is the computed segmentation mask and $\hat{h}_p$ is the correspondent region in human expert segmentation mask, operators $\oplus$ and $\wedge$ are bitwise XOR and AND respectively. We also defined a multi-regions accuracy function (MRA) as the ratio between number of matched regions (when overlapping area is greater than 80%) and total number of regions (RTN) marked by the human expert.

Many approaches to fuzzy rules learning have been proposed in the genetic framework. Knowledge base (KB) of FIS is constituted by two different information levels. The first one is the data base (DB) containing the sets of linguistic

terms considered in the linguistic rules and the membership functions defining
the semantics of the linguistics labels . The second one is the rule base (RB) that
comprises a collection of linguistic rules that are joined by the also operator so
multiple rules can fire for the same input. A genetic algorithm can compose the
rule set with fixed DB, the data base with fixed rules or both.

Genetic learning of rule base, is generally achieved by two approaches. The
first one is referred as Pittsburg approach and consists on a representation of
the entire RB with a single chromosome whereas the latter considers a single
rule for each chromosome. Methods assuming chromosome as single rule are
conversely referred as Michigan approach, many others were developed from
that such as iterative learning (ITL) and cooperative competitive methods[5].
this mechanism, A pure Pittsburg approach is choosen for this work for avoiding
the complexity required by the systems based on Michigan method, obtaining a
simpler and easily implementable algorithm. We also assumed a predefined set of
membership functions, considering an input range represented with 5 partitions
is sufficiently descriptive for the considered features

The algorithm is articulated as shown in fig.3. Each locus in the chromosome
is codified with an integer between 0 and 5, inclusive. The values from 1 to
5 are used to represent the corresponding fuzzy partition labels for the current
input in the rule, number 0 is used as "wild card" symbol, representing all possible
membership labels in that input, therefore effectively ignoring that input feature.
Thus chromosome is composed by $Q \times 8$ genes, as each rule has 7 input genes
and one output value. A stochastic universal sampling algorithm is chosen for
the fitness evaluation for the characteristics of this method which fulfill zero
bias and minimum spread[1]. For analogous reasons a shuffle crossover operator
is used[3] with a per gene mutation rate of 0.01.

The fitness of each individual χ is evaluated as the sum of SRA on all the N_s
images of the training set TS:

$$\Phi(\chi) = \frac{1}{N_s} \sum_{p=1}^{N} SRA_p \tag{5}$$

Best individuals are selected in accordance to an universal stochastic sampling
with a generation gap (value 0.75) and deferred to the reproduction phase. The
best individuals are selected to replace worst ones using an elitist strategy. A
population of 60 rule sets is used in all the experiments. Stopping criterion is
max iteration reached defined with a trial and error process.

4   Experiments and Results

The algorithm has been tested on three data-set of different origin. The first one
(BreastDS) is composed by 35 images (Fig. 4) collected during standard screen-
ing protocols of the mammary glands executed by expert radiological operators
at Senologic Unit - Azienda Ospedaliero Universitaria Careggi (AOUC), Firenze,
Italy with 8 MHz linear probes from MyLab™ 50 (Esaote S.p.A., Firenze).

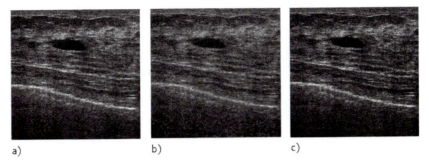

a)   b)   c)

Fig. 4. Breast neoplastic lesion. a) Original bioimage. b) Partitions generated and deferred to the evaluation and selection stage. c) Resulting selected region.

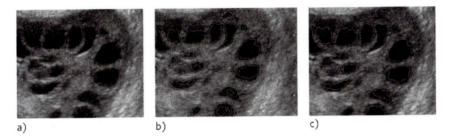

a)   b)   c)

Fig. 5. Segmentation of follicles. a) Original bioimage. b) Partitions generated and deferred to the evaluation and selection stage. c) Resulting selected region. Particular image of PCOS from [4].

The second (FolliclesDS) is composed of a set of transvaginal ultrasound images (Fig. 5) of ovarian follicles in different conditions, 18 images extracted from databases of PCOS [4] for use in human in-vitro fertilization assisted ultrasonography and a 50 images set of different medical institutions origins and ultrasound equipments. The third (NerveDS) is composed of 100 images from ultrasound videos (Fig. 6) employed in ultrasound-guided axillary brachial plexus block in Centro Traumatologico Ortopedico, AOUC. Used apparatus is a MyLab™ 50 with a 13 MHz probe. Each dataset has been manually segmented by an expert radiologist with an image editing software and mouse or pen tablet.

Training sets are set of images extracted from each datasets. They were choosen representing the different morphologies of structures to be identified with the aid of medical experts. They contains 15, 10 and 15 images from BreastDS, FolliclesDS and NerceDS respectively. The optimal number of rules has been evaluated by examining the total error rate with different values of Q. Results, reported in Fig. 7, show a decrease of the error rate with the increase of Q, until we obtain a substantial stabilization of the error with rule sets greater than an optimal number of rules and a slow down of the convergence performance of the learning process. The value of Q is, respectively for the BreastDS, FollicleDS and NerveDS, 11 rules, 8 rules and 9 rules.

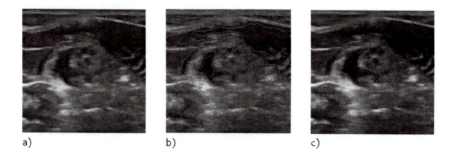

a)   b)   c)

Fig. 6. Brachial nerve during brachial plexus block anesthetic injection. a) Original image. b) Partitions generated and deferred to the evaluation and selection stage. c) Resulting selected region.

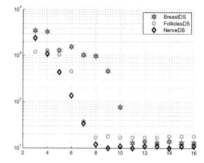

Fig. 7. Total segmentation error with different values of Q for the three sets

The MRA index is high for all data sets, its maximum assessed over the entire data-sets is respectively 0.99, 0.97 and 0.98. Mean values of SRA index are 0.97, 0.98, 0.98, respectively, while its minimum values are 0.89, 0.88, 0.87, respectively. Convergence is always reached in less than 80 generations during learning in the three datasets (Fig. 8).

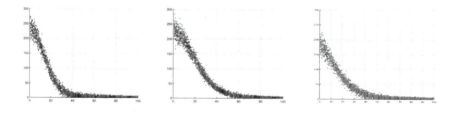

Fig. 8. Total segmentation error, mean on population, 30 runs of the learning algorithm with selected number of compact-form fuzzy rules. From left to right: BreastDS, FolliclesDS, NerveDS.

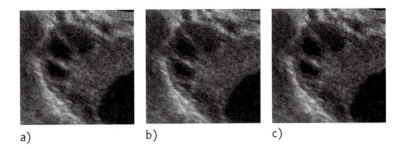

a)   b)   c)

Fig. 9. Different segmentation attempts of ovarian follicles[4] segmentation. a) Active contours as shown in [8] with same parameters for all follicular regions. b) Simple thresholding with manually fixed level. c) Proposed algorithm.

5   Conclusions

Reported results show the proposed approach is able to correctly adapt to different segmentation tasks in ultrasound imaging. The application of a fuzzy inference system is shown to be useful to refine behavior of a generalist method of segmentation.The use of a FIS representation is shown to be efficent for the purpose of region selection. Selection of connected components by means of a scoring function avoids the issues, typical of other segmentation algorithms such as active contours, where contours can be highly dependent on the choice of the seeding points and the parameters of the model so they necessitate an ad-hoc tuning for each region. Under-segmentation such as regions containing part of adjacent follicles as shown in fig. 9 is quite improbable with proposed method wich discards contours with attributes of shape unlikely for a human follicle. Mandani rules assures a complete understandability of automatically generated rules which is agreable for clinicians. The learning phase for the generation of the taylored RB for each of the particular medical applications required less than 45 minutes in every described data set. Current implementation is based on interpreted code executed on a 32 bit OS desktop computer with a standard dual core processor at 2.81 GHz. The processing time required by the final segmentation algorithm (with learned rules) is under 500 ms for each image, thus reasonably low, while the accuracy of the obtained segmentation is high.

Acknowledgments. We wish to thank the medical operators of the radiology and anesthesiology departments of AOUC - University Hospital of Careggi, Florence, for their support in providing the images and the necessary medical expertise, and in particular Dr. J. Nori, Dr. R. Deodati, and Dr. A. Di Filippo.

References

1. Baker, J.E.: Reducing Bias and Inefficiency in the Selection Algorithm. In: Proceedings of the Second International Conference on Genetic Algorithms and their Application, pp. 14–21. Lawrence Erlbaum, Hillsdale (1987)

2. Bocchi, L., Rogai, F.: Segmentation of Ultrasound Breast Images: Optimization of Algorithm Parameters. In: Di Chio, C., Cagnoni, S., Cotta, C., Ebner, M., Ekárt, A., Esparcia-Alcázar, A.I., Merelo, J.J., Neri, F., Preuss, M., Richter, H., Togelius, J., Yannakakis, G.N. (eds.) EvoApplications 2011, Part I. LNCS, vol. 6624, pp. 163–172. Springer, Heidelberg (2011)
3. Caruana, R., Eshelman, L.: Representation and hidden bias II: Eliminating defining length bias in genetic search via shuffle crossover. In: Ridharan, N.S. (ed.) Proceedings of the 11th International Joint Conference on AI, pp. 750–755. Morgan Kaufmann (1989)
4. Chizen, D., Pierson, R.: Global library of women's medicine (2010)
5. Cordón, O.: Genetic fuzzy systems: Evolutionary tuning and learning of fuzzy knowledge bases, vol. 19. World Scientific Pub. Co. Inc. (2001)
6. Driankov, D., Hellendoorn, H., Reinfrank, M., Ljung, L., Palm, R., Graham, B., Ollero, A.: An Introduction to Fuzzy Control. Springer, Heidelberg (1996)
7. Hernandez, G., Herrmann, H.J.: Cellular Automata for Elementary Image Enhancement. Graphical Models and Image Processing 58(1), 82–89 (1995)
8. Gritti, F., Giannotti, E., Nori, J., Bocchi, L.: Active contour segmentation for breast cancer detection using ultrasound images. In: II Congresso Nazionale di Bioingegneria (2010)
9. Huang, Y.L., Chen, D.R.: Watershed segmentation for breast tumor in 2-D sonography. Ultrasound in Medicine & Biology 30(5), 625–632 (2004)
10. Konouchine, V., Vezhnevets, V.: Interactive image colorization and recoloring based on coupled map lattices. Computer, 231–234 (2006)
11. Mamdani, E., Assilian, S.: An Experiment in Linguistic Synthesis with a Fuzzy Logic Controller. International Journal of Man-Machine Studies 7(1), 1–13 (1975)
12. Nicolae, M., Moraru, L., Onose, L.: Comparative approach for speckle reduction in medical ultrasound images. Romanian J. Biophys. 20(1), 13–21 (2010)
13. Vezhnevets, V., Konouchine, V.: GrowCut: Interactive multi-label ND image segmentation by cellular automata. In: Proc. of Graphicon, pp. 150–156. Citeseer (2005)
14. Zadeh, L.A.: Outline of a New Approach to the Analysis of Complex Systems and Decision Processes. IEEE Transactions on Systems, Man, and Cybernetics SMC-3 (January 1973)
15. Chi, Z., Yan, H., Pham, T.: Fuzzy Algorithms: With Applications to Image Processing and Pattern Recognition. World Scientific (1996)

Object Recognition with an Optimized Ventral Stream Model Using Genetic Programming

Eddie Clemente[1,2], Gustavo Olague[1], León Dozal[1], and Martín Mancilla[1]

[1] Proyecto EvoVision,
Departamento de Ciencias de la Computación, División de Física Aplicada,
Centro de Investigación Científica y de Estudios Superiores de Ensenada,
Carretera Ensenada-Tijuana No. 3918, Zona Playitas, Ensenada, 22860, B.C., México
{eclemen,ldozal,olague}@cicese.edu.mx
http://cienciascomp.cicese.mx/evovision/
[2] Tecnológico de Estudios Superiores de Ecatepec. Avenida Tecnológico S/N,
Esq. Av. Carlos Hank González, Valle de Anáhuac, Ecatepec de Morelos

Abstract. Computational neuroscience is a discipline devoted to the study of brain function from an information processing standpoint. The ventral stream, also known as the "what" pathway, is widely accepted as the model for processing the visual information related to object identification. This paper proposes to evolve a mathematical description of the ventral stream where key features are identified in order to simplify the whole information processing. The idea is to create an artificial ventral stream by evolving the structure through an evolutionary computing approach. In previous research, the "what" pathway is described as being composed of two main stages: the interest region detection and feature description. For both stages a set of operations were identified with the aim of simplifying the total computational cost by avoiding a number of costly operations that are normally executed in the template matching and bag of feature approaches. Therefore, instead of applying a set of previously learned patches, product of an off-line training process, the idea is to enforce a functional approach. Experiments were carried out with a standard database and the results show that instead of 1200 operations, the new model needs about 200 operations.

Keywords: Evolutionary Artificial Ventral Stream, Complex Designing System, Heterogeneous and Hierarchical Genetic Programming.

1 Introduction

The human brain is the best example of a purposeful system that transforms numerous complex signals into a set of complex actions. Today, the exact way in which the brain organizes and controls its actions remains a mystery. The endeavour of understanding the inner working of the brain is challenged by several communities grouped into the field of neuroscience that includes the following disciplines: psychology, neurology, psychiatry, cognitive science, cybernetics, computer science, and philosophy, to mention but a few. The advent of

C. Di Chio et al. (Eds.): EvoApplications 2012, LNCS 7248, pp. 315–325, 2012.

computer technology starts a new age in which the brain is modeled as an information processing system giving raise to a field known as computational neuroscience. Although the complexity of the brain is recognized within the domain of evolutionary computation; there is no meaningful work on the development of algorithms modeling the ventral stream and their application to the solution of complex problems [5]. The research described in the present paper aims to develop a new research area in which evolutionary computing is combined with previous proposals from computational neuroscience to assess the validity of algorithmic representations for problem solving. In particular, the goal of the present paper is to illustrate the necessary steps for evolving an artificial ventral stream applied to the problem of object recognition.

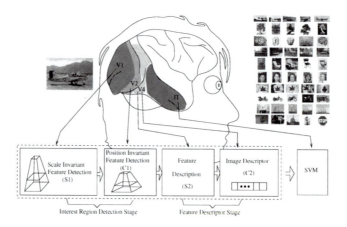

Fig. 1. Analogy between the ventral stream and the proposed computational model

Figure 1 depicts the classical model of the ventral stream that is known as the standard model for object recognition. In this model, the ventral stream begins with the primary visual cortex, $V1$, goes through visual area, $V2$, then through visual area, $V4$, and finally to the inferior temporal cortex. Therefore, the ventral stream is said to be organized as a hierarchical and functionality specialized processing pathway [6]. The idea exposed in this paper is to evolve an artificial occipitoparietal pathway in such a way of defining an interest region stage; as well as, a feature description stage. The proposed model starts with a color image that is decomposed into a set of alternating "S" and "C" layers, which are named after the discovery of Hubel and Wiesel of the simple and complex cells [7]. This idea was originally implemented by Fukushima in the neocognitron system [4]. This system was further enhanced by other authors including the convolutional networks [8] and the HMAX model by Riesenhuber and Poggio [9]. In all these models the simple layers apply local filters in order to compute higher-order features and the complex layers increase invariance by combining units of the same type.

1.1   Problem Statement

Despite powerful paradigms for object recognition developed in the last decades; it is acknowledged that a solution remains elusive. The problem studied in this paper is the recognition of object categories and this is solved with a biological inspired model that is optimized through an evolutionary algorithm. The goal is to search for the best expressions that are grouped into a hierarchical structure by emulating the functionality of the ventral stream using genetic programming. The major result that is presented in this work is the simplification of the computational process that brings a significant economy in the final algorithm.

The remainder of the paper is organised as follows: section 2 describes the artificial ventral stream divided in two parts known as detection and description; section 3 presents the hierarchical and heterogeneous genetic programming that is used as a way of solving the computational problem; section 4 provides experimental results, and section 5 draws the conclusions and discusses possible future work.

2   An Artificial Ventral Stream

The aim of this section is to describe an artificial ventral stream (AVS) with the goal of solving the problem of object class recognition. In this way, an artificial ventral stream is defined as a computational system, which mimics the functionality of the visual ventral pathway by replicating the main characteristics that are normally found in the natural system. In this sense, previous research developed by computer scientists and neuroscientists such as: [4,1,12,11,10] follow a line of research where the natural and artificial systems are explained through a data-driven scenario. Thus, the idea is to extract a set of features from an image database using a hierarchical representation. The ventral stream is modeled as a process that replicates the functionality of the primary visual cortex $V1$, the over extrastriate visual areas $V2$ and $V4$, and the inferotemporal cortex IT. Thus, the image is transformed into a new representation where: bars, edges, and gratings, are outlined and the whole information is combined into an output vector that represents the original image. This process is characterized by the application of *a priori* information in the form of image patches, which are normally used during the training of the proposed model. In this way, the artificial ventral stream is evaluated by a classification system that is implemented with a support vector machine.

Contrary to previous research, the idea developed in this paper is to propose a function driven scenario based on the genetic programming paradigm. This section proposes a way in which an artificial ventral stream could be evolved in order to emulate key functions that are used to describe the human ventral stream; specifically the standard model. These functions, called Evolutionary Visual Operators (EVO), are optimized in order to render an improved design of the whole visual stream. In particular, the aim is to recognize the building blocks that are used in the solution of a multi-class object recognition problem. Thus, the search functions highlight the set of suitable features that point out the properties of the object such as: color, form, and orientation. The main advantage of the proposed system is reflected on the lower amount of computations that provided a significant economy without sacrificing the overall quality. Next, we describe both detection and description stages following the hierarchical model of the ventral stream.

2.1   Interest Region Detection

The interest region detection stage is depicted in Figure 2. The input color image $I(R, G, B, x, y)$ is decomposed into a pyramid of 12 scales $(I_{\sigma_1}, ..., I_{\sigma_{12}})$, each being smaller than the last one within a factor of $2^{\frac{1}{4}}$; while maintaining its aspect ratio and the information of the corresponding color bands (R, G, B). In this way, the pyramid could be seen as a multidimensional structure that is used to introduce scale invariance information along the multiple bands and at the same time integrating the multiple color bands R, G, B. In this stage, the idea is to apply an EVO to the image pyramid in order to simplify the amount of information. The EVO should be understood as a general concept that is applied to the artificial ventral stream; in such a way, that for each step of the information processing there are specialized programs that fit the problem in an optimal way.

For example, during the stage devoted to the scale invariant feature detection, an interest region operator (IRO) is evolved in order to be adapted to this specific function, see Figure 2. Hence, the IRO can be seen as an specialized operator designed with a GP process that extract special or unique features such as: borders, lines at different orientations, corners; and finally, others that do not need to be human readable. Moreover, one property of genetic programming is the characteristic of being a white box, which is something of great value in the approach that is presented here. In this work, the IRO's domain is defined by the color and orientation at 12 scales $I_{\sigma_1}, ..., I_{\sigma_{12}}$ and its codomain is the

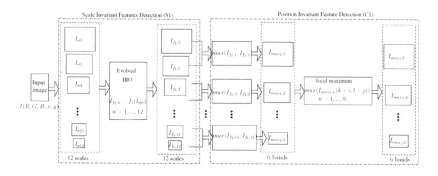

Fig. 2. Flowchart of the interest region detection stage

resulting pyramid of images $I_{f_1,1}, ..., I_{f_1,12}$ that are obtained after applying a suitable IRO. These steps have the functionality of replicating the layer V1 that consists of a set of simple cells. Note, that the structure of the IRO is built from the terminals and functions provided in Table 1. Here, the terminals not only include the RGB color space; but also, the C, M, Y, H, S, I, K that are obtained from the corresponding transformation between color spaces.

Next, in order to enhance the data a *maximum* operation is applied over the local regions, $max(I_{(f_1,2n-1)}, I_{(f_1,2n)})$ with $n = 1, .., 6$, between each pair of consecutive images of the 12 scale pyramid. Then, another maximum filter is applied in order to substitute each sample of the 6 bands $(I_{max1,1}, ..., I_{max1,6})$ by the maximum within an interval around a region ϵ of size $i \times j$ around the sample's position (k, l):

$$I_{max2,n} = \max[I_{max1,n}(k + i, l + j)] \tag{1}$$

In this particular case, $i = j = 9$ and k and l move at steps of 5 elements for each band. These two operations improve the position and scale invariance within a larger region and it also reduces the information into fewer bands: $(I_{(max2,1)}, ..., I_{(max2,6)})$.

This process emulates the first stage of a simple feedforward hierarchical architecture that is composed of a set of simple cells, modeled here with the IRO, and the cortical complex cells which bring some tolerance respect to small changes in position and scale. Therefore, in the proposed model, the layers $S1$ and $C1$ have the purpose of detecting features that are invariant to scale and position. The next section explains how to describe image regions containing the detected features.

Table 1. Set of terminals and functions

Terminals IRO:	$R, G, B, C, M, Y, H, S, I, K, D_x(R), D_x(G), D_x(B),$ $D_x(C), D_x(M), D_x(Y), D_x(H), D_x(S), D_x(I), D_x(K), D_y(R),$ $D_y(G), D_y(B), D_y(C), D_y(M), D_y(Y), D_y(H), D_y(S), D_y(I),$ $D_y(K), D_{xx}(R), D_{xx}(G), D_{xx}(B), D_{xx}(C), D_{xx}(M), D_{xx}(Y),$ $D_{xx}(H), D_{xx}(S), D_{xx}(I), D_{xx}(K), D_{yy}(R), D_{yy}(G), D_{yy}(B),$ $D_{yy}(C), D_{yy}(M), D_{yy}(Y), D_{yy}(H), D_{yy}(S), D_{yy}(I), D_{yy}(K)$
Functions IRO:	$+, -, /, *, |-|, |+|, (\cdot)^2, log(\cdot), D_x(\cdot), D_y(\cdot), D_{xx}(\cdot), D_{xy}(\cdot)$ $D_{yy}(\cdot), Gauss_{\sigma=1}(\cdot), Gauss_{\sigma=2}(\cdot), 0.05(\cdot)$
Terminals IDO:	$C1, D_x(C1), D_{xx}(C1), D_y(C1), D_{yy}(C1), D_{xy}(C1)$
Functions IDO:	$+, -, /, *, |-|, |+|, \sqrt{\cdot}, (\cdot)^2, log(\cdot), D_x(\cdot), D_y(\cdot), D_{xx}(\cdot)$ $D_{xy}(\cdot), D_{yy}(\cdot), Gauss_{\sigma_1}(\cdot), Gauss_{\sigma_2}(\cdot), 0.05(\cdot)$

2.2 Feature Description

Once that all regions have been highlighted the next step is to describe such important regions. The typical approach is based on template matching between the information obtained in the previous section and a number of prototype patches. The goal is to learn a set of prototypes that are known as a universal dictionary of features and which are used by all object categories. Hopefully, the SVM can recognize the prototypes that correspond to a specific image of a single category. On the other hand, in this paper the functionality of layer $S2$ is evolved in order to enhance the set of prominent features that was highlighted by the interest region detector. It should be noted that each evolved function is a composite function that is capable of substituting several prototype features; thus, reducing significantly the total number of operations needed to define all object features that are used to describe and classify the input images. According to Figure 3, the information provided by $C1$ is feedforward to $k-1$ operators that emulate a set of lower order hypercomplex cells. In other words, for each input image $I_{max2,n}$ with $n = 1, ..., 6$, a set of functions $f_i(I_{max2,n})$ with $i = 2, ..., k$, are applied in order to highlight the necessary characteristics that recognize each object class. Note, that each of these functions is an EVO built by the GP from the terminals and functions shown in Table 1, and which performs the patch information descriptive operation (IDO) during this second stage. Hence, this set of functions replaces the universal dictionary proposed by [12,10,11] and it could be said that it corresponds to a function driven approach.

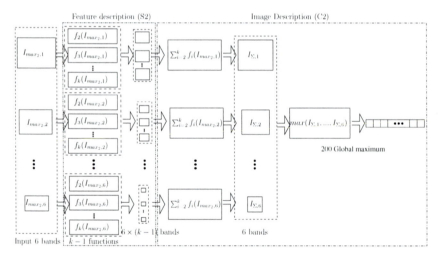

Fig. 3. Flowchart of the feature description stage

Finally, the methodology includes a layer $C2$ for which the outputs of the $k-1$ EVOs are combined and feedforwarded into $I_{\Sigma,n} = \sum_{i=2}^{k} f_i(I_{max2,n})$ with $n = 1, ..., 6$, resulting into a new pyramid of 6 bands. This step is different to the traditional layer $C2$ where an Euclidean norm is applied to identify the best patches. In this way, the approach proposed here requires only to add the $k-1$ functions' responses. Thus, the image description vector is built by selecting the 200th highest values from the image pyramid that is sort out of the $C2$ layer.

3   Heterogeneous and Hierarchical GP

This section describes the heterogeneous and hierarchical genetic programming (HHGP) that was implemented to optimize the artificial ventral stream. Figure 4 depicts the main steps in the search of an optimal AVS using a mixture of tree-based representations organized similarly to a linear genetic programming structure. The representation that is proposed ensures the development of complex functions, while freely increasing the number of programs. In this way, the structure can growth in number and size of its elements. It is important to remark that each individual should be understood as the whole AVS and it is therefore not only a list of tree-based programs but the whole processing depicted in Figures 2 and 3. Thus, the algorithm executes the following steps. First, it randomly creates an initial population of 30 AVS, where each one is represented as a string of heterogeneous and hierarchical functions called chromosome. In this way, each function corresponds to a gene and is represented as a tree with a maximum depth of 7 levels. Also, each string has a maximum length of 10 genes or functions. Thus, the initial population is initialized with a ramped half and half technique for each gene and the size of the whole chromosome is randomly created. The variation is performed with four different operators that work at the chromosome and gene levels and all operations are selected based on fitness following the scheme proposed by Koza in which the probability of selecting all genetic operations sum to one. Hence, the probability of crossover at chromosome and gene levels is 0.4 respectively and the mutation at chromosome and gene levels is 0.1 for each operation. In this way, the evolutionary loop start the execution of each AVS by computing its fitness using a SVM that is

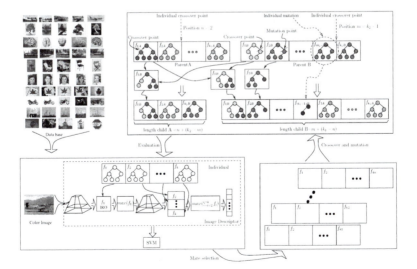

Fig. 4. General flowchart of the methodology to synthesize an artificial ventral stream

used to calculate the accuracy for solving a multiclass recognition problem. The algorithm considers 10 classes and 30 images per class and it uses the Caltech 101 database with the following objects: cars, brains, airplanes, bonsai, chairs, faces, leaves, schooner, motorcycles, and stop signals. Next, an AVS is selected from the population with a probability based on fitness using a roulette-wheel selection to participate in the genetic recombination; while, the best AVS is retained for further processing. In this way, a new population is created from the selected AVS by applying only one genetic operator; for example, the crossover or mutation operation at chromosome or gene levels. As in genetic algorithms our HHGP program executes the crossover between two selected AVS at the chromosome level by applying a "cut and splice" crossover. Each parent string has a separate choice of crossover point; for example, in Figure 4 the position $n = 2$ of Parent A with length k_1 and the position $m = k_2 - 1$ from Parent B with length k_2. Thus, all data beyond the selected crossover point in either AVS string is swapped between both parents A and B. Hence, the resulting child A has a length of $n + (k_2 - m)$ genes that in this case is 3; and the child B has a length of $m + (k_1 - n)$ genes. Moreover, The result of applying a crossover at the gene level is performed by randomly selecting two parent AVS based on fitness in order to execute a subtree crossover between both selected genes. Note, that the IRO can only be selected for a subtree crossover between two parents f_{1A} and f_{1B}. Thus, the f_{1A}'s subtree is replaced with the f_{1B}'s subtree and vice versa to create two new AVS genes. On the other hand, the mutation at the chromosome level leads the selection of a random gene of a given parent to replace such function by a new randomly mutated gene; for example, the position $k_2 - 1$ at parent B; see Figure 4. Moreover, the mutation at the gene level is calculated over an AVS by applying a subtree mutation to a probabilistically selected gene; in other words, a mutation point is chosen at a selected gene and the subtree up to that point is removed and replaced with a new subtree as is illustrated in Figure 4, where the tree f_{2B} is mutated. Finally, the evolutionary loop is terminated until an acceptable classification is reached; i.e., the accuracy is equal to 100% or the total number of generations $N = 50$ is reached.

4   Experimental Results

This section describes the results of evolving the AVS with the multi-class problem in a Dell Precision T7500 Workstation, Intel Xeon 8 Core, NVIDIA Quadro FX 3800 and running Linux OpenSUSE 11.1. The SVM implementation was developed with the libSVM [2]. The best result of the HHGP gives a classification accuracy of 78% in training. In order to compare and validate the performance of the evolved AVS a test against the original HMAX implementation written in Matlab [9]; as well as, the CUDA version of the HMAX model is provided here. The test consists on the comparison of the number of convolutions, speed and performance. Tables 2 and 3 provide the classification results for testing the HMAX and the evolved AVS using 15 images per class. Note, that the HMAX implementations use gray scale images, while the evolved AVS was programmed

Table 2. This table shows the confusion matrix obtained during the testing of the HMAX and its classification accuracy = 71.33% (107/150 images)

	Airplanes	Bonsai	Brains	Cars	Chairs	Faces	Leaves	Motorcycle	Schooner	Stop Signal
Airplanes	11	1	0	1	0	0	0	2	0	0
Bonsai	0	10	3	1	0	1	0	0	0	0
Brains	0	1	10	0	1	0	1	1	0	1
Cars	1	1	0	11	0	1	0	0	1	0
Chairs	0	0	1	1	11	0	0	1	1	0
Faces	0	2	1	0	0	10	1	0	0	1
Leaves	0	0	1	0	0	1	12	0	0	0
Motorcycle	2	0	0	2	0	0	0	11	1	0
Schooner	0	0	1	0	0	2	0	2	10	0
Stop Signal	0	0	1	0	0	2	1	0	0	11

Table 3. This table shows the confusion matrix obtained during the testing of the AVS and its classification accuracy = 80% (120/150 images)

	Airplanes	Bonsai	Brains	Cars	Chairs	Faces	Leaves	Motorcycle	Schooner	Stop Signal
Airplanes	3	1	0	6	0	1	0	3	1	0
Bonsai	0	13	0	0	0	0	0	0	2	0
Brains	1	0	13	0	0	0	0	1	0	0
Cars	0	0	0	14	0	0	0	1	0	0
Chairs	0	2	0	0	10	3	0	0	0	0
Faces	0	0	0	0	0	15	0	0	0	0
Leaves	0	0	0	0	0	0	15	0	0	0
Motorcycle	0	0	0	3	0	0	0	12	0	0
Schooner	0	1	0	0	0	1	0	1	12	0
Stop Signal	0	0	0	0	0	1	0	0	1	13

using the color space. Table 5 shows the number of convolutions per function to illustrate that a significant number of computations was reduced with our proposal, even considering the application of color space. Hence, the AVS applies 216 convolutions while the HMAX model uses 1248 convolutions using a universal dictionary of 200 patches. Therefore, the performance of the AVS process was improved significantly since the total number of convolutions is reflected on a lower computational time, see Table 4.

Table 4. This table shows the total running time

Image size	HMAX MATLAB	HMAX CUDA	Artificial V. S.	
896 × 592	34s	3.5s	2.6s	
601 × 401	24s	2.7s	1.25s	
180 × 113	9s	1s	0.23s	

Table 5. Number of convolutions (NC) for each function of the best individual

$f_1 = 0.05D_x($ $D_y(I))$	$f_2 = \frac{log(D_{xx}(C1))}{log(D_x(D_x(C1)-C1))}$	$f_3 = (|D_{xx}(C1) - D_x($ $D_{yy}(C1))|)(|D_y(C1)|)$	$f_4 = log(D_{xx}(C1))$	$f_5 = D_{yy}(C1)$ $+D_x(D_y(C1))$
$NC = 24$	$NC = 24$	$NC = 36$	$NC = 12$	$NC = 24$
$f_6 = D_{yyy}(C1)$	$f_7 = (log(D_x(D_y(C1))))$ $(C1 \cdot Gauss_{\sigma=2}(C1))$	$f_8 = |D_y(C1)$ $-D_y(C1)|$	$f_9 = D_y(D_x(D_y($ $C1))) - log(D_y(C1))$	$f_{10} = 0.05(D_{yy}(C1)$ $-D_x(D_y(C1)))$
$NC = 18$	$NC = 18$	$NC = 12$	$NC = 24$	$NC = 24$

5 Conclusions

The goal of this paper was to develop an approach based on GP to solve an object recognition problem using as model the ventral stream. The proposal follows a functional approach with several genetic programs being evolved in a hierarchical structure. All programs use different elements within the terminal and function sets according to the particular stage of the artificial ventral stream. The main result is a simplification of the overall structure that provides a lower computational cost. In future work we would like to explore other models.

Acknowledgments. This research was founded by CONACyT, Project 155045 - "Evolución de Cerebros Artificiales en Visión por Computadora".

References

1. Bartlet, W.: SEEMORE: Combining Color, Shape, and Texture Histogramming in a Neurally Inspired Approach to Visual Object Recognition. Neural Computation 9, 777–804 (1997)
2. Chih-Chung, C., Chih-Jen, L.: LIBSVM: a library for support vector machines. ACM Transactions on Intelligent Systems and Technology 2(27), 1–27 (2011) Software available at, http://www.csie.ntu.edu.tw/~cjlin/libsvm
3. Fei-Fei, L., Fergus, R., Perona, P.: Learning generative visual models from few training examples: an incremental Bayesian approach tested on 101 object categories. In: IEEE Workshop on Generative-Model Based Vision, CVPR 2004 (2004)
4. Fukushima, K.: Necognitron: A Self-Organizing Neural Network Model for a Mechanism of Pattern Recognition Unaffected by Shift in Position. Biological Cybernetics 36, 193–202 (1980)
5. Holland, J.H.: Complex Adaptive Systems. A New Era in Computation 121(1), 17–30 (1993)
6. Hubel, D., Wiesel, T.: Receptive Fields of Single Neurones in the Cat Striate Cortex. J. Physiol. 148, 574–591 (1959)
7. Hubel, D.: Exploration of the Primary Visual Cortex. Nature 299, 515–524 (1982)
8. LeCun, Y., Bottou, L., Bengio, Y., Haffner, P.: Gradient-based Learning applied to Document Recognition. Proceedings of the IEEE (1998)
9. Riesenhuber, M., Poggio, T.: Hierarchical Models of Object Recognition in Cortex. Nature Neuroscience 2(11), 1019–1025 (1999)

10. Mutch, J., Lowe, D.: Object Class Recognition and Localization Using Sparse Features with Limited Receptive Fields. International Journal of Computer Vision, IJCV (2008)
11. Serre, T., Wolf, L., Bilechi, S., Riesenhuber, M., Poggio, T.: Robust Object Recognition with Cortex-Like Mechanisms. IEEE Transactions on Pattern Analysis and Machine Intelligence 29(3), 411–426 (2007)
12. Ullman, S., Vidal-Naquet, M., Sali, E.: Visual features of intermediate complexity and their use in classification. Nature Neurosciencie 5(7), 682–687 (2002)
13. Ungerleider, L., Haxby, J.: "'What' and 'where' in the Human Brain". Current Opinion in Neurobiology 4, 157–165 (1994)

Evolving Visual Attention Programs through EVO Features

León Dozal[1], Gustavo Olague[1], Eddie Clemente[1,2], and Marco Sánchez[1]

[1] Proyecto EvoVision,
Departamento de Ciencias de la Computación, División de Física Aplicada,
Centro de Investigación Científica y de Estudios Superiores de Ensenada,
Carretera Ensenada-Tijuana No. 3918, Zona Playitas, Ensenada, 22860, B.C., México
{eclemen,ldozal,olague}@cicese.edu.mx
http://cienciascomp.cicese.mx/evovision/
[2] Tecnológico de Estudios Superiores de Ecatepec. Avenida Tecnológico S/N,
Esq. Av. Carlos Hank González, Valle de Anáhuac, Ecatepec de Morelos

Abstract. Brain informatics (BI) is a field of interdisciplinary study covering topics such as attention, memory, language, computation, learning and creativity, just to say a few. The BI is responsible for studying the mechanisms of human information processing. The dorsal stream, or "where"stream, is intimately related to the processing of visual attention. This paper proposes to evolve *VAPs* that learn to attend a given object within a scene. Visual attention is usually divided in two stages: feature acquisition and feature integration. In both phases there are specialized operators in the acquisition of a specific feature, called *EVOs*, and on the fusion of these features, called *EFI*. In previous research, those referred operators were established without considering the goal to be achieved. Instead of using established operators the idea is to learn and optimize them for the visual attention task. During the experiments we used a standard database of images for visual attention. The results provided in this paper show that our approach achieves impressive performance in the problem of focus visual attention over complex objects in challenging real world images on the first try.

Keywords: Evolutionary visual attention, organic genetic program, evolved visual operators, evolved feature integration, artificial dorsal stream.

1 Introduction

Until recently, it was widely believed that humans construct a complete representation of the visual field [5]. This has been amply refuted by a large amount of vison research. Visual attention is one of the most important mechanisms in the visual system, since the brain or visual cortical areas are unable to process all the information within the entire visual field. Thus, there are two basic phenomena that define the problem of visual attention. The first basic phenomenon is the limited capacity of information processing. At any given time, only a small

C. Di Chio et al. (Eds.): EvoApplications 2012, LNCS 7248, pp. 326–335, 2012.

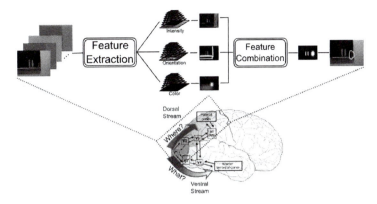

Fig. 1. Dorsal stream and visual attention computational model

amount of information, available on the retina, can be processed and used in the control of visual behavior. The second basic phenomenon is selectivity or the ability to filter unwanted information [4].

In this way, it is said that visual attention is controled by both cognitive, or top-down, factors, such as knowledge, expectation and current goals; bottom-up factors that reflect sensory stimulation [3]. The low level mechanisms for feature extraction act in parallel on the entire visual field to provide the "bottom-up" signals that highlights the image. Finally, it is said that attention is focused sequentially on those highlight regions of the image to analyze them in more detail ([12], [9]).

1.1   Problem Statement

Visual attention is a skill, which allows the creature, living or artificial, to direct their gaze rapidly towards the objects of interest in the visual environment [9]. The objects of interest refers to those objects or regions in the environment, which contain important information at a given time. Since the late 19th century the visual attention mechanism has been studied by researchers from different scientific disciplines; such as: neurologists, physiologists, psichologists, and in the last three-decades by people working on computer vision. This last community has studied the problem of visual attention as a feasible way to reduce the complexity of visual information processing. In this paper we address the "top-down" visual attention problem with a biologically inspired model that is evolved and optimized through evolutionary computation. The main goal is to obtain a set of visual attention programs that are trained to attend a given object in a scene. This is achived through the implementation of what we call organic genetic programming, which is a modified version of classic GPs that consists of individuals with several tree structures, each tree with a different set of functions and terminals. The results show that our methodology performs training process with excellent performance.

The remainder of this paper is organized as follows: Section 2 describes the visual attention program aproach that is divided in two parts: acquisition of early visual features and feature-integration for attention; Section 3 presents the organic genetic programming that is used as a way of solving the complex optimization of visual attention programs for visual task solution; Section 4 provides experimental results and their analysis, and Section 5 draws the conclusions and discusses possible future work.

2   Evolving Visual Attention Programs

There is a growing interest in applying evolutionary computation within computer vision to solve difficult problems and to improve the traditional vision algorithms; as well as, to propose new ones. On the other hand, from a biological perspective, it has been found that the development of specific visual mechanisms in the primate brain is product of evolution, specifically this is linked to natural selection [2] as explained in the evolutionary theory. The main idea explained in this paper is to derive the necessary steps in the search of a computer vision methodology by evolving visual attention programs.

2.1   Visual Attention Programming

The visual attention program (VAP) is a function driven approach that consider the biological visual process from the standpoint of its functionality, paying special attention to its aim, rather than considering the particular visual characteristics and the manner in which they are obtained; an approach that some authors refer as data driven. The brain and visual system can be extremely complex and despite rapid scientific progress, much about how they work remains a mistery. However, we know that vision is useful for accomplishing certain tasks, although very little is known on how vision performs such tasks. VAP exploits the knowledge about a given task and the intrinsic characteristics of the scene, to create complex programs based on functions, called visual operators, specialized in the extraction of physical visual features from the observed scene, such as: color, borders, and intensity. This paper explores a new way about, how these features are combined, inhibited or excited, to highlight the necessary visual information for the task at hand. In particular, we claim that genetic programming provides suitable tools necessary to implement this approach in highly creative ways. Next, we list some useful concepts.

Definition 1. *Let f be a function $f : U \subset \mathbb{R}^2 \to \mathbb{R}$. The graph or image I of f is the subset of $\mathbb{R}^3$ that consist of the points $(x, y, f(x, y))$, in which the ordered pair (x, y) is a point in U and $f(x, y)$ is the value at that point. Symbolically, the image $I = \{(x, y, f(x, y)) \in \mathbb{R}^3 | (x, y) \in U\}$.*

Color digital images are composed of three images at different wavelengths of light that are red, green and blue. Note that it is possible to convert an image that is represented in RGB into another color space. In this way, we say that a color image is the set of images named $I_{color} = \{I_r, I_g, I_b, I_c, I_m, I_y, I_k, I_h, I_s, I_v\}$.

Therefore, the input of a *VAP* is an I_{color}. Moreover, it is said that the outcome of a *VAP* is an optimized saliency map, *OSM*. An *OSM* is an image whose pixel values represent the prominence of the visual features being considered. Thus, given the representation of the scene and the outcome of a *VAP*, we can define the above concepts in a more formal way as follows:

Definition 2 (Visual Attention Program). *Let v be the function $v : I_{color} \to$ OSM that represents a* VAP. *The domain I_{color} consist of a set of images that characterise a real scene, and the codomain is the* OSM. *We may say that v induces an* OSM *over I_{color}, written as $v \overset{I_{color}}{\Rightarrow}$ OSM.*

Therefore we can distinguish two different feature processing stages within the *VAP* known as: acquisition and integration. In the same vein, it is necessary to say that a conspicuity map (CM) is an intermediate representation of the prominence of each separate visual feature, as shown in Figure 2. In this sense, we specify the *VAP* as an evolved feature composition (EFC) function as follows.

Definition 3 (Evolved Feature Composition). *The function v is represented as a composite function $v = FI \circ FA : I_{color} \to$ OSM. *The function $FA : I_{color} \to CM$ is known as the feature acquisition stage; and the function $FI : CM \to$ OSM *is known as the feature integration stage. Here, CM represents the set of conspicuity maps.*

The output of a *VAP* is an *OSM* that is characterized by a set of prominent regions that indicate the position of objects in scene where attention will be directed. Our proposed system is based on the psychological model for visual attention introduced by Treisman[12]; as well as, the evolutionary algorithms that are inspired by biological evolution. In the following two subsections, the functions that conform *VAP* are explained in detail.

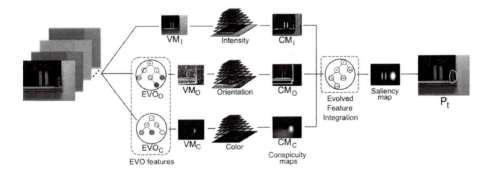

Fig. 2. Visual attention program

2.2   Acquisition of Early Visual Features

Three visual operators are applied in a separated way to emphasize: intensity, color, and orientation attributes. In biologically plausible models as [8], these operators are established according to the knowledge in neuroscience about how these features are obtained in the visual cortex of the brain, the explanation follows a data driven approach. The operation of the dorsal stream, is a product of the evolutionary process. For this reason we decided to use evolutionary computation to obtain these evolved visual operators (EVO) as depicted on Figure 2. Below, the EVO features used in the VAP are defined.

Feature Orientation. A feature orientation is a funtion $EVO_O : I_{color} \rightarrow VM_O$ that is evolved to optimize the extraction or rejection of edges present in the image based on a top-down process. The result of this operation is a visual map VM_O in which the pixel value represents the feature prominence, in such a way, that the larger the pixel value, the greater the prominence of the feature. Then, VM_O can be seen as a function $VM_O : I_{color} \rightarrow I$, that is obtained with a high level visual operation, $VM_O = EVO_O(I_{color})$. The evolutionary method uses a set of funtions and terminals that allows an EVO_O to cooperate with the VAP to accomplish a purpose; such functions and terminal are provided in Table 1. The notation used is as follows. I_{T_O} can be any of the terminals in T_O; as well as, the output of any of the functions in F_O; D_u symbolizes the image derivatives along direction $u \in \{x, y, xx, yy, xy\}$; G_σ are Gausssian smoothing filters with $\sigma = 1$ or 2.

Table 1. Set of functions and terminals used by EVO_O

$F_O = \{+, -, \times, \div, \lvert+\rvert, \lvert-\rvert, \sqrt{I_{T_O}}, I_{T_O}^2, log_2(I_{T_O}), G_{\sigma=1}, G_{\sigma=2},$ $\lvert I_{T_O} \rvert, \dfrac{I_{T_O}}{2}, D_x, D_y\}$
$T_O = \{I_r, I_g, I_b, I_c, I_m, I_y, I_k, I_h, I_s, I_v, G_{\sigma=1}(I_r), G_{\sigma=2}(I_r),$ $D_x(I_r), D_y(I_r), D_{xx}(I_r), D_{yy}(I_r), D_{xy}(I_r), ... \}$

Table 2. Set of functions and terminals used by EVO_C

$F_C = \{+, -, \times, \div, \lvert+\rvert, \lvert-\rvert, \sqrt{I_{T_C}}, I_{T_C}^2, log_2(I_{T_C}), Exp(I_{T_C},$ $Complement(I_{T_C})\}$
$T_C = \{I_r, I_g, I_b, I_c, I_m, I_y, I_k, I_h, I_s, I_v\}$

Feature Color. A feature color is a function $EVO_C : I_{color} \rightarrow VM_C$ that is evolved to optimize the extraction or rejection of colors, which are presented, in the objects that appear in the image. The result is a visual map VM_C containing the prominence of the feature. Then, VM_C can be seen as a function $VM_C : I_{color} \rightarrow I$, that is obtained with a high level visual operation, $VM_C = EVO_C(I_{color})$. In the same way, as in EVO_O, the evolutionary process uses a set of functions and terminals provided in Table 2. The notation is summarized as follows, I_{T_C} can be any of the terminals in T_C, as well as the output of any of the functions in F_C; function $Complement(I_{T_C})$ symbolizes a negative image that is represented as the total inversion of an image.

Finally, as in previous work, to obtain the intensity of an input image I_{color}, the red, green and blue values of each pixel are averaged. The formula is developed as a function $VM_I : I_{color} \rightarrow I$, that is obtained with the following formula $VM_I = \frac{I_r + I_g + I_b}{3}$.

Conspicuity Maps. The conspicuity maps (CMs) are obtained by means of a center-surround function $CM : VM \rightarrow I$, that is applied in order to simulate the center-surround receptive fields. This natural structure allows the ganglion cells to measure the differences between firing rates in center (c) and surroundings (s) of ganglion cells. First, a pyramid $VM_l(\alpha)$ of nine spatial scales $S = \{1, 2, ..., 9\}$ is created for each of the three resulting VMs. Afterwards, an across-scale substraction $\ominus$ is performed, resulting in a center-surrond map $VM_l(\omega)$ in which the value of the pixel is augmented as the contrast along their neighbors at different scales is higher. Finally, the $VM_l(\omega)$ maps are added using an across-scale addition $\oplus$ in order to obtain conspicuity maps CM_l. At this point, we have three CMs, one for each feature, as shown in Figure 2. The CM is obtained by performing as in Walther and Koch model [13]. Immediately, CMs are combined to obtain a single saliency map as shown below.

2.3   Feature-Integration for Visual Attention

The following step is the fusion of conspicuity maps CMs into a single map of salience. This is a difficult problem because the CMs belong to different and unrelated visual modalities. In neuroscience, an exact description about how the brain makes this integration or where is located the saliency map in the brain is unknown. In this work, the problem statement considers that the problem must be addressed regarding the task to be performed. In other words, since the task needs to accomplish a purpose; then the main criterion should be the one that guides the suitable combination of characteristics. In this sense, genetic programming is very useful since it provides a methodology to address the problem. Therefore, we decided to evolve the integration of CMs and we called this function, evolved feature integration (EFI). As a result of the EFI, the structure of the $VAPs$ becomes dynamic since the fusion considers different combinations of the CMs throughout the process. The definition of the EFI function is as follows:

Definition 4 (Evolved Feature Integration). *Let* EFI *be a function* $EFI : CM_l \rightarrow OMS$. *The domain* CM_l, *with* $l \in \{O, C, I\}$, *consists of a set of images or conspicuity maps, and the codomain is the* OSM, *then*

$$EFI : CM_l \rightarrow I, \text{ and the } OSM \text{ is obtained with an image operation,}$$
$$OSM = EFI(CM_l) \quad \therefore \quad OSM = I$$

The evolutionary method uses the set of funtions and terminals listed in Table 3 in order to create a fusion operator that highlights the features of the object of interest.

Hence, an OSM is characterized through a proto-object P_t or a secuence of proto-objects $\{P_1, P_2, \ldots, P_i, \ldots, P_t\}$ [11]. This structures provide local descriptions of scene.

Table 3. Set of functions and terminals used by *EFI*

$F_{fi} = \{+, \ -, \ \times, \ \div, \ \| + \|, \ \| - \|, \ \sqrt{I_{T_{fi}}}, \ I_{T_{fi}}^2, \ Exp(I_{T_{fi}}),$ $G_{\sigma=1}, G_{\sigma=2}, \|I_{T_{fi}}\|, \ D_x, \ D_y\}$
$T_{fi} = \{CM_I, \ CM_O, \ CM_C, \ D_x(CM_I), \ D_y(CM_I), \ D_{xx}(CM_I),$ $D_{yy}(CM_I), \ D_{xy}(CM_I), \ ... \ \}$

Definition 5 (Proto-Object). *A P_t is defined as a proto-object or salient region of the* OSM, *which is being attended at time t.*

In the next section we explain the evolutionary process used to obtain the *VAPs*.

3   Organic Genetic Programming

In this section, we describe the main aspects for the evolution of *VAPs* through the use of what we called the organic genetic programming (OGP). All changes introduced in the OGP embody an organic motivation, in a sense of describing an organ or tissue of a living organism and their complexity. We tried to introduce this changes in order to deal with the evolution of complex structures. These changes are explained below.

The first phase of the OGP is the training. In this phase the OGP learns to focus a prominent object using an image database for training. Algorithm 1, list the steps that the OGP performs in order to obtain the *VAPs*. In this work, we propose a *VAPs* genotype that is robust because it is capable of encoding in a better way the phenotype of the dorsal stream. More specifically, the genotype consists of a triplet of trees representation; each having a different and specialized operation. Hence, each tree has its own independent set of functions and terminals. Unlike the classic GP that only has one tree representation using a unique set of functions and terminals. In this way, the functions and terminals sets are listed in Tables 1, 2, and 3 for orientation, color and feature integration, respectively. *VAPs* genotype is conformed as follows: the 1st tree is an EVO_O; the 2nd tree is an EVO_C, and the 3rd tree is an *EFI*; each one with a maximum depth of 9 levels. The first one encodes the orientation, or the operation of the orientation-sensitive cells in V1 [7]. The second one encodes the color, or operation of photoreceptor cells and color-sensitive cells present in the layers V1 and V4 of the visual cortex. Finally, the third one encodes the way in which the features are combined to obtain the saliency map, or operation of the posterior parietal cortex [6]. The algorithm initialize the population of 50 *VAPs* with a ramped half-and-half technique.

After initialization, the OGP needs a well-posed fitness function. In this work, we propose to use the F-measure as the fitness function in order to compare and select among several *VAPs*. This measure has already been used in previous works as evaluation in applications related to computer vision such as [10] and [1]. The calculation of *VAPs*'s fitness can be explained as a comparison between a P_t attended by the *VAPs* and a manual location, attention and segmentation of an object, considered as an ideal visual attention criterion. A target is considered attended if a

---

Algorithm 1. Organic Genetic Programming Algorithm

Randomly create an initial population of *VAPs*.

repeat

    Execute each *VAPs*, using the training images database, and compute its fitness.

    Select one or two $VAP(s)$ from the population with a probability proportional to their fitness to participate in genetic recombination.

    Create new $VAP(s)$ by applying genetic operations with specific probabilities.

until An acceptable solution is found or some other stopping condition is met (e.g., a maximum number of generations is reached).

return The best *VAPs* up to this point.

---

subset, not empty, of pixels that conform the object intersect the proto-object P_t. Another difference arises from the existence of two-level complexity in the structure of the genotype. Consequently, the gene level, as in classic GP, considers the tree as the unit where the genetic operation is performed. In addition, the OGP manages the chromosome level that recognize the whole genotype, composed of a tripet of trees, as a unit where it have to work. Therefore, the OGP allows the creation of new genetic operators inspired by gene and chromosomal biological mutations, each one operating at a different level. These genetic operators are selected based on a probability that is set following the scheme proposed by Koza, for which each operation is computed independently and their addition of probabilities is one. Hence, the probability of crossover at gene and chromosome levels is 0.4 and the mutation probability at both levels is 0.1.

The next step is concentrated on the selection of one or two *VAPs* using the roulete-wheel approach. Thus, the best *VAP* is kept in the following generation and the genetic recombination is repeated until a new population is created. Finally, the evolutionary loop finish until a total number of generations, N=50, is reached. Therefore, once the training stage ends and the *VAP* with the best fitness is obtained; then, the testing stage starts. Hence, the fittest *VAP* is tested using a different image database, known as the testing database.

4   Experiments and Results

This section is an overview about the specific details of the experiments; also, the results of evolving the *VAPs* are showed and discussed. Experiments were performed in a Dell Precision T7500 Workstation, Intel Xeon 8 Core, NVIDIA Quadro FX 3800 and Linux OpenSUSE 11.1 operating system. Next, the obtained *VAPs* and their performance are presented considering one for each target object that are: red can and triangle sign. The first evolved VAP_{Can}, was obtained at generation 5 and it was the individual number 27. The Figure 3 shows the final structure surrounded by a red-dotted line. Note, that the *EFI* employs the EVO_O as the unique relevant information. Therefore, intensity and EVO_C were not used for the VAP_{Can}. Also, in the upper left corner of the same figure are listed the percentage of images where the red can was successfully focused using the VAP_{Can}. The *EVOs* and *EFI* of the VAP_{Can} are shown below.

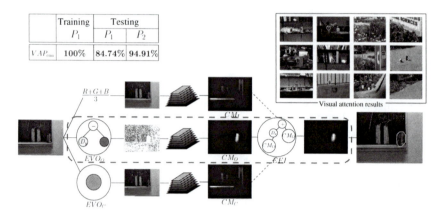

Fig. 3. Evolved structure of VAP_{Can} obtained through the OGP to attend the red can in the images

$$EVO_O = I_m - D_y(I_y), \ EVO_C = I_g, \ EFI = D_x(CM_O) + CM_O$$

The second example is named $VAP_{Triangle}$, which was achived at generation 2 and it was the individual number 24. The Figure 4 shows the final structure surrounded by a red-dotted line. Note, that the EFI employs the EVO_O and EVO_C as relevant information. Therefore, intensity was not used for the VAP_{Can}. Also, in the upper left corner of the same figure are listed the percentage of images where the triangle signal was successfully focused using the $VAP_{Triangle}$. The $EVOs$ and EFI of the $VAP_{Triangle}$ are shown below.

$$EVO_O = (D_y(I_m))^2, \ EVO_C = \exp(I_b), \ EFI = \frac{D_x(CM_O)}{CM_C}$$

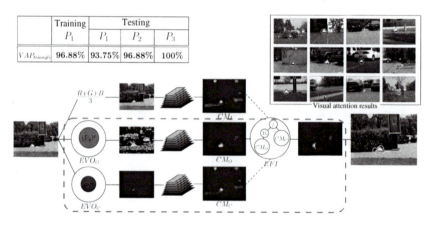

Fig. 4. This figure shows the VAP obtained through the GP to attend the signal in the images

5 Conclusions

The aim of our research was to show that the OGP is a powerful methodology that is capable of obtaining *VAPs*s that can be seen as "top-down" models of the visual attention system that is capable of solving the visual attention problem. This work follows a function driven approach that leads to the creation of different programs that optimize the use of available resources in order to solve a particular task. As a conclusion, for some task it's not necessary to compute all features; thus, simplifying the *VAPs* final structure. In the future, we would like to integrate new EVOs on the evolution as well as the "bottom-up" models of visual attention.

Acknowledgments. This research was founded by CONACyT through the Project 155045 - "Evolución de Cerebros Artificiales en Visión por Computadora".

References

1. Atmosukarto, I., Shapiro, L.G., Heike, C.: The use of genetic programming for learning 3D craniofacial shape quantifications. In: Proceedings of the 2010 20th International Conference on Pattern Recognition, ICPR 2010 pp. 2444–2447 (2010)
2. Barton, R.A.: Visual Specialization and Brain Evolution in Primates. Proceedings of the Royal Society of London Saries B-Biological Sciences 265(1409), 1933–1937 (1998)
3. Corbetta, M., Shulman, G.L.: Control of goal-directed and stimulus-driven attention in the brain. Nature Reviews Neuroscience 3, 201–215 (2002)
4. Desimone, R., Duncan, J.: Neural Mechanisms of Selective Visual Attention. Annual Reviews 18, 193–222 (1995)
5. Feldman, J.A.: Four frames suffice: A provisional model of vision and space. Behavioral and Brain Sciences 8, 265–289 (1985)
6. Gottlieb, J.: From thought to action: the parietal cortex as a bridge between perception, action, and cognition. Neuron 53, 9–16 (2007)
7. Hubel, D., Wiesel, T.: Receptive Fields of Single Neurones in the Cat Striate Cortex. J. Physiol. 148, 574–591 (1959)
8. Itti, L., Koch, C.: Feature Combination Strategies for Saliency-Based Visual Attention Systems. Journal of Electronic Imaging 10(1), 161–169 (2001)
9. Kosh, C., Ullman, S.: Shifts in selective visual attention: towards the underlying neural circuitry. Human Neurobiology 4, 219–227 (1985)
10. Perez, C.B., Olague, G.: Learning Invariant Region Descriptor Operators with Genetic Programming and the F-Measure. In: 19th International Conference on Pattern Recognition, ICPR 2008 (2008)
11. Rensink, R.A.: Seeing, sensing, and scrutinizing. Vision Research 40, 1469–1487 (2000)
12. Treisman, A.M., Gelade, G.: A Feature-Integration Theory of Attention. Cognitive Psichology 12(1), 97–136 (1980)
13. Walther, D., Kosh, C.: Modeling Attention to Salient Proto-Objects. Neural Networks 19, 1395–1407 (2006)

Evolutionary Purposive or Behavioral Vision for Camera Trajectory Estimation

Daniel Hernández[1], Gustavo Olague[1,*], Eddie Clemente[1,2], and León Dozal[1]

[1] Proyecto EvoVision,
Departamento de Ciencias de la Computación, División de Física Aplicada,
Centro de Investigación Científica y de Estudios Superiores de Ensenada,
Carretera Ensenada-Tijuana No. 3918, Zona Playitas, Ensenada, 22860, B.C., México
{dahernan,olague,eclemen,ldozal}@cicese.edu.mx
http://cienciascomp.cicese.mx/evovision/
[2] Tecnológico de Estudios Superiores de Ecatepec. Avenida Tecnológico S/N,
Esq. Av. Carlos Hank González, Valle de Anáhuac, Ecatepec de Morelos

Abstract. Active, animate, purposive or behavioral vision are all understood as a research area where a seeing system interacts with the world in such a way of creating a balance between perception and action. In particular, it is said that a selective perception process in combination with a specific motion-action works as a unique complex system that accomplishes a visuomotor task. In the present work, this is understood as a visual behavior. This work describes a real-working system composed of a camera mounted on a robotic manipulator that is used as a research platform for evolving a visual routine specially designed in the estimation of specific motion-actions. The idea is to evolve an interest point detector with the goal of simplifying a well-known simultaneous localization and map building system. Experimental results shows as a proof-of-concept that the proposed system is able to design a specific interest point detector for the case of a straight-line displacement with the advantage of eliminating a number of heuristics.

Keywords: Evolutionary Visual Behavior, Multiobjective Evolution, Purposive Vision, SLAM.

1 Introduction

Active vision is a research area where perception meets action, in such a way that the processes of visual information adapt their behavior to the task; as a result, of the interaction between the camera and the given scenario. For example, in active vision a robotic system is able to manipulate the attitude of a camera to achieve some task or purpose related to the observation of the environment where the robot acts [3]. A distinctive characteristic of active vision is the idea that the observer is capable of engaging into some kind of activity whose purpose is to change the visual parameters according to the environmental conditions [7].

---

* Corresponding author.

C. Di Chio et al. (Eds.): EvoApplications 2012, LNCS 7248, pp. 336–345, 2012.

In this way, purposive vision as an information process does not function in isolation but as a part of a bigger system that interacts with the world in highly specific ways ([1],[2]). The idea studied in this work is related to the fact that a purposive visual system is part of a complex system whose interaction with the world is developed not generally but in specific ways. Therefore, the aim of purposive or behavioral vision is to evolve a visual routine via an evolutionary algorithm, whose overall goal is to adapt the visual program to a specific purposeful task.

Figure 1 illustrates the problem that we would like to approach using a camera mounted at the end of a robotic manipulator. The idea is that a visual behavior requires specific information related to the task that is being confronted. On the left-side of the figure, the visual behavior performed by a person is related to the extraction of visual information that is needed in order to be able to read a map; as well as, the mental activity that is applied to extract the information needed to find an object within a scene. In this way, the person performs a set of actions including a visual behavior that needs to accomplish a number of aspects such as: visual perception and self-localization, in order to solve the problem of being lost. The figure on the right-side, describes the necessary steps in a self-localization and map building (SLAM) process. This process is normally modeled through an estimation process where the visual perception executes an action in order to achieve the purpose of auto-localization. Here, the visual process is carried out by an interest point detector that is applied to the input image sequence, and the action is realized by an SLAM method. Thus, the idea is to adapt the visual routine to accomplish an specific movement evaluated by the SLAM system; in order to achieve a desired task. Therefore, the goal of this paper is to show that it is possible to evolve a visual routine with respect to an specific camera displacement in such a way to avoid the use of a number of explicit heuristics that are normally incorporated to the SLAM system.

This paper is organized as follows: first the evolutionary purposive vision system is described in three parts including the interest point detection, the simultaneous localization and map building, and the camera trajectory estimation using genetic

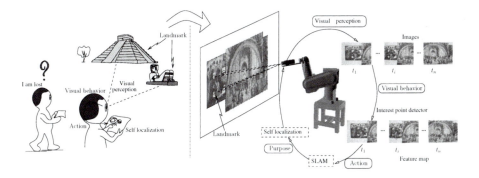

Fig. 1. Visual behavior is characterized by the sequence of actions that an organism performs in response to the environment where is located

programming. Then, the multiobjective visual behavior algorithm is explained. Finally, results of a real working system are presented followed by a conclusion.

2   Evolutionary Purposive Vision

The main idea of this paper is the evolution of a visual behavior through a visual routine. The instance of this visual routine is an interest point detector, which will be employed inside an SLAM system in order to solve the camera pose estimation problem, leading to the trajectory estimation. Therefore, this section we described the three main aspects of the work, the definition of interest point detection, the description of the SLAM system and the relationship through evolution of theses two concepts.

2.1   Interest Point Detection

An interest point is a small region on an image which contains visually prominent information. The right side of Figure 1 depicts a robotic system in a hand-eye configuration that was used to evaluate the evolved interest point detectors. The idea is to create a visual behavior in which the visual routine is able to extract the most relevant visual information for the development of an specific task; like the evaluation of a straight line in the pose estimation approach of SLAM. In particular, the value of importance of a pixel in an image is the result of a mapping $K : \mathbb{R}^+ \to \mathbb{R}$, this transformation is known as the *operator*, which should not be confused by the *detector*. The first one is applied to an image to obtain the importance of each pixel, and the latter is an algorithm that extract all the interest points or regions in an image. In this way, most interest point detectors work as follows:

1. Apply the operator K to the input image in order to obtain the *interest image I^**.
2. Perform a non-maximum suppression operation.
3. Establish a threshold to determine if a given maximum should be considered an interest point.

It has been shown that it is possible to evolve a general purpose interest point detector through genetic programing [10]. The idea in this work is to evolve a visual behavior, in such a way of maintaining general purpose characteristics, such as: *repeatability* and *dispersion*; while adapting the visual information according to an specific trajectory. In summary, the following properties are the desired characteristics for the evolved detectors:

1. *Repeatability*, the expected interest points should be robust to environmental conditions and geometric transformations.
2. *Dispersion*, the detector should be able to find points over the whole image.
3. *Trajectory estimation*, the detector must find useful information that simplifies the camera pose estimation computed through an SLAM system.

2.2   Simultaneous Localization and Map Building

The SLAM implementation used to test the visual behaviors was the one presented by [4], an schematic representation of this system can be seen in Figure 2, it presents the SLAM problem as an estate estimation process, done through Kalman filtering, and using the concept of interest point detection to find visual landmarks in the environment, in order to build a feature based map for the SLAM problem [8]. The SLAM system is divided into three main stages: initialization stage, the state propagation (done through Kalman filtering) and the map update.

Initialization stage. In order to estimate the trajectory of the camera, it is necessary to calculate its position over time; hence, the state of the camera is defined by the camera's pose and its displacement with respect to time. This is calculated through a Kalman filtering process, where the attitude of the camera $\hat{x}_v$ is given by its position r and orientation q, and the information about its motion, is defined by the linear and angular speeds, v and ω. On the other hand, it is also necessary to maintain a map of the environment; a task performed in this case through a sparse set of visual landmarks. Therefore, the system must also track the spacial position for each feature y_i. Thus, the state that the system needs to estimate is s_t which is composed of the information about the camera and the features. The state estimation known as $\hat{s}_t$ is coupled with the estimation uncertainty $\mathbf{S}_t$. This system is calibrated meaning it has a known initial state; in other words the camera position is known along with a starting map, composed of four landmarks. The state s_t and its covariance $\mathbf{S}_t$ have the following structure

$$\mathbf{s}_t = \begin{pmatrix} \hat{x}_v \\ \hat{y}_1 \\ \hat{y}_2 \\ \vdots \end{pmatrix} \quad \mathbf{S}_t = \begin{bmatrix} S_{xx} & S_{xy_1} & S_{xy_2} & \cdots \\ S_{y_1x} & S_{y_1y_1} & S_{y_1y_2} & \cdots \\ S_{y_2x} & S_{y_2y_1} & S_{y_2y_2} & \cdots \\ \vdots & \vdots & \vdots & \end{bmatrix}$$

State Propagation. The state is estimated using a Kalman filtering method, which is divided in two stages, prediction and measurement. The future states of the camera are assumed to change according to a dynamic model of the form

$$\mathbf{s}_t = \mathbf{A}\mathbf{s}_{t-1} + \mathbf{w}_t$$

where $\mathbf{A}$ is called the state transition matrix, and $\mathbf{w}_t$ represents the process noise. At each step, the Kalman filter makes a prediction of the current state

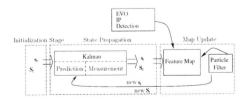

Fig. 2. The stages of the SLAM system along the evolutionary method integrated in the map update stage

denoted as $\mathbf{s}_t^-$, also known as *a priori* state, along with its error covariance matrix $\mathbf{S}_t^-$. This step is known as *prediction stage*, and it is normally calculated using the dynamic model

$$\mathbf{s}_t^- = \mathbf{A}\mathbf{s}_{t-1}$$
$$\mathbf{S}_t^- = \mathbf{A}\mathbf{S}_{t-1}\mathbf{A}^T + \mathbf{\Lambda}_w$$

where $\mathbf{\Lambda}_\mathbf{w}$ is the process covariance noise, which is assumed to be a white Gaussian noise. The *measurement stage* consists in using the information captured by the sensor to ameliorate the state estimation. In this way, the measurements $\mathbf{z}_t$, about the image location at time t of some visual landmarks, are assumed to be related to the current state as follows

$$\mathbf{z}_t = \mathbf{C}\mathbf{s}_t + \mathbf{v}_t$$

where $\mathbf{v}_t$ is the measurement noise, and $\mathbf{C}$ relates the camera pose to the considered image features. This stage is tackled from an active search approach, using the predicted camera position $\mathbf{s}_t^-$ to determine the visible landmarks y_i and their possible image position $\mathbf{z}_t^-$ defined as

$$\mathbf{z}_t^- = \mathbf{C}\mathbf{s}_t^-$$

along with the corresponding uncertainty $\mathbf{\Lambda}_z$ as follows

$$\mathbf{\Lambda}_z = \mathbf{C}\mathbf{S}_t^-\mathbf{C}^\mathbf{T} + \mathbf{\Lambda}_\mathbf{v}$$

these predictions are useful to restrict the search for visual landmarks to a region of the image. The measurement model is then used to calculate an *a posteriori* state estimate $\hat{\mathbf{s}}_t$ and its covariance matrix $\mathbf{S}_t$ by incorporating the measurements $\mathbf{z}_t$ as follows

$$\hat{\mathbf{s}}_t = \mathbf{s}_t^- + \mathbf{G}_t(\mathbf{z}_t - \mathbf{C}\mathbf{s}_t^-)$$
$$\mathbf{S}_t = \mathbf{S}_t^- - \mathbf{G}_t\mathbf{C}\mathbf{S}_t^-$$

where $\mathbf{G}_t$ is the Kalman gain, computed as

$$\mathbf{G}_t = \mathbf{S}_t^-\mathbf{C}^T(\mathbf{C}\mathbf{S}_t^-\mathbf{C}^T + \mathbf{\Lambda}_v)^-1$$

with $\mathbf{\Lambda}_v$ being the measurement covariance. In this way, the *a posteriori* state, the estimation at time t is complete; thus, the state $\hat{\mathbf{s}}_t$ together with its uncertainty $\mathbf{S}_t$ will be used on the next cycle.

Map Update. After having a good estimation of the camera position, the system must extend the feature map. Thus, in this step the system employs the interest point detector to find information in the environment that may improve future state estimations. In the original system, it used the Shi-Tomasi [9] detector, along with some heuristics such as:

- *Selection of interest regions.* The system would only look for interest points in regions of the image where there are no known landmarks; while, trying to keep the feature map dispersion.

– *Elimination of bad features.* This process would delete landmarks that are unstable, meaning visual features that did not appear in the predicted image region $\mathbf{z}_t^-$; on the other hand, eliminating landmarks whose depth could not be easily estimated, even though these points were considered as "interesting" by the detector.

An specialized detector should be able to cope with those problems. In this way, the designed detector bring implicitly these characteristics on the interest points, by highlighting stable landmarks that are dispersed all over the environment. The final step for the map update calculates the depth of the visual landmarks. Since, the system uses monocular vision, it is said that it is impossible to calculate the depth of a point using a single image; therefore, the system uses a particle filter to find the depth of a visual landmark. Once, the depth is know for a certain point, the corresponding feature is added to the map for further estimations.

2.3   Camera Trajectory Estimation Using Genetic Programming

The main idea of this paper is to describe the adaptation of the visual behavior, through the evolution of visual routines as interest point detectors. Such a detector is used inside the SLAM system, see section 2.2. The scenario can be seen in Figure 3, where the camera is moving on a straight line, parallel to a wall. The SLAM system estimates the camera position while it moves, using the evolved detector within the feature map building. Then, the position estimations are applied to evaluate the detector using a trajectory adjustment. The evolved detectors should have the properties mentioned in section 2.1. Due to the amount of work required for the evolutionary algorithm, a camera motion sequence was captured in order to be use in the offline evaluation of the evolved detectors. The hypothesis for this work is that a good detector for the camera's trajectory estimation is one with high repeatability and high point dispersion. Which lead to stable landmarks and a disperse feature map. The following functions were use to evaluate the evolved detectors, seeking to minimize them:

Repeatability. The average repeatability $r_K(\epsilon)$ is calculated for the operator K by evaluating the repeatability between two consecutive images, using a

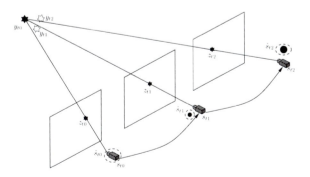

Fig. 3. Testing environment for the visual routines

neighborhood of size ϵ. It is important to note that the repeatability is calculated using the position of the camera produced by the highly-accurate robot movement, instead of using the homographies between the images [11].

$$rI_i(\varepsilon) = \frac{|R_{I_i}(\varepsilon)|}{min(\gamma_{i-1}, \gamma_i)}$$

where $\gamma_{i-1} = |\{x_{i-1}^c\}|$ and $\gamma_i = |\{x_i^c\}|$ are the number of points detected in images I_{i-1} and I_i. $R_{I_i}(\varepsilon)$ is a set of pairs of points (x_{i-1}^c, x_i^c) that were found in two consecutive images within a region of radius ε:

$$f_1 = \frac{1}{r_K(\epsilon) + c_1}$$

where c_1 is a constant to avoid invalid division.

Dispersion. $\mathcal{D}_p(K)$ is the average dispersion of the located points within the image sequence using operator K; where $c_2 = 10$ is a normalization constant.

$$f_2 = \frac{1}{e^{\mathcal{D}_p(K) - c_2}}$$

The point dispersion in image I_i is calculated using the points' entropy $D(I, X) = -\sum P_j \cdot log_2(P_j)$ where X is the set of detected points and P_j is approximated using a histogram.

Trajectory Adjustment. This fitness is measured using the χ^2 adjustment of the estimated trajectory using a real straight-line trajectory.

$$f_3 = \sum_{i=1}^{M} \frac{[y_i - f(x_i; p_1, p_2, ..., p_M)]^2}{\sigma_i}$$

3   Multiobjective Visual Behavior Algorithm

A Multiobjective (MO) approach allows us to incorporate several optimization criteria [5]. An important characteristic of a MO algorithm is that it searches for a set of Pareto optimal solutions, instead of a single best solution [6]. In our problem, each individual represents an operator to be used inside an interest point detector within the SLAM system, which is built using the functions and terminals detailed in Table 2. The evolutionary process was executed using a Unibrain Fire-i camera mounted on a Stäubli RX-60 robot with six degrees of freedom. The camera captures monochromatic images with a resolution of 320×240 pixels. The trajectory used for the experiments was a straight-line, parallel to a wall rich in visual information, the length of the path is 70 centimeters, the camera moves at a speed of 7 cm/second, with a shutter speed of 15 images/second. This results on an average of 150 sequence of images for the trajectory run. The parameters for the evolutionary algorithm can be found in Table 1. Thus, each individual is evaluated within the SLAM system. The resulting image sequence is employed to evaluate the individual's repeatability and dispersion. Follow by the population operations, using the SPEA2 [12] algorithm for parent selection, one-point crossover and sub-tree mutation for creating the offspring.

Table 1. Parameters used in the Multi-objective GP for synthesis of image operators for interest point detectors

Parameters	Description
Population	30 individuals
Generations	30 iterations
Initial population	Ramped Half-and-Half
Genetic operations probabilities	Crossover $p_c = 0.85$, Mutation $p_\mu = 0.15$
Max three depth	6 levels
File size (SPEA2)	15
Parent selection	15

Table 2. Functions and terminals used to built the population

Functions	$F = \{+, \lvert + \rvert, -, \lvert - \rvert, \lvert I_{out} \rvert, *, \div, I_{out}^2, \sqrt{I_{out}}, log_2(I_{out}),$ $EQ(I_{out}), k \cdot I_{out}, \frac{\delta}{\delta x} G_{\sigma_D}, \frac{\delta}{\delta y} G_{\sigma_D}, G_{\sigma=1}, G_{\sigma=2}\}$
Terminals	$T = \{I, L_x, L_{xx}, L_{xy}, L_{yy}, L_y\}$

4   Experiments

The resulting Pareto front can be seen in Figure 4, which shows the distribution of the individuals along the fitness space. The Table 3 lists eight non-dominated individuals produced with the system just described. The resulting operators are called *Interest Point operators specialized in the SLAM problem, IPSLAM.*

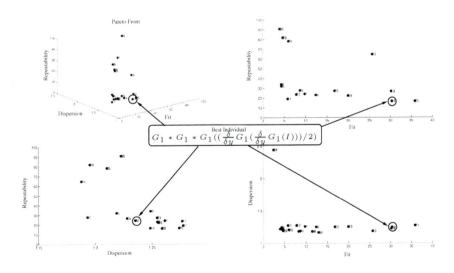

Fig. 4. Resulting Pareto front for the evolutionary process

Table 3. Best individuals found after the evolutionary process. We chose individuals from different areas of the Pareto front.

Name	Operator	Fitness
IPSLAM1	$G_1 * \lvert \frac{\delta}{\delta x} G_1(\frac{\delta}{\delta x} G_1(I)) - \frac{\delta}{\delta y} G_1(I) \rvert$	$f_1 = 9.76$ $f_2 = 1.26$ $f_3 = 24.14$
IPSLAM2	$\lvert \lvert \lvert \lvert G_1 * I - I \rvert^2 + I \rvert \rvert + \frac{I^2}{\frac{\delta}{\delta x} G_1(I)} \rvert$	$f_1 = 3.93$ $f_2 = 1.22$ $f_3 = 90.68$
IPSLAM3	$G_1 * G_1 * (\frac{\delta}{\delta x} G_1(\frac{\delta}{\delta y} G_1(I)) - G_1(I))$	$f_1 = 8.68$ $f_2 = 1.19$ $f_3 = 27.81$
IPSLAM5	$G_1 * G_2 * \frac{\delta}{\delta y} G_1(\frac{\delta}{\delta y} G_1(I))$	$f_1 = 4.33$ $f_2 = 1.21$ $f_3 = 32.02$
IPSLAM6	$G_1 * \lvert \frac{\delta}{\delta x} G_1(\frac{\delta}{\delta y} G_1(I)) - \frac{\delta}{\delta y} G_1(\frac{\delta}{\delta y} G_1(I)) \rvert$	$f_1 = 4.30$ $f_2 = 1.24$ $f_3 = 34.02$
IPSLAM11	$\frac{\delta}{\delta y} G_1(\frac{\delta}{\delta y} G_1(I)) \times (G - 1 * I - G_1 * G_2(\frac{\delta}{\delta x} G_1(\frac{\delta}{\delta y} G_1(I)) - G_1 * I))$	$f_1 = 5.77$ $f_2 = 1.27$ $f_3 = 19.13$
IPSLAM19	$G_1 * G_1 * G_1((\frac{\delta}{\delta y} G_1(\frac{\delta}{\delta y} G_1(I)))/2)$	$f_1 = 30.30$ $f_2 = 1.24$ $f_3 = 16.62$
IPSLAM25	$\frac{\delta}{\delta y} G_1(\frac{\delta}{\delta y} G_1(I))$	$f_1 = 35.86$ $f_2 = 1.27$ $f_3 = 16.57$
IPSLAM27	$Log((\frac{\delta}{\delta y} G_1(I) \times \frac{\delta}{\delta y} G_1(\frac{\delta}{\delta y} G_1(I))) + \lvert G_2 * G_1(I) + \frac{\delta}{\delta y} G_1(\frac{\delta}{\delta y} G_1(I)) \rvert)$	$f_1 = 7.79$ $f_2 = 1.27$ $f_3 = 23.56$

Note, that the most used operation in these functions is the image derivate in the y direction. The use of this operation is related to the selected movement applied for the evolutionary process. Since, the camera is moving horizontally, the vertical features should be more stable. Once, we had several good solutions, we made a comparison of the individuals being used within the SLAM system. Figure 5 shows the result for the individual $IPSLAM19$ that achieves the better results. The sequence on the left corresponds to the images captured by the

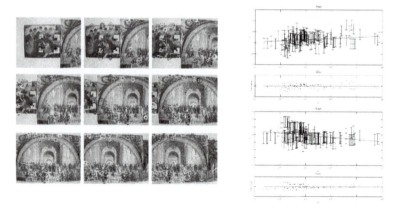

Fig. 5. System execution using the $IPSLAM19$ individual

camera, where the ellipses on the images represent the computed landmarks on the map. The right side of the figure shows the error graph for the XY and XZ planes of the estimated trajectory.

5 Conclusion

The aim of this paper was to adapt a visual behavior in a purposive way through a multiobjective evolutionary computation approach. The specialized behavior is based on solving a specific task that was able to simplify a real-world SLAM system by eliminating two heuristics around the interest point detector applied in the map building. This paper illustrates that it is coherent to see an evolutionary algorithm as a purposive process.

Acknowledgments. This research was founded by CONACyT through the Project 155045 - "Evolución de Cerebros Artificiales en Visión por Computadora". First author supported by scholarship 267339/220773 from CONACyT. This research was also supported by TESE through the project DIMI-MCIM-004/08.

References

1. Aloimonos, J., Weiss, I., Bandyopadhyay, A.: Active vision. In: Proceedings of the First International Conference on Computer Vision, pp. 35–54 (1987)
2. Aloimonos, Y.: Active Perception, p. 292. Lawrence Erlbaum Associates (1993)
3. Ballard, D.: Animate vision. Artificial Intelligence Journal 48, 57–86 (1991)
4. Davison, A.J.: Real-time simultaneous localisation and mapping with a single camera. In: Proceedings of the Ninth IEEE International Conference on Computer Vision, vol. 2, pp. 1403–1410. IEEE Computer Society, Washington, DC, USA (2003)
5. Dunn, E., Olague, G.: Pareto optimal camera placement for automated visual inspection. In: Proceedings of IEEE/RSJ International Conference on Intelligent Robots and Systems (2005)
6. Dunn, E., Olague, G., Lutton, E., Schoenauer, M.: Pareto optimal sensing strategies for an active vision system. In: Proceedings of IEEE Congress on Evolutionary Computation, vol. 1, pp. 457–463 (2004)
7. Fermüller, C., Aloimonos, Y.: The Synthesis of Vision and Action. In: Landy, et al. (eds.) Exploratory Vision: The Active Eye, ch. 9, pp. 205–240. Springer, Heidelberg (1995)
8. Lepetit, V., Fua, P.: Monocular model-based 3d tracking of rigid objects: A survey. In: Foundations and Trends in Computer Graphics and Vision, vol. 1, pp. 1–89 (2005)
9. Shi, J., Tomasi, C.: Good features to track. In: Proceedings of Computer Vision and Pattern Recognition, pp. 593–600 (1994)
10. Trujillo, L., Olague, G.: Automated design of image operators that detect interest points. Evolutionary Computation 16, 483–507 (2008)
11. Trujillo, L., Olague, G., Lutton, E., Fernández, F.: Multiobjective design of operators that detect points of interest in images. In: Proceedings of Genetic and Evolutionary Computation Conference, pp. 1299–1306 (2008)
12. Zitzler, E., Laumanns, M., Thiele, L.: Spea2: Improving the strength Pareto evolutionary algorithm. Tech. rep., Evolutionary Methods for Design (2001)

On Evolutionary Approaches to Unsupervised Nearest Neighbor Regression

Oliver Kramer

Fakultät II, Department for Computer Science,
Carl von Ossietzky Universität Oldenburg,
26211 Oldenburg, Germany
oliver.kramer@uni-oldenburg.de

Abstract. The detection of structures in high-dimensional data has an important part to play in machine learning. Recently, we proposed a fast iterative strategy for non-linear dimensionality reduction based on the unsupervised formulation of K-nearest neighbor regression. As the unsupervised nearest neighbor (UNN) optimization problem does not allow the computation of derivatives, the employment of direct search methods is reasonable. In this paper we introduce evolutionary optimization approaches for learning UNN embeddings. Two continuous variants are based on the CMA-ES employing regularization with domain restriction, and penalizing extension in latent space. A combinatorial variant is based on embedding the latent variables on a grid, and performing stochastic swaps. We compare the results on artificial dimensionality reduction problems.

1 Introduction

In many scientific disciplines structures in high-dimensional data have to be detected, e.g., in face and speech recognition, gesture recognition, and in genome data processing tasks. Dimensionality reduction methods reduce the dimensionality of the data to make them easier to process, e.g., for visualization, and for post-processing with other machine learning techniques that may suffer from the curse of dimensionality effect [3].

Given a set of N high-dimensional patterns $\mathbf{y}_1, \ldots, \mathbf{y}_N \in \mathbb{R}^d$ in the d-dimensional data space $\mathbb{R}^d$, many dimensionality reduction methods compute low-dimensional representations $\mathbf{x}_1, \ldots, \mathbf{x}_N \in \mathbb{R}^q$ with $q < d$ of the corresponding high-dimensional patterns. The variables $\mathbf{x}_i$ are called latent points, latent variables, or embeddings in the latent space $\mathbb{R}^q$.

The low-dimensional representations should capture the most important characteristics of their high-dimensional pendants, e.g., they should maintain neighborhood relations and distances (neighbored points in data space should be neighbored in latent space). This is sometimes referred to as *intrinsic structure* of the data. In this paper we concentrate on unsupervised nearest neighbor regression, a method we recently proposed [6]. UNN is based on unsupervised regression, a framework for dimensionality reduction, which has originally been

C. Di Chio et al. (Eds.): EvoApplications 2012, LNCS 7248, pp. 346–355, 2012.
© Springer-Verlag Berlin Heidelberg 2012

introduced by Meinicke [9]. The idea is to reverse the regression formulation such that low-dimensional data samples in latent space optimally reconstruct high-dimensional output data. UNN regression fits nearest neighbor regression into this unsupervised setting. The spatial formation of the low-dimensional vectors during training is induced by the regression method, and is supposed to represent important information about the high-dimensional space. In [6] we argued that the UNN optimization problem (see Section 3) is difficult to solve. Although the two proposed iterative strategies allow the construction of a fast solution, some open questions remain: how close do the UNN heuristics come to the optimal solution, and is an evolutionary treatment of the problem a reasonable alternative to the UNN heuristics. How close can stochastic variants come to the iterative heuristics, and how many dimensions can be managed.

Section 2 reviews related work. In Section 3 we revisit the UNN optimization problem, and a constructive heuristic. Section 4 concentrates on the continuous perspective with two regularization variants. Section 5 introduces the combinatorial variant, while Section 6 presents an experimental comparison of the proposed methods. The paper closes with a summary of the main results in Section 7.

2   Related Work

Perhaps the most famous dimensionality reduction method is the principal component analysis (PCA), which assumes linearity of the manifold [4,12]. Further famous approaches are Isomap by Tenenbaum *et al.* [15], and locally linear embedding (LLE) by Roweis and Saul [13]. The work on unsupervised regression for dimensionality reduction starts with Meinicke [9], who introduced the algorithmic framework for the first time. In this line of research early work concentrated on non-parametric kernel density regression, i.e., the counterpart of the Nadaraya-Watson estimator [10] denoted as unsupervised kernel regression (UKR). Klanke and Ritter [5] introduced an optimization scheme based on LLE, PCA, and leave-one-out cross-validation (LOO-CV) for UKR. Kramer and Gieseke [7] employed the CMA-ES to replace the complicated optimization scheme of UKR by a simple stochastic one. Carreira-Perpiñán and Lu [2] argue that training of non-parametric unsupervised regression approaches is quite expensive, i.e., $\mathcal{O}(N^3)$ in time and $\mathcal{O}(N^2)$ in memory. Parametric methods can accelerate learning. In this line of research unsupervised regression approaches were introduced, e.g., based on radial basis function networks (RBFs) [14], and Gaussian processes [8].

3   Unsupervised KNN Regression

An UNN regression manifold is defined by N latent variables $\mathbf{x} \in \mathbb{R}^q$ organized in an $N \times q$ matrix $\mathbf{X}$ with the unsupervised formulation:

$$\mathbf{f}_{UNN}(\mathbf{x}; \mathbf{X}) = \frac{1}{K} \sum_{i \in \mathcal{N}_K(\mathbf{x}, \mathbf{X})} \mathbf{y}_i \tag{1}$$

for a K that defines the neighborhood size of KNN regression. The matrix $\mathbf{X}$ contains the latent points $\mathbf{x}$ that define the manifold, i.e., the low-dimensional representation of the data $\mathbf{Y}$. Parameter $\mathbf{x}$ is the location where the function is evaluated. With the help of Equation (1) we can define the UNN regression optimization problem. An optimal UNN regression manifold minimizes the data space reconstruction error (DSRE), i.e., the function:

$$E(\mathbf{X}) = \frac{1}{N} \|\mathbf{Y} - \mathbf{f}_{UNN}(\mathbf{x}; \mathbf{X})\|_F^2 \tag{2}$$

that measures the differences between the KNN-mapping from latent space, and the target patterns in data space[1]. An optimal UNN manifold is a set of latent points $\mathbf{X}$ that minimizes the reconstruction of the data points $\mathbf{Y}$ employing KNN regression. Minimizing Equation (2) is a hard optimization problem. First, the problem dimensionality scales with size N of the data set. Second, without further constraints on the latent space, the search space of the optimization problem is very large. In [6] we introduced iterative strategies based on one-dimensional fixed latent point topologies (topological sorting). The first iterative strategy UNN 1 tests all intermediate latent positions, and places the latent points at locations that achieve the lowest DSRE:

1. Choose one element $\mathbf{y} \in \mathbf{Y}$,
2. test all $\hat{N} + 1$ intermediate positions of the $\hat{N}$ embedded points $\hat{\mathbf{Y}}$ in latent space,
3. choose the latent position that minimizes $E(\mathbf{X})$, and embed $\mathbf{y}$,
4. remove $\mathbf{y}$ from $\mathbf{Y}$, add $\mathbf{y}$ to $\hat{\mathbf{Y}}$, and repeat from Step 1 until all elements have been embedded.

Figure 1(a) illustrates the $\hat{N} + 1$ possible embeddings of a data sample into an existing order of points in latent space (yellow/bright circles). The position of element $\mathbf{x}_3$ results in a lower DSRE with $K = 2$ than the position of $\mathbf{x}_5$, as the mean of the two nearest neighbors of $\mathbf{x}_3$ is closer to $\mathbf{y}$ than the mean of the two nearest neighbors of $\mathbf{x}_5$. Figure 1(b) shows an example of a UNN 1 embedding of the 3D-S (upper part shows *unsorted* S, lower part shows colorization w.r.t. the embedding), similar colors correspond to neighbored positions in latent space, i.e., a meaningful neighborhood preserving embedding took place. UNN 2 searches for the closest embedded pattern $\mathbf{y}^*$ and places the latent points on the site of the neighborhood that achieves the lowest DSRE.

---

[1] with Frobenius norm

$$\|\mathbf{A}\|_F = \sqrt{\sum_{i=1}^{d} \sum_{j=1}^{N} |a_{ij}|^2} \tag{3}$$

with matrix elements a_{ij} of $\mathbf{A}$.

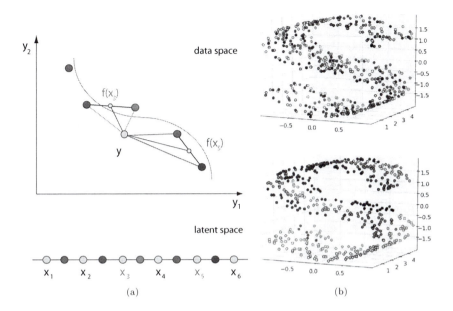

Fig. 1. Left: Illustration of UNN 1 embedding of a low-dimensional point to a fixed latent space topology w.r.t. the DSRE testing all $\hat{N} + 1$ positions [6]. Right: Example of UNN 1 result of a 3D-S before (upper right) and after embedding (lower right) with UNN 1 and $K = 10$.

4   Continuous Perspective

In the following, we employ a continuous optimization perspective on UNN. For this sake we allow optimization of neighborhood topologies in (a subset of) $\mathbb{R}^q$. We choose the following representation: A candidate solution $\mathbf{X} \in \mathbb{R}^{q \times N}$ is the matrix of latent vectors $\mathbf{x} \in \mathbb{R}^q$, i.e., a vector of scalars for $q = 1$. For one fitness evaluation $f(\mathbf{x})$ the overall DSRE of the whole embedding is computed. Similar to UKR we have to regularize the continuous UNN model. An unconstrained UNN regression formulation would allow the latent points to move infinitely apart from each other, which may complicate the optimization process. Two kinds of regularization approaches will be compared: restriction of latent space to a hypercube $[0, 1]^q$, and penalizing extension with a summand $\lambda \|\mathbf{X}\|$.

4.1   CMA-ES

As optimization approach we employ the CMA-ES by Hansen and Ostermeier [11]. The CMA-ES belongs to the class of evolution strategies (ES) [1]. In each generation t, a population of λ points $\mathbf{X}_i^t$, $i = 1, \ldots, \lambda$ is produced with the multivariate normal distribution:

$$\mathbf{X}_i^t = \mathbf{m}^t + \sigma^t \mathcal{N}_i(0, \mathbf{C}^t), \text{ for } i = 1, \ldots, \lambda. \tag{4}$$

with $t \in \mathbb{N}$. The variables define sequences, i.e., $\mathbf{m}^t$ with $t \in \mathbb{N}$ defines the sequences of mean values of the Gaussian distribution generated by the CMA-ES (corresponding to the estimate of the optimum), while $\sigma^t, t \in \mathbb{N}$ defines the sequence of step sizes, and $\mathbf{C}^t, t \in \mathbb{N}$ of covariance matrices, respectively. Individuals are ranked according to their fitness f:

$$f(\mathbf{x}_{1:\lambda}^t) \leq \ldots \leq f(\mathbf{x}_{\mu:\lambda}^t) \leq \ldots f(\mathbf{x}_{\lambda:\lambda}^t), \tag{5}$$

following the notation that $\mathbf{x}_{i:\lambda}^t$ is the i-th best individual. The mean $\mathbf{m}^t$ is updated with the μ best solutions in each generation:

$$\mathbf{m}^{t+1} = \sum_{i=1}^{\mu} \omega_i \mathbf{x}_{i:\lambda}^t \tag{6}$$

with positive and normalized weights ω_i. Core of the CMA-ES is the update of the covariance matrix $\mathbf{C}^t$, which is adapted to the local fitness conditions. The update rule for $\mathbf{C}^t$, $\mathbf{m}^t$, and σ^t can be found in Hansen *et al.* [11]. The population sizes are chosen as $\lambda = 4N$, and $\mu = \lambda/2$.

4.2   Restriction of Latent Space to $[0, 1]^q$

Regularization is a technique in machine learning to avoid complex models and over-fitting. In our optimization process it is undesirable to let the latent variables move infinitely apart from each other. All neighborhood relations can be instantiated within a restricted part of latent space. First, we restrict the latent space to the unit hypercube $\mathbf{x} \in [0, 1]^q$:

$$\min E(\mathbf{X})_r \text{ subject to } x_{ij} \in [0, 1] \tag{7}$$

The constraint forces the latent points to stay in the unit hypercube in latent space. To handle the interval constraint, we penalize deviations from the interval employing a quadratic penalty:

$$p(\mathbf{X}) = \sum_{i,j} \epsilon_{ij} \text{ with } \epsilon = \begin{cases} (x_{ij} - 1)^2 & \text{if } x_{ij} > 1 \\ x_{ij}^2 & \text{if } x_{ij} < 0 \\ 0 & \text{else} \end{cases} . \tag{8}$$

In Section 6 the latent space restriction approach will be compared to the penalty approach experimentally.

4.3   Penalizing Extension in Latent Space with $\lambda \|\mathbf{X}\|$

The second variant for regularization of UNN that we employ is penalizing extensions in latent space:

$$\min E(\mathbf{X})_p := \min \left(E(\mathbf{X}) + \lambda \|\mathbf{X}\| \right) \tag{9}$$

Also this technique has been applied to regularize UKR models, and is a more frequent way to regularize models in machine learning. The penalty limits the sum of lengths of all latent variables.

To understand the influence of the parameter λ we test various assignments for regularization parameter λ, and different neighborhood sizes K for the 3D-S data set[2]. We repeat each experiment 25 times. Table 1 shows the median DSRE of the experiments. The CMA-ES terminates after 1,000 generations. It can be observed that the DSRE can be reduced significantly in comparison to the initial state with the CMA-ES for small settings of λ. For large settings the optimization process first primarily concentrates on reduction of the extension, and neglects optimization of the DSRE in the 1,000 generations we employ. Hence, we set $\lambda = 10^{-2}$ in the remainder of this work.

Table 1. Analysis of regularization parameter λ for two neighborhood sizes $K = 2, 10$ on the 3D-S data set with two data set sizes $N = 30, 50$

N	K	init	10^{-2}	10^{-1}	10^0	10^1	10^2
30	2	34.82	18.64	21.76	32.06	36.23	36.25
30	10	51.39	36.99	38.95	51.33	51.14	50.33
50	2	60.50	39.516	52.01	56.01	63.63	58.08
50	10	86.17	67.44	75.14	81.19	84.91	82.10

5   Combinatorial Perspective on UNN Regression

The continuous perspective allows an infinite number of positions for latent points, only the regularization approach restricts the number of possible solutions. However, the number of possible K-nearest neighborhoods, and consequently of different fitness values is much lower, i.e. $\binom{N}{K}$. For large N this is still an intractable number of solutions, and makes it impractical to test all possible latent variable neighborhoods. Techniques are required to reduce the search space like the iterative embedding strategies UNN 1 and 2, see Section 3. In the following, we propose an evolutionary discrete search strategy that randomly selects two points in latent space, and swaps their positions, see Algorithm 1 for the corresponding pseudocode.

Let $\mathbf{X} = (\mathbf{x}_1, \ldots, \mathbf{x}_N)$ be a solution, i.e., an (initially random) order of the high-dimensional patterns $\mathbf{y}_i$, while $\mathbf{x}_i$ specifies the position of pattern $\mathbf{y}_i$. The $(1+1)$-EA randomly selects two positions $p_1, p_2 \in \mathbb{N}^+$, swaps the corresponding patterns $\mathbf{x}_{p_1}$ and $\mathbf{x}_{p_2}$, and accepts the new order, iff the DSRE is decreased. This process is repeated until a termination condition is fulfilled (e.g., no DSRE change for a defined number of generations).

The swap is accepted, if the DSRE can be reduced (Line 4). Otherwise, it is rejected and the latent points are reset to their original position. It is reasonable

---

[2] The 3D-S data set consists of $d = 3$ dimensional patterns arranged to an S with and without hole, see [6].

Algorithm 1. $(1+1)$-EA FOR UNN

Require: data set $\mathbf{Y}$, **Request:** embedding $\mathbf{X}$
1: initialization: random order of $\mathbf{X} = (\mathbf{x}_1, \ldots \mathbf{x}_N)$.
2: **repeat**
3:    choose two points $p_1, p_2 \in \mathbb{N}$
4:    change $\mathbf{X}$ to $\mathbf{X}'$ by swapping $\mathbf{x}_{p_1}$ and $\mathbf{x}_{p_2}$
5:    replace $\mathbf{X}$ by $\mathbf{X}'$ if $f(\mathbf{X}') \leq f(\mathbf{X})$
6: **until** termination condition
7: **return** embedding $\mathbf{X}$

---

to let the stochastic process terminate, if the DSRE does not change for $\kappa \in \mathbb{N}^+$ iterations. Figure 2 shows a 1-dimensional example grid. Assume the left blue (dark) latent point $\mathbf{x}_1$, and the right yellow (bright) latent point $\mathbf{x}_2$ have randomly been selected to swap their positions. The DSRE is computed for their old and the novel neighborhood. The swap is accepted, as their own colors are more similar to colors of the novel neighborhoods than the colors of the previous neighborhoods.

Fig. 2. Linear latent space (grid with $q = 1$) before (left), and after a swap (right) that leads to a lower DSRE for $K = 2$. Similar colors correspond to low distances in latent space.

6   Experimental Analysis

In the following, we analyze the evolutionary variants experimentally.

6.1   Comparison

Table 2 shows the experimental results of 25 CMA-ES runs for the data sets 3D-S and 3D-S with hole, each with $N = 30$ patterns. The figures show the resulting DSRE at the beginning (init), in comparison to UNN 2 [6], the CMA-ES with both regularization strategies, and the $(1 + 1)$-EA that works on the grid representations. To allow a comparison to one of the state-of-the-art methods in dimensionality reduction, we also compare to locally linear embedding (LLE) [13].

First, the experimental results show that UNN 2 achieves a lower DSRE than LLE for $K \geq 5$. This result shows the strength of the iterative heuristic. Concerning the evolutionary optimization approaches we can observe that the $(1+1)$-EA achieves a lower DSRE than UNN 2. The continuous approaches and UNN 2 achieve similar results: On 3D-S for $K = 2$ the penalized variant, for $K = 5$ the

Table 2. Comparison of DSRE for initial data set, embedding of UNN 2, the evolutionary approaches, and LLE for $q = 1$

K	3D-S			3D-S hole		
	2	5	10	2	5	10
init	34.8 ±0	46.8 ±0	51.3 ±0	28.3 ±0	40.5 ±0	41.2 ±0
UNN 2	23.4 ±0	31.3 ±0	43.3 ±0	15.6 ±0	20.8 ±0	28.3 ±0
CMA, $[0,1]^q$	24.5 ±11.6	27.6 ±8.4	36.3 ±15.0	24.0 ±11.7	20.4 ±20.4	29.1 ±15.4
CMA, $\lambda\|\mathbf{X}\|$	22.2 ± 6.1	31.6 ± 10.8	41.4 ± 8.3	14.7 ± 10.9	24.5 ± 11.8	33.8 ± 21.3
$(1+1)$-EA	**13.3 ± 1.3**	**24.4 ± 2.5**	**31.1 ± 1.7**	10.8 ± 0.4	**17.6 ± 1.4**	**24.7 ± 0.4**
LLE	13.7 ±0.0	34.1 ±0.0	49.6 ±0.0	**10.4** ±0.0	23.6 ±0.0	31.2 ±0.0

constrained variant, and for $K = 10$ both variants achieve a lower DSRE than UNN 2. On 3D-S with hole UNN 2 and the continuous variants achieve similar results. We want to point out that the evolutionary variants have a budget of $1,000$ fitness evaluations (computing the overall DSRE), i.e., $1,000 \cdot N$ complete DSRE computations (30,000 for $N = 30$), while UNN 2 solves the problem with a budget of $0.5 \cdot N \cdot (N+1)$ DSRE computations (465 for $N = 30$). At the same time the $(1+1)$-EA shows better results than the CMA-ES. On both data sets the restricted CMA-ES shows better results than the penalized CMA-ES in two of the three cases. But due to the high standard deviations, a statistical significance cannot be reported. In contrast, the $(1+1)$-EA is quite robust with small standard deviations, and consequently statistically significant improvements.

6.2   Curse of Dimensionality

The number of patterns defines the dimensionality of the UNN optimization problem. The question arises: How does the dimensionality influence the success of the evolutionary optimizers, i.e., the DSRE? We try to answer this question in the following by showing the DSRE achieved depending on the number of patterns. Figure 3(a) shows the DSRE for UNN 1, UNN 2, both CMA variants, and the $(1+1)$-EA on the 3D-S data set (without hole). We can observe that the stochastic variants are slightly better at the beginning of the optimization process, but fail for higher dimensions. The continuous variants do not scale well, while the $(1+1)$-EA is still able to approximate the optimum. But the heuristics UNN 1 and UNN 2 scale much better, in particular UNN 1 turns out to be the best optimizer for larger problem sizes. Figure 3(b) shows the corresponding outcome of the experiments for the 3D-S data set with hole, where similar observations can be made. Here, it is interesting that the two continuous variants show a very similar behavior.

For these two cases we show the relative fitness improvement with variances in the following. For the 3D-S data set with hole, Figure 4 shows the mean relative DSRE improvement E_s/E_e, with E_s being the DSRE at the start (initialization), and E_e the DSRE in the end (after termination). The relative error is plotted with variance for the constrained (left), and the continuous (right) CMA-ES-based optimization. The relative improvement gets worse with

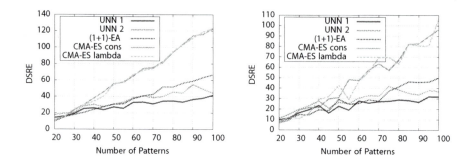

Fig. 3. Curse of dimensionality for evolutionary UNN variants on (a) 3D-S, and (b) 3D-S with hole

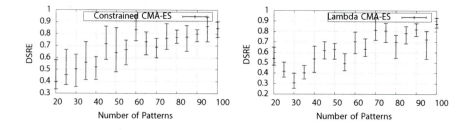

Fig. 4. Relative error E_s/E_e depending on the number of patterns for the 3D-S data set with hole for the constrained CMA-ES ($[0,1]^q$, left), and the regularized ($\lambda\|\mathbf{X}\|$, right)

increasing problem size. We can observe that the worst runs on larger problem sizes (as of $N = 50$) show very bad results, with almost no improvement. The constrained variant shows higher variances for small data sets.

7   Conclusions

In this work we have introduced evolutionary variants to compute UNN regression embeddings. It turns out that the evolutionary variants achieve higher accuracies than their heuristic pendants on small data sets. However, the high accuracy has to be paid with a slow approximation speed. But for data sets with $N >> 50$ the optimization problem becomes difficult to solve, in particular for the continuous variants with the CMA-ES. The $(1 + 1)$-EA scales better w.r.t. the data set sizes, as the restriction to a grid decreases the solution space. In comparison, the UNN heuristics are much faster, in particular for high dimensions, as they can compute a solution in $\mathcal{O}(N^2)$ (or $\mathcal{O}(N \log N)$ employing k-d-trees). As a conclusion we can recommend a stochastic approach for small data sets, but recommend to employ the iterative heuristics for data set sizes larger than 50.

References

1. Beyer, H.G., Schwefel, H.P.: Evolution strategies - A comprehensive introduction. Natural Computing 1, 3–52 (2002)
2. Carreira-Perpiñán, M.Á., Lu, Z.: Parametric dimensionality reduction by unsupervised regression. In: Conference on Computer Vision and Pattern Recognition (CVPR), pp. 1895–1902 (2010)
3. Hastie, T., Tibshirani, R., Friedman, J.: The Elements of Statistical Learning. Springer, Berlin (2009)
4. Jolliffe, I.: Principal component analysis. Springer Series in Statistics. Springer, New York (1986)
5. Klanke, S., Ritter, H.: Variants of unsupervised kernel regression: General cost functions. Neurocomputing 70(7-9), 1289–1303 (2007)
6. Kramer, O.: Dimensionality reduction by unsupervised k-nearest neighbor regression. In: Proceedings of the International Conference on Machine Learning and Applications (ICMLA), pp. 275–278. IEEE Computer Society Press (2011)
7. Kramer, O., Gieseke, F.: A stochastic optimization approach for unsupervised kernel regression. In: Genetic and Evolutionary Methods (GEM), pp. 156–161 (2011)
8. Lawrence, N.D.: Probabilistic non-linear principal component analysis with gaussian process latent variable models. Journal of Machine Learning Research 6, 1783–1816 (2005)
9. Meinicke, P.: Unsupervised Learning in a Generalized Regression Framework. Ph.D. thesis, University of Bielefeld (2000)
10. Meinicke, P., Klanke, S., Memisevic, R., Ritter, H.: Principal surfaces from unsupervised kernel regression. IEEE Trans. Pattern Anal. Mach. Intell. 27(9), 1379–1391 (2005)
11. Ostermeier, A., Gawelczyk, A., Hansen, N.: A derandomized approach to self adaptation of evolution strategies. Evolutionary Computation 2(4), 369–380 (1994)
12. Pearson, K.: On lines and planes of closest fit to systems of points in space. Philosophical Magazine 2(6), 559–572 (1901)
13. Roweis, S.T., Saul, L.K.: Nonlinear dimensionality reduction by locally linear embedding. Science 290, 2323–2326 (2000)
14. Smola, A.J., Mika, S., Schölkopf, B., Williamson, R.C.: Regularized principal manifolds. J. Mach. Learn. Res. 1, 179–209 (2001)
15. Tenenbaum, J.B., Silva, V.D., Langford, J.C.: A global geometric framework for nonlinear dimensionality reduction. Science 290, 2319–2323 (2000)

Evolutionary Regression Machines
for Precision Agriculture[*]

Heikki Salo, Ville Tirronen, and Ferrante Neri

Department of Mathematical Information Technology,
P.O. Box 35 (Agora), 40014 University of Jyväskylä, Finland
{heikki.salo,ville.tirronen,ferrante.neri}@jyu.fi

Abstract. This paper proposes an image processing/machine learning system for estimating the amount of biomass in a field. This piece of information is precious in agriculture as it would allow a controlled adjustment of water and fertilizer. This system consists of a flying robot device which captures multiple images of the area under interest. Subsequently, this set of images is processed by means of a combined action of digital elevation models and multispectral images in order to reconstruct a three-dimensional model of the area. This model is then processed by a machine learning device, i.e. a support vector regressor with multiple kernel functions, for estimating the biomass present in the area. The training of the system has been performed by means of three modern meta-heuristics representing the state-of-the-art in computational intelligence optimization. These three algorithms are based on differential evolution, particle swarm optimization, and evolution strategy frameworks, respectively. Numerical Results derived by empirical simulations show that the proposed approach can be of a great support in precision agriculture. In addition, the most promising results have been attained by means of an algorithm based on the differential evolution framework.

1 Introduction

Food production and agriculture have been transformed from a solar based industry into one relying on fuel, chemicals, sensors and technology. The use of chemicals and fuel increased dramatically in 60s and 70s and several concerns were stated about the effect of this increase to our health and the health of our environment. This concern and the advances in imaging technology resulted in the development of the Precision Agriculture (PA), see [10]. PA is a farming technique based on observing and responding to intra-field variations. Clearly, the observation of variations in the field is crucially important to promptly apply a countermeasure.

A fundamentally important entity to monitor within a field is the produced biomass since an accurate map of field biomass is necessary for crop yield estimation and optimal field management, see [13]. If an exact inventory of plant

---

[*] This research is supported by the Academy of Finland, Akatemiatutkija 130600, Algorithmic Design Issues in Memetic Computing. A special thank to Antti-Juhani Kaijanaho for the useful discussions.

C. Di Chio et al. (Eds.): EvoApplications 2012, LNCS 7248, pp. 356–365, 2012.

mass is known, more careful economical planning can be done. Furthermore, if some parts of the field fall behind in growth, intervention methods, such as fertilization or additional irrigation, can be used.

Field biomass mapping systems are often image based, where spectral or false color images are acquired from satellites, aeroplanes, and devices mounted on tractors and other field equipment. Field map creation is based on machine vision techniques that include a wide variety of machine learning elements where the biomass estimation is based on features and models built from the images. For example, in [14], an estimation scheme using several different vegetation indices based on relationships of multispectral images is proposed. In [6], biomass estimation is performed by means of stereoscopic vision techniques used to construct the so called Digital Elevation Models (DEM), i.e. 3-D representations of the terrain surface, from sets of ordinary aerial photographs. In [17], the combination of both multispectral images and digital elevation based measurements of the biomass is successfully proposed.

In order to estimate the biomass in a field, multiple images and measurements must be taken and the images must be processed. Thus, the problem can be presented as a non-parametric regression, which is further complicated by the large variability in images.

In this paper, we propose a chain of operations that extracts suitable information from image data and creates a non-parametric estimator for biomass using a machine learning technique for performing the non-parametric regression. This technique, is based on Support Vector Regression (SVR) (see [3]). and ensemble learning (see [11] and [1,2]).

The training of the SVM ensemble is obtained by means of three modern computational intelligence optimization algorithms, based on Evolution Strategies (ES), Differential Evolution (DE) and Particle Swarm Optimization (PSO), respectively.

The remainder of this paper is organized as follows. In Section 2 we introduce the chain of operations and the support vector regression and ensemble learning techniques. Section 3 shows the performance comparison of the three meta-heuristics considered in this study. Finally, Section 4 gives the conclusions of this work.

2   Intelligent System for Biomass Estimation

The proposed chain is schematically represented in Fig. 1. A set of images is taken by an Unmanned Aerial Vehicle (UAV). These images are processed into of DEM and multispectral images, as shown in [17], thus producing a set of data which is processed by a machine learning technique to associate to each portion of land (patch) with a biomass value.

The first step in the chain (1) is collating the images acquired by an unmanned aerial vehicle into a digital elevation model (see Figure 2) and an orthographic map of the field. In our case this is done by the UAV operator using image correlation and stereoscopic vision techniques (see [12] for a survey of this topic). This phase is not parametrised in this paper.

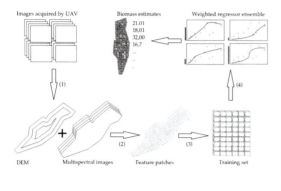

Fig. 1. General workflow of an biomass estimation system

Fig. 2. A Digital Elevation Model (DEM) of a field

In the second step of the process (2), the field area is divided into test patches. Each patch contains a specific sample with a different biomass. For each patch we calculate the following features, which are used to train the estimation system:

1. A cumulative histogram of the elevation values: $\text{cdf}_i^{DEM} = \sum_{j=0}^{i} \frac{n_j}{n}$, where $0 < i/leq5$, and n_j is number of elevation samples where the elevation is between $\min + j(\min - \max)$ and $m + (j+1)(\min - \max)$, n is the number of elevation samples within the patch and min and max and are the minimum and the maximum elevations in the patch respectively.
2. The average (μ) of the elevations measured in the patch.
3. The variance (s_{DEM}^2) of the elevations measured in the patch.
4. The variance (s_{NIR}^2) of the NIR channel responses measured in the patch.
5. A Cumulative histogram of the NIR channel responses: $\text{cdf}_i^{NIR} = \sum_{j=0}^{i} \frac{n_j}{n}$, where $0 < i/leq2$, and n_j is number of NIR channel responses in range $\min + j/(\min - \max)$ and $m + (j+1)/(\min - \max)$, n is the number of responses within the patch and min and max and are the minimum and the maximum responses in the patch respectively.

These features must be preprocessed to equalize features with different ranges. In this paper, we consider preprocessing with both simple scaling and the principal component analysis (PCA), which can be used to reduce the feature vector dimensionality. In the case of PCA a proper ratio of dimension reduction T_{pca} must be selected properly.

These features are paired with the physically measured dry-biomass values in the step (3) to produce the training set from which we build the biomass estimator using regression analysis in the step (4). Regression analysis is the science of determining the relationship between dependent and independent variables and it is used in devising automatic prediction and forecasting tools. When the relationship between the parameters is unknown, the problem is named non-parametric regression. SVM, following the example given in [3], are used here to perform the non-parametric regression tasks. In order to construct a SVR from a set of point and value pairs, $\{(x_i, y_i)\} \subset \mathbb{R}^n \times \mathbb{R}$, we must find a function $f : \mathbb{R}^n \to \mathbb{R}$ such that f deviates at most an ϵ amount from the training points:

$$|y_i - f(x_i)| \le \epsilon \tag{1}$$

$$\tag{2}$$

while f should be as simple as possible.

In our case, the points x_i represent the features of the terrain acquired by the imaging system and the values y_i represent the biomass measures associated with the corresponding terrain features. The resulting function f will be the biomass estimator for the non-measured parts of the terrain. In order to model this problem, it is enough to consider a linear function $f(x) = w \cdot x + b$ and equate simplicity to flat slope, which can later on be generalized to a non-linear estimator by using a non-linear mapping of the data points. This results in the following optimization problem:

$$\text{minimize} \quad \|w\|^2 + C \sum_{i=1}^{l} (\xi_i + \xi_i^*) \tag{3}$$

$$\text{subject to} \quad y_i - w \cdot x + b \le \epsilon + \xi_i \tag{4}$$

$$w \cdot x + b - y_i \le \epsilon + \xi_i^*. \tag{5}$$

Many datasets contain noise and other deviations that make it impossible to meet this constraint at all or without giving up the simplicity requirement. To properly handle these cases, the slack variables ξ_i and ξ_i^* are added to the constraint for additional flexibility and the fitness in penalized according to the parameter C. The latter parameter determines the trade-off between flatness of the function and the deviations from the estimate. By transforming inequality into equality constraints, the optimization problem is reformulated in the following way:

$$\text{maximize} \quad \frac{1}{2} \sum_{i,j=1}^{l} (\alpha_i - \alpha_i^*)(\alpha_j - \alpha_j^*)(x_i \cdot x_j)$$

$$- \epsilon \sum_{i=1}^{l} (\alpha_i - \alpha_i^*) + \sum_{i=1}^{l} y_i (\alpha_i - \alpha_i^*)$$

$$\text{subject to} \quad \sum_{i=1}^{l} (\alpha_i - \alpha_i^*) = 0$$

$$\text{and} \quad \alpha_i, \alpha_i^* \in [0, C[,$$

where α_i and α_i^* are Lagrange multipliers. It can be observed that $w = \sum_{i=1}^{l} (\alpha_i - \alpha_i^*) x_i$ and $f(x) = \sum_{i=1}^{l} (\alpha_i - \alpha_i^*) x_i \cdot x + b$. Thus, the function f is entirely characterized by the scalar product between the training points. Then, this optimization problem can be efficiently solved using quadratic programming techniques [16].

In addition, this characterization via scalar products allows an easy extension from the linear case to the non-linear one by applying a suitable non-linear mapping θ to the data prior to training the model. Although such mappings can be computationally demanding, for some θ there exist such functions k that $k(x, y) = \theta(x) \cdot \theta(y)$, which allow the efficient calculation of scalar products in the codomain of θ. These functions k are called kernel functions. Three popular kernel functions are considered in this paper: 1) Linear $k(x, y) = x \cdot y$; 2)Radial Basis $k(x, y) = e^{-\sigma||x-y||^2}$; 3) Sigmoid $k(x, y) = \tanh\left(\gamma x \cdot y + c_0\right)$.

In order to build up an efficient intelligent system it is fundamental to properly select the parameters ϵ, C, to design the preprocessing scheme and the related parameters as well as the kernel function k with its corresponding parameters. In this study, we propose an alternative for finding the proper estimator by combining the outputs of several, differently modelled, weaker estimators as an ensemble. Such ensemble methods have been found to be very effective tools for various machine learning tasks in a survey in [8]. In this study we model the selection of each of the regression tasks sub-components, scaling, feature reduction, and regressor kernel selection by assigning them weights. Each weight represents the selection probability of the component whose weight is associated. Thus, our problem consists of finding the optimal weights for each sub-component along with their related parameters.

Ensembles are constructed according to the optimization based scheme in Fig. 3. First, Bagging (bootstrap aggregation) is done to avoid overfitting estimators to fitness dataset and subset of 20 samples are selected from the overall training set for each regressor. Then, the components for the regressors are selected according to weights given by the optimization process and their respective parameters are picked according to Table 1. The trained regressors

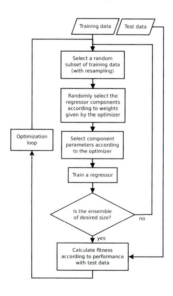

Fig. 3. General overview of the proposed system

are then tested with the test set samples and their average error is passed to the optimizer, which then proceeds to search for better set of parameters.

The parameters to be selected and their respective range of variability are listed in Table 1.

The related goal is to find the set of parameters listed in Table 1 such that the median error from the actual biomass values of the samples is minimized. A training set is used to perform the machine learning while test set is used to calculate the fitness.

Table 1. Parameters for the optimization problem

Variable	effect	range
x_1	C	$[2^{-5}, 2^{15}]$
x_2	ϵ	$[2^{-15}, 2^3]$
x_3	γ for RBF kernel	$[2^{-15}, 2^3]$
x_4	σ for Sigmoid kernel	$[2^{-15}, 2^3]$
x_5	coefficient for Sigmoid kernel	$[0, 1]$
x_6	T_{pca}	$[0.0000001, 0.5]$
x_7	Weight for Linear SVR kernel	$]0, 1]$
x_8	Weight for RBF SVR kernel	$]0, 1]$
x_9	Weight for Sigmoid SVR kernel	$]0, 1]$
x_{10}	Weight for using no reduction	$]0, 1]$
x_{11}	Weight for using PCA reduction	$]0, 1]$

3   Numerical Results

This study was conducted in Agrifood Research Finland (MTT)'s experimental Hovi crop field, which is situated in Vihti, Finland. For this study, MTT arranged a test season where growth between plots were varied using different seed and pesticide amounts during sowing.

The data consists of 91 test plots which were imaged using NIR capable UAV drone. The images were then postprocessed into an ortophotograph and a Digital Elevation Model, which describes the terrain height. DEMs are commonly used as basis for building maps and geographic information systems and can be constructed from sets of plain 2D images using image correlation and stereoscopic vision techniques (see, [12] for survey of this topic). For our application we have acquired a DEM of the target field using the stereoscopic vision techniques. The test plot locations and reference points for DEM calculation were measured using Real Time Kinematic GPS. Currently, our spectral data consists of of the DEM and near-infrared part of the spectrum, due to our UAVs occasional and catastrophic ineptitude in being aerial.

The test plots, which were randomly divided in the training and test sets, used in the training of the estimator, plus the validation set, which is used to evaluate the resulting ensembles. Each set consists of 30 samples. The target attribute in this study is the total dry biomass of the test plots and the reference values were acquired by manually collecting samples from the test plots and oven drying and weighting them.

The proposed model of the ensemble learner is trained using the three following optimization algorithms:

1. Proximity based Differential Evolution (Prox-DE) [4]
2. Frankenstein-Particle Swarm Optimization (F-PSO) [9]
3. Covariance Matrix Adaptation Evolution Strategy CMA-ES) according the implementation given in [7]

The Prox-DE algorithm is a Differential Evolution scheme which, instead of randomly (with uniform distribution) selecting the individuals undergoing mutation, employs a probabilistic set of rules for preferring the selection of solutions closely located to each other. The F-PSO algorithm employs a Particle Swarm Optimization structure and a set of combined modifications, previously proposed in literature in order to enhance the performance of the original paradigm. The CMA-ES is a well-known algorithm based on Evolution Strategy employing the so called maximum likelihood principle, i.e. it attempts to increase the probability of successful candidate solutions and search steps. The distribution of the solution and their potential moves tend to progressively adapt to the fitness landscape and take its shape.

For each algorithm, 75 simulation runs were run with a budget of 55 000 fitness evaluations. The parameters of the optimization algorithms are taken from the original articles in literature and are: for the Prox-DE F= 0.7, Cr= 0.3, S_{pop} = 60; for F-PSO v_{max} = 1, w_{min} = 0.4, w_{max} = 0.9, wt_{max} = 360, S_{pop} = 60, topology$_k$ = 2000, topology update period = 11; for CMA-ES σ = 0.5.

Table 2 shows the performance of each algorithm. The first three columns give numerical values for average, standard deviation and the best value of distribution over 75 simulations for each algorithm. The last column visualizes the distribution using sparkline histograms [15]. The distribution shows that Proximity Based Differential Evolution is both the most stable and the best performing of the algorithms in this problem. The algorithm seems to produce a good average result but is lucky in finding few extraordinarily good values during the test. The CMA-ES produces the second best values for this problem, but is less stable than the other algorithms, while the Frankenstein-PSO fails to produce a competitive result. This observed ordering of the algorithms is statistically significant (Mann-Whitney U-test,$p \le 0.005$, see [5]). The experiment was not repeated in order to achieve the significance.

A short analysis of the convergence speed and required iterations can be made in Figure 4, which shows that CMA-ES converges slightly faster than the other algorithms.

The trained regressor ensemble is tested on the validation data not present during the training and the results are summarized in the Table 3. The table shows the average prediction error on the validation set and the sparkline histogram of the estimation error.

Table 2. Fitness values achieved with 75 runs.

Algorithm	Average±Std.Dev	Best	Histogram
Prox-DE	2.546±0.248	1.401	
F-PSO	3.400±0.230	2.989	
CMA-ES	3.020±0.314	2.546	

1.401   3.568

Table 3. Estimation errors on the training data

Algorithm	Average±Std.Dev	Histogram
Prox-DE	5.482±4.600	
F-PSO	7.863±6.159	
CMA-ES	5.883±4.516	

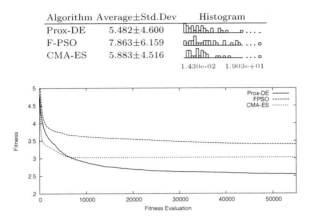

Fig. 4. Convergence of the algorithms

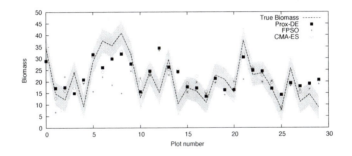

Fig. 5. Performance of the estimators with the validation dataset

Discarding the failure of the F-PSO based ensemble, the result values are within $500g$, which makes the effective difference between the training methods minor. Also, the convergence behaviour suggests that if more computational resources were at hand, it could be beneficial to allow Prox-DE to run for a longer time in hope of improving the solution. When contrasting the performance to the complexity of the algorithm, it is clear that the very simple Prox-DE is the most cost efficient choice for the task. In the practical side of things, the achieved accuracy is enough to control a tractor towed fertilizer dispensers, which have relatively few dispensation settings.

4   Conclusions

A biomass estimation system based on image analysis and machine learning has been proposed in order to support precision agriculture. This system collects multiple images taken by a UAV, reconstructs a 3-D model of the field, and extracts biomass information by means of a support vector regressor. The machine learning structure coordinates an ensemble of components and related parameters using the weights representing the activation probability of the corresponding components. The training has been performed by means of Prox-DE, F-PSO, and CMA-ES showed that the proposed machine learning system is realiably capable to detect the biomass present in the field by subsequent operations on the images. Amongst the three optimization algorithms considered in this study for performing the learning of the support vector regressor, Prox-DE appears to be the most reliable choice as the other two meta-heuristics seem to detect, on average, solutions characterized by a mediocre performance and then to be unable to improve upon them.

Possibilities of remote sensing applications for precision agriculture have been studied before. However, the presented estimation results in this paper are hard to compare since the premises and the target attributes in the studies differ.

Interesting areas for future studies would be to compare results with differently produced orthophotographs and Digital Elevation Model. Also, with a larger set of extracted features it would be interesting to shift the focus particularly to optimizing methods for dimension reduction.

References

1. Breiman, L.: Bagging predictors. Machine Learning 24(2), 123–140 (1996)
2. Breiman, L.: Random Forests. Machine Learning 45(1), 5–32 (2001)
3. Drucker, H., Burges, C.J.C., Kaufman, L., Smola, A., Vapnik, V.: Support vector regression machines. In: Advances in Neural Information Processing Systems, pp. 155–161 (1997)
4. Epitropakis, M.G., Tasoulis, D.K., Pavlidis, N.G., Plagianakos, V.P., Vrahatis, M.N.: Enhancing differential evolution utilizing proximity-based mutation operators. IEEE Transactions on Evolutionary Computation 15(1), 99–119 (2011)
5. Fay, M.P., Proschan, M.A., et al.: Wilcoxon-Mann-Whitney or t-test? On assumptions for hypothesis tests and multiple interpretations of decision rules. Statistics Surveys 4, 1–39 (2010)
6. Gimelfarb, G.L., Haralick, R.: Terrain reconstruction from multiple views. Computer Analysis of Images and Patterns (1997)
7. Hansen, N., Müller, S.D., Koumoutsakos, P.: Reducing the time complexity of the derandomized evolution strategy with covariance matrix adaptation (CMA-ES). Evolutionary Computation 11(1), 1–18 (2003)
8. Maclin, R., Opitz, D.: Popular ensemble methods: An empirical study. Journal of Artificial Intelligence Research 1(11), 169–198 (1999)
9. Montes De Oca, M., Stutzle, T., Birattari, M., Dorigo, M.: Frankenstein's PSO: A Composite Particle Swarm Optimization Algorithm. IEEE Transactions on Evolutionary Computation 13(5), 1120–1132 (2009)
10. Moran, M.S., Inoue, Y., Barnes, E.M.: Opportunities and limitations for image-based remote sensing in precision crop management. Remote Sensing of Environment 61(3), 319–346 (1997)
11. Schapire, R.E.: The strength of weak learnability. Machine Learning 5(2), 197–227 (1990)
12. Scharstein, D.: A taxonomy and evaluation of dense two-frame stereo correspondence algorithms. International Journal of Computer Vision 47(1), 131–140 (2002)

13. Serrano, L., Filella, I.: Remote sensing of biomass and yield of winter wheat under different nitrogen supplies. Crop Science 40(3), 723–731 (2000)
14. Thenkabail, P.S., Smith, R.B., De Pauw, E.: Hyperspectral vegetation indices and their relationships with agricultural crop characteristics. Remote Sensing of Environment 71(2), 158–182 (2000)
15. Tirronen, V., Weber, M.: Sparkline Histograms for Comparing Evolutionary Optimization Methods. In: Proceedings of 2nd International Joint Conference on Computational Intelligence. pp. 269–274 (2010)
16. Vapnik, V.N.: The Nature of Statistical Learning Theory, Statistics for Engineering and Information Science, vol. 8. Springer, Heidelberg (1995)
17. Zebedin, L., Klaus, A., Grubergeymayer, B., Karner, K.: Towards 3D map generation from digital aerial images. ISPRS Journal of Photogrammetry and Remote Sensing 60(6), 413–427 (2006)

A Generic Approach to Parameter Control

Giorgos Karafotias, S.K. Smit, and A.E. Eiben

Vrije Universiteit, Amsterdam, Netherlands
g.karafotias@vu.nl,{sksmit,gusz}@cs.vu.nl

Abstract. On-line control of EA parameters is an approach to parameter setting that offers the advantage of values changing during the run. In this paper, we investigate parameter control from a generic and parameter-independent perspective. We propose a generic control mechanism that is targeted to repetitive applications, can be applied to any numeric parameter and is tailored to specific types of problems through an off-line calibration process. We present proof-of-concept experiments using this mechanism to control the mutation step size of an Evolutionary Strategy (ES). Results show that our method is viable and performs very well, compared to the tuning approach and traditional control methods.

1 Introduction

When defining an evolutionary algorithm (EA) one needs to configure various settings: choose components (such as variation and selection mechanisms) and set numeric values (e.g. the probability of mutation or the tournament size). These configurations largely influence performance making them an important aspect of algorithm design.

The field of evolutionary computing (EC) traditionally distinguishes two approaches for setting parameter values[4]:

- *Parameter tuning*, where parameter values are fixed in the initialization stage and do not change while the EA is running.
- *Parameter control*, where parameter values are given an initial value when starting the EA and undergo changes while the EA is running.

The capability of parameter control to use adequate parameter values in different stages of the search is an advantage, because the run of an EA is an intrinsically dynamic process. It is intuitively clear –and for some EA parameters theoretically proven– that different values may be optimal at different stages of the evolution. This implies that the use of static parameters is inherently inferior to changing parameter values on-the-fly.

The conceptual distinction between tuning and control can be lifted if we consider the control mechanism as an integral part of the EA. In this case the EA and its parameter control mechanism (that may be absent) are considered as one entity. Furthermore, this

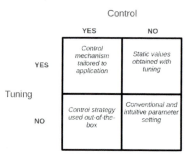

Fig. 1. Tuning vs control

C. Di Chio et al. (Eds.): EvoApplications 2012, LNCS 7248, pp. 366–375, 2012.

composed entity may or may not be tuned before being applied to a new problem. The resulting matrix of four options is shown in Figure 1.

The combinations in the top row have the advantage of enhanced performance at the cost of the tuning effort [11]. The options in the left column offer the benefits of time varying parameter values mentioned above with a tradeoff of increased complexity.

In this paper we introduce a generic control mechanism that is located in the top-left corner of the matrix, i.e. it combines on-line parameter adjustment (control) and off-line configuration (tuning). This means that the evolutionary algorithm incorporates a parameter control mechanism (seen as a black box for the moment) but this controller itself has certain parameters that can be configured for a problem through an off-line tuning process. Such a method has both the advantage of enhanced performance, and the possibility to be tuned to a specific problem (and is therefore most appropriate for repetitive problems). On the other hand, this mechanism also has the disadvantages of being complex, and the need for computational time dedicated to problem tailoring. Hence, the questions we want to address in this paper are:

- Is such an approach viable?
- What is the added value of this approach, when compared to traditional approaches such as static parameter-values, the 1/5th rule and self-adaptation?
- On what kind of feedback from the search process should such a parameter control mechanism base its decisions?

2   Related Work

Parameter control is an increasingly popular topic in the field of evolutionary algorithms[5]. The outline of the most commonly used methods is quite similar: one of the parameter values is altered based on some specific evidence. Most often these methods are designed for specific parameters. The most popular parameter-specific control methods focuses on mutation probability [6], mutation step size [12] and operator selection [17] but methods also exist for the selection pressure [18], the population-size[16], the fitness function [9] and the encoding [13].

Some generic control methods for numeric parameters also exist. In [19] an adaptive mechanism is proposed that works in alternating epochs, first evaluating parameter values in a limited set and then applying them probabilistically. In the end of every such pair of epochs the set of possible parameter values is updated according to some heuristic rule. In Lee and Takagi [8] an adaptive control mechanism based on fuzzy logic is proposed. Instantiation of the rule set of the controller is achieved through an off-line calibration process using a GA. Lee and Takagi concluded that such an approach was very beneficial, and led to a much better performance than using the fixed parameter values. However, the fixed values used in this study were the ones commonly used at that time, based on the early work of DeJong, rather than found using parameter tuning.

A two-layer approach to control is presented in [3] and [10]: the lower layer adaptively controls EA parameters driven by an upper level that enforces a user-defined schedule of diversity or exploration-exploitation balance (though these are not parameters per se). The algorithm in [10] includes a built-in learning phase that calibrates the controller to the EA and problem at hand by associating parameter values to diversity and mean fitness using random samples. In [3], the lower control level is an adaptive operator selection method that scores operators according to the diversity-fitness balance they achieve as compared to a balance dictated by the upper level user defined schedule. However, neither of the two make a comparison against static parameter-values found using parameter tuning.

Extensive literature reviews on parameter control can be found in [5] and [2].

3   Parameter Control Roadmap

In this section we present a simple framework for parameter control mechanisms. The purpose of this framework is not to provide any theoretical grounding or proofs but to serve as a roadmap that helps in designing and positioning one's mechanism.

We define a parameter control mechanism as a combination of three components:

1. A choice of parameters (i.e. *what* is to be controlled).
2. A set of observables that will be the input to the control mechanism (i.e. what *evidence* is used).
3. An algorithm/technique that will map observables to parameter values (i.e. *how* the control is performed).

These components are briefly described in the following paragraphs. However, they are based on the definition of the state of an evolutionary algorithm, which is therefore introduced first.

EA State. We define the state S_{EA} of an evolutionary algorithm as:

$$S_{EA} = \{G, \bar{p}, \mathcal{F}\} \tag{1}$$

where G is the set of all the genomes in the population, $\bar{p}$ is the vector of current parameter values, and $\mathcal{F}$ is the fitness function.

A triple S_{EA} uniquely specifies the state of the search process for a given evolutionary algorithm (the design and specific components of the EA need not be included in the state since they are the same during the whole run) in the sense that S_{EA} fully defines the search results so far and is the only observable factor that influences the search process from this point on (though not fully defining it, given the stochastic nature of EA operators). Time is not part of S_{EA} as it is irrelevant to the state itself; it introduces an artificial uniqueness and a property that is unrelated to the evolution. Of course, state transitions are not deterministic.

3.1   Parameters

The starting point when designing a control mechanism is the parameter to be controlled (as well as choices such as when and how often the parameter is updated). The importance of various parameters and the effect or merit of controlling each of them are subjects that will not be treated here (we refer to [2]). Instead, here we will only distinguish between *numeric* (e.g. population size, crossover probability) and *symbolic* (e.g. recombination operator) parameters.

3.2 Observables

The observables are the values that serve as inputs to the controller's algorithm. Each observable must originate from the current state S_{EA} of the EA since, as defined above, it is the only observable factor defining how the search will proceed from this point on.

However, the raw data in the state itself are unwieldy: if we were to control based on state S_{EA} directly, that would imply that the control algorithm should be able to map every possible S_{EA} to proper parameter values. Consequently, preprocessing is necessary to derive some useful abstraction, similar to the practise of dataset preprocessing in the field of data mining. We define such an observable derivation process as the following pipeline:

$$Source \rightarrow (Digest) \rightarrow (Derivative) \rightarrow (History)$$

Parentheses denote that steps can be bypassed.

i. *Source*: As stated above, the source of all observables is the current state of the EA, i.e. the set of all genomes, the current parameter values and the fitness function.

ii. *Digest*: A function $D(S_{EA}) = v$ that maps an EA state to a value, e.g. best fitness or population diversity.

iii. *Derivative*: Instead of using directly a value v we might be more interested in its speed or acceleration (e.g. to make the observable independent to the absolute values of v or to determine the effect of the previous update as the change observed in the most recent cycle).

iv. *History*: The last step in defining an observable is maintaining a history of size W of the value received from the previous step. This step includes a decision on the sliding window size W and the definition of a function $F_H(v_1, v_2, ..., v_W)$[1] that, given the last W values, provides a final value or vector (e.g. the minimum value, the maximum increase between two consecutive steps, the whole history as is etc.).

The above observable derivation is meant to be a conceptual framework and not an implementation methodology. For example, the current success ratio (in the context of Rechenberg's 1/5 rule) can in theory be derived from a state S_{EA} by applying the selection and variation operators to G and calculating the fitnesses of the results though obviously that would be a senseless implementation.

3.3 Algorithm

Any technique that maps a vector of observable values to a vector of parameter values can be used as an algorithm for the control mechanism, e.g. a rule set, an ANN or a function are all valid candidates. The choice of the proper algorithm seems to bear some resemblance to choosing an appropriate machine learning technique given a specific task or dataset. Whether EA observables display specific characteristics that make certain biases and representations more suitable is a question that needs to be investigated. In any case, it is obvious that given the type of parameter controlled (i.e. numeric or nominal) different techniques are applicable.

---

[1] Notice that indices have no relation to time but merely indicate a sequence of W elements.

Here we distinguish between two main categories of control techniques, regardless the algorithm and representation used, based on a fundamental characteristic of the controller: whether it is static or it adapts itself to the evolutionary process.

i. *Static*: A static controller remains fixed during the run, i.e. given the same observables input it will always produce the same parameter values output. In other words, the values produced only depend on the current observables input:

$$p = c(o) \quad and \quad o_1 = o_2 \Rightarrow c(o_1) = c(o_2)$$

where $o \in O, p \in P$ are the vectors of observables and parameter values respectively and $c : O \mapsto P$ is the mapping of the controller.

ii. *Dynamic*: A dynamic controller changes during the run, i.e. the same observables input can produce different parameter values output at different times. This implies that the controller is stateful and that the values produced depend on both the current observables input and the controller's current state:

$$p = c_p(o, S_C) \quad and \quad S_C^{t+1} = c_S(o_t, S_C^t)$$

where $o \in O, p \in P$ are the vectors of observables and parameter values respectively, $S_C \in S$ is the state of the controller and $c_p : O \times S \mapsto P, c_S : O \times S \mapsto S$ are the mappings of the controller.

According to this classification, a time-scheduled mechanism is a trivial case of a dynamic controller; it maintains a changing state (a simple counter) but is "blind" to the evolutionary process since it does not use any observables. It should be noted that we do not consider control mechanisms necessarily as separate and distinct components, e.g. we classify self-adaptation in ES as a dynamic controller since it implicitly maintains a state influenced by the evolutionary process.

4   Experimental Setup

The experiments presented here are designed as a proof of concept for the viability of a generic, EA-independent and parameter-independent control mechanism that is instantiated through an off-line tuning process and targeted to repetitive applications. That means that the present configuration belongs in the upper left square of Fig 1, i.e. it combines on-line control of the EA parameters and off-line tuning of the controller to a specific kind of problem.

The parameter we chose for our initial control experiments is the mutation step size σ in evolution strategies. The specific parameter may seem a trivial choice given the number and efficiency of existing control techniques but its simplicity and the existence of related theory and practical experience make σ a parameter suitable for analysis.

4.1   Evolutionary Algorithm

The EA used, is a $(10 + \lambda)$ ES with Gaussian mutation. It has no recombination and uses uniform random parent selection.

4.2 Parameter

As stated above, the controlled parameter is the mutation step size σ. In this experiment, σ will be updated in every generation from the start of the run.

4.3 Observables

The observables that act as input to the controller, are based on the current parameter values, diversity and fitness. The first input, the current σ, is input directly without going through the digest, derivative and history stages of the pipeline. The second input is the diversity, using the Population Diversity Index (PDI) [15] as the digest function and bypassing derivatives and history. Finally, we use two different fitness-based observables: (i) f_N uses a digest of the best fitness normalized in $[0, 1]$ and no history, (ii) Δf uses a best fitness digest (f_B) and a history with length W and the history function

$$F_H(f_B^1, ..., f_B^W) = \frac{f_B^W - f_B^{W/2}}{f_B^W - f_B^1}$$

We choose to use this formula to measure change instead of a derivative to make the controller robust to shifting and stretching of the fitness landscape.

We compare the two fitness-based observables on their own as well as paired with the diversity observable. The Δf observable is combined with the current σ observable following the intuition that if changes in fitness are observed then changes in the parameter value should be output, thus the old value must be available. This yields four sets of observables: $\{f_N\}$, $\{f_N, PDI\}$, $\{\Delta f, \sigma\}$ and $\{\Delta f, PDI, \sigma\}$.

4.4 Control Method

As a control method we chose a neural network (NN) as a generic method for mapping real valued inputs to real valued outputs. We use a simple feed-forward network without a hidden layer. The structure of the nodes is fixed and the weights remain static during an EA run and are set by the off-line tuning process. All inputs are, by definition, in the range $[0, 1]$. The weights are tuned $w \in [-1, 1]$. We tested three different activation functions (Table 1) with the output $o \in [0, 1]$. We also tested limiting the range of σ by multiplying the output with 0.1 (testing traditional practise of keeping $\sigma < 0.1$).

All six activations were combined with all four observable sets. These combinations were tested against six standard test problems[1]: Sphere, Corridor, Rosenbrock, Ackley, Rastrigin and Fletcher & Powell . This setup was repeated for two generation gaps: $\lambda = 70$ as a standard in numeric optimization and $\lambda = 1$ motivated by robotics applications [7]. The total number of control-problem instances is 288, see Table 1.

In all cases, the search for good controllers (NN weights) was performed using Bonesa [14]. Bonesa is an iterative model-based search procedure based on an intertwined searching and learning loop. The search loop is a generate-and-test procedure that iteratively generates new vectors, pre-assesses their quality using a surrogate model of the performance landscape, and finally tests the most promising vectors by executing an algorithm run with these specific values. In its turn, the learning loop uses the information obtained about the quality of the tested vectors to update the model of the performance surface. Furthermore, it uses a kernel filter to reduce the noise caused by the stochasticity of the algorithm.

For each combination of problem, observables set and activation, a good set of weights was found with an off-line tuning using Bonesa. After this tuning session, the EA using the best found controller instance was run 100 times to validate its performance. These performances are then compared to outcomes of 100 runs with:

- a static σ, that is also tuned using Bonesa, using the same computational budget as for finding the controller instance
- the Rechenberg's 1/5 rule applied to a global σ using the success ratio of all mutations in the population
- a self-adaptation approach using a single σ
- the theoretical optimum derived by Rechenberg [12] (applicable only to the Sphere and Corridor functions)

Table 1. Experimental Setup

λ	1, 70
Observables	$\{f_N\}$, $\{f_N, PDI\}$, $\{\Delta f, \sigma\}$, $\{\Delta f, \mathrm{PDI}, \sigma\}$
Activations	$a_1(x) = x$ $a_2(x) = 1/(1 + e^{-6x})$ $a_3(x) = \tanh 3x$
Problems	Sphere, Corridor, Ackley, Rosenbrock, Rastrigin, Fletcher& Powell
EA	$(10 + \lambda)$-ES with: Gaussian mutation, no recombination and uniform random parent selection, limit of 10000 evaluations
Instantiation	Bonesa with a budget of 3000 tests to calibrate weights $w_i \in [-1, 1]$

5 Results

The performance results of the calibrated control mechanisms and the benchmarks are presented in Table 2.

First, we consider the performance of control over the test problems by examining the results column-wise. We can observe that the calibrated control mechanisms are able to perform well (or comparably to the benchmarks) on all problems. Comparing our control to the tuned static approach, we find that on all problems except F&P, there are multiple control settings that are significantly better. For the F&P function there are control settings with no significant difference in average. Compared to self-adaptation, our control mechanism is overwhelmingly better on the Corridor, Ackley and Rastrigin problems while, for the other test functions, there are several settings that result in a tie. Control is always able to outperform the theoretical optimum (notice that this optimum was derived for a $(1 + 1)$ES).

Second, we consider control settings separately (by examining the results row-wise) to determine what observables and activation functions are better, and if there is a combination that consistently performs well. From this perspective, using a hypertangent

Table 2. Performances results. For each problem, we mark the control mechanisms that are significantly better than static (underlined), significantly better than self-adaptation (bold) and not significantly different than self-adaptation (italic). All problems are to be minimized.

	$\lambda = 1$						$\lambda = 70$					
	Sphere	Corridor	Rsnbk	Ackley	Rsgn	FletchP	Sphere	Corridor	Rsnbk	Ackley	Rsgn	FletchP
{f_N,PDI}												
lin	0.01504	3.174	7.906	1.639	35.35	9355	0.01525	3.175	7.898	1.004	36.36	9141
lin.1	0.0326	9.124	5.001	14.25	61.08	1.008e+04	0.04045	9.133	11.62	16.18	57.73	1.251e+04
sig	0.0963	1.694	6.626	2.081	34.73	9880	0.1009	1.696	7.219	2.104	35.28	1.09e+04
sig.1	0.0126	9.124	5.969	5.634	61.41	8969	0.04751	8.938	6.666	7.454	59.9	1.002e+04
tanh	0.03605	1.682	4.709	0.07069	35.79	8618	0.04408	1.684	7.129	0.3	35.09	1.013e+04
tanh.1	0.07144	9.124	5.56	4.755	57.09	8538	0.07844	8.938	6.698	7.758	57.16	1.054e+04
{f_N}												
lin	9.704e-07	1.514	57.22	9.114e-07	46.62	8481	9.003e-06	1.638	181.8	8.978e-07	49.22	1.589e+04
lin.1	0.01152	9.22	371.5	7.964	60.97	2.704e+04	1.438	9.135	869.5	11.7	59.86	7.54e+04
sig	0.538	1.593	11.18	4.908	36.21	7669	0.472	1.701	11.45	13.52	36.25	6675
sig.1	0.05019	9.124	5.27	9.725	60.87	1.016e+04	0.05714	8.938	7.228	7.091	58.87	1.087e+04
tanh	9.223e-07	1.512	18.75	9.156e-07	20.93	7714	8.889e-07	1.602	74.1	8.789e-07	23.64	1.026e+04
tanh.1	9.903e-07	9.13	150.9	2.573	57.89	1.572e+04	0.09544	8.864	469.4	4.896	59.55	3.421e+04
{Δf, PDI, σ}												
lin	0.003837	6.519	6.283	2.608	35.2	7992	0.02674	6.519	8.136	3.467	46.99	8631
lin.1	0.02484	9.124	5.382	18.48	61.07	9401	0.04189	9.217	19.49	18.43	58.27	1.276e+04
sig	0.1092	1.787	7.007	7.269	35.46	9850	0.1147	1.792	7.022	5.597	36.75	8892
sig.1	0.01506	9.124	6.031	8.065	61.02	1.104e+04	0.03794	9.218	6.706	10.7	58.4	9969
tanh	0.007184	2.147	6.834	0.4822	34.35	6159	0.003973	2.15	7.499	0.8447	38.72	8217
tanh.1	0.009213	9.124	5.645	9.314	58.3	8622	0.06944	8.938	6.86	14.35	60.08	1.008e+04
{Δf, σ}												
lin	0.0027	6.24	5.379	1.4	49.95	8757	0.002984	6.24	8.11	7.87	36	7489
lin.1	0.01787	9.129	5.369	18.51	60.48	1.081e+04	0.04558	9.22	12.19	18.45	58.82	1.182e+04
sig	0.06929	1.79	7.485	6.75	32.44	9914	0.06655	1.792	7.895	8.002	37.46	8917
sig.1	0.01255	8.938	5.966	6.929	61.2	9806	0.04192	9.217	6.651	11.44	59.57	1.032e+04
tanh	0.002631	1.963	5.715	0.762	32.2	6467	0.0008254	1.964	6.282	0.9197	36.67	6759
tanh.1	0.002881	8.938	5.726	8.473	57.99	8137	0.05975	8.938	6.8	11.41	59.08	1.006e+04
static	0.01721	4.659	6.721	9.383	39.31	1.073e+04	0.06903	3.378	7.254	9.234	37.07	9313
15rule	6.414	9.221	1212	18.44	85.84	1.85e+05	0.5263	9.124	34.34	18.28	56.29	1.828e+04
sssa	9.256e-07	8.938	7.203	15.04	57.65	8740	8.887e-07	8.938	7.458	15.65	55.34	6781
rechopt	0.002939	4.659	21.49	7.887	39.2	8712	1.027	4.659	21.73	7.863	39.2	8631

activation is always the best option: performance is always significantly better than static for most problems, while it is significantly better than self-adaptation in three out of six problems and at least as good in four. The best choice of observables is either $\{f_N\}$ or $\{\Delta f, \sigma\}$.

Though choosing between f_N or Δf is mostly a matter of feasibility (calculating normalized fitness is not possible if the bounds of the fitness values are not known), using diversity as an observable or not, is a more fundamental question. Contrary to our initial expectations, adding diversity to the input does not always offer an advantage even for multimodal problems (including the irregular and asymmetric Fletcher&Powell). Keeping all other factors fixed, using diversity as an input produces a significant improvement in 23.6% of all cases and only in 8.3% of the cases that are using the hypertangent activation. This advantage of using diversity combined with the hypertangent activation is only observed for the Rosenbrock problem that is asymmetric but still unimodal. Our assumption is that the ineffectiveness of observing diversity is due to the survivor selection mechanism. Diversity could be useful as feedback if σ control aimed at maintaining a diverse population. However, $(\mu + \lambda)$ selection negates such an effort since any new "explorative" individuals will have inferior fitness and will be immediately discarded. Thus, a strong selective pressure could be the reason observing diversity is ineffective.

6 Conclusions and Future Work

Based on our results, the questions stated in the introduction can be answered. The main conclusion that can be drawn is that the generic approach taken in this paper is viable and fruitful. In contrast to previous work, we were able to find a problem-tailored control mechanism that outperforms the tuned (but static) parameter values for each of the problems. More specific, the combination of $\{f_N\}$, or $\{\Delta f, \sigma\}$, as observables and a hypertangent function of the neural network, outperforms the tuned parameter values in most problems. Even more promising is the fact that this combination performed equally well or better than traditional methods for controlling a single σ, such as the 1/5th rule and self-adaptation. However, in contrast to these methods, our generic approach has the possibility to be extended to other parameters, possibly leading to increased performance. Although this comes with the added cost of specific problem-tailoring, the benefits, especially in case of repetitive problems, can be significant.

Inevitably, this conclusion depends on the experimenters design decisions. In our case, the most important factors (beyond the underlying algorithm itself) are:

- The observables chosen as inputs to the control strategy.
- The parameters to be controlled
- The technique that maps a vector of observable values to a vector of parameter values

Changing either of these can, in principle, lead to a different conclusion. With respect to the observables chosen, we can conclude that these indeed highly influence the results. Remarkably, adding diversity as an input appears to have hardly any added value for controlling σ, possibly due to the selection mechanism used. The normalized fitness and $\{\Delta f, \sigma\}$ appear to be the best choices for input. Note that calculating the normalized fitness is not always possible, and is most probably the less robust choice.

Regarding the future, we expect more studies along these ideas, exploring the other possible implementations of the most important factors. Most enticing is the possibility of applying it to the other parameters of the evolutionary algorithm. This has the prospect of delivering high quality control methods that can adapt the underlying algorithm to the different stages of the search process.

References

1. Bäck, T.: Evolutionary Algorithms in Theory and Practice. Oxford University Press, Oxford (1996)
2. De Jong, K.: Parameter Setting in EAs: a 30 Year Perspective. In: Lobo, F., Lima, C., Michalewicz, Z. (eds.) Parameter Setting in Evolutionary Algorithms. SCI, vol. 54, pp. 1–18. Springer, Heidelberg (2007)
3. di Tollo, G., Lardeux, F., Maturana, J., Saubion, F.: From Adaptive to More Dynamic Control in Evolutionary Algorithms. In: Hao, J.-K. (ed.) EvoCOP 2011. LNCS, vol. 6622, pp. 130–141. Springer, Heidelberg (2011)

4. Eiben, A.E., Hinterding, R., Michalewicz, Z.: Parameter Control in Evolutionary Algorithms. IEEE Transactions on Evolutionary Computation 3(2), 124–141 (1999)
5. Eiben, A.E., Michalewicz, Z., Schoenauer, M., Smith, J.E.: Parameter Control in Evolutionary Algorithms. In: Lobo, F., Lima, C., Michalewicz, Z. (eds.) Parameter Setting in Evolutionary Algorithms. SCI, vol. 54, pp. 19–46. Springer, Heidelberg (2007)
6. Fogarty, T.C.: Varying the probability of mutation in the genetic algorithm. In: Proceedings of the Third International Conference on Genetic Algorithms, pp. 104–109. Morgan Kaufmann Publishers Inc., San Francisco (1989)
7. Karafotias, G., Haasdijk, E., Eiben, A.E.: An algorithm for distributed on-line, on-board evolutionary robotics. In: Proceedings of the 13th Annual Conference on Genetic and Evolutionary Computation, GECCO 2011, pp. 171–178. ACM (2011)
8. Lee, M.A., Takagi, H.: Dynamic control of genetic algorithms using fuzzy logic techniques. In: Proceedings of the Fifth International Conference on Genetic Algorithms, pp. 76–83. Morgan Kaufmann (1993)
9. Majig, M., Fukushima, M.: Adaptive fitness function for evolutionary algorithm and its applications. In: International Conference on Informatics Research for Development of Knowledge Society Infrastructure, pp. 119–124 (2008)
10. Maturana, J., Saubion, F.: On the Design of Adaptive Control Strategies for Evolutionary Algorithms. In: Monmarché, N., Talbi, E.-G., Collet, P., Schoenauer, M., Lutton, E. (eds.) EA 2007. LNCS, vol. 4926, pp. 303–315. Springer, Heidelberg (2008)
11. Nannen, V., Smit, S., Eiben, A.E.: Costs and Benefits of Tuning Parameters of Evolutionary Algorithms. In: Rudolph, G., Jansen, T., Lucas, S., Poloni, C., Beume, N. (eds.) PPSN 2008. LNCS, vol. 5199, pp. 528–538. Springer, Heidelberg (2008)
12. Rechenberg, I.: Evolutionstrategie: Optimierung Technisher Systeme nach Prinzipien des Biologischen Evolution. Fromman-Hozlboog Verlag, Stuttgart (1973)
13. Schraudolph, N.N., Belew, R.K.: Dynamic parameter encoding for genetic algorithms. Machine Learning 9, 9–21 (1992)
14. Smit, S., Eiben, A.E.: Multi-problem parameter tuning using bonesa. In: Hao, J., Legrand, P., Collet, P., Monmarché, N., Lutton, E., Schoenauer, M. (eds.) Artificial Evolution, pp. 222–233 (2011)
15. Smit, S.K., Szláavik, Z., Eiben, A.E.: Population diversity index: a new measure for population diversity. In: GECCO (Companion), pp. 269–270 (2011)
16. Smith, R., Smuda, E.: Adaptively resizing populations: Algorithm, analysis and first results. Complex Systems 9(1), 47–72 (1995)
17. Spears, W.M.: Adapting crossover in evolutionary algorithms. In: Proceedings of the Fourth Annual Conference on Evolutionary Programming, pp. 367–384. MIT Press (1995)
18. Vajda, P., Eiben, A.E., Hordijk, W.: Parameter Control Methods for Selection Operators in Genetic Algorithms. In: Rudolph, G., Jansen, T., Lucas, S., Poloni, C., Beume, N. (eds.) PPSN X 2008. LNCS, vol. 5199, pp. 620–630. Springer, Heidelberg (2008)
19. Wong, Y.-Y., Lee, K.-H., Leung, K.-S., Ho, C.-W.: A novel approach in parameter adaptation and diversity maintenance for genetic algorithms. Soft Computing - A Fusion of Foundations, Methodologies and Applications 7, 506–515 (2003)

Applying (Hybrid) Metaheuristics to Fuel Consumption Optimization of Hybrid Electric Vehicles

Thorsten Krenek[1], Mario Ruthmair[2], Günther R. Raidl[2], and Michael Planer[1]

[1] Institute for Powertrains and Automotive Technology,
Vienna University of Technology, Vienna, Austria
{thorsten.krenek,michael.planer}@ifa.tuwien.ac.at
[2] Institute of Computer Graphics and Algorithms,
Vienna University of Technology, Vienna, Austria
{ruthmair,raidl}@ads.tuwien.ac.at

Abstract. This work deals with the application of metaheuristics to the fuel consumption minimization problem of hybrid electric vehicles (HEV) considering exactly specified driving cycles. A genetic algorithm, a downhill-simplex method and an algorithm based on swarm intelligence are used to find appropriate parameter values aiming at fuel consumption minimization. Finally, the individual metaheuristics are combined to a hybrid optimization algorithm taking into account the strengths and weaknesses of the single procedures. Due to the required time-consuming simulations it is crucial to keep the number of candidate solutions to be evaluated low. This is partly achieved by starting the heuristic search with already meaningful solutions identified by a Monte-Carlo procedure. Experimental results indicate that the implemented hybrid algorithm achieves better results than previously existing optimization methods on a simplified HEV model.

Keywords: hybrid metaheuristic, genetic algorithm, downhill-simplex, particle-swarm-optimization, hybrid electric vehicles, driving cycles.

1 Introduction

Due to the requirement of lower fuel consumption and emissions it is necessary that the automotive industry comes up with new approaches. One of these are hybrid electric vehicles (HEV) which have a much higher flexibility concerning operation strategies and components compared to conventional vehicles utilizing only a combustion engine. The propulsion system of HEVs consists of a conventional combustion engine and electric machines. With the assistance of electric machines it is possible to achieve higher efficiency, in particular by providing energy recuperation in deceleration phases.

Nowadays engines and vehicles can be numerically simulated with high accuracy, which makes it easier to analyze different operation strategies and the consequences of their modification. Our aim is to minimize the fuel consumption in

C. Di Chio et al. (Eds.): EvoApplications 2012, LNCS 7248, pp. 376–385, 2012.

exactly specified driving cycles of such HEV computer models. The vehicle is simulated by the software GT-SUITE[1] using physics-based one-dimensional modeling thus being able to calculate the fuel consumption and the battery state of charge (SOC) for a specific driving cycle. Depending on the duration of the driving cycle, this can take several minutes on current hardware. In general, the fuel consumption is influenced by a large number of adjustable parameters from which we preselected a meaningful subset for optimization: velocities at which the vehicle switches from parallel to series hybrid mode and vice versa, the SOC operating limits and the gear shifting strategy. In parallel mode the internal combustion engine (ICE) and/or the electric machines are used for propulsion while in series mode only electric propulsion is provided utilizing the ICE to power the electric generator. A detailed parameter description is given in Section 5. All n parameters $p = (p_1, \ldots, p_n)$ of the HEV model are real-valued and have individual lower and upper bounds $[p_i^{\min}, p_i^{\max}]$, $\forall i = 1 \ldots n$. The battery SOC is required to be nearly identical at the beginning and the end of a driving cycle in order to guarantee a fair comparison to other vehicles. So we considered the quadratic deviation between the SOC at the beginning (SOC_{begin}) and at the end (SOC_{end}) of the driving cycle. The objective function to be minimized is:

$$f(p) = w_{\text{cons}} \cdot cons(p) + w_{\text{sdev}} \cdot (SOC_{\text{begin}}(p) - SOC_{\text{end}}(p))^2$$

The fuel consumption is denoted by $cons(p)$ and constants $w_{\text{cons}} \geq 0$ and $w_{\text{sdev}} \geq 0$ are used for weighting the individual terms appropriately. A solution p^* is optimal if $f(p^*)$ is minimal, so $f(p^*) \leq f(p)$, $\forall p$. A direct determination of proven optimal parameter settings is practically impossible due to the high complexity of f, even obtaining the objective for one set of parameters by simulation is quite time-consuming. So the goal was to find a heuristic optimization strategy making it possible to reliably find a solution that is close to optimal only requiring a limited number of simulations. Beginning with standard optimization techniques diverse in most cases more efficient algorithms than Design Of Experiments (DOE) [11], which is included in GT-SUITE, have been developed by considering special properties of the problem. A genetic algorithm (GA) [9], a downhill-simplex method [12], and an algorithm based on swarm intelligence (PSO) [5] provided, after some specific tailoring, in preliminary experiments the best results. Major features are: Starting solutions are not initialized randomly but by a Monte Carlo search procedure to reduce the number of required iterations. In the GA's recombination operator the choice which value is passed on depends on the deviation of the parameter values from the two parent solutions to the best solutions in the population. The simplex reduction in the downhill simplex method is not applied here because it re-calculates all points of the new simplex and this mostly ends up in worse objective function values due to possibly unbalanced SOCs. The best solution from the PSO algorithm is additionally improved by a surface-fitting algorithm. Finally, the individual metaheuristics are combined to a hybrid optimization approach taking into account the strengths and weaknesses of the single procedures.

---

[1] GT-SUITE is a software by Gamma Technologies, Inc., http://www.gtisoft.com

For a model of an existing HEV with complex operation strategy a fuel saving of about 33% compared to a related conventionally powered vehicle could be achieved. The part our hybrid optimization algorithm contributes is about five percent in comparison to setting the parameters by the methods implemented in GT-SUITE. These standard optimization methods in particular have problems with the high number of parameters. Furthermore, we are able to show that our proposed algorithm achieves better results on another simplified HEV benchmark model too, see Section 5.

The following Section discusses related work, Section 3 presents the individual metaheuristics which are then combined in Section 4 to a hybrid algorithm, Section 5 shows experimental results, and Section 6 concludes the article.

2   Related Work

In GT-SUITE a *Design of Experiments* optimization method is implemented. Here the search space is typically approximated by a quadratic or cubic polynomial function based on a large number of simulated parameter sets distributed in the search space. The minimum of this function is then derived analytically. In [7] and [14] several optimization algorithms are applied to HEV models and the authors state that the considered search space is highly non-linear with non-continuous areas. Similarly to our problem, the goal is to minimize the fuel consumption for a given driving cycle. As additional constraint they consider a minimum requirement on vehicle dynamics. As simulation software ADVISOR[2] is used and the applied optimization algorithms are taken from iSIGHT[3], VisualDOC[4] and MATLAB[5]. As optimization procedures *fmincon* from MAT-LAB, VisualDOC's DGO and RSA, as well as the search strategies Sequential Quadratic Programming (SQP) [13], DIviding RECTangle (DIRECT) [1] and a GA are applied. Unfortunately, there is no information given about the implementation and configuration of the used algorithms, in particular concerning the GA. The best result is achieved by the DIRECT method, the gradient strategies can only find rather poor local optima.

In [2] and [3] among others the simulation software PSAT[6] and its DIRECT optimization algorithms, a GA, Simulated Annealing (SA) and PSO are applied to a HEV model whereas SA and DIRECT are the most successful approaches. The objective is the same as in [7] and [14].

Furthermore, in [3] a hybrid algorithm combining SQP with DIRECT is presented but only applied on a simpler test function. However, in few iterations the global optimum is found in most cases. In [4] and [10] a multi-objective GA

---

[2] ADVISOR (Advanced Vehicle Simulator) is a software from AVL, http://www.avl.com

[3] iSIGHT is a software from Simulia, http://www.simulia.com

[4] VisualDOC is a software from VR & D, http://www.vrand.com

[5] MATLAB is a software from MathWorks http://www.mathworks.de

[6] PSAT (Powertrain System Analysis Toolkit) was developed by Argonne National Laboratory, http://www.transportation.anl.gov/modeling_simulation/PSAT

is successfully applied to a HEV model, considering fuel consumption and emissions minimization. Comparisons with other methods are not presented. In [15] and [16] a PSO algorithm was proposed for a HEV model for improving a given operation strategy. ADVISOR is used as simulation software. The SOC deviation on the defined driving cycles is integrated in the objective function. Given the characteristics of the vehicle the operation strategy is optimized resulting in an improvement compared to the strategy before. How the original strategy has already been optimized before is not stated. Compared to GT-SUITE parts of the objective function can be calculated much faster in ADVISOR and PSAT by directly solving mathematical functions. As a consequence, such models can be simulated significantly faster and gradient strategies can be applied. However, the benefit of using GT-SUITE is the much higher accuracy of the HEV model. In the mentioned related work not only the operation strategy but also other criteria, e.g. the battery capacity and the number of battery cells, are optimized. The requirement of a balanced SOC is either considered as a side constraint or by adding the difference to a balanced SOC as a penalty term to the objective function. In the first case a large number of infeasible solutions are possibly calculated, mainly by methods like DOE [11].

Most previous works use standard optimization methods from existing libraries without problem-specific adaptations, and different articles report different optimization methods to work best. Unfortunately, a direct comparison of these approaches is hardly possible since only few algorithmic details are available. Thus, it is difficult to draw general conclusions about appropriate methods for the optimization of HEVs. GT-SUITE provides a DOE-based optimization method too, however, in our studies we recognized that DOE can only handle up to five parameters in reasonable time for our HEV models.

3   Metaheuristics

We now describe the new metaheuristic approaches we developed. For more details, in particular also deeper studies of the individual algorithms' performances and influences of strategy parameters, we refer to the first author's master thesis [6], on which this article is based.

Monte-Carlo Search Method. The Monte-Carlo method [8] is primarily used to generate manifold initial solutions for the other algorithms. The initial range of values for each parameter is set to the entire range of possible values. Consequently, in a first step only random solutions are generated. After each iteration the parameter range is reduced by a factor and moved towards the best known solution. Due to the fact that the algorithm mainly generates initial solutions subject to further improvement we choose a factor between 0.8 and 0.9 and keep the number of computed solutions constant.

Downhill-Simplex Method. This method [12], also known as Nelder-Mead method, is based on a v-simplex, which is a polytope of dimension v defined by $v + 1$ points spanning the convex hull. Each point corresponds to a particular set of parameters together with its objective function value. By comparing the different function values the tendency of the values and gradient directions are approximated. In each iteration, the point with the worst value is replaced by a newly derived one. In our implementation we omit the otherwise usual shrinking of the whole simplex because it would be very time-consuming to re-calculate the objective values of all points of the simplex. Furthermore, these new points are likely to have an unbalanced SOC.

Genetic Algorithm (GA). In our GA [9] each individual is directly represented by a vector of real parameter values. The selection of solutions from the population for pairwise recombination occurs uniformly at random. To recombine two solutions p^1, p^2, for each parameter $i = 1 \ldots n$, either p_i^1 or p_i^2 is adopted. The choice which value is passed on considers the average deviation to the d best solutions q^j, $j = 1 \ldots d$, in the population:

$$dev_i(p^k) = \frac{1}{d} \sum_{j=1}^{d} |q_i^j - p_i^k| \qquad \forall k \in \{1,2\}, \ \forall i = 1 \ldots n$$

The probability of adopting the i-th parameter from parent p^k is then defined as $P_i^{\text{comb}}(p^k) = 1 - dev_i(p^k)/(dev_i(p^1) + dev_i(p^2))$. Furthermore, each parameter is mutated with a small probability P^{mut} by assigning it a new random value within its bounds. Once an offspring solution p' has been generated and its objective value $f(p')$ determined via simulation, a solution r is randomly selected from the population and replaced with probability $P^{\text{rep}} = (f(r) - c)/(f(p') + f(r) - 2c)$. The correction value $c \in [0, \min\{f(p'), f(r)\})$ is used to control the influence of the objective values: the higher c the higher the probability of a new solution with better objective value being chosen as new member of the population.

Particle-Swarm-Optimization (PSO). This optimization method was originally derived from the behavior of birds and shoals of fish [5]. Each solution p^j, $j = 1 \ldots m$, corresponds to an individual of a swarm of size m moving within the search space. The motion depends both on the best known solution of the individual and the best solution of the entire swarm. First, m solutions are randomly selected from the solution set of the Monte-Carlo search procedure to form the initial population. For each individual j the so-far best "local" solution $p^{L,j}$ encountered on its path is stored. Moreover, p^G denotes the overall best known solution. In each iteration the parameter set of each individual is modified depending on both the local and global best solutions. For each individual s^j a velocity vector $v^j \in [-1,1]^n$ is defined and updated as follows:

$$v_i^j \leftarrow v_i^j + \alpha^{\text{L}} \cdot \frac{p_i^{\text{L},j} - p_i^j}{p_i^{\max} - p_i^{\min}} + \alpha^G \cdot \frac{p_i^G - p_i^j}{p_i^{\max} - p_i^{\min}} + rand \qquad \begin{array}{l} \forall j = 1 \ldots m, \\ \forall i = 1 \ldots n. \end{array}$$

Constants $\alpha^L, \alpha^G \geq 0$, with $\alpha^L + \alpha^G = 1$, control the influence of the local and global best solutions, respectively, and $rand$ is a random value uniformly distributed in $[-0.1, 0.1]$. The positions (solutions) of the individuals are then updated by $p_i^j \leftarrow p_i^j + v_i^j \cdot (p_i^{\max} - p_i^{\min})/\delta$, where $\delta \geq 1$ controls the step size. If a parameter steps out of its corresponding range, it is set to the corresponding limit. The algorithm terminates after a specified number of iterations.

Surface-Fitting. We use surface-fitting to improve the best solution obtained by the PSO algorithm in our hybrid metaheuristic approach, see Section 4. In each iteration $e \geq 6$ solutions are derived from the so far best solution by varying two randomly selected parameters p_1, p_2 slightly. The range of the variation is limited by the following factors: factor $area$ is initialized with 1 and increases by 1 after every fourth solution. The factors (fit_1, fit_2) are continuously assigned the values $(-1, -1)$, $(1, -1)$, $(-1, 1)$ and $(1, 1)$. The constant rad denotes the step size relative to the range of feasible parameter values. For the chosen parameters $i = 1 \dots 2$ the parameter values are calculated by $p_i = p_i + area \cdot fit_i \cdot rad \cdot (p_i^{\max} - p_i^{\min})$. The new solutions are evaluated and the objective function is approximated by function $c_1 + c_2 \cdot p_1 + c_3 \cdot p_2 + c_4 \cdot p_1^2 + c_5 \cdot p_2^2 + c_6 \cdot p_1 \cdot p_2$. Coefficients $c_1 \dots c_6$ and the minimum of the approximation function are calculated using the GNU Scientific Library and finally evaluated by simulation.

4   Hybrid Meta-heuristic (PSAGADO)

Each presented method has its own strenghts and weaknesses. On average the GA was able to achieve the best results since by mutation it was possible to escape from unpromising areas of the search space. However, rather good solutions often could not be further improved. The results of the PSO and the downhill simplex method are highly dependent on the chosen initial solutions. If only the PSO is applied, the solutions have to be broader distributed in the search space and should have nearly a balanced SOC. Our hybrid approach (*Particle-Swarm And Genetic Algorithm with Downhill-simplex Optimization*, PSAGADO) combines the previously presented algorithms trying to exploit their strengths. Initial solutions are determined by the Monte-Carlo search method and stored in a solution pool. As not much is known about the search space the PSO is well suited to be the central algorithm, since it is a robust method considering solutions with high diversity. After a certain number of iterations the best solution of the PSO is improved by the surface-fitting procedure if possible. Surface-fitting is only applied to the best solution because of runtime considerations. The GA is applied next using the final swarm of the PSO as initial population. If most of the individuals are similar, the GA still can lead to new best solutions by increasing diversity by mutation. If the solutions are well distributed in the search space recombination is frequently able to combine two good parameter sets to a better one. After recombination two solutions are randomly chosen from the population. If the new solution is better than both selected, one solution is replaced by the new solution and the other one by a random solution from the initial solution pool to restrict similar solutions in the pool. Otherwise only the chosen solution with the lower objective value will be replaced by the new solution. If the GA is able to find a new best solution, half of the solutions closest

---

Algorithm 1. PSAGADO

---

1 execute Monte-Carlo search and store all solutions as initial pool
2 **while** *termination criterion not met* **do**
3 | execute PSO
4 | apply surface-fitting on the best solution of PSO
5 | execute GA
6 | **if** *new best solution found* **then** replace half of the solutions closest to best
7 | **else**
8 | | execute downhill-simplex
9 | | **if** *no new best solution found* **then** replace all solutions

---

Table 1. Algorithm settings

Monte-Carlo	$resize = 0.89$
SIMPLEX	$v = 15$
SURFACE-FITTING	$e = 12$, $rad = 0.02$
PSO	$\alpha^L = 0.3$, $\alpha^G = 0.7$, $m = 30$, $\delta = 10$
GA	$c = \min\{f(p'), f(r)\} - 2$, $d = 10$, $P^{\mathrm{mut}} = 10\%$

to the best solution are replaced by random solutions from the pool to increase diversity and prevent too much focus on the best solution. The distance $D(p)$ of parameter set p to the best solution p^{best} is calculated by

$$D(p) = \sum_{i=1}^{n} \left(\frac{|p_i - p_i^{\mathrm{best}}|}{p_i^{\mathrm{max}} - p_i^{\mathrm{min}}} \right)^2 .$$

If the GA is not able to achieve any improvement, the simplex method is applied. This usually occurs when most of the PSO solutions are very similar. Although this could mean that most solutions are near the global optimum bad solutions may still exist possibly resulting in a shift of the simplex and leading to a new best solution. If the simplex method leads to an improvement, the process continues with the PSO. However, if most solutions are quite similar and the PSO and GA cannot achieve new best solutions then the simplex method usually results in no improvement, too. In this case a restart is performed by replacing all solutions but the so-far best with solutions from the initial pool and continuing with the PSO. Algorithm 1 shows the implementation of PSAGADO.

5   Experimental Results

We applied PSAGADO to a complex real-world and a simplified benchmark HEV model. Unfortunately we are not allowed to publish details for the real-world model due to a non-disclosure agreement with the manufacturer. Overall, a fuel saving of about 33% compared to a related conventionally powered vehicle could be achieved, and the remarkable part PSAGADO contributes is about five percent in comparison to the parameter setting found by DOE integrated in GT-SUITE. As simplified benchmark HEV model we used the "parallel-series" example supplied by GT-SUITE and compare PSAGADO to the integrated DOE and the individual metaheuristics. To further reduce simulation times a shorter

Table 2. Final objective values of PSAGADO, DOE, GA, SIMPLEX and DOE

	Runs	Sol. p. Run	Worst	Best	Average	Std.Dev.
PSAGADO	10	3600	207.52	206.69	206.92	0.23
PSO	10	3600	229.93	207.22	212.43	12.85
GA	10	3600	208.64	206.98	207.23	0.25
SIMPLEX	10	3600	230.57	207.94	215.93	14.10
DOE	10	3600	210.87	210.19	210.40	0.23

driving-cycle is used here altogether leading to an evaluation time for one parameter set of about 30 seconds. Thus, the runtime of the optimization algorithms can be neglected compared to the simulation times. Important algorithm specific settings are shown in Table 1. The Monte-Carlo search method calculates 35 solutions at each of 15 total iterations. The population size for the PSO and GA is 25. In each optimization cycle the PSO is iterated ten times, the surface-fitting method is applied five times and in the GA 60 new solutions are derived. In case of no improvement, the simplex will be updated 15 times. The constants in the objective function are set to $w_{cons} = 3.6$ and $w_{sdev} = 9$. All parameter values have been determined in preliminary tests to fit the limited number of simulations. The fuel consumption $cons$ is measured in mg, the SOC in percent. The parameters to be optimized are the gearshift strategy defined by $gear1up$ to $gear4up$, the charging limits of the battery SOC_{min}, SOC_{max} and hybrid mode thresholds $hev1, hev2$ specifying the velocities switching from parallel to series mode and vice versa. DOE uses the latin-hypercube method to select the parameter sets and approximates the mathematical model by a cubic replacement function. Results obtained from 10 runs with 3600 evaluated solutions per run for each considered algorithm are summarized in Table 2.

In the optimization progress we observed several local optima from which one cannot escape by changing only one parameter. If the Monte-Carlo method leads to a poor local optimum it may take some time until PSAGADO gets out of it mainly because of the low diversity of the initial solution pool. To prevent this the range reduction factor could be increased or the number of iterations in the Monte-Carlo search procedure could be reduced. Another possibility would be to entirely skip the Monte-Carlo method and use only random solutions. However, since the number of simulations is strictly limited we decided to initially restrict the search space even if there is a risk of getting stuck in a local optimum. DOE often fails because of an inaccurate model approximation in the relevant areas containing good solutions which can be explained by the rather naive uniform sampling strategy. Table 3 shows the best solutions obtained by the individual algorithms; notable are the remarkably strong differences in the parameter values. Among PSO, SIMPLEX and the GA, the GA performed best, using mutation to escape from unfavorable areas of the search space. The results of the PSO strongly depend on the diversity and the SOC balance of the initial solutions. In the downhill-simplex method it is necessary to start with solutions with almost balanced SOC otherwise it is difficult to find good solutions. Characteristic optimization progresses of all methods are shown in Fig. 1, where worst-case scenarios of downhill-simplex method and PSO are shown together in one curve.

Table 3. Obtained best parameter sets of PSAGADO, PSO, SIMPLEX, GA and DOE

Parameter	Boundaries	PSAGADO	PSO	SIMPLEX	GA	DOE
$hev1$ [km/h]	65–100	65.00	65.00	65.04	65.02	100.00
$hev2$ [km/h]	10–60	60.00	60.00	59.95	59.82	60.00
SOC_{max}	0.7–0.9	0.79	0.70	0.78	0.73	0.90
SOC_{min}	0.1–0.7	0.50	0.57	0.47	0.55	0.10
$gear1up$ [km/h]	12–30	29.93	12.93	27.08	29.62	25.87
$gear2up$ [km/h]	32–50	47.84	42.01	38.45	47.19	44.69
$gear3up$ [km/h]	52–70	57.53	57.87	59.11	57.52	53.72
$gear4up$ [km/h]	72–100	72.00	72.10	87.63	72.00	76.13

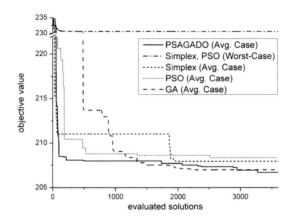

Fig. 1. Characteristic optimization progresses

6   Conclusions and Future Work

We considered the problem of optimizing diverse control strategy parameters of hybrid vehicles in order to minimize fuel consumption over a given driving-cycle. This problem is characterized by the relatively large number of real-valued parameters, the multi-modality and discontinuity of the search space, and in particular the expensive simulations required for evaluating solutions. Consequently, we investigated diverse heuristic strategies including Monte Carlo and Downhill-Simplex approaches, a specifically adapted GA, and a PSO. Considering the individual properties of these methods, we finally combined them in the hybrid PSAGADO. Results on a complex real-world scenario were remarkable, with PSAGADO's solution leading to a reduction of the fuel consumption of about five percent in comparison to a standard optimization strategy provided by the GT-SUITE simulator. As we are not allowed to give more details here on these results, a simplified benchmark model was further used for comparison, also indicating the superiority of PSAGADO over the individual metaheuristics and GT-SUITE's DOE.

In future work more testing is necessary and the search space should be studied in more detail in order to possibly exploit certain features in the optimization in better ways. A promising idea seems to be to approximate the objective function with a neural network which is refined at the same time as the optimization is performed.

References

1. Finkel, D.: Direct optimization algorithm user guide. North Carolina State University, Center for Research in Scientific Computation 2 (2003)
2. Gao, D., Mi, C., Emadi, A.: Modeling and simulation of electric and hybrid vehicles. Proceedings of the IEEE 95(4), 729–745 (2007)
3. Gao, W., Porandla, S.: Design optimization of a parallel hybrid electric powertrain. In: IEEE Conference on Vehicle Power and Propulsion, pp. 6–12. IEEE (2005)
4. Huang, B., Wang, Z., Xu, Y.: Multi-objective genetic algorithm for hybrid electric vehicle parameter optimization. In: IEEE/RSJ International Conference on Intelligent Robots and Systems, Beijing, China, pp. 5177–5182 (2006)
5. Kennedy, J., Eberhart, R.: Particle swarm optimization. In: Proceedings of the IEEE International Conference on Neural Networks, vol. 4, pp. 1942–1948. IEEE (1995)
6. Krenek, T.: Verbrauchsminimierung eines Hybridfahrzeuges im Neuen Europäischen Fahrzyklus. Master's thesis, Vienna University of Technology, Institute of Computer Graphics and Algorithms, Vienna, Austria (2011)
7. Markel, T., Wipke, K.: Optimization techniques for hybrid electric vehicle analysis using advisor. In: Proceedings of the ASME International Mechanical Engineering Congress and Exposition, New York, USA, pp. 11–16 (2001)
8. Meywerk, M.: CAE-Methoden in der Fahrzeugtechnik. Springer, Heidelberg (2007)
9. Michalewicz, Z.: Heuristic methods for evolutionary computation techniques. Journal of Heuristics 1(2), 177–206 (1996)
10. Montazeri-Gh, M., Poursamad, A., Ghalichi, B.: Application of genetic algorithm for optimization of control strategy in parallel hybrid electric vehicles. Journal of the Franklin Institute 343(4-5), 420–435 (2006)
11. Myers, R., Montgomery, D., Anderson-Cook, C.: Response surface methodology: process and product optimization using designed experiments. Wiley (2009)
12. Nelder, J., Mead, R.: A Simplex Method for Function Minimization. Oxford Journals - The Computer Journal, British Computer Society 7(4), 308–313 (1965)
13. Nocedal, J., Wright, S.: Numerical optimization. Springer, Heidelberg (2006)
14. Wipke, K., Markel, T., Nelson, D.: Optimizing energy management strategy and degree of hybridization for a hydrogen fuel cell SUV. In: Proceedings of 18th Electric Vehicle Symposium, Berlin (2001)
15. Wu, J., Zhang, C., Cui, N.: PSO algorithm-based parameter optimization for HEV powertrain and its control strategy. International Journal of Automotive Technology 9(1), 53–59 (2008)
16. Wu, X., Cao, B., Wen, J., Bian, Y.: Particle swarm optimization for plug-in hybrid electric vehicle control strategy parameter. In: IEEE Conference on Vehicle Power and Propulsion, Harbin, China, pp. 1–5 (2008)

Improved Topological Niching
for Real-Valued Global Optimization

Mike Preuss

Chair of Algorithm Engineering, Computational Intelligence Group,
Dept. of Computer Science, Technische Universität Dortmund, Germany
mike.preuss@tu-dortmund.de

Abstract. We show how nearest-better clustering, the core component
of the NBC-CMA niching evolutionary algorithm, is improved by ap-
pyling a second heuristic rule. This leads to enhanced basin identification
for higher dimensional (5D to 20D) optimization problems, where the
NBC-CMA has previously shown only mediocre performance compared
to other niching and global optimization algorithms. The new method is
integrated into a niching algorithm (NEA2) and compared to NBC-CMA
and BIPOP-CMA-ES via the BBOB benchmarking suite. It performs
very well on problems that enable recognizing basins at all with reasonable effort (number of basins not too high, e.g. the Gallagher problems),
as expected. Beyond that point, niching is obviously not applicable any
more and random restarts as done by the CMA-ES are the method of
choice.

1 Introduction

The idea to apply niching in *evolutionary optimization* (EC) is almost as old as
EC itself, starting with Sharing and Crowding in the 1970s. The general scheme
is to host multiple explicitly or implicitly separated populations in one optimization run and by driving these into the better regions of a multimodal problem
landscape, provide more than one good solution at once. However, they can also
be considered global optimizers as on complex landscapes, it is necessary to retrieve as many good local optima as possible to determine the global one (or
at least a very good one). Premature convergence has never been completely
ruled out and probably will never be, otherwise one would have to safely recognize where the global optimum is before actually approaching it. We believe
that while it is possible to identify some basins of attraction before too many
function evaluations are spent, trying to detect which of these would host the
global optimum with some safety is impossible without actually applying several
restricted (or local) searches.

Generally, niching relies on geometrical properties of the populations (as distances), which makes it less and less applicable if the number of search space
dimensions rises. Most methods are experimentally tested in 2D and maybe 3 to
5D, rarely on 10 or more dimensions. [9] gives a good overview over the "classical" niching strategies and some of their newer variants, also taking different

C. Di Chio et al. (Eds.): EvoApplications 2012, LNCS 7248, pp. 386–395, 2012.
© Springer-Verlag Berlin Heidelberg 2012

Clearing procedures into account. Today, there are still new works in the niching context every year, some of the most recent being [8], [10], and [5]. And we can assess some development from fixed niche radii towards either adaptive radii or even shapes [8], while the topological methods do completely without radii, with [10] and without [7] employing additional evaluations. An interesting related approach that follows more the parallelization ideas of island models [3] is the Particle Swarm CMA-ES [4]. It is especially designed for multi-funnel landscapes (arguably second-order multimodal problems) and shows good results on the CEC 2005 test set even for 10 to 50 dimensions.

As documented in [10], many niching methods perform quite well for the lower dimensional classical test problems as 6-hump camel back, but extensive testing on recent benchmarks as the BBOB 2009/2010 test set[1] is rarely seen. To our knowledge, the *nearest-better clustering cumulative matrix adaptation evolution strategy* (NBC-CMA-ES) [5] is the only niching algorithm which has been investigated on this benchmark yet. In this work, we enhance the NBC-CMA by two strong modifications:

a) We add a second heuristic rule to the basin identification mechanism that is known as nearest-better clustering that is especially tasked at higher dimensional search spaces.
b) We change the local search scheme of the NBC-CMA from a parallel one to a more sequential one that is more robust against basin identification mismatches.

The resulting algorithm is termed NEA2 and we experimentally investigate how each of the modifications increases performance either concerning basin identification or by comparing to the NBC-CMA on the BBOB problem collection. However, we start with giving a very rough general picture of the functioning of the NBC-CMA in order to enable understanding the changes.

2   NBC-CMA General Scheme

The algorithm starts with setting out a larger start population (the default size is 40D) as evenly as possible in the search space. This can be done by employing a space-filling method as *Latin Hypercube Sampling* (LHS) but in principle, a uniform random distribution may also be used, this just lowers the basin identification quality gradually. On this sample, the NBC (see next section) is run and returns a split into separate populations. These are set up as separate CMA-ES runs, respecting the default population size of the CMA-ES according to the search space dimensionality: $\lambda = 4 + \lfloor 3ln(D) \rfloor$ (CMA-ES default parameters are collected nicely here [2]). These separate CMA-ES instances are then run for one generation, the resulting individuals collected, and the basin identification mechanism applied again as described above, starting the loop again. Learned step-sizes and covariance-matrices are stored also in the individuals and after

---

[1] http://coco.gforge.inria.fr/doku.php

clustering and redistributing individuals into populations, each population uses the values found in the best individual (otherwise these values could not be adapted over time). The CMA-ES restart conditions are also applied, so that in case of stagnation the whole algorithm is started over.

As the NBC returns a previously unknown number of basins, and taking into account that it is a heuristic and can fail (if so often by recognizing several basins where there is only one), it is necessary to limit the maximum number of niches pursued at the same time. Of the identified niches, only the best $nich_{max}$ ones are considered, where $nich_{max}$ has a default value of 20. This is connected to a major problem for the NBC-CMA: If too many basins have been erroneously detected, many evaluations are wasted because several populations follow the same peak and eventually progress to the same optimum. This downgrades the performance enormously on very simple functions (e.g. the sphere function). Thus, we clearly have a strong motivation to increase the reliability of NBC, the basin identification method.

3   Nearest Better Clustering with Rule2

This basin identification method has been introduced in [6]. It works by connecting every search point in the population to the nearest one that is better and cutting the connections that are longer than $2\times$ the average connection. The remaining connections determine the found clusters by computing the weakly connected components. The reasoning behind cutting the longest connection is that they are very likely to reach out into another basin; these points seem to be locally optimal, at least considering the given sample. The scheme has huge advantages compared to other clustering methods as no additional evaluations beside the initial scan are needed, and neither shape nor size of the basins is predefined but recognized from the sample. It works very well for a reasonably large populations in two or three dimensions, but increasingly fails if the number of dimensions increases.

Therefore, we now add a second additional cutting rule: For all search points that have at least 3 incoming connections (it is the nearest better point for at least 3 others), we divide the length of its own nearest-better connection by the median of its incoming connections. If this is larger than a precomputed correction factor b, the outgoing connection is cut (and we have one additional cluster). Both rules are applied in parallel, that is, the edges to cut due to rule 2 must be computed before actually cutting due to rule 1. Edges cannot be cut more than once, so if both rules apply, this is not specially treated. Algorithm 1 presents the updated NBC method containing both rules. As the basin identification capability of rule 1 in 2D seems to be sufficient, rule 2 is only applied in at least 3 dimensions.

The motivation for rule 2 was that in a sufficiently large samples (at least around $40 \times D$), often points with several incoming connections are found whose outgoing edge is not cut with rule 1 because it is longer than all the incoming ones but not one of the longest of the the whole sample. However, determination

---

Algorithm 1. Nearest-better clustering (NBC) with rule2

---

1 compute all search points mutual distances;
2 create an empty graph with num(*search points*) nodes;
   `// make spanning tree:`
3 forall the *search points* **do**
4   ⌊ find nearest search point that is better; create edge to it;

   `// cut spanning tree into clusters:`
5 RULE1: delete edges of length $> \phi \cdot$ mean(*lengths of all edges*);
6 RULE2: forall the *search points with at least 3 incoming and 1 outgoing edge*
   do
7   ⌈ **if** *length(outgoing edge)/median(length(incoming edges))* $> b$ **then**
8   ⌊ ⌊ cut outgoing edge;

   `// find clusters:`
9 find connected components;

---

of the correction factor b for rule 2 is not as trivial as setting the cutting factor to 2 for rule 1. It is experimentally derived and presumed to depend on D and the sample size S. As we want to recognize only one cluster on unimodal problems and ideally two or more on multimodal problems, we employ two extreme test functions, namely the sphere (2) and a deceptive function (1) with 2^D optima, located in the corners of the hypercube, restricting the search space to $[0, 1]^D$.

$$dec(\boldsymbol{x}) = \sum_{i=1}^{D} 1 - 2 * abs(0.5 - x_i) \tag{1}$$

$$sphere(\boldsymbol{x}) = \sum_{i=1}^{D} (x_i - 0.5)^2 \tag{2}$$

Figure 1 shows the recognized number of clusters on our two problems while varying b for up to $20D$ and sample sizes of 300. This already provides a good idea how to set b in order to prevent obtaining more than 1.1 clusters on average for the sphere, and it also shows that with these values of b, we can still expect to obtain at least 2 clusters on average on the deceptive function (white areas). However, the corridor is shrinking for higher dimensions. By applying a simple interval search method for finding the right b for every combination of (S, D), using at least 100 repeats for every measurement, we obtain a sequence of points we can use for a linear regression over $log10(S)$. This makes sense as the figure already shows a near linear structure (in logarithmic scaling). The resulting formula is given in (3).

$$b(S, D) = (-4.69 * 10^{-4} * D^2 + 0.0263 * D + 3.66/D - 0.457) * log10(S)$$
$$+ 7.51e - 4 * D^2 - 0.0421 * D - 2.26/D + 1.83 \tag{3}$$

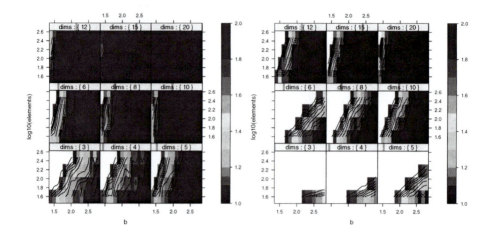

Fig. 1. Number of clusters found over up to 300 elements in dimensions 3 to 20, by applying different values for correction factor b. Left: Sphere (we must not obtain more than 1 cluster here), right: Deceptive test function with 2^D basins. b has been chosen to result in at most 1.1 clusters on the sphere. The gap to the corresponding value for the deceptive function shows that still at least 2 clusters can be found reliably here up to around $20D$.

3.1   Experiment: Effectiveness of Rule 2

In order to assess if rule 2 helps to identify more (meaningful) clusters than before especially in higher dimensions we compare the average number of obtained clusters and the pairwise accuracy of all points on the sphere and the deceptive problem with and without applying rule 2. The *pairwise accuracy* (*pa*) counts the fraction of the points belonging to the same basin of attraction of all pairs of points (without ordering) that have been put into the same cluster by the clustering method. Note that the *pa* measure gets very small rapidly if the number of clusters is much lower than the one of the existing basins (some clusters cover several basins), but it may still be a good first orientation concerning the quality of a clustering.

Setup. We run the clustering with and without rule 2 in 3,4,5,6,8,10,12,15,20 dimensions, on the sphere and the deceptive function, respectively, with 50 repeats. The sample sizes are always $40 \times D$.

Task. To assess a successful improvement (due to rule 2), we demand that a) the average number of obtained clusters on the deceptive function is significantly higher (Wilcoxon rank-sum test at 5%) than the number of clusters with rule 1 alone, b) the cluster numbers on the sphere function do not surpass 1.1 on average, and b) that the pairwise accuracy for the case with rule 2 is significantly (same test) better than the one without at least up to 10 dimensions

(10 dimensions means 1024 optima, as the pa has quadratic nature, the values will become nearly 0 quickly if D increases).

Results/Visualization. The mean numbers of clusters and pairwise accuracies on the deceptive function are given on the right side of fig. 2 (the left side gives an example in 3D), on the sphere function the method always returns 1 without rule 2, the mean cluster numbers with rule 2 are: $1.06, 1, 1.1, 1.08, 1.1, 1.08, 1.16, 1.14, 1.18$ in dimensions $3, 4, 5, 6, 8, 10, 12, 15, 20$. The difference in cluster numbers on the deceptive function is significant for all dimensions, and for the accuracies up to dimension 15.

Observations. Cluster numbers recognized with rule 2 increase up to around 6 for ten dimensions, then fall again. Without rule 2, values are practically 1 for $D > 3$. The accuracy values for rule 2 are always around double the size than without.

Discussion. It is clear that the deceptive function resembles a more or less ideal case, real problems may be much more difficult. Nevertheless, it is important that the improved clustering method at least works well under these conditions. It clearly does so, as seen from the figures, for $D > 3$ the difference is enormous. Required statistical significance has also been achieved, so that we can expect to improve also the performance of a niching method built on the improved basin identification method we deliver. However, the pa measure is a bit difficult to handle as its values get very small with increasing number of basins. One may think of an alternative.

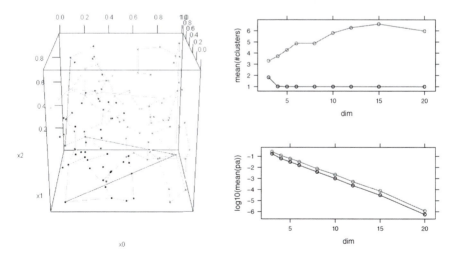

Fig. 2. Left: Example clustering on the 3D deceptive test function, the 120 points are colored per cluster, gray lines are the nearest better edges, blue lines represent the edges cut due to rule 1, red lines the ones cut due to rule 2. Right: Average number of recognized clusters (top) and pairwise accuracies (bottom) for $40 \times D$ points, with rule 2 (red) and without (blue). Note the logarithmic scaling in the lower figure, Absolute values for red are constantly about factor 2 higher.

4   NEA2: Parallel Niching and Sequential Local Optimization

The NBC-CMA as suggested in [5] had its difficulties with performing adequately on unimodal functions. The reason is that as the NBC basin identification mechanism is a heuristic, it is prone to erroneously building more than one cluster where this is unnecessary. The new variant of the algorithm we suggest here, NEA2, overcomes this problem by switching from a BFS-like to a DFS-like search in which the clusters are treated sequentially sorted according to their best members (best first). Should the problem be less multimodal then detected, (e.g. unimodal), NEA2 would perform very similar to the CMA-ES (without heuristic population enlargement as e.g. implemented by the IPOP- [1] and BIPOP-CMA) as every start point leads to the same optimum.

Another advantage of a sequential method is that it is much easier to detect if we approach an already found local optimum again because we already know the local optimum approximations resulting from the previous "local searches". However, this is currently not exploited. We have to admit that changing from parallel to sequential searches is unavoidable because many more clusters are now found also in higher dimensions, where NBC-CMA detected only one. Summarizing, the whole NEA2 method works as given in algorithm 2.

Algorithm 2. NEA2 (with updated NBC)

---

1 distribute an evenly spread sample over the search space;
2 apply the NBC to separate the sample into populations according to detected basins;
3 **forall the** *populations* **do**
4 | run a CMA-ES until restart criterion is hit;

    // start all over:
5 **if** *!termination* **then**
6 | goto step 1

---

5   Experimental Comparison

It is clear that the BBOB test set of 24 functions does not resemble the ideal test bed for niching methods as one would never employ one if it is highly likely that the treated problem is unimodal. However, we compare the new variant NEA2 to its ancestor NBC-CMA and also to the BIPOP-CMA-ES (winner of the BBOB 2009 competition) on this benchmark suite because a) there is a considerable amount of data and knowledge on the performance of different algorithm types generated during the last two BBOB competitions, and b) there are at least some multimodal functions without strong global structure (which would be the setting niching algorithms are targetted at). These functions are the one in the last group, f20 (Schwefel), f21 (Gallagher 1), f22 (Gallagher 2), f23 (Katsuuras), and f24 (Lunacek). While f20 can be considered a deceptive problem, f21/f22 are moderately multimodal, f23/f24 are very highly multimodal and f24 additionally possesses a funnel structure.

Setup. NEA2 is run with a maximum of 300,000 function evaluations as was done with NBC-CMA in [5]. This may not enough for a full picture, but enables a first comparison. NEA2 employs an initial sample of $40 \times D$ that is spread over the search space by means of an LHS (Latin Hypercube Sampling). All CMA-ES specific parameters (used inside the NEA2) are set to their defaults, the initial step size is 1.5. NBC-CMA had used an initial sample of 100, a fixed population size of $\mu = 5, \lambda = 10$, and a maximum of 20 concurrent populations. Runs are immediately stopped if the BBOB frameworks signals hitting the global optimum. We run over the dimensions $2, 3, 5, 10, 20$, all 15 instances provided by the BBOB set.

Task. We require that we see improvement of NEA2 in comparison to the NBC-CMA in many cases, and few performance losses (over the 5 functions and 5 different dimensions. This is not a very formal criterion, but the benchmark set is a bit too small to do final decisions anyway.

Results/Visualization. Performance pictures as generated by the BBOB tools are provided in figure 3. Counting the number of improvements (over problems and dimensions, together 25) results in 11 improvements, 3 losses, and 11 cases of equal (or no) performance. The log files indicate that the number of basins identified is usually much higher for the NEA2 than for the NBC-CMA, especially in $5D$ and up. Note that small differences should not be over-interpreted as we have only 15 repeats on different instances, so a considerable amount of noise in the result can be expected.

Observations. On the Schwefel problem, not much difference between NBC-CMA and NEA2 is visbible, even on the Gallagher problems, the difference is small. However, on f23 and f24, NEA2 is clearly better than NBC-CMA (although only in small dimensions).

Discussion. The very similar behavior of NBC-CMA and NEA2 (while still being better than the BIPOP-CMA-ES) on the Gallagher problems is a bit disappointing. Obviously, a better clustering did not help to speed up search here. However, on f24 and especially f23, there is an unexpected clear difference. As these functions are highly multimodal, better clustering should be of limited importance. We are currently not able to give a good explanation for this, but it seems clear that the basin identification plays some role because in $D > 3$, especially on f23, it obviously gets unproportionally more difficult to obtain the global optimum for NEA2.

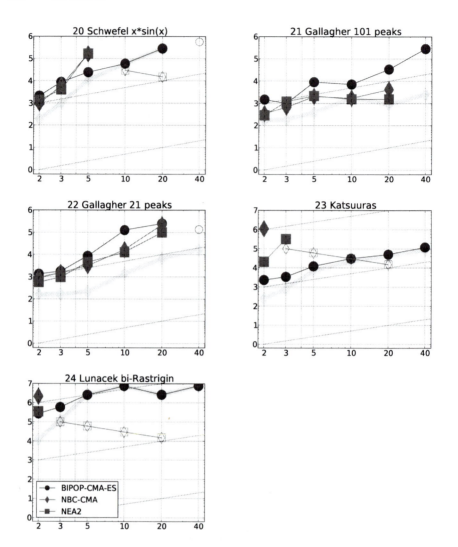

Fig. 3. Performance comparison of NEA2, NBC-CMA and BIPOP-CMA-ES on the 5 multimodal problems without strong global structure the BBOB test set provides. X-axis: problem dimension, y-axis: evaluations in log10-scale. Note that NEA2 and NBC-CMA have been allowed a maximum of 3×10^5 evaluations only.

6   Summary and Conclusions

We have shown that by adding a second heuristic rule to the nearest-better clustering algorithm (NBC), its performance in basin identification is greatly improved. However, more data on multimodal test functions with known basins or at least local optima is needed to see how large this improvement is on functions that have not been taken into account while designing the heuristic.

As found in sec. 3.1, it may also make sense to think of another accuracy measure to rate different clusterings when the true basins are known. Concerning the comparison of the proposed NEA2 algorithm to the NBC-CMA, we can attest slight improvements, and at the same time we obtain many more clusters to start with, which could be interesting for deriving some knowledge about the treated problem quickly (e.g. its degree of multimodality). However, on the BBOB set it is currently not possible to check the accuracy of the obtained clustering easily. This deserves some further testing, also employing different problem generators which provide more support in this respect (but unfortunately much less in others as automated visualization).

References

1. Auger, A., Hansen, N.: A restart CMA evolution strategy with increasing population size. In: Proceedings of the IEEE Congress on Evolutionary Computation, CEC 2005, Edinburgh, UK, September 2-4, pp. 1769–1776. IEEE Press (2005)
2. Hansen, N.: The CMA Evolution Strategy: A Tutorial, http://www.lri.fr/~hansen/cmatutorial.pdf (version of June 28, 2011)
3. Martin, W.N., Lienig, J., Cohoon, J.P.: Island (migration) models: evolutionary algorithms based on punctuated equilibria. In: Handbook of Evolutionary Computation, pp. pp. C6.3:1–C6.3:16. Institute of Physics Publishing, Bristol (1997)
4. Müller, C.L., Baumgartner, B., Sbalzarini, I.F.: Particle swarm CMA evolution strategy for the optimization of multi-funnel landscapes. In: Proceedings of the Eleventh Congress on Evolutionary Computation, CEC 2009, pp. 2685–2692. IEEE Press (2009), http://dl.acm.org/citation.cfm?id=1689599.1689956
5. Preuss, M.: Niching the CMA-ES via nearest-better clustering. In: Proceedings of the 12th Annual Conference Companion on Genetic and Evolutionary Computation, GECCO 2010, pp. 1711–1718. ACM (2010)
6. Preuss, M., Schönemann, L., Emmerich, M.: Counteracting genetic drift and disruptive recombination in $(\mu + /, \lambda)$-EA on multimodal fitness landscapes. In: Beyer, H.G., et al. (eds.) Proc. Genetic and Evolutionary Computation Conf. (GECCO 2005), Washington D.C, vol. 1, pp. 865–872. ACM Press, New York (2005)
7. Preuss, M., Stoean, C., Stoean, R.: Niching foundations: basin identification on fixed-property generated landscapes. In: Proceedings of the 13th Annual Conference on Genetic and Evolutionary Computation, GECCO 2011, pp. 837–844. ACM (2011)
8. Shir, O.M., Emmerich, M., Bäck, T.: Adaptive Niche Radii and Niche Shapes Approaches for Niching with the CMA-ES. Evolutionary Computation 18(1), 97–126 (2010)
9. Singh, G., Deb, K.: Comparison of multi-modal optimization algorithms based on evolutionary algorithms. In: Proceedings of the 8th Annual Conference on Genetic and Evolutionary Computation, GECCO 2006, pp. 1305–1312. ACM (2006)
10. Stoean, C., Preuss, M., Stoean, R., Dumitrescu, D.: Multimodal optimization by means of a topological species conservation algorithm. IEEE Transactions on Evolutionary Computation 14(6), 842–864 (2010)

Towards a Deeper Understanding of Trade-offs Using Multi-objective Evolutionary Algorithms

Pradyumn Kumar Shukla, Christian Hirsch, and Hartmut Schmeck

Karlsruhe Institute of Technology – Institute AIFB
76128 Karlsruhe, Germany

Abstract. A multi-objective optimization problem is characterized by multiple and conflicting objective functions. The conflicting nature of the objectives gives rise to the notion of trade-offs. A trade-off represents the ratio of change in the objective function values, when one of the objective function values increases and the value of some other objective function decreases. Various notions of trade-offs have been present in the classical multiple criteria decision making community and many scalarization approaches have been proposed in the literature to find a solution satisfying some given trade-off requirements. Almost all of these approaches are point-by-point algorithms. On the other hand, multi-objective evolutionary algorithms work with a population and, if properly designed, are able to find the complete preferred subset of the Pareto-optimal set satisfying an a priori given bound on trade-offs. In this paper, we analyze and put together various notions of trade-offs that we find in the classical literature, classifying them into two groups. We then go on to propose multi-objective evolutionary algorithms to find solutions belonging to the two classified groups. This is done by modifying a state-of-the-art evolutionary algorithm NSGA-II. An extensive computational study substantiates the claims of the paper.

Keywords: Multi-objective optimization, NSGA-II, Trade-offs.

1   Introduction

Many real-world, mathematical and economical problems are characterized by the presence of several objective functions which are (at least partially) conflicting (see [3,12]). These problems are called multi-objective optimization problems (MOPs). Solving a problem results in a set of Pareto-optimal solutions, where, without loss of generality, we consider minimization of all the objective functions. A solution is termed as Pareto-optimal if a *decrease* in one of the objective functions can only happen at the expense of an *increase* in another objective function. Trade-offs are a basic tool in decision making and we find various notions in the classical multiple criteria decision making (MCDM) community (see [12]).

Many scalarization approaches ([5,16]) have been proposed in the literature to find a solution satisfying some given trade-off requirements. Almost all of these approaches are point-by-point algorithms and one solution is obtained at the end of the algorithm [5,12]. On the other hand, evolutionary multi-objective

C. Di Chio et al. (Eds.): EvoApplications 2012, LNCS 7248, pp. 396–405, 2012.

algorithms work with a population and, if properly designed, are able to find the complete preferred subset of the Pareto-optimal set satisfying an a priori given bound on trade-offs. Although over the last decades evolutionary multi-objective optimization algorithms have been successfully applied in finding the complete set (or a representative subset, in case of continuous problems) of Pareto-optimal solutions, we find quite less work in designing evolutionary multi-objective algorithms for finding the complete preferred subset of Pareto-optimal solutions. Among the many reasons for this is a proper lack of understanding of the various trade-off concepts that have been prevalent in the MCDM literature. Trade-offs require comparison between two points, and hence population based algorithms, like evolutionary algorithms, are a natural choice for finding solutions that satisfy a priori bounds on the trade-offs.

In this paper, we analyze and put together various notions of trade-offs that we find in the classical literature, classifying them into two groups. This classification is based on domination ideas. We then go on to propose multi-objective evolutionary algorithms to find solutions belonging to the two classified groups. This is done by modifying a state-of-the-art evolutionary algorithm NSGA-II. Moreover, an extensive computational study is used to test the proposed algorithms on a large variety of difficult test problems.

This paper is divided into five sections of which this is the first. The next section presents (an inexhaustive list of) various notions of trade-offs that we find in literature. These are classified into two groups in the same section. Section 3 discusses some (existing and new) multi-objective evolutionary algorithms for finding solutions belonging to the two groups. These algorithms are tested on a variety of test-problems in Section 4. Finally, conclusions and outlook are presented in the last section.

2   Trade-offs and Their Classification

Let $F_1, \ldots, F_m : \mathbb{R}^n \to \mathbb{R}$ and $X \subseteq \mathbb{R}^n$ be given. Consider the following multi-objective optimization problem (MOP):

$$\min \mathbf{F}(\mathbf{x}) := (F_1(\mathbf{x}), F_2(\mathbf{x}), \ldots, F_m(\mathbf{x})) \qquad \text{s.t. } \mathbf{x} \in X.$$

A central optimality notion for the above problem is that of Pareto-optimality. A point $\mathbf{x}^* \in X$ is called *Pareto-optimal* if no $\mathbf{x} \in X$ exists so that $F_i(\mathbf{x}) \leq F_i(\mathbf{x}^*)$ for all $i = 1, \ldots, m$ with strict inequality for at least one index i. Let $X_p(\mathbf{F}, X)$ denote the set of Pareto-optimal points of the above MOP. A criticism of Pareto-optimality is that it allows unbounded trade-offs. To avoid this, starting with the classical work of Geoffrion [7], various stronger optimality notions, known as proper Pareto-optimality, have been defined. Different classes of properly Pareto-optimal exist [12], and the notion of trade-off in them is inherent.

Definition 1 (Geoffrion M-proper Pareto-optimality [15]). *Let $M > 0$ be given. Then, a point $\mathbf{x}^* \in X$ is M-proper Pareto-optimal if $\mathbf{x}^* \in X_p(\mathbf{F}, X)$ and if for all i and $\mathbf{x} \in X$ satisfying $F_i(\mathbf{x}) < F_i(\mathbf{x}^*)$, there exists an index j such that $F_j(\mathbf{x}^*) < F_j(\mathbf{x})$ and moreover $(F_i(\mathbf{x}^*) - F_i(\mathbf{x}))/(F_j(\mathbf{x}) - F_j(\mathbf{x}^*)) \leq M$.*

This definition of proper Pareto-optimal solutions can be practically modified in a number of ways (see details in [9]):

1. The constant M could in general be a function $\mathcal{M}$ of both $\mathbf{x}$ and $\mathbf{x}^*$ [12]. It could also depend on the objective functions F_i and F_j.
2. Based on a (not necessarily disjoint) partition of the index set $\mathcal{I} := \{1, \ldots, m\}$ into two sets $\mathcal{I}_1$ and $\mathcal{I}_2$, we could require that the indices i and j in Definition 1 belong to the sets $\mathcal{I}_1$ and $\mathcal{I}_2$, respectively.
3. The requirement that there exists an index j such that $F_j(\mathbf{x}^*) < F_j(\mathbf{x})$ could be modified to that we restrict ourselves only to those indices j for which $F_j(\mathbf{x}^*) + \Delta < F_j(\mathbf{x})$, where $\Delta > 0$ is a given threshold.

Definition 2 (Marginal rate of substitution [12]). *The marginal rate of substitution* $t_{ij} = t_{ij}(\mathbf{x}^*)$ *at the point* $\mathbf{x}^*$ *represents the amount of gain in the i-th objective that compensates a loss of one unit of the j-th objective, while the other objectives remain unaltered.*

Definition 3 (Allowable trade-off [17]). *An allowable trade-off between criteria i and j, with* $i, j = 1, \ldots, m$, *denoted by* a_{ij}, *is the largest amount of decay in criterion i considered allowable to the decision maker to gain one unit of improvement in criterion j. Also,* $a_{ij} \geq 0$ *for all i and j,* $i \neq j$.

If $a_{ij} = 0$, then the decision maker's preference model is based on the classical Pareto-cone domination structure [12]. A trade-off between two criteria incurred along a direction $\mathbf{d}$ is called *directional trade-off*.

Definition 4 (Directional trade-off [17]). *A directional trade-off between criteria i and j, with* $i, j = 1, \ldots, m$, $i \neq j$, *denoted by* $t_{ij}(\mathbf{d})$, *is defined as follows:*

$$t_{ij}(\mathbf{d}) = 0, \quad \text{if } d_i \leq 0 \text{ and } d_j \leq 0$$

$$t_{ij}(\mathbf{d}) = \frac{d_i}{-d_j}, \quad \text{if } d_i > 0 \text{ and } d_j < 0$$

$$t_{ij}(\mathbf{d}) = \infty, \quad \text{if } d_i \geq 0 \text{ and } d_j \geq 0, \mathbf{d} \neq \mathbf{0}.$$

A direction $\mathbf{d} \in \mathbb{R}^m$ *is an attractive direction if* $t_{ij}(\mathbf{d}) \leq a_{ij}$ *for every pair of criteria* $i, j = 1, \ldots, m$, $i \neq j$.

Based on the above definition, Wiecek et al. [17] construct a model where they assume that the decision maker allows one criterion i to decay only if all the other criteria $j \neq i$ improve. The values a_{ij} come from the decision maker. It may be of interest to repeat the process with more than one selection of criteria and the model includes that. Let P_i, for a given criterion i, be the set of all attractive directions. All the attractive directions are appended with $-\mathbb{R}^m_{\geq}$ so as to obtain a set (which is a cone) of attractive directions given by

$$P := \bigcup_i P_i \cup (-\mathbb{R}^m_{\geq}). \tag{1}$$

Here, we consider the MOP with a domination structure given by the cone P. For an m-dimensional problem, P can be represented with the help of an $m(m-1)$ by m matrix A (see [17]). Since P is the union of all P_i's, in general the cone P_1 might be non-convex.

Engineering applications of many of the above trade-offs can be found in [2,3,10]. Corresponding to an arbitrary but fixed trade-off notion $\mathcal{T}$, let $X_{\mathcal{T}}(\mathbf{F}, X)$ denote the set of those Pareto-optimal solutions that satisfy the trade-off bounds specified by $\mathcal{T}$. The trade-off bounds can have different forms depending on the definitions involved. For example, the bound is M for $\mathcal{T}$ corresponding to Definition 1 and it is a_{ij} for all i, j corresponding to Definitions 3 and 4.

Definition 5. *A trade-off notion $\mathcal{T}$ will be called* Pareto compatible *if there exists a $k \times m$ matrix $\mathcal{A}$, for some $k \in \mathbb{N}$, such that $X_{\mathcal{T}}(\mathbf{F}, X) = X_p(\mathcal{A}\mathbf{F}, X)$, otherwise we call $\mathcal{T}$ to be* Pareto incompatible.

The distinction above is not only for a better understanding of the various trade-off notions but has also great algorithmic and computational implications. These issues will be further discussed in the next two sections.

Lemma 1. *The bounds $\mathcal{T}$ corresponding to Definition 1 and all their practical modifications are Pareto incompatible. The bounds $\mathcal{T}$ corresponding to Definitions 2, 3, and 4 are Pareto compatible.*

Proof: We provide a brief sketch of the proof as a rigorous mathematical proof is outside the scope of this paper. A closer look at Definitions 2, 3, and 4 will reveal that the bounds on the trade-off are pairwise, without consideration of the other objectives. Based on this, we can construct an appropriate matrix $\mathcal{A}$ in a way similar to the matrix A. For a description of the structure of A, please refer to [17]. Definition 1 and the practical modifications are Pareto incompatible as the existence of a conflicting objective (j) is assumed and it could be *any* such objective. This will not provide the elements of the matrix $\mathcal{A}$. $\square$

3   Algorithms

In this section, we introduce algorithms for finding a representative subset of the complete preferred front $X_{\mathcal{T}}(\mathbf{F}, X)$ corresponding to a given trade-off notion $\mathcal{T}$. For this, we take a state-of-the-art algorithm, the non-dominated sorting genetic algorithm NSGA-II [3], and we tailor it for finding the solutions from the set $X_{\mathcal{T}}(\mathbf{F}, X)$. The idea of non-dominated sorting is to sort the entire population into different fronts (set of solutions), such that the solutions in the first front are not dominated by any other solution, the solutions in the second front are only dominated by solutions in the first front and so on.

In the recent years, evolutionary algorithms have shown their potential in including user preferences directly in their search mechanism (see [1] and the references therein). Trade-off based preferences have been incorporated in multi-objective evolutionary algorithms for some Pareto compatible [1] and some Pareto incompatible [15] notions.

One way to develop algorithms for Pareto compatible notions is to change the objective function from $\mathbf{F}$ to $\mathcal{A}\mathbf{F}$ (see Definition 5). A special feature of this class of algorithms is that they work on a k-dimensional objective function instead of the original m dimensions (as $\mathbf{F}$ changes to $\mathcal{A}\mathbf{F} \in \mathbb{R}^k$). Usually, k can be (much) larger than m. Although the approach is simple, in the sense that any existing algorithm could be used and the modification is just in the objective function, the increase in the number of objectives also causes an increase in the non-dominated sorting complexity (see [4,11]).

The situation is more difficult for the case of Pareto incompatible notions as the problem cannot be simply changed, rather the algorithms have to be redesigned. One such approach uses a general framework for finding properly Pareto-optimal points [14]. It finds M-proper solutions among the set of non-dominated solutions and uses these solutions to change the ranking of the population. Another approach [15] is to introduce additional constraints that take into account the amount of violation of M-proper Pareto optimality.

On the basis of the discussion in the above two paragraphs, we propose the following NSGA-II based algorithms for finding the set $X_\mathcal{T}(\mathbf{F}, X)$ corresponding to Pareto-compatible trade-off notions.

TNSGA-IIR: This algorithm uses the general framework from [14]. TNSGA-IIR finds the set of solutions that satisfy the bound specified by $\mathcal{T}$ from among the non-dominated solutions. This set found is given a better rank in the population.

TNSGA-IIG: This algorithm modifies the objective function to $\mathcal{A}\mathbf{F}$ and runs the original NSGA-II on this modified objective function. This is the extension of the standard guided domination based NSGA-II [3] to general Pareto compatible $\mathcal{T}'s$.

When $\mathcal{T}$ is Pareto incompatibe, we can think of the following three algorithms.

MNSGA-IIR: This algorithm again uses the general framework from [14] in order to find the set of solutions that satisfy the bound specified by $\mathcal{T}$ from among the non-dominated solutions. Like in TNSGA-IIR, this set found is given a better rank in the population.

MNSGA-IIC: This algorithm uses the constraint formulation to penalize those solutions that violate the trade-off bound specified by $\mathcal{T}$. Although there could be many ways to introduce the constraints, here we generalize the formulation from [15].

MNSGA-IICR: This algorithm uses both the ranking as in MNSGA-IIR and also the constraint formulation of MNSGA-IIC.

It is easy to see that TNSGA-IIGR will be the same as TNSGA-IIG. The algorithms presented above can be used for general $\mathcal{T}$'s and cover a wide spectrum of possibilities that could be used to find the set $X_\mathcal{T}(\mathbf{F}, X)$. That includes changing the problem to take into account the matrix $\mathcal{A}$, changing the ranking to *favor* good solutions, introducing constraints to *penalize* bad solutions, or a combination of these. Working with a population is a great advantage of evolutionary algorithms (in terms of using the trade-off information efficiently), and we will show the computational efficacy of these algorithms in the next section.

4   Simulation Results

We test the five algorithms on a number of test problems of varying complexity. The test problems are taken from various test suits proposed in the multiobjective community. Specifically, we include two problems from the CEC-2007 competition (SZDT1, SZDT2), two from the ZDT suite (ZDT3, ZDT4) [3], one from the DTLZ family (DTLZ4-3D) [8], four from the WFG suite (WFG1, WFG2, with both 2 and 3 objectives) [8] and one from the CTP family [3]. For all problems solved, we use a population of size 100 and set the maximum number of function evaluations to 20,000 (i.e., 200 generations). Moreover, we use a standard real-parameter SBX and polynomial mutation operator with $\eta_c = 15$ and $\eta_m = 20$, respectively [3].

For Pareto incompatible notions, we use Definition 1 together with $M = 1.5$ and 5.0. In the case of Pareto compatible notions, we use the matrix $\mathcal{A} = (a_{ij})$, with $a_{11} = a_{22} = 1.0$ and $a_{12} = a_{21} = 0.5$. Note that these values restrict the efficient front. This is the preferred efficient front that needs to be found. For all problems, we compute a well-distributed approximation of the preferred front as follows. Corresponding to the problem, first we generate 10,000 well-diverse points on the efficient front. Then, we calculate the preferred points, i.e., the points that satisfy the $\mathcal{T}$ requirement (with M-proper Pareto-optimality criteria or values from the matrix). We use the Inverted generational distance (IGD) and Generational distance (GD) metrics in order to evaluate the results. For statistical evaluation, we use the attainment surface based statistical metric [6]. We run each algorithm for 101 times and the median (50%) attainment surface (51st) is plotted. The source code of all the algorithms is made available[1], and the data files for all the 101 runs of all the problems are available on request.

Fig. 1 shows the attainment surfaces of MNSGA-II$_R$, MNSGA-II$_C$, and MNSGA-II$_{CR}$ on ZDT3. It can be seen that MNSGA-II$_R$ performs the best in this case. Although the constraint based algorithm MNSGA-II$_C$ is able to come close to the preferred front on ZDT3, we find that on the multi-modal test problem ZDT4 this algorithm performs poorly (see Fig. 2). MNSGA-II$_C$ and MNSGA-II$_{CR}$ penalize the whole population, and in this case we see that the population converges prematurely to a local front. MNSGA-II$_R$ on the other hand, splits only the non-dominated set, and this approach seems to work well for multi-modal problems. Figs. 3, 4, and 6 show similar results on three dimensional problems. Nevertheless, all the approaches work well on the nonlinearly constrained test problem CTP7 (see Fig 5).

Figs. 7 and 8 show simulation runs of TNSGA-II$_R$ and TNSGA-II$_G$ on the test problem WFG1-3D. The black points are the preferred front, and in this case we see that both algorithms seem to be away from the preferred front. As WFG1-3D is a very difficult problem, convergence to the front is not possible after 20,000 function evaluations. Nevertheless, we see that both algorithms approach the preferred front in a focused way so as not to explore the entire feasible region in

---

[1] http://www.aifb.kit.edu/web/TNSGA-II/en

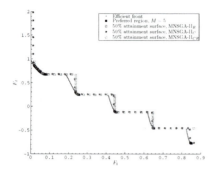

Fig. 1. Attainment surface plot of the algorithms on ZDT3 ($M = 5.0$)

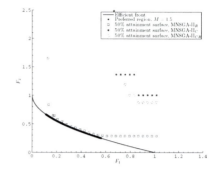

Fig. 2. Attainment surface plot of the algorithms on ZDT4 ($M = 1.5$)

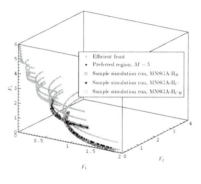

Fig. 3. Preferred front and sample run of the algorithms on WFG2-3D ($M = 5.0$)

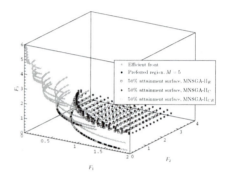

Fig. 4. Attainment surface plot of the algorithms on WFG2-3D ($M = 5.0$)

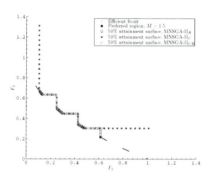

Fig. 5. Attainment surface plot of the algorithms on CTP7 ($M = 1.5$)

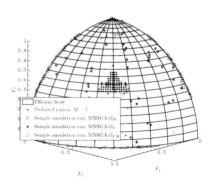

Fig. 6. Preferred front and sample run of the algorithms on DTLZ4 ($M = 5.0$)

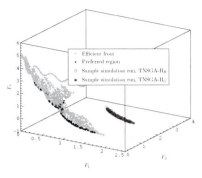

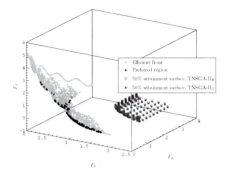

Fig. 7. Preferred front and sample run of the algorithms on WFG1-3D ($t = 0.5$)

Fig. 8. Attainment surface plot of the algorithms on WFG1-3D ($t = 0.5$)

the objective space. This is important for higher dimensional problems as then the search effort is not wasted in the entire space.

Looking at Tables 1 and 2, we see that _R algorithms perform very well for finding a representative subset of $X_\mathcal{T}(\mathbf{F}, X)$ for Pareto incompatible notions but not for Pareto compatible notions. This is interesting and shows that modifying the entire population (either by using constraints or by using guided approaches) is beneficial only for Pareto compatible trade-offs. Pareto compatibility is an additional information that is global in the sense of a problem being changed completely (to $\mathcal{AF}$). Hence, for these notions, _R algorithms are not a good choice. In this case, an algorithm like _G is useful as it works with a modified problem (that is also equivalent to using a different domination concept, see [3] for details). _R

Table 1. Median and interquartile range for Pareto incompatible notions

IGD ($M = 1.5$)	MNSGA-II$_R$	MNSGA-II$_C$	MNSGA-II$_{CR}$
SZDT1	$0.000762_{0.000304}$	$0.001651_{0.000560}$	$0.001647_{0.000464}$
SZDT2	$0.028343_{0.691251}$	$0.831794_{0.061160}$	$0.833084_{0.048942}$
ZDT3	$0.001138_{0.000484}$	$0.106242_{0.043848}$	$0.085041_{0.042644}$
ZDT4	$0.000368_{0.000327}$	$0.013582_{0.013215}$	$0.010992_{0.010310}$
WFG1-2D	$0.239695_{0.016413}$	$0.330208_{0.005360}$	$0.332137_{0.005831}$
WFG1-3D	$0.716829_{0.011457}$	$0.709812_{0.032873}$	$0.691462_{0.032658}$
WFG2-2D	$0.186246_{0.001344}$	$0.190389_{0.004472}$	$0.191518_{0.005499}$
WFG2-3D	$0.041108_{0.138842}$	$0.224028_{0.193395}$	$0.225610_{0.191548}$
DTLZ4	$0.000037_{0.000103}$	$0.000009_{0.000007}$	$0.000009_{0.000010}$
CTP7	$0.000031_{0.000002}$	$0.000033_{0.000003}$	$0.000031_{0.000004}$

GD ($M = 5.0$)	MNSGA-II$_R$	MNSGA-II$_C$	MNSGA-II$_{CR}$
SZDT1	$0.001068_{0.000247}$	$0.002602_{0.002424}$	$0.001289_{0.000196}$
SZDT2	$0.001554_{0.000356}$	$0.016176_{0.042619}$	$0.144943_{0.260740}$
ZDT3	$0.000116_{0.000016}$	$0.005647_{0.005665}$	$0.000153_{0.000040}$
ZDT4	$0.000493_{0.000353}$	$0.002799_{0.012020}$	$0.001698_{0.003546}$
WFG1-2D	$0.093332_{0.013041}$	$0.090901_{0.004792}$	$0.094342_{0.014087}$
WFG1-3D	$0.119694_{0.002628}$	$0.135277_{0.003272}$	$0.119546_{0.003465}$
WFG2-2D	$0.003064_{0.001949}$	$0.005497_{0.003660}$	$0.005279_{0.004595}$
WFG2-3D	$0.007298_{0.012241}$	$0.013697_{0.001827}$	$0.008344_{0.011168}$
DTLZ4	$0.006325_{0.000955}$	$0.011298_{0.000618}$	$0.007000_{0.001151}$
CTP7	$0.000013_{0.000013}$	$0.000011_{0.002246}$	$0.000010_{0.000001}$

Table 2. Median and interquartile range for Pareto compatible notions

	TNSGA-II$_R$ (GD)	TNSGA-II$_G$ (GD)	TNSGA-II$_R$ (IGD)	TNSGA-II$_G$ (IGD)
SZDT1	$0.000695_{0.000165}$	$0.000679_{0.000149}$	$0.000491_{0.000194}$	$0.000470_{0.000164}$
SZDT2	$0.007413_{0.002875}$	$0.000848_{0.000233}$	$0.062127_{0.000094}$	$0.062138_{0.000092}$
ZDT3	$0.000048_{0.000003}$	$0.000049_{0.000002}$	$0.057771_{0.000150}$	$0.057757_{0.057667}$
ZDT4	$0.000373_{0.000299}$	$0.000388_{0.000336}$	$0.000282_{0.000198}$	$0.000264_{0.000239}$
WFG1-2D	$0.194169_{0.065272}$	$0.077383_{0.006835}$	$0.217352_{0.003753}$	$0.213599_{0.004660}$
WFG1-3D	$0.118367_{0.003043}$	$0.117167_{0.003405}$	$0.093049_{0.002028}$	$0.092257_{0.002960}$
WFG2-2D	$0.011109_{0.009349}$	$0.006193_{0.004448}$	$0.105864_{0.005186}$	$0.105376_{0.004505}$
WFG2-3D	$0.016094_{0.012247}$	$0.016376_{0.009669}$	$0.064790_{0.021157}$	$0.048527_{0.034876}$
DTLZ4	$0.006010_{0.000517}$	$0.005998_{0.000632}$	$0.010273_{0.000900}$	$0.010009_{0.001633}$
CTP7	$0.000007_{0.000000}$	$0.000010_{0.000001}$	$0.000021_{0.000001}$	$0.000046_{0.000003}$

algorithms only change the ranking based on the non-dominated set (or the best front), and this methodology is useful for Pareto incompatible trade-offs. For these trade-offs, there is not a modified problem, and hence, together with the theoretical results from [14], changing the ranks based on splitting the non-dominated solutions might be a viable way to compute $X_{\mathcal{T}}(\mathbf{F}, X)$. Moreover, as the simulation results show, _C and _CR algorithms are not a good option for Pareto incompatible notions. The constraints in the _C and _CR algorithms are very restrictive and lead to premature convergence, especially on multi-modal and difficult problems.

5   Conclusions

This study is towards a deeper understanding of various trade-off notions that are present in the multi-objective community. Trade-offs are a basic tool in decision making and have been a subject of active research since the late sixties, starting with the seminal work of Geoffrion [7]. Multi-objective evolutionary algorithms have been used to find a representative subset of the complete efficient front and could be properly designed to find a preferred region of the efficient front as well. This is due to the population based advantage that they have (see [13] for more on this interesting aspect). Trade-offs compare solutions with conflicting objectives from among the solutions in a population and in this way are ideally suited for a population based algorithm. With this idea in mind, we characterized the various notions of trade-offs that we find in literature. Based on this characterization, we investigated various algorithms for finding the solutions that satisfy a given trade-off bound (depending on the notion). We found that Pareto compatible notions require an algorithm that works on a modified problem or works with a global change of domination. Pareto incompatible notions, on the other hand, are best handled by algorithms that only change the ranking of a subset of the entire population, that is, of the non-dominated solutions. This paper sheds adequate light on the notion of trade-offs and we hope to see more population based algorithms for this task.

Acknowledgements. This work was supported by the German Federal Ministry of Economics and Technology (MeRegio - Minimum Emission Regions, Grant 01ME08001A).

References

1. Branke, J., Deb, K., Miettinen, K., Slowinski, R. (eds.): Multiobjective Optimization, Interactive and Evolutionary Approaches [outcome of Dagstuhl seminars]. LNCS, vol. 5252. Springer, Heidelberg (2008)
2. Coello, C.A.C., Christiansen, A.D.: Multiobjective optimization of trusses using genetic algorithms. Computers and Structures 75(6), 647–660 (2000)
3. Deb, K.: Multi-objective optimization using evolutionary algorithms. Wiley (2001)
4. Fang, H., Wang, Q., Tu, Y., Horstemeyer, M.F.: An efficient non-dominated sorting method for evolutionary algorithms. Evol. Comput. 16, 355–384 (2008)
5. Fischer, A., Shukla, P.K.: A Levenberg-Marquardt algorithm for unconstrained multicriteria optimization. Oper. Res. Lett. 36(5), 643–646 (2008)
6. Fonesca, C.M., Fleming, P.J.: On the Performance Assessment and Comparison of Stochastic Multiobjective Optimizers. In: Ebeling, W., Rechenberg, I., Voigt, H.-M., Schwefel, H.-P. (eds.) PPSN 1996. LNCS, vol. 1141, pp. 584–593. Springer, Heidelberg (1996)
7. Geoffrion, A.M.: Proper efficiency and the theory of vector maximization. Journal of Mathematical Analysis and Applications 22, 618–630 (1968)
8. Huband, S., Hingston, P., Barone, L., White, L.: A review of multi-objective test problems and a scalable test problem toolkit. IEEE Transactions on Evolutionary Computation 10(5), 280–294 (2005)
9. Kaliszewski, I.: Soft computing for complex multiple criteria decision making. Springer, New York (2006)
10. Kalsi, M., Hacker, K., Lewis, K.: A comprehensive robust design approach for decision trade-offs in complex systems design. Journal of Mechanical Design 123(1), 1–10 (2001)
11. Kung, H.T., Luccio, F., Preparata, F.P.: On finding the maxima of a set of vectors. Journal of the Association for Computing Machinery 22(4), 469–476 (1975)
12. Miettinen, K.: Nonlinear Multiobjective Optimization. Kluwer, Boston (1999)
13. Prügel-Bennett, A.: Benefits of a population: Five mechanisms that advantage population-based algorithms. IEEE Transactions on Evolutionary Computation 14(4), 500–517 (2010)
14. Shukla, P.K., Hirsch, C., Schmeck, H.: A Framework for Incorporating Trade-Off Information Using Multi-Objective Evolutionary Algorithms. In: Schaefer, R., Cotta, C., Kołodziej, J., Rudolph, G. (eds.) PPSN XI. LNCS, vol. 6239, pp. 131–140. Springer, Heidelberg (2010)
15. Shukla, P.K.: In Search of Proper Pareto-optimal Solutions Using Multi-objective Evolutionary Algorithms. In: Shi, Y., van Albada, G.D., Dongarra, J., Sloot, P.M.A. (eds.) ICCS 2007. LNCS, vol. 4490, pp. 1013–1020. Springer, Heidelberg (2007)
16. Shukla, P.K.: On the Normal Boundary Intersection Method for Generation of Efficient Front. In: Shi, Y., van Albada, G.D., Dongarra, J., Sloot, P.M.A. (eds.) ICCS 2007. LNCS, vol. 4487, pp. 310–317. Springer, Heidelberg (2007)
17. Wiecek, M.M.: Advances in cone-based preference modeling for decision making with multiple criteria. Decis. Mak. Manuf. Serv. 1(1-2), 153–173 (2007)

OpenCL Implementation of Particle Swarm Optimization: A Comparison between Multi-core CPU and GPU Performances

Stefano Cagnoni[1], Alessandro Bacchini[1], and Luca Mussi[2]

[1] Dept. of Information Engineering, University of Parma, Italy
cagnoni@ce.unipr.it, alessandro.bacchini@studenti.unipr.it
[2] Henesis s.r.l., Parma, Italy
luca.mussi@henesis.eu

Abstract. GPU-based parallel implementations of algorithms are usually compared against the corresponding sequential versions compiled for a single-core CPU machine, without taking advantage of the multi-core and SIMD capabilities of modern processors. This leads to unfair comparisons, where speed-up figures are much larger than what could actually be obtained if the CPU-based version were properly parallelized and optimized.

The availability of OpenCL, which compiles parallel code for both GPUs and multi-core CPUs, has made it much easier to compare execution speed of different architectures fully exploiting each architecture's best features.

We tested our latest parallel implementations of Particle Swarm Optimization (PSO), compiled under OpenCL for both GPUs and multi-core CPUs, and separately optimized for the two hardware architectures.

Our results show that, for PSO, a GPU-based parallelization is still generally more efficient than a multi-core CPU-based one. However, the speed-up obtained by the GPU-based with respect to the CPU-based version is by far lower than the orders-of-magnitude figures reported by the papers which compare GPU-based parallel implementations to basic single-thread CPU code.

Keywords: Parallel computing, GPU computing, Particle Swarm Optimization.

1 Introduction

Particle Swarm Optimization (PSO), the simple but powerful algorithm introduced by Kennedy and Eberhart [4], is intrinsically parallel, even more than other evolutionary, swarm intelligence or, more in general, population-based optimization algorithms.

Because of this, several parallel PSO implementations have been proposed, the latest of which are mainly based on GPUs [2,3,8,9]. It is very hard to fairly compare the results of different implementations, on different architectures or

C. Di Chio et al. (Eds.): EvoApplications 2012, LNCS 7248, pp. 406–415, 2012.

compilers, or even of the same programs run on different releases of software-compatible hardware. Execution time is generally the only direct objective quantitative parameter on which comparisons can be based. Therefore, this is the approach that most authors are presently adopting. After all, obtaining a significant speed-up in the algorithm's execution time is obviously the main reason for developing parallel versions of the algorithm, considering also that PSO is also intrinsically one of the most efficient stochastic search algorithm, for the simplicity and compactness of the equations on which it is based.

Undoubtedly, the GPU-based parallelization of PSO has produced impressive results. Most papers on the topic report speed-ups of orders of magnitude with respect to single-core CPU-based versions, especially when large-size problems or large swarms are taken into consideration. This may lead to a misinterpretation of the results, suggesting that CPUs[1] are overwhelmingly outperformed by GPUs on this task. However, most comparisons have actually been made between accurately-tuned GPU-based parallel versions and a sequential version compiled for a single-processor machine. Thus, while being absolutely objective and informative, they do not reflect the actual disparity in the top performances which can be offered by the two computational architectures. In fact, they totally ignore the parallel computation capabilities of modern CPUs, both in terms of number of cores embedded by the CPU architecture and of CPU's SIMD (Single Instruction Multiple Data) computation capabilities.

Papers comparing GPUs and CPUs "at their best" have started being published only recently [5]. This has been mainly justified by the lack of a handy environment for developing parallel programs on CPUs. As new environments or libraries for parallel computing supporting both GPUs and CPUs, like OpenCL and Microsoft Accelerator, have been released, this gap has been filled. In particular, OpenCL allows one to develop parallel programs for both architectures, offering the chance to either compile the same code for execution on both CPUs and GPUs, or, more importantly, to develop parallel implementations of the same algorithm, specifically otpimized for either computing architecture.

We have previously presented two GPU implementations of PSO [6,7], a synchronous and an asynchronous version, developed both on CUDA, nVIDIA's environment for GPU computing using its cards. Our implementations were aimed at maximizing execution speed, disregarding other limitations, such as the maximum number of particles of which a swarm could be composed. Thus, our best-performing GPU-based parallel implementation could only manage up to 64 particles, depending on hardware capabilities. In such a work, we also have made a comparison with the single-thread sequential implementation of a standard PSO (SPSO2006 [1]), mainly to allow for a comparison with other, previously published, results.

In this paper we try to make the fairest possible comparison between computing performances which can be obtained by GPU-based and CPU-based parallelized versions of PSO, developed within the OpenCL environment. On the one

---

[1] From here onwards, the term CPU will refer to a multi-core CPU, if not differently specified.

hand, we have slightly modified our most efficient GPU-based PSO algorithm, allowing for swarms of virtually any size to be run, at the price of a slight reduction in performances. On the other hand, we have also developed a parallel OpenCL version of PSO, whose structure and parameters are optimized for running on a CPU. This way, we have obtained two implementations having similar limitations, usable for a fair comparison between the actual performances which can be obtained by the two different architectures.

We report results obtained on a set of classical functions used for testing stochastic optimization algorithms, by five different GPUs and CPUs which are presently offered in typical medium/high-end desktop and laptop configurations.

In the following sections, we first describe our parallel algorithm and the slight differences between the CPU-oriented and GPU-oriented versions. We then report results obtained in the extensive set of tests we have performed to compare their performances. Finally, we close the paper with a discussion of the results we obtained, and draw some conclusions about the efficiency and applicability of the two implementations.

2   PSO Parallelization

The parallel versions of PSO developed in our previous work are quite similar, both being fine-grained parallelizations down to the dimension level, i.e., we allocate one thread per particle's dimension in implementing the PSO update equations. As long as enough resources are available, this means that it is virtually possible to perform a full update in a single step. The other common feature is the use of a ring topology of radius equal to one for the particles' neighborhoods, which minimizes data dependencies between the executions of the update cycles of each particle.

The main difference between the two implementations is related with the update of the particles' social attractors, i.e., the best-performing particle in each particle's neighborhood or in the whole swarm, depending on whether a "local-best" or "global-best" PSO is being implemented, respectively. Our earlier-developed version [6] implements the most natural way of parallelizing PSO, as regards both task separation and synchronization between particles: thus, it was implemented as three separate kernels: i. position/velocity update; ii. fitness evaluation; iii. local/global best(s) computation. Each kernel also represented a synchronization point, so the implementation corresponded to the so-called "synchronous PSO", in which the algorithm waits for all fitness evaluations to be over before computing the best particles in the swarm or in each particle's neighborhood. In a later-developed version [7] we relaxed this synchronization constraint, obtaining an "asynchronous PSO" by letting a particle's position and velocity be updated independently of the computation of its neighbors' fitness values. This allowed us to implement the whole algorithm as a single kernel, minimizing the overhead related to both context switching and data exchange via global memory, as the whole process, from the first to the last generation, is scheduled as a single kernel call. Therefore, while, in the synchronous version, the status of each particle needed to be saved/loaded after each kernel, in the asynchronous version the parallelization occurs task-wise, with no data exchange in global memory before the whole process terminates. While this feature is optimal for speed, it introduces a severe drawback related with resource availability. No partial inter-generation results can be saved during it to allow another batch of particles to be run within the same generation; therefore, a swarm can only comprise up to N_{max} particles, N_{max} being the number of particles which the resources available in the GPU permit to be managed at the same time.

To allow for a virtually unlimited size of the swarm, we have turned back to a synchronous version, which saves the partial results of each generation back to the global memory, while still using a single kernel for running a generation of a swarm of up to N_{max} particles. Therefore, if a swarm of $N > N_{max}$ is to be run, a full update of the swarm can be obtained by running a batch of $\lceil N/N_{max} \rceil$ instances of the kernel sequentially.

The three stages of the previous synchronous version of PSO have been merged and synchronization between stages removed: the only synchronization occurs at the end of a generation. This avoids some accesses to the global memory because each particle loads its state at the beginning of each generation and writes it back to memory only at the end of it. While this, obviously, has a price in terms of execution speed, it has the advantage of limiting the delay between the times of update between particles to no more than one generation.

3   Open Computing Language

The evolution both of parallel processors and of the corresponding programming environments has been extremely intense but, until recently, far from any standardization. The Open Computing Language, OpenCL^{TM}, released at the end of 2008, is the first open standard framework for cross-platform and heterogeneous resources programming. It permits to easily write parallel code for modern processors and obtain major improvements in speed and responsiveness for a very wide range of applications, from gaming to scientific computing.

On the one hand, to cope with modern computers which often includes one or more processors and one or more GPUs (and possibly also DSP processors and other devices), the OpenCL Platform Model definition (see Figure 1) includes one "host device" controlling one or more "compute devices". Each compute device is composed of one or more "compute units" each of which is further divided into one or more "processing elements". Usually host and compute devices are mapped onto different processors, but may also be mapped onto the same one. This is the case for the Intel OpenCL implementation where host and device can be the very CPU.

On the other hand, the OpenCL Execution Model defines the structure of an OpenCL application which runs on a host using classical sequential code to submit work to the available compute devices. The basic unit of work for an OpenCL device is referred to as "work item" (something like a thread) and its code is called "kernel" (something similar to a plain C function). Work items

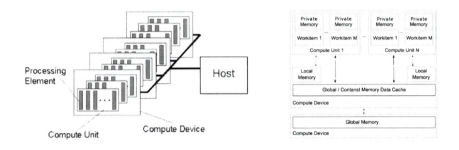

Fig. 1. OpenCL definition for the Platform Model (left) and the Memory Hierarchy (right)

are grouped under many "work groups": all the work items belonging to the same work group actually run in parallel, while different work groups can be scheduled sequentially based on the available resources. To achieve the best performances, each device has its own memory hierarchy: the global memory is shared across all work groups, the local memory is shared across all work items within a work group, and private memory is only visible to a single work item. Local and private memory usage usually greatly improve performances for the so called memory-bounded kernels. Finally, in addition to the memory hierarchy, OpenCL defines some barriers to synchronize all the work items within a work group: work items belonging to different work groups cannot be synchronized while a kernel is running.

The OpenCL standard also permits the compilation of OpenCL code at execution time, so the code can be optimized on place to execute the requested operation as fast as possible. Despite this, to achieve best performances on different kinds of devices, it is still necessary to write slightly different versions of the same kernel to achieve top performances: for example, a kernel must explicitly use vector types (*float4*, *float8*, ...) to fully exploit the SIMD instructions of modern CPUs.

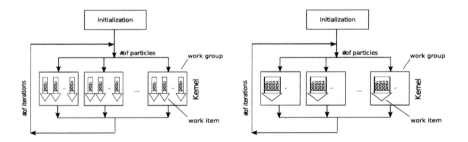

Fig. 2. Workflow organization for the GPU version (left) and the GPU version (right) of PSO

4   PSO Implementation within OpenCL

This new OpenCL implementation of PSO roughly inherits the same structure of our previous versions developed for CUDA [7]. The three main steps of PSO are implemented as a single kernel scheduled many times in a row to simulate the generational loop. This organization introduces a fixed synchronization point for all particles at the end of every generation, realizing a synchronous PSO variant.

Each particle is simulated by a work group comprising a work item for each dimension of the problem to optimize. At the beginning of the kernel, the particles' positions, velocities and best positions are read from the memory: this step severely limits the kernel performances because of the accesses to the global memory, the slowest operation for an OpenCL device. Then velocities and positions are updated, but no bottlenecks are present here because only simple arithmetic operations are required. Fitness evaluation comes next and is usually the most complex stage of the kernel because of the possible presence of transcendental functions and parallel reductions used to compute the sum of the many addends calculated by each work item. The last stage updates the particles' best fitness values, positions, best positions and velocities and stores them into the global memory: some further waste of time occurs here, due to the high latency of global memory write operations.

The OpenCL standard allows for the compilation of OpenCL kernels at execution time, so it is actually possible to define preprocessor constants to save some parameter passing. This feature is used to embed all PSO parameters into the kernel code, still allowing the user to specify everything when launching the application. This technique is especially advantageous when dealing with parallel reductions: in our case knowing in advance the number of problem dimensions (i.e., the number of work items inside each work group) allows us to minimize the number of synchronization barriers inside each work group and improve overall performances. Another way to reduce the execution time of the kernel is using the native transcendental functions ("fast math" enabled) which are usually implemented in hardware and are an order of magnitude faster than non-native ones. As already mentioned, to obtain the best performances, it is obviously necessary to write the code taking into account which device will run the kernel, in order to exploit its peculiar parallel instructions. Accordingly, we developed two versions of the same kernel for the two architectures under consideration.

4.1   GPU-Based Implementation

This version is oriented to massively parallel architectures, like GPUs, for whom each work item maps to a single thread. Indeed, nVidia GPUs have hundreds of simple ALU that process only a single instruction at a time: in this case it is usually better to run thousands of light threads instead of hundreds of heavy ones. Moreover, global and local memory on GPU devices are located in distinct areas and have different performances: local memory is usually one order of magnitude faster than global memory. The best practice is hence to use local memory as much as possible and minimize the number of accesses to global memory. Finally, the OpenCL implementation by AMD and nVidia ensure that work items are scheduled in groups of at least 32 threads at a time which, in some cases, makes it possible to avoid the use of synchronization barriers. For example, this is the case for the final steps of parallel reductions.

4.2   CPU-Based Implementation

CPUs[2] are architecturally different from GPUs and the above-mentioned optimizations are not always the best option for this kind of devices. CPUs have a very small number of cores compared to GPUs, but each core includes a large amount of cache and a complex processing unit: branch prediction, conditional instructions and misaligned memory accesses are usually more efficient on CPUs than on GPUs. The problem of having a limited number of cores is also overcome by SIMD (Single Instruction, Multiple Data) instructions, an extension to the standard x86 instruction set that allows each ALU to perform parallel operations over a set of 4 or 8 homogeneous values. SSE-SIMD instructions allow for parallel operations on 4 floats/ints or 2 doubles at the same time, while new AVX-SIMD instructions allow for parallel operations on 8 floats or 4 doubles. The OpenCL standard natively supports SIMD instructions via vector data type: int4, int8, float4, float8, . . . For these reasons our version of the PSO kernel was rewritten for CPUs, grouping the work items four by four or eight by eight so that each work item takes care of four or eight dimension of the problem under optimization. It is easy to see how this introduces big improvements, considering that the OpenCL standard requires that all the work items belonging to the same work group be executed on the same compute unit (one CPU core) sequentially.

5   Test and Results

We compared the performance of our parallel CPU-based and GPU-based PSO implementation on a set of "classical" functions used to evaluate stochastic optimization algorithms.

Our goal was to fairly compare the performances of our GPU and CPU implementations in terms of speed. We also verified the correctness of our implementations (results not reported, see [7] for the tests on the previously developed algorithms) by checking that the minima found by the parallel algorithms did not differ significantly from the results of the sequential SPSO [1]. We kept all algorithm parameters equal in all tests, setting them to the standard values suggested for SPSO: $w = 0.72134$ and $C1 = C2 = 1.19315$. The test were performed on different processors: two Intel CPUs (i7-2630M and i7-2600K) having 4 physical and 8 logical cores; two nVidia GPUs (GT-540M and GTX-560Ti) having 96 and 384 cores, respectively; and an ATI Radeon HD6950 GPU (1408 cores). These can be considered typical examples of the processors and GPUs which presently equip medium/high-end laptop and desktop computers.

Table 1 lists the functions used as benchmark with the corresponding ranges within which the search was constrained.

---

[2] We consider typical x86 CPUs here.

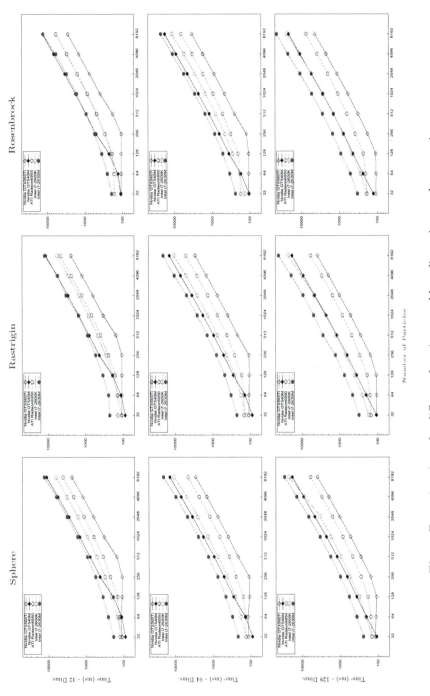

Fig. 3. Execution times for different functions, problem dimensions and swarm sizes

Table 1. Function set used for the test and corresponding search ranges

Function	Search Range
Sphere	$[-100, +100]^N$
Elliptic	$[-100, +100]^N$
Rastrigin	$[-5.12, +5.12]^N$
Rosenbrock	$[-30, +30]^N$
Griewank	$[-600, +600]^N$

For each function, we tested the scaling properties with respect to problem dimension and swarm size by running 20 repetitions of each test. Figure 3 reports the graphs of the average results obtained in all tests for three out of the five functions we considered. Similar results were obtained for the other two functions.

The relative performances of the five processors are quite regular and repeatable, with the GPUs outperforming CPUs by a speed gain ranging from about 1 to no more than 5-6 for the largest swarm sizes. For the smallest swarm sizes, depending on the resources available on each processor, and therefore more evidently for the most powerful processors, after a flat segment, a "knee" appears when the resources avaiable are no more sufficient for running the algorithm at the same time, after which the graph (in log-log scale) becomes smoothly linear. Only some minor peculiarities can be observed. One is the apparently surprising performances of the nVidia GT540M (a GPU dedicated to mobile systems, with a limited number of resources) with small swarm sizes. In some tests, for swarms of 32 particles, it has even exhibited the top performances. This can be easily explained, considering that the processor clock frequency of the GT540M is slightly higher than the corresponding frequency in the GTX560Ti so it can perform better than the latter if the function is not too memory-intensive (memory bandwidth is much lower in the GT540M). Another peculiarity is the relative performance of the desktop processor by Intel with respect to the mobile version, which shows a narrowing of the performance gap between the two processors in "extreme" conditions (large swarms and high-dimensional functions). We have no explanation for this, except for the possible influence of other components of the PC which may affect its global performances.

6 Final Remarks

We have assessed the performances of basically the same version of PSO implemented on a set of five CPUs and GPUs, taking advantage of the opportunity, offered by OpenCL, to develop, and optimize, the code for different architectures within the same environment.

The main goal of our work was to compare GPU and CPU performances using PSO code which had been optimized for both computing architectures, going beyond the usual comparison between parallel GPU code and single-thread sequential CPU code, where GPUs outperform CPUs by orders of magnitude.

From this point of view, we showed that, while GPUs still outperform CPUs in this task, the performance gap is not so large as the usual "unfair" comparison tends to suggest, as the speed gain within processors belonging to comparable market segments never gets even close to the order of magnitude. One should also consider that PSO, as well as the target functions used in the test, is probably one of the algorithms that is most suitable for parallelization on massively parallel architectures.

From the practical point of view of the development cost, however, one should consider that the comparison has been fair also in this regard. In fact, developing the parallel PSO using OpenCL from scratch has the same cost for both architectures. So, especially for larger optimization problems to be tackled by PSO, it makes little sense to use the CPU, if a well-performing GPU is available, unless, of course, one needs to produce graphics at the same time. What this work mainly suggests is that the range of problems in which a GPU clearly outperforms a CPU is possibly much smaller than a superficial interpretation of the results usually available on the topic might induce one to imagine.

References

1. (2006), http://www.particleswarm.info/Standard_PSO_2006.c
2. Cadenas-Montes, M., Vega-Rodriguez, M.A., Rodriguez-Vazquez, J.J., Gomez-Iglesias, A.: Accelerating particle swarm algorithm with GPGPU. In: 19th Euromicro International Conference on Parallel, Distributed and Network-based Processing (PDP), pp. 560–564. IEEE (2011)
3. de P. Veronese, L., Krohling, R.A.: Swarm's flight: Accelerating the particles using C-CUDA. In: Proc. IEEE Congress on Evolutionary Computation (CEC 2009), pp. 3264–3270. IEEE (2009)
4. Kennedy, J., Eberhart, R.: Particle swarm optimization. In: Proc. IEEE International Conference on Neural Networks, vol. IV, pp. 1942–1948. IEEE (1995)
5. Lee, V.W., Kim, C., Chhugani, J., Deisher, M., Kim, D., Nguyen, A.D., Satish, N., Smelyanskiy, M., Chennupaty, S., Hammarlund, P., Singhal, R., Dubey, P.: Debunking the 100X GPU vs. CPU myth: an evaluation of throughput computing on CPU and GPU. In: Proc. 37th International Symposium on Computer Architecture (ISCA), pp. 451–460. ACM (2010)
6. Mussi, L., Daolio, F., Cagnoni, S.: Evaluation of parallel particle swarm optimization algorithms within the CUDA architecture. Inf. Sciences 181(20), 4642–4657 (2011)
7. Mussi, L., Nashed, Y.S.G., Cagnoni, S.: GPU-based asynchronous Particle Swarm Optimization. In: Proceedings of the 13th Annual Conference on Genetic and Evolutionary Computation (GECCO), pp. 1555–1562. ACM (2011)
8. Papadakis, S.E., Bakrtzis, A.G.: A GPU accelerated PSO with application to Economic Dispatch problem. In: 16th International Conference on Intelligent System Application to Power Systems (ISAP 2011), pp. 1–6. IEEE (2011)
9. Zhou, Y., Tan, Y.: GPU-based parallel particle swarm optimization. In: Proc. IEEE Congress on Evolutionary Computation, CEC 2009, pp. 1493–1500. IEEE (2009)

A Library to Run Evolutionary Algorithms in the Cloud Using MapReduce

Pedro Fazenda[1,2], James McDermott[2], and Una-May O'Reilly[2]

[1] Institute for Systems and Robotics, IST, Lisbon, Portugal
[2] Evolutionary Design and Optimization Group, CSAIL, MIT

Abstract. We discuss ongoing development of an evolutionary algorithm library to run on the cloud. We relate how we have used the Hadoop open-source MapReduce distributed data processing framework to implement a single "island" with a potentially very large population. The design generalizes beyond the current, one-off kind of MapReduce implementations. It is in preparation for the library becoming a modeling or optimization service in a service oriented architecture or a development tool for designing new evolutionary algorithms.

Keywords: MapReduce, cloud computing, Hadoop, evolutionary algorithms.

1 Introduction

We think that the cloud may allow avoiding potential edge conditions arising from inadequate population sizes. It could also prompt the evolutionary computation community to return to natural evolution for another round of inspiration where there are fewer limitations imposed by the scale of an algorithm's computing resources. In this paper we relate our progress in designing an evolutionary algorithm (EA) library to run on the cloud. To date we use the Hadoop open-source MapReduce style distributed data processing framework [1] and implement genetic programming (GP) with a single island.

The library has capabilities that are not presently addressed in evolutionary computation precedents. There are no MapReduce GP precedents to our knowledge. Pushing beyond the one-off kind of MapReduce implementations of genetic algorithms (GAs) and Differential evolution algorithms that hard code their fitness evaluation and representation-based operators into mappers and reducers, its use of serialization allows these elements to be plugged in like they usually are in EA implementations. The library also provides the programmer with an abstraction of GP which hides the distributed implementation (which uses MapReduce). Specificly:

- An *evolve* method runs evolution on the island. The programmer does not have to be concerned that, within this method, fitness evaluation, selection, and variation occur on different computation partitions known as mappers and reducers.

C. Di Chio et al. (Eds.): EvoApplications 2012, LNCS 7248, pp. 416–425, 2012.
© Springer-Verlag Berlin Heidelberg 2012

- The programmer can introduce alternate genome representations, fitness functions and variation operators which can be called by the **evolve** method, without having to be concerned that MapReduce distributes them.
- We provide the programmer with an abstraction for collecting statistics for an island which hides the underlying MapReduce implementation.

Implicit in our goals is execution efficiency. We also want an extensively validated and extended version of our library to eventually operate as a modeling or optimization service in a service oriented architecture. For this we are designing for eventually supporting "application customization" which would allow a representation, fitness function and set of genetic operators to be passed to the library for use. The library would eventually inter-operate with a browser-based application (or, eventually, iPad or mobile phone application). We will eventually support the uploading of software files with problem-specific fitness function, representation and operators plus training data from the browser. Our vision extends to the application providing online execution monitoring plugins which access data flowing from the cloud execution. For this, library data stream and storage formats will have to be cloud-standard so that off-the shelf visualization tools can be used.

We intend to make best use of Hadoop open software libraries and software infrastructure to eliminate reinventing the wheel. We intend to release a public version of our library once it is sufficiently mature.

In this paper, our contributions are: to describe aspects of the library's design to date, to preliminarily evaluate the library's GP module's scaling properties and to explore efficient cluster configurations using the Amazon Elastic Compute Cloud ($EC2$).

We proceed as follows: in Section 2, we discuss our motivation for a cloud-scale EA library: it is a prerequisite to fulfilling our vision for a research project we alternately call FlexEA (sometimes also called FlexGP). This section includes related work. In Section 3 we describe our library design. In Section 4 we present a very preliminary investigation. In Section 5 we consider the design advantages and disadvantages of using MapReduce. Section 6 concludes and lists future work.

2   Motivations and Precedents

Motivation: Evolution is an overwhelmingly complex process. Yet today's EA designs are motivated mostly by its fundamentals: populations, selection and genetic inheritance, not its complexities. Our project, FlexEA, is influenced by our impatience to transfer, for regular use, more of the complex mechanisms and phenomena of natural evolution, such as speciation, complex migration and phenotype development to our algorithms. Because we believe massive computing power will be required, we are in the process of examining how well cloud computing will serve our purposes. We also aspire to a set of scalable, on-demand, application-customizable competencies, e.g. optimization and modeling, fulfilled by evolutionary computation, available as a utility, cloud-based software service component in a service oriented architecture (SOA).

FlexEA is currently focused on advances in genetic programming but our agenda will eventually also include discrete and continuous evolutionary optimization. We are also committed to solving real world applications. For these uses, we need a general purpose, cloud-scale, EA library which can support the introduction of different representations, operators and selection mechanisms with eventual transition to becoming a SOA component.

As a cloud programming framework we are pursuing using Hadoop MapReduce because of its general advantages. Hadoop is an open-source MapReduce distributed data processing framework which offers a set of cloud infrastructure services. Its distributed file system is fault tolerant through distributed data replication and presented to the software engineer via an API with a single file system abstraction. It optimizes data communication by scheduling computations near the data. Its master node-client worker architecture naturally load balances because it uses a global task queue for task scheduling. It reduces data transfer overheads by overlapping data communication with computations when map and reduce steps are involved. It offers fault tolerance features, scalability and is claimed to be easy to use. It duplicates executions of slower tasks and handles failures by rerunning the failed tasks using different workers. These features allow it to offer automatic parallelization and an efficient implementation of MapReduce [2].

Precedents: There are precedents for using MapReduce in machine learning, e.g. the Apache MAHOUT project [3]. Precedents in evolutionary computation include [4, 5, 6, 7, 8, 9]. A first result concurred with the intuition that a non-iterative framework such as MapReduce is ill suited to the iterative nature of EAs [4]. However, this was later refuted by [5](pg 9) whose approach is to "hammer the GAs to fit into the MapReduce model" [5](pg 9). Each generation is configured as a MapReduce job and the algorithm runs in non-overlapping generations (see Figure 1). Keys of mappers are randomized to ensure a randomized shuffle staging to reducers. A sliding window-based selection algorithm design which allows reducers to approximate how tournament solution selects the best of a tournament is described. Job overhead is incurred but in [10] a means of breaking the MapReduce stage barrier has a constant 15% benefit with increasing dataset size. The authors of [7] demonstrate a simple (OneMax) GA MapReduce implementation scaling well with large dimensions ($10^4...10^5$) and population sizes up to 16 million.

In [9] the same MapReduce design is used for a GA that substitutes a domain specific representation, fitness evaluation and crossover to competently solve real world problems from the domain of job shop scheduling problems. A co-evolutionary approach in [8] which requires a local fitness evaluation for each candidate solution plus population-relative fitness evaluation also uses MapReduce. The local fitness evaluation is partitioned to mappers then reducers compute the population-relative fitness.

3   Library Design

Packages, Class Hierarchy: Our library, called **EDO-Lib**, has a core package named *evodesign.core*. It has general classes to support EAs. It has a `Reporter` class and an `IslandModel` class. Sub-packages include *core.individual*, *core.fitness*, and *core.operator*. These are appropriately specialized to a set of general purpose sub-packages used in most GA and GP design such as: *ga.fitness*, *ga.individual*, *ga.operator* and *gp.fitness*, *gp.fitness.casefeeder*, *gp.individual*, plus *gp.operator*. Specialized packages extend *evodesign.core* to execute core functionality on different platforms. For example, *flexea.mapreduce* is a package with our implementation of a cloud-based EA using the Hadoop MapReduce framework.

The *islandmodel* package is a sub-package of *evodesign.core* and contains an `Island` class and a means of initializing multiple islands with migration. An island's central method is ***evolve(nbrGenerations)***. The class `HIsland` specializes it for a Hadoop MapReduce island. The ***evolve*** method of `HIsland` starts by writing serialized versions of the `Initialize`, `FitnessEvaluator`, `Crossover` and `Mutation` objects to the Hadoop Distributed File System (HDFS) for mappers and reducers to retrieve them. It initializes the sequence of jobs (one per generation) by setting up the map and reduce classes. It then dispatches the jobs for execution and waits for them to terminate. The `HIsland` ***evolve*** method can invoke the classes and methods of the *evodesign.core* package and inherit their specializations. Thus we are able to design the GP and GA classes and methods at an algorithmic abstraction which is isolated from the MapReduce framework wherever the design is general. When an implementation is specific to MapReduce, such as a monitoring component that extends `Reporter` to write to the HDFS or a `FitnessCaseFeeder` that reads from it to access fitness cases for GP symbolic regression, we develop in a specialized package *flexea.mapreduce*.

EA Design: For our actual implementation with MapReduce, we make a number of design decisions which follow [5] and others. Specifically, we set up a MapReduce job to run for each generation. Each generation, multiple mappers execute fitness evaluation of the population. Each sends the fitness result for a candidate solution to a random reducer. Each reducer runs a tournament and creates offspring with genetic operators before writing batches of the new generation to HDFS. See Figure 1. To initialize the population, we first execute another MapReduce job with identity reducers. It leaves data optimally placed for the subsequent generations like [9]. We randomize the keys of mappers to ensure a randomize shuffle staging to reducers per the example of [5]. We also use the sliding window-based selection algorithm design of [5] which allows reducers to approximate how tournament solution selects the best of a tournament. Other library design decisions involved novel considerations:

Fitness Function Serialization: The fitness function `FitnessEvaluator` is procedural and also a first class object. It needs to be compiled and dispatched to the cluster before the algorithm executes. In a non-Hadoop context, i.e. when using *evodesign.core* package, this is not necessary because the `FitnessEvaluator` can be set by injection, e.g. `island.setFitnessEvaluator(fitnessEvaluator)`.

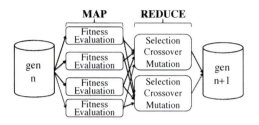

Fig. 1. Precedent-based mapping of EA generation to MapReduce framework

Then it can be used by the **evolve** method. In Hadoop, injection will not work because mappers and reducers are already defined and initialized across the Hadoop cluster and an injection only would have affect on the Master node. The library therefore serializes `FitnessEvaluator` and other functions which are also first class objects such as `Initialize`, `Crossover` and `Mutate` of the `Island`, i.e. writes them to HDFS so they can be retrieved when needed. For example, `Initialize` is retrieved by the initialization mappers, `FitnessEvaluator` by the `EAMapClass` mappers that will perform fitness evaluation and `Crossover` by the `EAReduceClass` reducers that will perform crossover.

Reporters: A core `reporter` allows any component to publish data to a sink such as MySQL database, web service, HDFS or Amazon S3. To aggregate distributed execution monitoring data, we specialize this core class. A specialized `Reporter` injected into the `Island` or one of its operators, e.g. `FitnessEvaluator`, sends to the sink providing a single point abstraction for data writing. A `Reporter` is serialized with its associated object through the HDFS. Thus, while the EA is executing, `Reporter` sinks can be accessed for monitoring at the browser. If the data is a MySQL database, the database can also be queried via a MySQL client. At the moment we are visualizing the data streams in real time using a MySQL database connector from Matlab.

Island Model and Migration: The library currently supports only one island. Later, to support multiple islands, it will have a channel class so that an island will be able to write emigrants to a channel and these can be read by multiple islands to pick up immigrants. The library implementation will initially implement a channel with the HDFS because its API provides a convenient single file system abstraction. This mechanism assures asynchronous island evolution and requires a minimal exchange of mailbox addresses. Conceptually the design is simple and efficient.

4   Library in Use

We develop our library on a single node Hadoop cluster and can unit test and debug with a small number of in-house 8 core servers. The results of this paper use Hadoop clusters we configured on Amazon $EC2$ [11]. We assign our Hadoop task master to one node and use each node as two mappers and reducers. Each node is a dual core Amazon large instance [11]. The large instance, costing $0.34/hr makes a large cluster relatively expensive. For example, given our initial timings, to solve this problem once on a 32 node cluster with a population size of 16384 running 100 generations will take approximately one hour, implying a set of 30 runs will cost $350. The cost of 30 runs with a population size of 1Million on 256 nodes would cost $8600! We expect to find optimizations to our setup to improve this by examining precisely why we need a large instance and hopefully only making just the task master a large instance. We also configure one node as a MySQL server to receive run data.

We consider a GP symbolic regression problem from [12] of dimensionality two:

$$f(x1, x2) = \frac{e^{-(x1-1)^2}}{1.2 + (x2 - 2.5)^2} \tag{1}$$

The fitness objective is to minimize the mean prediction error on 100 fitness cases. It executes in 4.0×10^{-3} seconds with essentially zero variance. Each mapper generates a random set of fitness cases at initialization.

To obtain the results in Table 1 for each cell we executed a single run of only 3 generations. We timed the initialization and final generation overhead for that run and added them to the averaged execution time of the three generations multiplied by 100 to estimate the time to execute 100 generations.

The function set is $x, y, +, *, -, \%,$ $exponent, square, cos, sin$, where % symbolizes protected division. The library has an implementation of standard GP tree crossover and biases internal node selection with a 90:10 ratio and uses maximum tree height of 17. Crossover probability is 1.0. Subtree mutation rate is 0.001. Candidate solutions in the initial population are generated with standard GP ramped half and half with a minimum tree depth of 2 and a maximum tree depth of 6. Tournament size (i.e. the moving window of the MapReduce approximation of tournament selection) is 7.

We estimate scaling in two dimensions: population size and cluster size. For a population size, we can examine the execution time of different cluster sizes. A cluster size trades off potential parallelism with communication overhead. Table 1 shows that while a 64 node cluster is ideal for population sizes 1024 and 4096, for some reason a 32 node cluster is fastest when population size increases to 16384. This result could be an artifact of our cursory estimation. When population size further increases, larger clusters become faster as would be expected: 128 nodes for population sizes 65536 and 262144, then 256 nodes for a population of approximately 1Million. Figure 2-right shows that allocating more resources to the problem at the same rate as increasing the population size provides a constant expected time for the run to complete. As well, for a cluster configuration, we can examine how execution time scales with increasing population sizes. See Figure 2, right. The result for a 256 node cluster for population size 1024 is probably due to some communication glitch within the cluster.

Table 1. Estimated Execution times for 100 generations of the benchmark on a Amazon $EC2$ cluster where the number of nodes doubles up from 2 to 256. Asterisk identifies cluster size offering shortest execution time.

Nodes/Cluster	2	4	8	16	32	64	128	256
Population Size								
1024	3654	3104	3152	3159	2903	2750*	2952	47309
4096	4822	4215	3864	3257	2956	2909*	3112	11866
16384	9721	4296	5474	4463	3913*	4569	6459	11724
65536	27916	17562	11090	7452	5887	7279	3903*	12603
262144		58184	34338	18792	11817	8432	6124*	10706
1048576				64567	34180	21284	13321	12130*

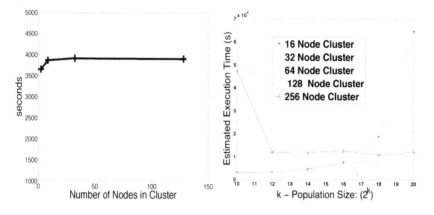

Fig. 2. Left: As population size is doubled concurrent with number of nodes in the cluster, estimated time to execute 100 generations is essentially constant. Right: Execution time scaling with increasing population size for different cluster sizes.

5   Design Discussion

Design Abstraction: An ideal library completely hides the MapReduce level and allows the programmer to focus upon EA design without spending too much effort considering the distributed implementations. We believe packaging and class hierarchy manage to fulfill this objective to a large extent. The true test will occur over the next year as more programmers in our group use the library. We will be able to observe how much they need to know about the MapReduce implementation and how much this influences their designs. When their design adds core functionality they will have to assure it can be implemented by MapReduce.

Design Scaling: The automatic scaling and parallelization of MapReduce is very compelling. The library is able to support much larger population sizes on an island than conventional MPI or socket implementations which map an island to a compute node. In MapReduce an island's population is distributed on as many nodes as desired. This implies the MapReduce could support the largest population sizes ever computed for a single island, even surpassing GPU implementations.

Design Complexity: A potential design complication is that the island level abstraction encourages the programmer to assume access to global knowledge of the population. However, it is not available because the MapReduce layer partitions out the computation. Consider tournament selection which can be thought of as global at the island abstraction but which we implement in the MapReduce layer in a partitioned manner. The MapReduce tournament selection design of [5], at large scale, faithfully emulates global tournament selection. It remains to be determined if other selection algorithms can support an underlying partitioning. One key unresolved example is multi-objective optimization where a pareto front of the population is used for selection.

Another potential complication is the assumption that fitness of each member of the population can be evaluated independently. This is not an assumption that holds for co-evolutionary dynamics where an individuals are compared pairwise or in bigger aggregations.

Job Overhead: MapReduce adds the overhead of job management to a EA run. We expect the overhead cost is best mitigated by using the MapReduce implementation for applications with computationally expensive fitness evaluation costs and/or which require large population sizes. Fitness functions with highly variable evaluation time may also be well suited to the Hadoop MapReduce implementation because Hadoop naturally load balances.

Performance Control: By using a programming framework like Hadoop MapReduce, we lose some transparency. In this situation, how can we tune for performance? We are concerned about the virtualization aspect of the cloud. Virtual host resource sharing will impact the consistency of resource availability. For example, Amazon uses XEN[1] and it offers only a small set of container types so that it can commission XEN hosts appropriately. However, we never know what VMs are active and sharing our resources when tuning. Ultimately Amazon, not our service, controls resource levels.

We will need to tune the population size with respect to generations and evaluation budget for a cluster configuration because the number of nodes can impact the performance. Sweeping the node size introduces a combinatorial problems. We are hopeful that we will find a configuration that is efficient for a span of population size to fitness evaluation cost ratios but we have not as yet had time to do so. We expect that we will have to eventually build a preliminary scaling investigation capability into the library. The EA service could be configured to run this automatically.

Development Effort: We estimate that it takes much longer to design a MapReduce implementation than it would to develop a socket or MPI model.

---

[1] http://xen.org/

There is a lot of software engineering overhead required to develop with Hadoop MapReduce libraries. In terms of ease of fit between the MapReduce programming framework and an EA control flow, there is no doubt that MapReduce's data flow, non-iterative model adds software engineering overhead in a sort of "square peg, round-hole" scenario. It has been somewhat complicated to map an iterative algorithm with an island level model to a non-iterative paradigm with parallelism inside the island. It also results in a code base which requires more effort to support and maintain which impacts research agility.

A mature, meticulous and conscientious software engineer has developed our library to date over the last 8 months working approximately 3 days weekly. Much of the time was spent understanding libraries and debugging. We think that the same island model implemented with sockets or MPI on the cloud would have taken much less time to develop because they involve much less library and infrastructure. However, this extra time, to us, is well invested because Hadoop MapReduce takes all responsibility for node failures and node resumption whereas we would have to look after this ourselves with MPI or sockets. As well, Hadoop MapReduce naturally load balances. It took us a week to engineer the Amazon $EC2$ and Hadoop infrastructure to set up a cluster that we can simply extend in terms of nodes. This has already brought great convenience. We obtain the sort of the automatic parallelization MapReduce offers. So, at the moment it seems that start up expense will be discounted well against these future benefits. We will have to see whether this infrastructure is easy to maintain, robust and stable.

6   Conclusions and Future Work

In summary, this paper described, from a practical software engineering vantage, using Hadoop MapReduce when providing a library or "service", not just an EA for one problem. It provided preliminary evaluation of using MapReduce programming framework for genetic programming.

Future performance analysis of the library will proceed hand in hand with performance optimization. We intend to use compression to improve effective data communication bandwidth. We also will revise our initialization to use mappers more efficiently. We intend to dig further into Hadoop to understand further performance tuning potential. We envision adding an auto-tuning component to the library which will, for a specific fitness function, investigate population size, node and mapper allocation to recommend an efficient setting for subsequent runs.

Acknowledgements. This material is based on work supported under a Portuguese National Science and Technology Foundation Graduate Research Fellowship, by FCT grant number **SFRH/BD/60481/2009**. The authors acknowledge the generous support of General Electric Global Research. Any opinions, findings, conclusions, or recommendations expressed in this publication are those of the author and do not necessarily reflect the views of the National Science and Technology Foundation, or the Portuguese government or General Electric Global Research.

References

1. Web resource: ApacheHadoop, `http://hadoop.apache.org/core`
2. Gunarathne, T., Wu, T.L., Qiu, J., Fox, G.: MapReduce in the clouds for science. In: 2010 IEEE Second International Conference on Cloud Computing Technology and Science, CloudCom (2010)
3. Web resource: MAHOUT, `http://mahout.apache.org/`
4. Jin, C., Vecchiola, C., Buyya, R.: MRPGA: An extension of MapReduce for parallelizing genetic algorithms. In: IEEE Fourth International Conference on eScience 2008, pp. 214–221. IEEE (2008)
5. Verma, A., Llora, X., Campbell, R., Goldberg, D.: Scaling genetic algorithms using MapReduce. Technical report, Illigal TR 2009007
6. Verma, A., Llora, X., Venkataraman, S., Goldberg, D., Campbell, R.: Scaling ECGA model building via data-intensive computing. In: 2010 IEEE Congress on Evolutionary Computation, CEC (2010)
7. Verma, A., Llora, X., Goldberg, D., Campbell, R.: Scaling genetic algorithms using MapReduce. In: Ninth International Conference on Intelligent Systems Design and Applications, ISDA 2009 (2009)
8. Wang, S., Gao, B.J., Wang, K., Lauw, H.W.: Parallel learning to rank for information retrieval. In: Proceedings of the 34th International ACM SIGIR Conference on Research and Development in Information, SIGIR 2011, pp. 1083–1084. ACM, New York (2011)
9. Huang, D.W., Lin, J.: Scaling populations of a genetic algorithm for job shop scheduling problems using MapReduce. In: 2010 IEEE Second International Conference on Cloud Computing Technology and Science, CloudCom (2010)
10. Verma, A., Zea, N., Cho, B., Gupta, I., Campbell, R.: Breaking the MapReduce stage barrier. In: 2010 IEEE International Conference on Cluster Computing (CLUSTER), pp. 235–244. IEEE (2010)
11. Web resource: Amazon EC2, `http://aws.amazon.com/ec2/`
12. Vladislavleva, E., Smits, G., Den Hertog, D.: Order of nonlinearity as a complexity measure for models generated by symbolic regression via pareto genetic programming. IEEE Transactions on Evolutionary Computation 13(2), 333–349 (2009)

A Fair Comparison of Modern CPUs and GPUs Running the Genetic Algorithm under the Knapsack Benchmark

Jiri Jaros[1] and Petr Pospichal[2]

[1] The Australian National University, ANU College of Engineering and Computer Science,
Canberra, ACT 0200, Australia
jiri.jaros@anu.edu.au
[2] Brno University of Technology, Faculty of Information Technology,
Bozetechova 2, 612 66 Brno, Czech Republic
ipospichal@fit.vutbr.cz

Abstract. The paper introduces an optimized multicore CPU implementation of the genetic algorithm and compares its performance with a fine-tuned GPU version. The main goal is to show the true performance relation between modern CPUs and GPUs and eradicate some of myths surrounding GPU performance. It is essential for the evolutionary community to provide the same conditions and designer effort to both implementations when benchmarking CPUs and GPUs. Here we show the performance comparison supported by architecture characteristics narrowing the performance gain of GPUs.

Keywords: GPU, multicore CPU, knapsack, performance comparison, CUDA.

1 Introduction

The Genetic Algorithms (GAs) have become a widely applied optimization tool since developed by Holland in 1975 [4]. Many researchers have shown GA abilities in real-world problems such as optimization, decomposition, scheduling and design. As the genetic algorithms are population based stochastic search algorithms, they often require hundreds of thousands test solutions to be created and evaluated.

One of the advantages of the genetic algorithms is their ability to be easily parallelized. During the last two decades, plenty of different parallel implementations have been proposed, such as island based or spatially structured GAs [19].

The trend over last few years has been to utilize Graphics Processing Units (GPUs) as general purpose co-processors. Although originally designed for rasterization and the game industry, their raw arithmetic power has attracted a lot of research [6], [15].

The evolutionary community has adopted this trend relatively quickly and a lot of papers have been presented in this area, collected e.g. at www.gpgpgpu.com. However, a developer experienced in computer architectures can shortly see that there is something amiss in the state of GA. Most of the papers compare the speedup of the GPU implementation against a sequential version, moreover, mostly implemented in the simplest possible way [8], [13], [16]. This is an exact contradiction with the way

C. Di Chio et al. (Eds.): EvoApplications 2012, LNCS 7248, pp. 426–435, 2012.

the speedup of the parallel processing is defined. G.A. Amdahl in 1967 stated the speedup as the performance of the parallel (GPU) version against the performance of the best known sequential version (an SSE/AVX multi-thread CPU one in the 21st century) [1]. What meaning would it make to accelerate the BubbleSort algorithm on GPUs and compare it with a sequential CPU one both with $O(n^2)$ when a parallel QuickSort with $O(n \log n)$ could be employed?

The main goal of this paper is to show a proper CPU and GPU implementation of the GA, written the ground up taking into account each the architectures features. This paper puts the reached speedups into relation with the architecture performance and discusses the validity of the results. We will simply convince ourselves that there is no way to reach speedups in order of 100 and beyond [7].

The well-known single-objective 0/1 knapsack problem is used as a benchmark. It is defined as follows: given a set of items (L), each with a weight $w[i]$ and a price $p[i]$, with $i = 1,..,L$. The goal is to pick such items that maximize the price of the knapsack and do not excess the weight limit (C) [18]. The solutions that break this limit are penalized according to amount of overweight and the peak price-weight item ratio.

As target architectures we have chosen the leading GPU and CPU on the market, namely the NVIDIA GTX 580 and the Intel Xeon X5650. It does not make any sense for the high performance computing community to compare with desktop CPUs providing that real-word problems require to be run on servers for a long time period.

2 Memory Layout of the GA

The section describes the memory layouts of the population, statistics and global data structures. It is crucial to allocate all host structures intended for host-device transfers by CUDA pinned memory routines. This makes it possible to use the Direct Memory Access (DMA) and reach the peak PCI-Express performance [11]. On the other hand, the host memory should be allocated using the `memalign` routine with 16B alignment when implementing the CPU only version. This helps CPU vector units to load chunks of data much faster and the compiler to produce more efficient code.

2.1 Population Organization

The population of GA has been implemented as a C structure consisting of two one-dimensional arrays. The first array represents the genotype while the second one represents fitness values. Assuming the size of the chromosome is L and the size of the population is N, the genotype is defined as an array$[N*L/32]$. As the knapsack chromosomes are based on the binary encoding, 32 items are packed into a single integer. This rapidly reduces the memory requirements as well as accelerates genetic manipulations employing logical bitwise operations. The fitness value array has the size of N.

Two different layouts of genotype can be found in literature [16]. The first one, referred to as chromosome-based, represents chromosomes as rows of a hypothetical

2D matrix implemented as a 1D array whilst the second one, referred to as gene-based, is a transposed version storing all genes with the same index in one row.

The chromosome-based layout simplifies the chromosome transfers in the selection and replacement phases as well as the host-device transfers necessary for displaying the best solution during the evolutionary process and in more advanced island-base models. In this case, multiple CUDA threads work on one chromosome to evaluate its fitness value. This layout should be preferred also for the CPU implementation in order to preserve data locality and enable the CPU to store chromosomes in the L1 cache and exploit modern prefetch techniques.

On the other hand, the gene-based representation allows working with multiple chromosomes at a time utilizing the SIMD/SIMT nature of CPUs and GPUs assuming there are no dependencies between chromosomes. However, evaluating multiple chromosomes at a time tends to run out of other resources such as registers, cache, shared memories, etc.

Taking into account architecture characteristics, the only thing that matters is to allow threads inside a warp to work on neighbor elements. Different warps can access different memory areas with only a small or no penalization. The chromosome-based layout seems to be the most promising layout enabling the warp to work with the genes of one chromosome, especially, if it is necessary for the fitness evaluation to read genes multiple times. The different warps can simply operate on different chromosomes. This reaches the best SIMD (SSE, CUDA) performance while reducing registers, share memory, and cache requirements. For this reason, the chromosome-based layout is used for both CPU and GPU.

2.2   GA Parameters Storage

A C data structure has been created to accommodate all the control parameters of the GA. Such parameters include the population and chromosome size, the crossover and mutation ratio, the statistics collecting interval, the total number of evaluated generations etc. Once filled in with command line parameters, the structure is copied to the GPU constant memory. This simplifies CUDA kernel invocations and saves memory bandwidth according to the CUDA C best practice guide [11].

2.3   Knapsack Global Data Storage

The knapsack global data structure describes the benchmark listing the price and weight for all items possible included in the knapsack. The structure also maintains the capacity of the knapsack and the item with the maximum price/weight rate. The prices and weights are stored in two separate 1D arrays. The benefit over an array of structures is data locality as all the threads first read prices and only then the weights.

The best memory area where to place this structure may seem to be the constant memory. Unfortunately, this area is too small to accommodate real-world benchmarks. Its capacity of 64KB allows solving problems up to 4K items. On the other hand, introducing L2 caches and a load uniform (LDU) instruction in Fermi cards [11] makes the benefits of constant memory negligible supposing all threads within a warp accesses the same memory location. As the result, the global data are stored in main GPU memory. The problem size (the chromosome size in bits) is always padded to a multiple of 1024 to prevent uncoalesced accesses.

3   Genetic Algorithm Routines

This section goes through the evolution process and comments on the genetic manipulation phase, fitness function evaluation, replacement mechanism and statistics collection. Each phase is implemented as an independent CUDA kernel to put global synchronization between each phase. The source codes can be downloaded from [5].

All the kernels of the GA has been carefully designed and optimized to exploit the hidden potential of modern GPUs and CPUs. It is essential for a good GPU implementation to avoid the thread divergence and to coalesce all memory accesses to minimize the required memory bandwidth. Thus, the key terms here are the **warp** and the **warp size** [11] . In order to write a good CPU implementation, we have to meet exactly the same restrictions. The warp size is now reduced to SSE or AVX width and coalescing corresponds to L1 (L2, L3) cache line accesses while GPU shared memory can be directly seen as the L1 cache.

As the main principles are the same, the CPU implementation follows the GPU one adding only an outer-most `for` cycle and the OpenMP `pragma omp parallel for` sections [2] to utilize all available CPU cores and simulate GPU execution.

3.1   Random Number Generation

As genetic algorithms are stochastic search processes, random numbers are extensively used throughout them. CUDA does not provide any support for on the fly generation of a random number by a thread because of many associated synchronization issues. The only way is to generate a predefined number of random numbers in a separate kernel [10]. Fortunately, a stateless pseudo-random number generator has recently been published based on hash functions [14]. This generator is implemented in C++, CUDA and OpenCL. The generator has been proven to be crush resistant with the period of 2^{128}. The generator is three times faster than the standard C `rand` function and more than 10x faster than the CUDA `cuRand` generator [12], [14].

3.2   Genetic Manipulation Phase

The genetic manipulation phase creates new individuals performing the binary tournament selection on the parent population and exchanging genetic material of two parents using uniform crossover with a predefined probability. Every gene of the offspring is mutated by the bit-flip mutation and stored in the offspring population.

The key for the efficient implementation of the genetic manipulation kernel is a low divergence and enough data to utilize all the CUDA cores. Each CUDA block is organized as two dimensional. The x dimension corresponds to the genes of a single chromosome while the y dimension corresponds to different chromosomes. The size of the x dimension meets the warp size of 32 to prevent lots of divergence within a warp. The size of y dimension of 8 is chosen based on the assumption that 256 threads per block is enough [15].

The entire grid is organized in 2D with the x size of 1, and the y size corresponding to the offspring population size divided by the double of the y block size (two offspring are produced at once). Since the x grid dimension is exactly the 1, the warps process the individuals in multiple rounds.

The selection is performed by a single thread in a warp. Based on the fitness values, two parents are selected by the tournament and their indices within the parent population are stored in shared memory.

Now, each warp reads two parents in chunks of 32 integer components (one integer per thread). As binary encoding enables 32 genes to be packed into a single integer, the warp effectively reads 1024 binary genes at once. Since this GA implementation is intended for use with very large knapsack instances, uniform crossover is implemented to allow better mixing of genetic material. Each thread first generates a 32b random number serving as the crossover mask. Next, logic bitwise operations are used to crossover the 32b genes. This removes all conditional code from the crossover except testing of the condition whether or not to do the crossover at all. This condition does not introduce any thread divergence as it is evaluated in the same way for the whole warp.

Mutation is performed in a similar way. Each thread generates 32 random numbers and sets the bit of the mask to 1 if the random number falls into the mutation probability interval. After that, the bitwise `xor` operation is performed on the mask and the offspring. This is done for both the offspring. Finally the warp writes the chromosome chunk to the offspring population and starts reading the next chunk.

3.3   Fitness Function Evaluation

The fitness function evaluation kernel follows the same grid and block decomposition as the genetic manipulation kernel. Evaluating more chromosomes at a time allows the GPU to reuse the matching chunk of global data and saves memory bandwidth.

Every warp processes one chromosome in multiple rounds handling a single 32b chunk at a time. In every round, the first warp of the block transfers the prices and weights of 32 items into shared memory employing coalesced memory accesses. After the barrier synchronization, every warp can read the knapsack data directly from shared memory. Now, every warp loads a single 32b chunk into shared memory. As all the threads within a warp access the same memory location (one integer), the L2 GPU cache is exploited. Every thread masks out an appropriate bit of the 32b chunk, multiplies it with the item price and weight, and stores the partial results into shared memory. When the entire chromosome has been processed, the partial prices and weights of the items placed in the knapsack have to be reduced to a single value. Since the chromosome is treated by a single warp a barrier-free parallel reduction can be employed. Finally, a single warp thread checks the total capacity of all the items and if the capacity has been exceeded, the fitness is penalized. Finally, the fitness value is stored in the global memory.

The CPU implementation evaluates chromosomes one by one provided that the global data can be easily stored in L3 cache. The evaluation process is distributed over multiple cores using OpenMP. The evaluation can be carried out immediately after a new offspring has been created which results in the chromosome being evaluated stored in L1 cache. This might also be possible for the GPU implementation, however, the kernel would run out of registers and shared memory resulting in poor GPU occupation and low performance.

3.4 Replacement Phase

The replacement phase employs the binary tournament over the parents and offspring to create the new parent population. The kernel and block decompositions are the same as in the previous phases. The only modification is that the kernel dimensions are derived from the parent population size.

Every warp compares a randomly picked offspring with the parent laying on the index calculated from the y index of the warp in the grid. If the offspring fitness value is higher than the parent one, the entire warp is used to replace the parent by the offspring. This restricts the thread divergence to the random number generation phase.

3.5 Statistics Collection

The last component of the genetic algorithm is the class collecting necessary statistics about the evolutionary process. It maintains the best solution found so far, and handing them over the CPU for saving into a log file.

The statistics collection consists of a kernel and statistics structure initialization. The GPU statistic structure maintains the highest and lowest fitness values over the population as well as the sum and the sum-of-squares over of fitness values. The last two values are necessary for calculating the average fitness value and the standard deviation. The last value is the index of the best individual.

The kernel is divided into twice as many blocks as the GPU has stream processors. Each block is decomposed into 256 threads based on the practice published in [15]. After the kernel invocation, the chunks of fitness values are distributed over the blocks. Each thread processes as many fitness values as necessary and stores the partial results into shared memory. After the barrier synchronization, the reductions over highest, lowest and two sum values are carried out. Finally, the first thread of each block uses a global memory lock to modify the global statistics.

After completion, the statistics structure is downloaded to host memory to compute average value and the standard deviation over the fitness values. Finally, the best solution is downloaded from GPU based on the index stored in the statistics structure.

The CPU implementation of the statistics collection has been left in a sequential form because the overhead of parallel execution would exceed the execution time provided by parallel processing.

4   Experimental Comparison of CPU and GPU Implementations

The goal of the experiments is to compare an optimized multicore CPU implementation with a well-designed GPU version and provide some insight into realistically achievable speedups. All the experiments were carried out on a dual Intel Xeon X5650 server equipped with a single NVIDIA GTX 580 running Ubuntu 10.04 LTS.

The knapsack benchmark with 10,000 items and a population size of 12,000 individuals were used. We chose such a big benchmark and large population to show the most optimistic results. The smaller the benchmark and population are, the slower a GPU will be compared to a multicore CPU. This is given by the massively parallel architecture of modern GPUs. Six thousand new individuals are created and evaluated every generation. The genetic algorithm works with tournament selections and replacement, a crossover ratio of 0.7, and a mutation ratio of 0.01. The statistics are collected after every generation. All the proposed codes were compiled using GNU C++ with the highest optimization level, SSE 4.2 support, the OpenMP library [2] and the CUDA 4.0 developer kit [11].

As the reference, we chose the GALib library [20] adopted by a lot of scientists. GALib is a comprehensive rapid prototyping library for evolutionary algorithms, however, the last version comes from 1997. Because of its age, the library cannot benefit from vector units (SSE/AVX) or multiple cores.

In order to validate the optimization abilities of the proposed implementations, we carried out 30 independent runs. The average highest fitness values reached after 100 generations as well as the standard deviation are plotted in Fig. 1. Although there is a statistically significant difference among the implementations, the practical impact on the result quality is negligible (lower than 0.1%).

The performance results are revealing. As a lot of researchers do not pay enough attention to the CPU implementation, the GALib is often compiled under default conditions without any optimization and with debugging support enabled. This degrades the performance (GALib-D) to 375 times slower than the GTX 580. Just a trivial modification of the GALib makefile (turning on appropriate compiler optimizations) can bring a huge performance gain for free. The GALib-O (Optimized) version is 221 times slower than the GPU (see Fig. 2).

Implementing the CPU version carefully rapidly decreases the execution time. The single thread (1T) implementation is 68 times slower than the GPU. This is similar to the speedup reported in many other studies, e.g. [17]. However, parallelization of the 1T version is trivial. It is only necessary to put three OpenMP pragmas in the entire code. The impact on the performance is significant! Running the GA on a single six-core processor reduces the speedup to 11.82 (5.78 faster than 1T). As common HPC servers are equipped with multiple CPU sockets, the dual Xeon5650 server takes only 6 times longer time to perform the task. This is appreciably different to the results reported speedups to 800x, 1000x, 2072x in [13], [16], [8], respectively.

The reason for such a big difference in the CPU codes is shown in Table 1. Three different CPU implementations were investigated using the PAPI performance counter library [9]. The key to the fast CPU code is to utilize cache memories properly.

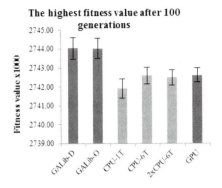

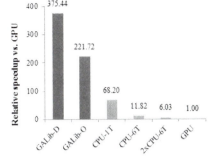

Fig. 1. Solution fitness value after 100 generations achieved by different implementations

Fig. 2. Speedup comparisons of GPU against different CPU implementations

The first line of the table clearly shows an awful L2 cache hit of the GALib. The problem lays in the way genotype and phenotype are organized, and the immense number of copy constructors employed virtually everywhere. Moreover, from the number of L2 cache accesses we can deduce, the L1 cache is also not exploited (caused by accesses with stride, e.g., when calculating the statistics in GALib). The custom CPU implementation reduces L2 access by factor of 26 (in case of the single thread implementation). The 12 thread implementation further benefits from data distribution over multiple L2 caches. In contrast, the L3 cache hit ratio seems to be better in the case of GALib. However, the number of L3 accesses of GALib is higher by 3 orders which leads to an enormous number of ALU stalls. On the other hand, interpreting the number of cache accesses in the table, the optimized CPU version spends more than 99.99% of time working within the L1 cache.

All these inefficiencies are projected to the CPU performance. GALib only reaches 1,099 MIPS (Million Instructions Per Second). On the other hand, the same six core Xeon running the optimized code reaches about 33,600 MIPS and the entire server can touch up to 67,248 MIPS. We have measured fixed point arithmetic instead of floating point FLOPs here because of the nature of GA encoding and the knapsack prices and weights being encoded as integers.

Given that fixed-point SIMD instructions are nearly as fast as single precision floating-point instructions on the CPU, we can calculate the efficiency of the CPU code. The dual Intel Xeon 5650 server reaches of 118 GFLOPs in the Intel optimized LINPACK benchmark. The overall efficiency of the CPU code is thus about 56%, which is pretty good. As the GPU is about 6 times faster, the peak GPU performance attacks 405 GFLOPS. Although this is roughly one fourth of the theoretical peak performance of the GTX580, it represents a very good result [7]. It should be clear we can never approach the peak performance because of many parallel reductions in the fitness function evaluation and statistics kernel, necessary synchronization and single thread operations inside the genetic manipulation, handing data over to the CPU to store statistics, and many other issues. The peak performance would require all the CUDA cores working in every clock cycle, which is not possible for such operations.

Table 1. PAPI performance counters profililng data of 100k knapack and 100 generations using GALib and the custom implementation on dual Intel Xeon X5650

	GALib-O	CPU-1T	2xCPU-6T
L2 cache hit	18.01%	98.81%	96.82%
L2 accesses	21 267M	800M	24M
L3 cache hit	99.04%	57.07%	81.3%
L3 accesses	17 278M	8M	12M
MIPS	1 099	5 392	67 248
Execution time	165.25s	48.52s	3.81s

In order to compare the 405 GFLOPs of this implementation with other CUDA applications, consider these values measured by SHOC benchmark [3] in single precision: FFT = 213 GFLOPS, GEMM = 529 GFLOPS, parallel reduction = 93 GFLOPS, parallel sort = 2 GFLOPS.

5   Conclusions

This paper points out the way many authors presents the speedups of the GPU implementation of the genetic algorithm against the CPU version. A lot of papers have used only a single thread implementation, [8], [13], [16], [17]. Such authors should have immediately divided their speedups by the factor of 6 at least. Do not forget there is nothing like a single thread CPU on the market any more. As we always need to perform multiple trials to produce good results, we can run as many trials as physical cores with negligible impact on the performance. The trials are embarrassingly parallel. The reason why some authors did not do so might lie in a foolish hunt for the highest speedup, or an attempt to hide the fact the performance gain by a GPU would have been so low that it would not have justified the amount of effort put it into.

The best GPU on the market has the peak performance of 1.5 TFLOPS while a typical server processor reaches the peak of 60 GFLOPs. Confronted with these architecture limits, it is not possible to report speedups of more than 100. Such speed-ups only show that CPU implementation is not well optimized. Fair comparison is to say how fast the implementation is in terms of GFLOPS, and what fraction of the peak performance has been achieved.

Modern GPUs have amazing computational power, and it is worth it porting computationally expensive applications such as evolutionary algorithms onto them. However, we must be careful about making the performance comparisons. We have clearly shown that a carefully implemented CPU version can be up to 30 times faster than a single thread default-compiled GALib. We have also shown that realistically a single NVIDIA GTX580 can outperform an Intel Xeon X5650 by a factor of around 12 while reaching an execution efficiency of 26% and performance of 405 GFLOPS. The proposed CPU and GPU implementations have been released as open source at [5].

Acknowledgement. This research has been partially supported by the research grant "Natural Computing on Unconventional Platforms", GP103/10/1517, Czech Science Foundation (2010-13), and the research plan "Security-oriented research in information technology", MSM 0021630528 (2007-13).

References

1. Amdahl, G.M.: Validity of the single processor approach to achieving large scale computing capabilities. In: Proceedings of the April 1820 1967 Spring Joint Computer Conference, vol. 23(4), pp. 483–485 (1967)
2. Chandra, R., Dagum, L., Kohr, D., et al.: Parallel programming in OpenMP. Morgan Kaufmann (2001)
3. Danalis, A., Marin, G., et al.: The Scalable HeterOgeneous Computing (SHOC) Benchmark Suite Categories and Subject Descriptors. In: Proceedings of the Third Workshop on General-Purpose Computation on Graphics Processors, GPGPU 2010 (2010)
4. Holland, J.H.: Adaptation in Natural and Artificial Systems. University of Michigan Press (1975)
5. Jaros, J.: Jiri Jaros's software website,
   http://www.fit.vutbr.cz/~jarosjir/prods.php.en
6. Kirk, D.B., Hwu, W.-M.: Programming Massively Parallel Processors: A Hands-on Approach. Morgan Kaufmann (2010)
7. Lee, V.W., Hammarlund, P., Singhal, R., et al.: Debunking the 100X GPU vs. CPU myth. In: Proceedings of the 37th Annual International Symposium on Computer Architecture, ISCA 2010, p. 451. ACM Press, New York (2010)
8. Luong, T.V.: GPU-based Island Model for Evolutionary Algorithms. Evaluation, 1089–1096 (2010)
9. Malony, A.D., Biersdorff, S., Shende, S., et al.: Parallel Performance Measurement of Heterogeneous Parallel Systems with GPUs. Performance Computing
10. NVIDIA: CUDA Toolkit 4. 0 CURAND Guide (2011)
11. NVIDIA: Cuda c best practices guide (2011)
12. NVIDIA: Math Library Performance CUDA Math Libraries (2011)
13. Pospichal, P., Schwarz, J., Jaros, J.: Parallel genetic algorithm solving 0/1 knapsack problem running on the gpu. In: 16th International Conference on Soft Computing MENDEL, pp. 64–70. Brno University of Technology, Brno (2010)
14. Salmon, J.K., Moraes, M.A., Dror, R.O., Shaw, D.E.: Parallel Random Numbers: As Easy as 1, 2, 3. In: Proceedings of 2011 International Conference for High Performance Computing, Networking, Storage and Analysis, SC 2011, pp. 16:1–16:12. ACM Press, New York (2011)
15. Sanders, J., Kandrot, E.: CUDA by Example: An Introduction to General-Purpose GPU Programming. Addison-Wesley (2010)
16. Shah, R., Narayanan, P., Kothapalli, K.: GPU-Accelerated Genetic Algorithms, cvit.iiit.ac.in
17. Simonsen, M., Pedersen, C.N.S., Christensen, M.H.: GPU-Accelerated High-Accuracy Molecular Docking using Guided Differential Evolution. In: Proceedings of the Genetic and Evolutionary Computation Confernce GECCO 2011. ACM Press (2011)
18. Simões, A., Costa, E.: An Evolutionary Approach to the Zero / One Knapsack Problem Testing Ideas from Biology. In: The Fifth International Conference on Artificial Neural Networks and Genetic Algorithms (ICANNGA 2001), April 22-25 (2001)
19. Tomassini, M.: Spatially Structured Evolutionary Algorithms. Springer, Heidelberg (2005)
20. Wall, M.: GAlib: A C ++ Library of Genetic Algorithm Components. Statistics (August 1996)

Validating a Peer-to-Peer Evolutionary Algorithm

Juan Luis Jiménez Laredo[1], Pascal Bouvry[1],
Sanaz Mostaghim[2], and Juan-Julián Merelo-Guervós[3]

[1] Faculty of Sciences, Technology and Communication,
University of Luxembourg, Luxembourg City L-1359, Luxembourg
{juan.jimenez,pascal.bouvry}@uni.lu
[2] Karlsruhe Institute of Technologie
Kaiserstrasse 89, Karlsruhe D-76133, Germany
sanaz.mostaghim@kit.edu
[3] University of Granada. ATC-ETSIIT
Periodista Daniel Saucedo Aranda s/n 18071, Granada, Spain
jmerelo@geneura.ugr.es

Abstract. This paper proposes a simple experiment for validating a
Peer-to-Peer Evolutionary Algorithm in a real computing infrastructure
in order to verify that results meet those obtained by simulations. The
validation method consists of conducting a well-characterized experiment
in a large computer cluster of up to a number of processors equal to the
population estimated by the simulator. We argue that the validation
stage is usually missing in the design of large-scale distributed meta-
heuristics given the difficulty of harnessing a large number of computing resources. That way, most of the approaches in the literature focus
on studying the model viability throughout a simulation-driven experimentation. However, simulations assume idealistic conditions that can
influence the algorithmic performance and bias results when conducted
in a real platform. Therefore, we aim at validating simulations by running
a real version of the algorithm. Results show that the algorithmic
performance is rather accurate to the predicted one whilst times-to-
solutions can be drastically decreased when compared to the estimation
of a sequential run.

1   Introduction

Given that most computer devices nowadays are connected to the Internet continuously, volunteer computing systems [2] have arisen as an alternative to su-
percomputers or wide-area grid systems. Volunteer computing systems usually
behave in a centralized fashion which might be a problem when huge numbers
of clients are simultaneously connected, posing a challenge to the server, or con-
verting it into a bottleneck in the case the systems become especially large.
Peer-to-Peer (P2P) systems, where no node has any special role, do take advantage of the nature of the Internet and take more advantage of the bandwidth
each node is connected with [14].

C. Di Chio et al. (Eds.): EvoApplications 2012, LNCS 7248, pp. 436–445, 2012.

These systems have received much attention from the scientific community within the last decade. In this context and under the term of P2P optimization, many optimization heuristics such as Evolutionary Algorithms (EAs), Particle Swarm (PSO) or Branch-and-bound have been re-designed in order to take advantage of such computing platforms [16,13,5,3]. The key issue here is that gathering a large amount of computational devices pose a whole set of practical problems. Therefore, and to the best of our knowledge, most of the approaches to P2P optimization –if not all– have been analyzed in simulators rather than in real environments. That way, this paper aims to go an step further and validate a P2P EA in a real large-scale infrastructure running up to 3008 parallel individuals.

To that aim, we consider the results published in [8] on the scalability of the Evolvable Agent model (i.e. a P2P EA model) in a simulated based environment. Such results are validated using an equally parametrized parallel version of the algorithm in a real environment[1]. In order to simplify the experimentation, the real platform consists of a cluster of homogeneous nodes. This allows the trace of computer failures and the minimization of asynchronous effects on the performance so that the characterization of the real environment mirrors the simulator settings.

In a first set of experiments, the algorithmic accuracy of the simulations is tested by running a medium size instance of trap functions [1]. We focus then on the fine-tuning of the population size and adjust both, simulation-based and parallel versions to their optimal sizes. Given that both approaches are equally parametrized, our validation proof relies on showing that the simulations require the same population size than the parallel version to induce the same progress in fitness. On the basis of previous results, a second set of experiments is conducted in order to prove the massive scalability of the approach. In this case, the computational performance of the model is assessed by tackling a larger problem instance for which the simulator points out large population size requirements. Here, results show that the parallel version is able to find the problem optimum in two and half hours in contrast with the estimation of hundred days of the sequential run.

The rest of the paper is organized as follow. Section 2 provides an overall description of the Evolvable Agent model. Section 3 explains the setup of the experiments. Results are analyzed in Section 4. Finally, we reach some conclusions and propose some future lines of work in Section 5.

2   Description of the Model

The Evolvable Agent (EvAg) model (proposed by Laredo et al. in [7]) is a fine-grained spatially structured EA in which every agent schedules the evolution

---

[1] In order to reproduce experiments, all the source-code –either the simulator or the parallel version of the algorithm– is available from our Subversion repository at http://forja.rediris.es/svn/geneura published under GNU public license.

process of a single individual and self-organizes its neighborhood via the newscast protocol. As explained by Jelasity and van Steen in [6], newscast runs on every node and defines the self-organizing graph that dynamically maintains some constant graphs properties at a virtual level such as a low average path length or a high clustering coefficient from which a small-world behavior emerges [15]. This makes the algorithm inherently suited for parallel execution in a P2P system which, in turn, offers great advantages when dealing with computationally expensive problems at the expected speedup of the algorithm.

Every agent acts at two different levels; the evolutionary level for carrying out the main steps of evolutionary computation (selection, variation and evaluation of individuals [4]) and the network level which defines P2P population structure.

The evolutionary level is depicted in Algorithm 1. It shows the pseudo-code of an $EvAg_i \in [EvAg_1 \ldots EvAg_n]$ where $i \in [1 \ldots n]$ and n is the population size. Despite the model not having a population in the canonical sense, neighbors EvAgs provide each other with the genetic material that individuals require to evolve.

Algorithm 1. Pseudo-code of an Evolvable Agent ($EvAg_i$)

Evolutionary level

$Ind_{current_i} \Leftarrow$ Initialize Agent
while not *termination condition* **do**
   $Pool_i \Leftarrow$ Local Selection($Neighbors_{EvAg_i}$)
   $Ind_{new_i} \Leftarrow$ Recombination($Pool_i, P_c$)
   Evaluate(Ind_{new_i})
   if Ind_{new_i} better than $Ind_{current_i}$ **then**
      $Ind_{current_i} \Leftarrow Ind_{new_i}$
   end if
end while

Local Selection($Neighbors_{EvAg_i}$)
$[Ind_{current_h} \in EvAg_h, Ind_{current_k} \in EvAg_k] \Leftarrow$ Random selected nodes from the newscast neighborhood

---

The key element at this level is the locally executable selection. Crossover and mutation never involve many individuals, but selection in EAs usually requires a comparison among all individuals in the population. In the EvAg model, the mate selection takes place locally within a given neighborhood where each agent selects the current individuals from other agents (e.g. $Ind_{current_h}$ and $Ind_{current_k}$ in Algorithm 1).

Selected individuals are stored in $Pool_i$ ready to be used by the recombination (and eventually mutation) operator. Within this process a new individual Ind_{new_i} is generated.

In the current implementation, the replacement policy adopts a replace if worst scheme, that is, if the newly generated individual Ind_{new_i} is better than the current one $Ind_{current_i}$, $Ind_{current_i}$ becomes Ind_{new_i}, otherwise, $Ind_{current_i}$ remains the same for the next generation. Finally, every EvAg iterates until a termination condition is met.

As previously mentioned, newscast is the canonical underlying P2P protocol in the EvAg model. It represents the network level of the model that conforms the population structure. Algorithm 2 shows the newscast protocol in an agent $EvAg_i$. There are two different tasks that the algorithm carries out within each node. The active thread which pro-actively initiates a cache exchange once every cycle and the passive thread that waits for data-exchange requests (the cache consists in a routing table pointing to neighbor nodes of $EvAg_i$).

Algorithm 2. Newscast protocol in $EvAg_i$

Active Thread
loop
    wait one cycle
    $EvAg_j \Leftarrow$ Random selected node from $Cache_i$
    send $Cache_i$ to $EvAg_j$
    receive $Cache_j$ from $EvAg_j$
    $Cache_i \Leftarrow$ Aggregate $(Cache_i, Cache_j)$
end loop

Passive Thread
loop
    wait $Cache_j$ from $EvAg_j$
    send $Cache_i$ to $EvAg_j$
    $Cache_i \Leftarrow$ Aggregate $(Cache_i, Cache_j)$
end loop

Every cycle each $EvAg_i$ initiates a cache exchange. It uniformly selects at random a neighbor $EvAg_j$ from its $Cache_i$. Then $EvAg_i$ and $EvAg_j$ exchange their caches and merge them following an aggregation function. In this case, the aggregation consists of picking the freshest c items (i.e. c is the maximum degree of a node. In this paper $c = 40$) from $Cache_i \cup Cache_j$ and merging them into a single cache that will be replicated in $EvAg_i$ and $EvAg_j$.

Within this process, every EvAg behaves as a virtual node whose neighborhood is self-organized at a virtual level with independence of the physical network. In this paper, we conduct experiments following the ideal case in which every computing core hosts a single EvAg.

3 Experimental Setup

The experimental setup in this paper is based on the simulations performed in [8] on the scalability of the Evolvable Agent model when tackling trap functions [1]. In order to validate the model, we will try to reproduce such results in a parallel infrastructure.

3.1 Simulation settings

A trap function is a piecewise-linear function defined on unitation (the number of ones in a binary string). There are two distinct regions in search space, one leading to a global optimum and the other leading to the local optimum (see Figure 1). In general, a trap function is defined by the following equation:

$$trap(u(\overrightarrow{x})) = \begin{cases} \frac{a}{z}(z - u(\overrightarrow{x})), if & u(\overrightarrow{x}) \le z \\ \\ \frac{b}{l-z}(u(\overrightarrow{x}) - z), & otherwise \end{cases} \qquad (1)$$

where $u(\overrightarrow{x})$ is the unitation function, a is the local optimum, b is the global optimum, l is the problem size and z is a slope-change location separating the attraction basin of the two optima.

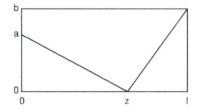

Fig. 1. Generalized *l-trap* function

For the following experiments, a 3-trap function was designed with the following parameter values: $a = l - 1$; $b = l$; $z = l - 1$. With these settings, 3-trap lies in the region between deception and non-deception. Scalability tests were then performed by juxtaposing m trap functions and summing the fitness of each sub-function to obtain the total fitness.

The bisection method [12] was used for each size m to determine the optimal population size P, that is, the lowest P for which 98% of the runs solve the

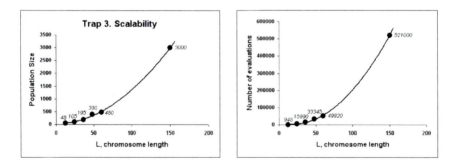

Fig. 2. Simulator estimated scalability of the Peer-to-Peer Evolutionary Algorithm [8] tackling different instances of the 3-trap problem [1]. On the left the estimated population sizes and the number of evaluations to solution on the right. Results are obtained for a selectorecombinative version of the algorithm (i.e. no mutation) and depicted as a function of the length of the chromosome, L.

traps functions. To find it, mutation rate is set to 0, so as to search a minimum population size such that using random initialization it is able to provide enough building blocks to converge to the optimum without other mechanism besides recombination and selection.

Figure 2 depicts the simulation-based results for increasing problem instances of the 3-trap problem (lengths of the chromosomes are $L = 12, 24, 36, 48, 60, 150$). As the problem scales, the P2P EA requires of both, a larger population size and a larger number of evaluations, to guarantee that the optimal solution is found with a probability of 0.98.

3.2   Parallel Version Settings

In order to run parallel experiments, we are going to consider two of the problem instances from previous simulations. The first instance of a medium size, i.e. $L = 48$ bits, and the second with $L = 150$ bits. Table 1 provides the simulator-based results for both instances that will be used as parameter inputs in the parallel runs. They characterize the settings of the algorithm to find the optimum 98 out of 100 times, e.g. in order to find the optimum in the $L = 48$ instance, the algorithm requires a population size of 390 individuals and a maximum of 140 generations.

Table 1. Simulator-based results for the population size and the respective number of generations in order to find the problem optimum

Instance	Population Size	Avg. n. of generations	Max. n. of generation
$L = 48$	390	85	140
$L = 150$	3000	173	250

The rest of parameter settings are summarized in Table 2.

Table 2. Parameters of the experiments

Trap instances

Size of sub-function (k)	3
Individual length (L)	48, 150

GA settings

Selection of Parents	Binary Tournament
Recombination	Uniform crossover, $p_c = 1.0$
Mutation	No mutation, $p_m = 0.0$

All experiments in this paper were conducted in the NEC Nehalem cluster at the HPC center of the University of Stuttgart (see http://www.hlrs.de/ systems/platforms/nec-nehalem-cluster for further details on the architecture). Here, it should be noted that P2P overlay networks behave independently of the underlying infrastructure they are running in, therefore, a cluster of homogeneous nodes can be consider a P2P system whenever it runs a P2P engine in every node. In that sense, using a cluster of homogeneous nodes has the advantage of simplifying the validation process. First, the side effects of asynchrony are minimized since every agent is scheduled by a single processor core running at the same frequency than the rest, and second, the lifetime and load of computers can be monitored so that we can ensure that there are no failures. Both effects, asynchrony and fault-tolerance to computer failures are left, therefore, as a future line of research.

4   Analysis of Results

In this section, we conduct two different sets of experiments. The first one focuses on verifying the results of the simulator in a real parallel platform. To that aim, a medium size instance of length $L = 48$ is considered. In a second experiment, we try to prove the massive scalability of the approach in terms of time speedups. Given that the goal of parallel Evolutionary Algorithms is to reduce times-to-solutions of expensive optimization problems, experiments were conducted on the largest problem instance of length $L = 150$.

4.1   Test-Case 1: Validating Results of the Simulator

In order to validate the results obtained by the simulator for the $L = 48$ instance, the P2P Evolutionary Algorithm was distributed in the NEC Nehalem cluster using a fine-grained parallelization in which every agent was scheduled in an independent thread, each running in its own processor. As described in [9], such settings stand for a worst-case scenario in which the fitness evaluations are computationally heavy. In advance, there are no restrictions limiting the number of agents per processor. Both algorithms (i.e. simulator-based algorithm and the parallel version) were equally parametrized and only differ on the population sizes P tested for the parallel approach.

Figure 3 depicts the average progress of the fitness convergence of the parallel runs for different population sizes $P = 50, 100, 200, 300, 400$. Given that the simulator predicts an optimal population size of $P = 390$, we aim at investigating the improvements on the fitness as the population size increases from $P = 50$ to $P = 400$. Results show that the smaller population sizes are not able to find the problem optimum (set to 48). However, for $P = 400$ the algorithm is able to track the optimal solution 8 times out of 10 as roughly estimated by the simulator (i.e. simulator actually predicts a success rate of 0.98 for $P = 390$). Taking into account the side effects of asynchrony and communications in the parallel version, we can conclude that the simulator-based results can be considered as a good estimate of the parallel performance of the algorithm.

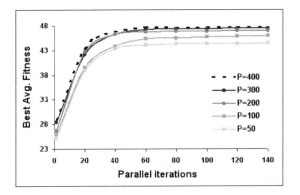

Fig. 3. Test-case 1. Best fitness convergence for the $L = 48$ instance using different population sizes (P). Results are averaged from 10 independent runs.

4.2   Test-Case 2: Testing the Massive Scalability of the Approach

In this experiment, we tackle the largest problem instance with a length of $L = 150$. The complexity of the instance is so high that the simulator estimates a population size of $P = 3000$ for the problem to be solved. However, and despite trap functions being algorithmically-complex problems (i.e. NP-hard), they are computationally lightweight. To emulate realistic time-consuming problems (e.g. the simulation guided optimization proposed by Ruiz et al. in [11] where the fitness function takes 6.5 seconds), we add a delay routine in every fitness evaluation that takes 16 seconds. Adding a delay routine aims reproducing heavy-loaded scenarios in which the ratio between communications and computation decreases. With these settings and according to the results in the simulator, the algorithm will require that 3000 individuals evolve during an average number of 173 generations to reach the optimum. In terms of time, that would translates into 100 days of sequential computation.

With the aim of reducing the time of convergence, the algorithm was parallelized using 3008 agents, each one running in a thread. The entire population was deployed in 188 computers, having 8 cores each and implementing hyper-threading with 2 threads per core. Note that the 8 extra individuals to the estimated 3000 are due to the composition of the architecture in which the parallelism extends to the microprocessor level –as McNairy and Bhatia describe in [10]– by adding several cores per processor and through hyper-threading technology.

Figure 4 depicts the convergence of the algorithm as a function of time. It shows how the algorithm is able to find optimality –the problem optimum is set to 150– after 2.5 hours of parallel processing which demonstrates that way the massive scalability of the approach.

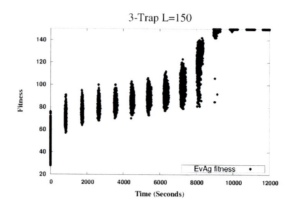

Fig. 4. Test-case 2. Fitness convergence for 3008 agents ($P = 3008$) in the $L = 150$ instance. The problem optimum is set to 150 and is found after 8900 seconds.

5   Conclusions

In this paper, we have conducted experiments for the validation of a Peer-to-Peer Evolutionary Algorithm in a cluster of homogeneous nodes. A common approach for designing such a kind of models is to use a simulation-driven experimentation given the difficulties of accessing a large amount of computers for testing. Therefore, model characterizations remain valid only under certain set of assumptions and the viability of the approaches is subject to the scope of the simulator.

In order to validate a model, we propose to reproduce results from simulations in a real-world system so that they can approach the predicted values. In that context, two simple experiments were conducted in a real infrastructure using a parallel version of the Peer-to-Peer Evolutionary Algorithm. The first experiment shows that the parallel version performs roughly the same than an equally-parametrized simulator-based run from which the validation of the model can be drawn. Specifically, the population size of the parallel version is adjusted to the same values of the simulations having an equivalent performance in fitness convergence. The second experiment focuses on determining the scalability of the parallel approach for large problem instances which, additionally, require of large population sizes. In this case, a massively parallel run is conducted in 3008 computing cores, each hosting an individual. The problem optimum is found after two and half hours of parallel execution in contrast to the estimate of one hundred days run if computed sequentially.

As future lines of work, we aim at investigating the parallel approach in a bigger set of scenarios taking into account the effects of heterogeneous computers on the algorithmic performance. We find that asynchrony, communication latencies and computer failures are the main issues to circumvent in order to deploy the algorithm in ad-hoc networks as they are Internet-based volunteer systems.

Acknowledgments. This work was supported by the Luxembourg FNR GreenIT Project (C09/IS/05), Spanish Ministry of Science Projects TIN2008-05941 and TIN2011-28627-C04 and Andalusian Regional Project P08-TIC-03903.

References

1. Ackley, D.H.: A connectionist machine for genetic hillclimbing. Kluwer Academic Publishers, Norwell (1987)
2. Anderson, D.P.: Boinc: A system for public-resource computing and storage. In: 5th IEEE/ACM International Workshop on Grid Computing, pp. 4–10 (2004)
3. Biazzini, M., Montresor, A.: Gossiping de: A decentralized heuristic for function optimization in p2p networks. In: ICPADS 2010, pp. 468–475 (2010)
4. Eibenand, A.E., Smith, J.E.: Introduction to Evolutionary Computing. Springer, Heidelberg (2003)
5. Guo, Y., Cheng, J., Cao, Y., Lin, Y.: A novel multi-population cultural algorithm adopting knowledge migration. Soft Comput. 15(5), 897–905 (2011)
6. Jelasity, M., van Steen, M.: Large-scale newscast computing on the Internet. Technical Report IR-503, Vrije Universiteit Amsterdam, Department of Computer Science, Amsterdam, The Netherlands (October 2002)
7. Laredo, J.L.J., Castillo, P.A., Mora, A.M., Merelo, J.J.: Exploring population structures for locally concurrent and massively parallel evolutionary algorithms. In: Proceedings of IEEE Congress on Evolutionary Computation (CEC2008), WCCI 2008, pp. 2610–2617. IEEE Press, Hong Kong (2008)
8. Laredo, J.L.J., Eiben, A.E., van Steen, M., Julián Merelo Guervós, J.: Evag: a scalable peer-to-peer evolutionary algorithm. Genetic Programming and Evolvable Machines 11(2), 227–246 (2010)
9. Laredo, J.L.J., Lombraña, D., de Vega, F.F., Arenas, M.G., Merelo, J.J.: A Peer-to-Peer Approach to Genetic Programming. In: Silva, S., Foster, J.A., Nicolau, M., Machado, P., Giacobini, M. (eds.) EuroGP 2011. LNCS, vol. 6621, pp. 108–117. Springer, Heidelberg (2011)
10. McNairy, C., Bhatia, R.: Montecito: a dual-core, dual-thread itanium processor. IEEE Micro. 25(2), 10–20 (2005)
11. Ruiz, P., Dorronsoro, B., Valentini, G., Pinel, F., Bouvry, P.: Optimisation of the enhanced distance based broadcasting protocol for manets. J. of Supercomputing. Special Issue on Green Networks, 1–28 (February 23, 2011), Online First^{TM}
12. Sastry, K.: Evaluation-relaxation schemes for genetic and evolutionary algorithms. Technical Report 2002004, University of Illinois at Urbana-Champaign, Urbana, IL (2001)
13. Scriven, I., Ireland, D., Lewis, A., Mostaghim, S., Branke, J.: Asynchronous multiple objective particle swarm optimisation in unreliable distributed environments. In: IEEE Congress on Evolutionary Computation, CEC 2008 (2008)
14. Steinmetz, R., Wehrle, K.: What is this Peer-to-Peer About? In: Steinmetz, R., Wehrle, K. (eds.) Peer-to-Peer Systems and Applications. LNCS, vol. 3485, pp. 9–16. Springer, Heidelberg (2005)
15. Watts, D.J., Strogatz, S.H.: Collective dynamics of "small-world" networks. Nature 393, 440–442 (1998)
16. Wickramasinghe, W.R.M.U.K., van Steen, M., Eiben, A.E.: Peer-to-peer evolutionary algorithms with adaptive autonomous selection. In: GECCO 2007, pp. 1460–1467. ACM Press, New York (2007)

Pool-Based Distributed Evolutionary Algorithms Using an Object Database

Juan-Julián Merelo-Guervós[1], Antonio Mora[1], J. Albert Cruz[2],
and Anna I. Esparcia[3]

[1] Departamento de Arquitectura y Tecnología de Computadores,
Universidad de Granada
{jmerelo,amorag}@geneura.ugr.es
http://geneura.wordpress.com
[2] Universidad de Ciencias Informáticas,
La Habana, Cuba
jalbert@uci.cu
[3] S2 Grupo, Valencia
aesparcia@s2grupo.es

Abstract. This work presents the mapping of an evolutionary algorithm to the CouchDB object store. This mapping decouples the population from the evolutionary algorithm, and allows a distributed and
asynchronous operation of clients written in different languages. In this
paper we present initial tests which prove that the novel algorithm design
still performs as an evolutionary algorithm and try to find out what are
the main issues concerning it, what kind of speedups should we expect,
and how all this affects the fundamentals of the evolutionary algorithm.

1 Introduction

Algorithms have traditionally been written with a single memory and CPU in
mind, current technological infrastructure includes a high variety of frameworks
and devices that make this paradigm be twisted and shifted in many different directions. Particularly in distributed evolutionary computation, traditional
notions of asynchronous, homogeneous and static computing systems have been
superseded by others in which one or all of these features are not present [1,10,2],
thus making the traditional distinction between master-slave and island based
models [13] fade by making them just two of all the possibilities that are created
along the different feature axes.

Using these new foundations for Evolutionary Algorithms (EAs) allows one to
take full advantage of the performance of modern CPUs and operating systems,
and, in some cases, opens up the possibility of using new devices for distributed
computing system, by making the participation in a distributed computation
experiment as easy as visiting a website [11].

This change in the computing framework might, and usually does, imply
changes in the algorithms themselves. A feature as common as threads makes
EAs escape the "sequential cage", and make us rethink how the biologically
inspired art of these algorithms [8] can be mapped to this new substrate.

C. Di Chio et al. (Eds.): EvoApplications 2012, LNCS 7248, pp. 446–455, 2012.
© Springer-Verlag Berlin Heidelberg 2012

For instance, database management systems are one pervasive technology in business computing, but it was not until 1999 that they were used as a base for persistent evolutionary algorithms by Bollini et al. [4]. They mention the fact that a database allows the simultaneous actuation of several clients, and change fundamentally the design of the EA from an *ab initio* strategy to an incremental one that makes use of the chromosomes that have been already created and evaluated, are stored, and can be efficiently retrieved, from the database.

Even database management systems change, and the last few years have seen the appearance of the so-called NoSQL, object or Key-Value stores [3]. These systems, beyond the obvious fact that they do not use the SQL language for accessing data stores, are characterized by the feature that they are key-value stores where the value is any kind of loosely structured document; in general, documents can include any data structure, although some of them (like, for instance, sets) might be present in only some cases. XML or JSON (JavaScript Object Notation,[6]) are commonly used as data description languages, while JavaScript has emerged as the most common data processing language. The languages are enhanced by functions that in some cases include *map/reduce* [17,7], which is an efficient way of working on large amounts of data without needing large amounts of memory. Map/reduce requests are structured in a *map* function that is applied to every element within the selection, creating a couple of data structures that are *reduced* by performing some operation on them. A *map* operation, for instance, might create an array with the values of a certain field; a *reduce* operation will create a hash that records how many times each value appears.

These features are usually accessed through a REST API, a lightweight way of interacting with HTTP based servers which uses the semantics and syntax of this protocol. Since in order to create a wrapper around a REST API the only requirement for a language is to be able to make web requests and build strings, NoSQL databases can be accessed either easily from the command line, from the address bar of a browser, or from libraries built in many different computer languages; in either case, it does not add much overhead to the raw request. Most of the overhead lies in the conversion from the native NoSQL format (usually JSON) to the data structures in the native language.

All these features make NoSQL databases an ideal candidate for creating the backoffice for a distributed computation experiment; even more so when most systems allow replication, so that a multi-star (that is, multiple and linked servers with clients *hanging* from each one) infrastructure can be created and single-points of failure avoided. This is what we have done in this paper: we have used *CouchDB* (a NoSQL DBMS) to set up a distributed evolutionary computing framework; after explaining how the EA was designed, we prove that first, those changes do not affect the essence of the algorithm by finding a solution using a number of evaluations comparable to a canonical EA; and then we measure how the system behaves in a distributed computation environment with several heterogeneous clients, trying to find out how many clients can be added in parallel.

The rest of the paper is organized as follows: next we will present the state of the art in similar systems; next, section 3 presents SofEA, the

CouchDB-based evolutionary algorithm and how evolution and the first proofs of concept; section 4 is devoted to show the results in more taxing environments, including certain speed-up measures, and finally we present the conclusions and future lines of work in section 5.

2 State of the Art

It is not usual to find pool-based implementations of EAs. In these methods, several nodes or *islands* share a *pool* where the common information is written and read. To work against a single pool of solutions is an idea that has been considered almost from the beginning of research in distributed evolutionary algorithms. Asynchronous Teams or A-Teams [15] were proposed in the early nineties as a cooperative scheme for autonomous agents. The basic idea is to create a work-flow on a set of solutions and apply several heuristic techniques to improve them, possibly including humans working on them. This technique is not constrained to EAs, since it can be applied to any population based technique, but in the context of EAs, it would mean creating different single-generation algorithms, with possibly several techniques, that would create a new generation from the existing pool.

The A-Team method does not rely on a single implementation, focusing on the algorithmic and data-flow aspects, in the same way as the Meandre [9] system, which creates a data flow framework, with its own language (called ZigZag), which can be applied, in particular, to EAs.

While algorithm design is extremely important, implementation issues always matter, and some (relatively) recent papers have concentrated on dealing with pool architectures in a single environment: Roy et al. [14] propose a shared memory multi-threaded architecture, in which several threads work independently on a single shared memory, having read access to the whole pool, but write access to just a part of it. That way, interlock problems can be avoided, and, taking advantage of the multiple thread-optimized architecture of today's processors, they can obtain very efficient solutions in terms of running time, with the added algorithmic advantage of working on a distributed environment. Although they do not publish scaling results, they discuss the trade off of working with a pool whose size will have a bigger effect on performance than the population size on single-processor or distributed EAs. The same issues are considered by Bollini and Piastra in [4], who present a design pattern for persistent and distributed EAs; although their emphasis is on persistence, and not performance, they try to present several alternatives to decouple population storage from evolution itself (*traditional* evolutionary algorithms are applied directly on storage) and achieve that kind of persistence, for which they propose an object-oriented database management system accessed from a Java client. In this sense, this approach is similar to AGAJAJ [11], since it uses for persistence a small database accessed through a web interface, but only for the purpose of interchanging individuals among the different nodes, not as storage for the whole population.

In fact, the efforts mentioned above have not had much continuity, probably due to the fact that there have been, until now, few (if any) publicly accessible online databases. However, given the rise of cloud computing platforms over the last few years, interest in this kind of algorithms has rebounded, with implementations using the public FluidDB platform [12] having been recently published. This implementation combines evolution with (possible) stigmergy (communication through the environment), the same as is proposed in this paper, since population is persistent, evolution is carried incrementally and the interaction among islands is only performed through the environment, in this case, the CouchDB DBMS. What we present here is similar; however, since it is a local database instead of a single copy of a global database accessed by all users, latency problems can be avoided, thus avoiding the scale problems we found in [12]. However, the results obtained in that work regarding packet size, and, in general, asynchronous organization of the algorithm, can also be applied to the present work.

This paper advances the state of the art by introducing a novel, more fine-grained parallelization technique, and also by testing it on a distributed and real world environment.

3 SofEA, a CouchDB-Based Evolutionary Algorithm

The first question is why choose CouchDB over other similar products, such as MongoDB or Redis. There have been several reasons for doing so: It is an open source product, which is available in most Linux distribution repositories. This means that it is very easy to create the infrastructure; second, it uses JavaScript as its query language, which makes learning to use it very easy, and it is introducing a new query language, UnQL[1], which is an hybrid between SQL and JSON. MongoDB can also be queried using JavaScript, but Redis cannot, using its own language instead; third, it uses persistent storage, which means stored procedures and data can be reused after reboot. Redis, on the other hand, handles everything in memory, which also implies a RAM consumption that is directly proportional to the amount of data you deal with, and finally, in the shape of DesktopCouch, CouchDB is a default install in most Linux desktops that include Gnome, although this does not imply CouchDB does not work in other operating systems. This means that the system can work *out of the box* (at least in Linux), without needing additional installs in most cases. Besides, CouchDB has as an advantage over MongoDB: a grater adherence to web standards (the primary interface to the data is through RESTful HTTP) which grant it an advantage for use in the Internet [16], and the possibility of using it in mobile platforms.

Mapping an EA to this system has to take into account its peculiar features and go with its grain to achieve maximum performance, locally in the server and globally on the system composed of server+clients.

The first step is to decouple population from the rest of the evolutionary algorithm. Usual EAs include population as a variable that is passed around together with operators and the rest of the algorithm; even distributed EAs

---

[1] http://www.unqlspec.org/display/UnQL/Home

encapsulate the population and the rest of the algorithm in a single problem. In this case, and following Bellini et al. [4], we decouple population storage and its processing. Population will be stored in CouchDB. A *document* will include a chromosome, a random number and the fitness value. Besides, CouchDB includes two other pieces of data into each document: the key (which will coincide with the chromosome) and a version number.

This version number, or *revision*, will be used to characterize the state of a chromosome in the population:

- Revision 1: newly created chromosome, no fitness computed yet
- Revision 2: chromosome with fitness
- Revision 3: *dead* chromosome.

Revisions are updated naturally by CouchDB; when a chromosome is updated with its fitness it is moved from revision 1 to 2; any further operation will take it to revision 3.

Since one of the strong points of CouchDB is its ability to cope with a high number of simultaneous requests, the EA itself has been divided into four different programs, which will operate independently and asynchronously.

- Initialization: will create a set of chromosomes in revision 1.
- Evaluation: will take packets of chromosomes in revision 1, compute their fitness, upgrading them to revision 2 (in traditional EA parlance, they would be part of *the population*).
- Reproduction: packets of chromosomes in revision 2 will be crossed over and mutated; newly generated chromosomes are obviously in revision 1.
- Elimination: the population (chromosomes in revision 2) is reduced down to a fixed number of chromosomes so that the less fit are progressively eliminated from it.

These last three components are run at the same time, although they can be started asynchronously. In fact, they can be run in any sequence. Since the population is *out there* any part, or all of them, can be run in different languages, operating systems, processes or machines. This allows also to optimize the implementation of each one of them by using the language that suits them better. This *horizontal* division of labor has been used before in papers by Castillo et al. [5], in this case to take computational load off a central server, which is used mainly as a clearinghouse for distributing the population; the full GA, however, is run on one of the clients, and there is a provision, in principle, for a single GA client, with possibly several evaluators.

The main problem with this configuration is the *starving* of the algorithms, that is, the lack of chromosomes for performing its task. Since operation is asynchronous, if reproduction is not fast enough the evaluator will run out of chromosomes to evaluate; if evaluator is not fast enough, the reproducer will not have a significant population to act on. The elimination phase is not so critical, but if it is not run frequently the reproductive population will grow out of proportion, thus reducing the exploitative ability of the whole algorithm. This is fixed, in part, by making components wait one second if there is not enough material to act on; however, this increases the number of useless requests to the server or servers we are using. The main handle we can use to act on this is introducing a slight delay when starting them and packet size. However, ultimately the key is to have enough chromosomes to evaluate, since the reproduction phase can create new ones (maybe with less efficiency) even with a few.

One of the main advantages of this configuration is the fact that every chromosome is evaluated just once. Since we use the chromosome string as a key, if the reproduction attempts to reintroduce a chromosome it will return an *already existing* error. This means that the reproduction phase (and, in fact, also the evaluation phase) becomes increasingly less successful with the ongoing algorithm, but also that every individual in *revision 2* is unique and thus diversity is always kept, no matter what kind of algorithm we introduce.

Since several evaluators and reproducers can act concurrently, we should issue them different chromosomes to work on. One of the possibilities would be to keep tabs on the server of the last one issued, but this is a problem since there is no guarantee that the result will be returned, and then it would also cause starvation if a slow client takes the last chromosomes to evaluate. So we included a random number in the document, which is used to sort the population and retrieve all the chromosomes after that first random number. There are two problems with this: if this number is too high, less chromosomes than the established packet will be returned. This could be avoided by issuing another request for the remaining number of chromosomes, however this is cumbersome and it is not really a big problem to have less chromosomes to operate on; the second one is that there is small probability (which increases with decreasing population size) that the same chromosome is returned twice to two different clients. This is not a problem with the reproduction phase, but it could be with the evaluation, causing a conflict. However, this only impacts on performance (some wasted CPU cycles) and in fact has been observed only a few times in experiments.

All these changes imply that whatever was proved for the old tried and trusted evolutionary algorithm no longer holds, so we will have to check whether, in fact, evolution proceeds and, in time, a solution is found. This is shown in Figure 3, applied to the classical ONEMAX problem with 128 bits. Block size, that is, the number of chromosomes requested in each step, for reproduction and evaluation was 64, and the base population was 128, after an initial non-evaluated population of 512. Since there is no *generation* here, in this figure we plot the fitness versus the number of evaluations. It can be seen that evaluation proceeds in more or less the same way in the five runs that we have plotted in this graph, reaching the maximum around 7000 evaluations. There are no big differences between them, and the time they take is also similar: around 150 seconds. In order to check what would the equivalent be in a generational algorithm, we run a simple evolutionary algorithm with the same code (when possible) and different population size (Figure 3, right); 100 runs were made for each population size; as it can be seen, that number of evaluations roughly corresponds to a population between 64 and 128, showing that, algorithmically, this evolutionary algorithm is

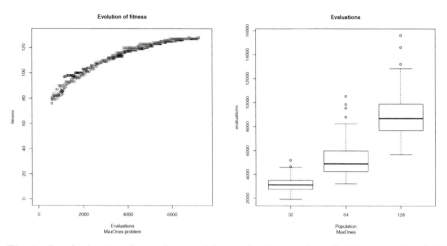

Fig. 1. Graph showing the evolution of fitness for 5 runs of the SofEA algorithm (left) and Boxplot of number of evaluations needed for the same problem in a canonical GA with different population sizes (right)

equivalent to the canonical GA with roughly the same population. However, Figure 2 shows the other problem these algorithms present: overpopulation. When the algorithm finishes, the number of non-evaluated chromosomes is close to 1000; since evaluation is slower than reproduction the gap only increases in time. In the next section we will see how parallelization avoids that problem. In this experiment we run the algorithm until 5000 evaluations were reached; packet size for evaluators and reproducers were set to 64 and base population size was 128.

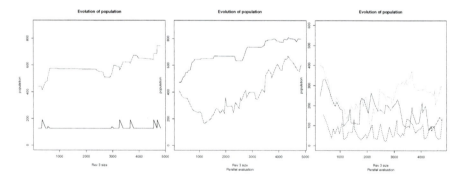

Fig. 2. Evolution or population in different states, plotted against the number of *dead* elements. Red represents the number of newly generated chromosomes, black those in revision 2 (left) and two parallel evaluators. Blue takes a block size of 32, red (bottom) 64 (middle); and finally, 3 and 4 parallel evaluators (dotted and dashed lines) and 3 evals + 1 reproducer (solid) (right).

4   Experiments and Results

Since the algorithm is separated in several modules, there are as many options to paralellize it by just multiplying its number, changing, if needed, the block size; the main factor, as in the *sequential* algorithm (that is, the single copy version shown above) is to avoid *starvation*, or lack of supply of individuals to work on, as well as the opposite effect, the glut or oversupply of chromosomes which wastes time by arriving to the end of the simulation with lots of chromosomes still to be evaluated.

We carried out several experiments in which we used two evaluators. To check whether an increment of speed was obtained just by the parallel evaluation or to the management of the oversupply of chromosomes,two packet sizes were tested: 32 and 64 for the evaluator. The first one got almost no time improvement, while the other obtained almost a 25% improvement in time. Since the number of evaluated chromosomes is independent of the block size, this improvement is simply due to the fact that the overall number of chromosomes created is less, as can be seen in Figure 2 (right). The number of non-evaluated chromosomes for block-size 32 is more or less the same as before (Figure 2, left). However, doubling the number of evaluations done in parallel effectively decreases the oversupply of chromosomes, resulting in a smaller number of total chromosomes generated and a better leveraging of the parallel evaluation.

To further improve solutions, another evaluator was added; within the same computer, a Javascript + JQuery evaluator running within the browser with block size equal to 32 was included. Having so many processes (all clients + server) running within the same computer did not result in a big overload, since it is an AMD 6-core computer with 8 GB of memory. We even managed to reduce approximately 1/6 the running time of the overall system. Effectively, looking at the solid red line in Figure 2 (right) we see that the oversupply of chromosomes is further reduced, which means that less effort is wasted generating chromosomes; besides, having more processes run in parallel improves the overall speed.

To see whether further improvement could be achieved we added another computer, which was running either an evaluator or an reproducer in the browser. In the first case, as shown by the black dotted line in 2 (right), we went from oversupply to starvation, and most of the time the evaluator did not have any chromosomes to work on; trying to avoid this by using another reproducer instead gave rise to the situation shown in 2 (right) with a dashed blue line: once again, the supply of chromosomes was not too high (compare it with Figure 2 – left), but neither of them resulted in an improvement in the time needed to perform 5000 evaluations, as was desired.

This implies that, for this particular configuration, the number of clients has reached a plateau and no more speedup is possible. However, the fact that we have achieved good speedups just by adding clients *within the same computer* encourages us to continue developing this system further.

5   Conclusions and Future Work

This paper describes the first experiments done with a CouchDB-based evolutionary algorithm, which experiments with a new form of parallelization and a new representation of the evolutionary algorithm that detaches storage from the process that work on them; this detachment is carried further by making a different process of every phase in an evolutionary algorithm: evaluation, reproduction and selection.

We have proved that the algorithm is able to work in this asynchronous and detached mode; further, that it is able to work in many different platforms by testing it with Perl and JQuery clients, all working on the same population asynchronously. Experiments have also shown firstly that one of the problems of this system is striking a balance with the supply of chromosomes to be evaluated and secondly, that clients cannot be added indefinitely. However, this is mainly related to the method we use for checking termination, and not to the algorithm itself: a rather expensive map/reduce request is done to the system involving retrieving all documents and checking their revision. The more clients are added the more requests are made; every packet evaluation or reproduction needs at least one of them. This is definitely a path for improvement: we could evaluate this only once by the reaper and store it in the database; this would save many requests and extend the number of possible clients.

Besides this improvement, currently the type of the client is decided beforehand; since the clients are served from the database, some intelligence could be added to it so that it was able to decide which clients were needed the most, even during the execution of the algorithm. If too many non-evaluated chromosomes were present, an evaluator could be served; else, a reproducer. The type of the client could even be changed at run time.

There is also some room for optimization of the CouchDB server by reducing the number of heavy-duty requests. Eventually, we expect to achieve speeds for the single clients system that are competitive with those achieved by a sequential system.

Acknowledgments. This work is supported by projects NEMESIS (TIN2008-05941) awarded by the Spanish Ministry of Science and Innovation and P08-TIC-03903 awarded by the Andalusian Regional Government.

References

1. Atienza, J., Castillo, P.A., García, M., González, J., Merelo, J.: Jenetic: a distributed, fine-grained, asynchronous evolutionary algorithm using Jini. In: Wang, P.P. (ed.) Proc. JCIS 2000 (Joint Conference on Information Sciences), vol. I, pp. 1087–1089 (2000); ISBN: 0-9643456-9-2
2. Bánhelyi, B., Biazzini, M., Montresor, A., Jelasity, M.: Peer-to-Peer Optimization in Large Unreliable Networks with Branch-and-Bound and Particle Swarms. In: Giacobini, M., Brabazon, A., Cagnoni, S., Di Caro, G.A., Ekárt, A., Esparcia-Alcázar, A.I., Farooq, M., Fink, A., Machado, P. (eds.) EvoWorkshops 2009. LNCS, vol. 5484, pp. 87–92. Springer, Heidelberg (2009)

3. Bartholomew, D.: SQL vs. NoSQL. Linux Journal 195, 4 (2010)
4. Bollini, A., Piastra, M.: Distributed and Persistent Evolutionary Algorithms: A Design Pattern. In: Langdon, W.B., Fogarty, T.C., Nordin, P., Poli, R. (eds.) EuroGP 1999. LNCS, vol. 1598, pp. 173–183. Springer, Heidelberg (1999)
5. Castillo, P.A., García-Arenas, M., Mora, A.M., Jiménez-Laredo, J.L., Romero, G., Rivas, V.M., Merelo-Guervós, J.J.: Distributed Evolutionary Computation using REST. CoRR abs/1105.4971 (2011)
6. Crockford, D.: JavaScript Object Notation (JSON) (July 2006), http://www.ietf.org/rfc/rfc4627
7. Dean, J., Ghemawat, S.: MapReduce: Simplified Data Processing on Large Clusters. Communications of the ACM 51(1), 107 (2008)
8. Goldberg, D.E.: Zen and the art of genetic algorithms. In: Schaffer, J.D. (ed.) ICGA 1995, June 4-7, pp. 80–85. George Mason University, Morgan Kaufmann, San Mateo, California (1989)
9. Llorà, X., Ács, B., Auvil, L., Capitanu, B., Welge, M., Goldberg, D.: Meandre: Semantic-driven data-intensive flows in the clouds. Tech. Rep. 2008103, Illinois Genetic Algorithms Laboratory (2008)
10. Gorges-Schleuter, M.: ASPARAGOS: An asynchronous parallel genetic optimization strategy. In: Schaffer, J.D. (ed.) Proceedings of the Third International Conference on Genetic Algorithms. Morgan Kaufmann Publishers (1989)
11. Merelo, J.J., Castillo, P., Laredo, J., Mora, A., Prieto, A.: Asynchronous distributed genetic algorithms with Javascript and JSON. In: Proceedings of WCCI 2008, pp. 1372–1379. IEEE Press (2008), http://atc.ugr.es/I+D+i/congresos/2008/CEC_2008_1372.pdf
12. Merelo, J.J.: Fluid evolutionary algorithms. In: IEEE Congress on Evolutionary Computation, pp. 1–8. IEEE (2010)
13. Nowostawski, M., Poli, R.: Parallel genetic algorithm taxonomy. In: Third International Conference on Knowledge-Based Intelligent Information Engineering Systems, pp. 88–92. IEEE (1999)
14. Roy, G., Lee, H., Welch, J., Zhao, Y., Pandey, V., Thurston, D.: A distributed pool architecture for genetic algorithms. In: IEEE Congress on Evolutionary Computation, CEC 2009, pp. 1177–1184 (May 2009)
15. Talukdar, S., Murthy, S., Akkiraju, R.: Asynchronous teams. International Series in Operations Research and Management Science, pp. 537–556 (2003)
16. Tiwari, S.: Professional NoSQL. John Wiley & Sons, Inc. (2011)
17. Yang, H., Dasdan, A., Hsiao, R., Parker, D.: Map-reduce-merge: simplified relational data processing on large clusters. In: Proceedings of the 2007 ACM SIGMOD International Conference on Management of Data, pp. 1029–1040. ACM (2007)

Migration and Replacement Policies for Preserving Diversity in Dynamic Environments

David Millán-Ruiz[1] and José Ignacio Hidalgo[2]

[1] Telefónica Digital, Distrito Telefónica, 28050, Madrid, Spain
[2] Complutense University of Madrid, Profesor José García Santesmases, 28040, Madrid, Spain

Abstract. This paper seeks to resolve the difficulties arising from the configuration and fine-tuning of a Parallel Genetic Algorithm (PGA) based on the Island Model, when the application domain is highly dynamic. This way, the reader will find a number of useful guidelines for setting up a PGA in a real, representative dynamic environment. To achieve this purpose, we examine different (existing and new) migration and replacement policies for three different topologies. Of course, there are many other factors that affect the performance of a PGA such as the topology, migrant selection, migration frequency, amount of migrants, replacement policy, number of processing nodes, synchronism type, configuration in the isolated islands, diversity of policies in different islands, etc which are also considered as a part of this study. The pivotal point of all the experiments conducted is the preservation of diversity by means of an appropriate balance between exploration and exploitation in the PGA's search process when the application domain is highly dynamic and strong time constraints arise. The experimental phase is performed over two problem instances from a real-world dynamic environment which is the multi-skill call centre.

Keywords: parallel genetic algorithms, migration policies, replacement policies, diversity maintenance, dynamic environments.

1 Yet Another Paper on Migration and Replacement Policies for PGAs?

Over the last years, an increasingly-growing interest in parallel and distributed computing has arisen in computer science. Specifically, this concern has recently guided most research activities on evolutionary computation towards areas such as parallel and distributed computational intelligence or parallel and distributed architectures and infrastructure. Truthfully, there exists a vast bibliography on parallel and distributed evolutionary approaches (see Section 2) although there are still paths to explore.

Additionally, we perceive a tendency to tackle gradually more complex problems and application domains which frequently entail the processing of extremely

C. Di Chio et al. (Eds.): EvoApplications 2012, LNCS 7248, pp. 456–465, 2012.

dynamic data flows. These demanding environments are usually hard to be efficiently handled by most of the existing, sequential techniques. In this context, parallel and distributed evolutionary algorithms do not only mitigate this drawback but also present several noteworthy characteristics such as robustness, traceability, problem simplification, adaptivity, scalability and speed-up.

Nevertheless, it is not always straightforward to control the internal dynamics of a PGA based on the island model, especially whether we seek to ensure a fair balance between exploration and exploitation in the search process within a dynamic environment. Too much exploration (high diversity in the population) can cause very slow convergence towards the optimum whereas an intense exploitation (low diversity) at the beginning can lead us to a premature convergence. Many authors have addressed this complex problem from many angles, producing a rich set of proposals (see Section 2).

This paper does not attempt to be the philosopher's stone for PGAs' configuration but to provide researchers and industrial professionals with some additional, useful guidelines on how to properly fine-tune the configuration of your parallel implementation when facing highly dynamic application domains (e.g. call centre management, datagrams routing, detection of mobility patterns, etc).

The main contribution of this work lies in the determination of the right setting-up for a PGA when it is applied to highly dynamic environments. We also propose new policies, which are inspired in other domain's solutions, to preserve a fair balance between exploration and exploitation in the search process. We also test out those policies in three different topologies in order to analyse their impact under two different scenarios which correspond to real data extracted from a multi-skill call centre.

The rest of this paper is organised as follows: Section 2 presents a survey of existing work from different points of view, considering commonalities with other problem domains. Section 3 describes the basic configuration of our PGA and all the parameters involved that can affect the maintenance of diversity during the experimental phase. Section 4 presents and analyses the experimental results derived from applying different policies for preserving diversity. Finally, Section 5 draws the main conclusions and provides some guidelines as future work.

2 State of the Art: Brief Survey of Existing Work

Determining the right configuration for a PGA is not truly a new issue as many authors have already worked on finding an appropriate setting-up as we are going to see along this section.

Roughly speaking, parallel versions of genetic algorithms can be categorised into coarse grained and fine-grained implementations (Cantú-Paz, 1998) [9]. Coarse-grained approaches maintain a population on each node where individuals are migrated according to a given policy. In contrast, fine-grained implementations keep an individual on each processing node which operates as a neighbour for selection and reproduction. As we have already introduced, we will investigate migration and replacement policies for PGAs based on the island model (coarse-grained approach) with special focus on those appropriate configurations for highly dynamic environments.

In 1987, Pettey (1987) [1] put forward a distributed model in which the best-fitted individuals of each node were migrated to each neighbour node in each generation, fully replacing the worst-fitted individuals of those neighbours. At the same time, Tanese (1987) [2] proposed a parallel implementation where each population was broken into a small number of subpopulations. Afterwards, each subpopulation was assigned to (and processed in) a different processing node within the system. The island model proposed in Cohoon (1987) [3] is an implementation of a distributed scheme where the idea of random migrant selection and replacement was put forward. In this proposal, each island was an isolated entity which was capable of selecting individuals, crossing them and evaluating their fitness value.

After that, Gordon (1992) [4] as well as Adamidis (1994) [5] reinforced the term of island model in their parallel proposals, while Collins (1992) [6] launched a grid model where individuals were placed in a node and interacted with their neighbours. In 1995, some authors went into the migrants selection in greater depth. This way, Belding (1995) [7] established an approach where the first n individuals were selected as migrants in relation to a predefined order.

Whitley (1997) [8] underlined that migration in parallel implementations caused additional selective pressure whereas Nowostawski (1999) [20] proposed a new taxonomy for PGAs based on a dynamic demes model. Special mention for Cantú-Paz (1998-2000) [9-11] who provided the most complete review of the state of the art on PGAs. Then, Alba (2001) [12] highlighted the importance of using asynchronous policies. In all the experiments conducted, asynchronous algorithms outperformed their equivalent synchronous counterparts in real time. Hu (2002) [13] described a model which is inspired by the stratified competition frequently seen in society and biology. The proposal defined stratified levels with fitness value limits. Individuals moved from low-fitness subpopulations to higher-fitness subpopulations whether they surpassed the fitness-based admission threshold of the receiving subpopulation. Higher fitness levels implied higher selection pressure (exploitation).

More recently, Lozano (2008) [14] put forward an explicit measure of diversity which entailed the replacement of existing individuals with lower values for the features being measured. The authors claimed to outperform existing replacement strategies presented to date, maintaining high levels of diversity. In contrast, Ruciński (2010) [15] examined the impact of the migration topology on the island model. This study compared different topologies and migration strategies in large networks. The authors concluded that the migration topology was a key factor to enhance the performance of a parallel global optimisation algorithm. Particularly, they recommended the use of ring topologies and suggested avoiding fully-connected and Barabasi-Albert topologies where a fast information spread over the entire network was allowed.

The most recent work on this topic, which went one step further, is Araujo (2011) [16]. The authors investigated, on a real parallel setup, a new strategy to enhance diversity in the island model. The proposal focused on the migrant selection phase of a genetic algorithm, taking into consideration the genotypic differences of the immigrant individual which was incorporated in a receiving subpopulation.

The key contribution of this paper is the determination of the right setting-up for a PGA, preserving a fair balance between exploration and exploitation in the search process, when the PGA is applied to a highly dynamic environment.

3   Setting-Up of the PGA

Same parameters of the PGA's configuration have been fixed (migration frequency, amount of migrants, synchronism type, number of processing nodes, same configuration in the isolated islands, and stopping condition) while others have been varied (topology and migration and replacement policies) in order to understand what policies perform best in dynamic environments.

3.1   Configuration of Each Island

All the islands have been set-up in the same way. Having a different configuration on each island obviously has an impact on the global performance as well as on the maintenance of diversity. Even more, the configuration of each island itself also makes high impact on the bottom line.

Bear in mind that the encoding and the fitness function as well as some genetic operators (due to problem constraints) are problem dependent. The setting of the simple GA existing in each island has been adapted to the problem being studied: the multi-skill call centre. The reader can find further information about the problem definition in Millán-Hidalgo (2010) [17].

The configuration of each island (steady-state GA) is similar to the one proposed in [17] although, in this study, we do not consider memetic algorithms, just simple genetic algorithms (Whitley, 1997) [19], and work over a different dataset. The exact configuration of our PGA is the following:

1. *Encoding*: We encode every solution as an array of integers whose indexes represent the available agents at a given instant and the array contents refer to the profile assigned to each agent.
2. *Fitness function*: We measure the service level resulting from the configuration of agents and incoming calls (see [17]).
3. *Population size*: The population of each island contains *30* different individuals encoded as hinted above.
4. *Initialisation*: The initial population is randomly generated.
5. *Selection*: Since the population needs to be bred each successive generation, we have chosen a binary tournament selection.
6. *Crossover*: The offspring inherits the common points in their parents and randomly receives the rest of genes from them.
7. *Mutation*: We apply a perturbation over each gene of the chromosome with a probability of *0.03*.

8. *Replacement policy*: We consider elitism with a probability of *0.93* to replace the worst-fitted individuals of the population in next generation. And, with a probability of *0.07*, a worse-fitted individual may be captured. Note that our basic GA relies on a steady-state scheme.

3.2   Common Framework

The following parameters have invariably been fixed in this study in order to analyse the impact of the migration and replacement policies, depending on the topology in which each configuration is studied.

1. *Migration frequency*: Each *60* seconds, all the islands are blocked for selection, migration and evaluation. If a generation is in process during the blockade in any of the islands, we take the previous stable population and go on with the process once the migrations have been carried out.
2. *Amount of migrants*: Each island sends a set of migrants according to the migration policy which represents the 10% of the size of the population (in practice, *3* individuals).
3. *Synchronism*: We have applied a synchronous scheme in which every island waits for every incoming set of migrants they have to evaluate.
4. *Number of processing nodes*: We consider 5 islands for every topology as this is the number of available processors which are fully utilised by the CPUs.
5. *Stopping condition*: The size of the time-frame considered is *300* seconds.

3.3   Topology

As previously mentioned, we have employed an island model in which every processing node is a steady-state GA. The topologies we have analysed are the following ones (see Figure 1 for a better understanding):

1. *Star topology*: We consider *4* subordinate islands which correspond to simple steady-state GAs. These islands are connected to a master island (another simple steady-state GA) which coordinates and synchronises the rest of subordinate islands (see Figure 1.A).
2. *Bidirectional ring topology*: each island sends and receives individuals from only other 2 islands, the previous one and the following one (see Figure 1.B).
3. *All-to-all topology*: every island is fully connected to the rest of islands (see Figure 1.C).

3.4   Migration and Replacement Policies

This section proposes different policies to define what individuals should be transferred to the neighbouring islands and which ones should be replaced in the receiving populations. The combinations of policies are listed below:

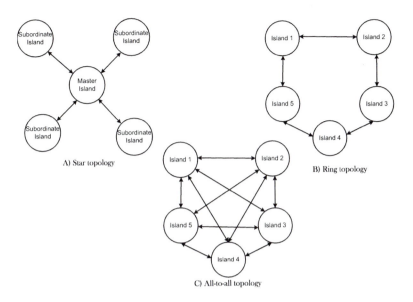

Fig. 1. Topologies being analysed: A) Star Topology, B) Bidirectional Ring Topology, C) All-to-all Topology

1. *Best-fitted individuals by worst-fitted individuals (BFI-WFI):* The best-fitted individuals from the source population replace the worst-fitted individuals from the receiving population. We substitute individuals who are "further" in terms of fitness value to the source ones.
2. *Best-fitted individuals by random individuals (BFI-RI):* The best-fitted individuals from the source population randomly replace individuals from the receiving population.
3. *Best-fitted individuals by best-fitted individuals (BFI-BFI):* The best-fitted individuals from the source population replace the best-fitted individuals from the receiving population. We replace individuals who are "closer" in terms of fitness value to the source ones.
4. *Best-fitted individuals by most different individuals (BFI-MDI):* The best-fitted individuals from the source population replace the most different individuals (according to the number of different genes) to them, existing in the receiving population.
5. *Best-fitted individual + "Annealing" by worst-fitted individuals (BFIA-WFI):* We select the best-fitted individual from the source population and a set of the following best-fitted individuals with probability η, proportional to the number of generations spent (the more generations are executed, the higher the probability is). This proposal is inspired by the simulated annealing approach which was pioneered by Kirkpatrick (1983) [18]. This way, the probability of not choosing the best-fitted individuals for migration is $1 - \eta$ (in this case, we randomly select another individual from the source population). Afterwards, the best-fitted individual from the source population and the set of "annealing" individuals replace the worst-fitted individuals from

the receiving population. Bear in mind that we always select the best-fitted individual of the source population to ensure a minimum of fast convergence as dynamic environments require prompt responses while also preserving diversity. Finally, we substitute individuals who are "further" in terms of fitness to the source ones.

4   Experiments

We now detail and analyse the experiments conducted over two different problem instances (medium and high difficulty, respectively).

4.1   Dataset Employed

This subsection describes the two problem instances (medium and high difficulty) that we have created to test out the configurations being studied. These two problem instances are composed by real data taken from our call centre during two different days at the same hour: a one-day campaign and a normal day. The size of each snapshot where each configuration has been executed is *300* seconds (*5* minutes). Note that around *800* incoming calls (n) simultaneously arrive during a normal day in such a time interval, whereas up to *2450* simultaneous incoming calls may arrive during this interval during a commercial campaign. The number of agents (m), for each time interval, oscillates between *700* and *2100*, having *16* different skills for each agent on average, grouped in skill profiles of *7* skills on average. The total number of call types considered for this study is *167*. When the workload (n/m) is really high, finding the right assignment among agents and incoming calls becomes fundamental.

4.2   Results of the Experimental Phase

For a fair comparison, every configuration has been run over the same problem instance *30* times. Table 1 shows the results obtained from the experimental phase for the medium-difficulty problem instance while Table 2 illustrates the respective ones for the high-difficulty problem instance. These tables show the best, worst and mean solution (and its standard deviation) out of the *30* executions performed. Ranking refers to the relative comparison between configurations, considering that the best setting-up represents the highest performance. We analyse the impact of each configuration of policies over the global performance of the PGA with regards to the topologies being studied in next section.

4.3   Discussion

We perceive that the bidirectional ring seems to be the most appropriate topology for dynamic environments, most likely because this topology allows for opportune convergence while preserving the required diversity. The star topology also entails high-quality outcomes but quickly gets stagnated. The reason is that the master island receives many migrants from the subordinate islands after some migrations (and it is even worse when there are many subordinate

Table 1. Results of each PGA configuration for each island topology in *30* executions (medium-difficulty problem instance). Values refer to the fitness values obtained by each combination of migration & replacement policies. The caption of each combination of policies is: Best-fitted individuals by worst-fitted individuals (*BFI-WFI*), Best-fitted individuals by random individuals (*BFI-RI*), Best-fitted individuals by best-fitted individuals (*BFI-BFI*), Best-fitted individuals by most different individuals (*BFI-MDI*) and Best-fitted individual + "Annealing" by worst-fitted individuals (*BFIA-WFI*).

Policy	Topology	Min	Max	Mean	SD	Ranking
BFI-WFI	Star	0.846698	0.847310	0.847092	0.0003	9
BFI-RI	Star	0.846744	0.847361	0.847102	0.0003	8
BFI-BFI	Star	0.846195	0.847068	0.846511	0.0004	12
BFI-MDI	Star	0.847119	0.847742	0.847471	0.0003	5
BFIA-WFI	Star	0.847119	0.847742	0.847489	0.0003	4
BFI-WFI	Ring	0.847141	0.848006	0.847535	0.0004	3
BFI-RI	Ring	0.846933	0.847908	0.847290	0.0004	7
BFI-BFI	Ring	0.847119	0.847742	0.847364	0.0003	6
BFI-MDI	Ring	0.853954	0.860611	0.858281	0.0031	2
BFIA-WFI	Ring	0.857322	0.861109	0.859702	0.0017	1
BFI-WFI	Hub	0.846149	0.847488	0.846856	0.0005	10
BFI-RI	Hub	0.846654	0.847201	0.846848	0.0002	11
BFI-BFI	Hub	0.834190	0.835465	0.834838	0.0005	14
BFI-MDI	Hub	0.831358	0.831984	0.831603	0.0003	15
BFIA-WFI	Hub	0.845520	0.846874	0.846378	0.0006	13

islands), implying that populations eventually become very similar. This intuitively involves a lack of diversity so that the gain of fitness gets fatally damaged. This phenomenon affects much more strongly to the hub topology as, being all the islands interconnected to each other, the diversity diminishes too much after one or two migrations.

A second key conclusion is that the replacement of individuals is another important feature to set-up. In this manner, replacing the worst-fitted individuals in the receiving population by the best-fitted individuals of the source population does not always behave better than taking the most different individuals. The process of analysing differences in the chromosomes in contrast implies that the PGA can run fewer generations (as it is a costly operation) but entails better fitness values in the end. The underlying principle may be that fitness-based comparisons can occasionally be misleading or deceptive, leading to the situation in which two close individuals in terms of genes in common may have associated very different fitness values, whereas two far chromosomes in terms of genes in common may have assigned close fitness values (Whitley, 1991). Another consequence of measuring gene differences as compared to gauging fitness values is that the lift of the fitness curve has a smoother slope in the first generations. Naturally, replacing the best-fitted individuals of the receiving population by the best-fitted ones of the source population implies a slower convergence in each processing node as we will find a larger percentage of less fitted individuals.

Table 2. Same as Table 1 but results now refer to the highly-difficult problem instance

Policy	Topology	Min	Max	Mean	SD	Ranking
BFI-WFI	Star	0.793660	0.793941	0.793796	0.0001	8
BFI-RI	Star	0.794102	0.794197	0.793561	0.0008	9
BFI-BFI	Star	0.791377	0.792888	0.792280	0.0007	12
BFI-MDI	Star	0.794265	0.794932	0.794693	0.0003	5
BFIA-WFI	Star	0.794288	0.795012	0.794688	0.0003	6
BFI-WFI	Ring	0.794610	0.795595	0.795223	0.0004	3
BFI-RI	Ring	0.794677	0.795216	0.794978	0.0002	4
BFI-BFI	Ring	0.794313	0.795221	0.794654	0.0004	7
BFI-MDI	Ring	0.792158	0.798497	0.796137	0.0028	2
BFIA-WFI	Ring	0.795679	0.798864	0.797696	0.0014	1
BFI-WFI	Hub	0.792373	0.792873	0.792669	0.0002	11
BFI-RI	Hub	0.791816	0.793589	0.792864	0.0008	10
BFI-BFI	Hub	0.790809	0.791874	0.791324	0.0004	14
BFI-MDI	Hub	0.790148	0.791492	0.790646	0.0006	15
BFIA-WFI	Hub	0.791097	0.791840	0.791566	0.0003	13

5   Conclusions and Future Work

We have seen that PGAs can deal with complex, real-world application domains
although they require specific tuning, depending on the nature of the problem being faced. This way, we have presented a basic configuration for PGAs based on
the island model and then performed several experiments over this initial setting-
up for two representative problem instances extracted from a real-world production environment. Same parameters were fixed (migration frequency, amount of
migrants, synchronism, number of processing nodes and stopping condition) while
others were changed (topology and migration and replacement policies) in order
to understand what policies perform best in dynamic environments. We have also
proved that the bidirectional ring topology seems to be the most suitable configuration for dynamic environments, most probably because this topology allows for convergence while maintaining diversity when it is required. Another key conclusion
has been that swapping the worst-fitted individuals in the receiving population by
the best-fitted individuals of the source population does not always perform better
than replacing the most different individuals in terms of gene differences, specially
when the migrants are not always the best-fitted ones (policy BFIA-WFI). The
best migration policy has been sending the best fitted-individual with some non-
necessarily best-fitted individuals (annealing set) as it provides diversity. As future
work, we propose to analyse in depth more migration and replacement policies and
more combinations of features as we have fixed many important "regulators" of diversity. We also plan to study other representative problem domains to generalise
the present conclusions.

Acknowledgements. We would like to thank José L. Vélez for his valuable
contribution to the optimisation and debugging of the parallel implementation.
This work has been partially supported by Spanish Government grants TIN
2008-00508 and Consolider CSD00C-07-20811.

References

[1] Pettey, C., Leuze, M., Grefenstette, J.: A parallel genetic algorithm. In: Proceedings of the Second International Conference on Genetic Algorithms and Their Applications. Lawrence Erlbaum Associates, Hillsdale (1987)

[2] Tanese, R.: Parallel genetic algorithms for a hypercube. In: Proceedings of the Second International Conference on Genetic Algorithms, pp. 177–183 (1987)

[3] Cohoon, J., Hegde, S., Martin, W., Richards, D.: Punctuated equilibria: a parallel genetic algorithm. In: Proceedings of the Second International Conference on Genetic Algorithms, Hillsdale, NJ, USA, pp. 148–154 (1987)

[4] Gordon, V.S., Whitley, D., Böhn, A.: Dataflow parallelism in genetic algorithms. Parallel Problem Solving from Nature 2, 533–542 (1992)

[5] Adamidis, P.: Review of Parallel Genetic Algorithms. Technical Report, Department of Electrical and Computer Engineering, Aristotle University, Thessalonik (1994)

[6] Collins, R.: Studies in Artificial Evolution. Ph.D. dissertation, Department of Computer Science, University of California at Los Angeles (1992)

[7] Belding, T.: The distributed genetic algorithm revisited. In: Proceedings of the Sixth International Conference on Genetic Algorithms, pp. 114–121 (1995)

[8] Whitley, D., Rana, S., Heckendorn, R.: Island Model Genetic Algorithms and Linearly Separable Problems. In: Corne, D.W. (ed.) AISB-WS 1997. LNCS, vol. 1305, pp. 109–125. Springer, Heidelberg (1997)

[9] Cantú-Paz, E.: A Survey of Parallel Genetic Algorithms. Calc. Parallelles (1998)

[10] Cantú-Paz, E.: Topologies, migration rates, and multi-population parallel genetic algorithms. In: Proceedings of GECCO, pp. 91–98 (1999)

[11] Cantú-Paz, E.: Efficient and Accurate Parallel Genetic Algorithms. Kluwer Academic Press (2000); ISBN: 0792372212

[12] Alba, E., Troya, J.M.: Analyzing synchronous and asynchronous parallel distributed genetic algorithms. Future Generation Comp. Syst., 451–465 (2001)

[13] Hu, J., Goodman, E.: The hierarchical fair competition (HFC) model for parallel evolutionary algorithms. In: Proceedings of the Congress on Evolutionary Computation, pp. 49–54. IEEE Press, Honolulu (2002)

[14] Lozano, M., Herrera, F., Cano, J.R.: Replacement strategies to preserve useful diversity in steady-state genetic algorithms. Inf. Sci. 178, 4421–4433 (2008)

[15] Ruciński, M., Izzo, D., Biscani, F.: On the impact of the migration topology on the Island Model. Parallel Computing 36, 555–571 (2010)

[16] Araujo, L., Merelo, J.J.: Diversity Through Multiculturality: Assessing Migrant Choice Policies in an Island Model. IEEE Transactions on Evolutionary Computation 15(4) (2011)

[17] Millán-Ruiz, D., Hidalgo, J.I.: A Memetic Algorithm for Workforce Distribution in Dynamic Multi-Skil Call Centres. In: Proceedings of the 10th European Conference on Evolutionary Computation in Combinatorial Optimisation, pp. 178–189 (2010)

[18] Kirkpatrick, S., Gelatt, C., Vecchi, M.P.: Optimization by Simulated Annealing. Science 220(4598), 671–680 (1983)

[19] Whitley, L.D.: Fundamental Principles of Deception in Genetic Search. Foundations of Genetic Algorithms 1, 221–241 (1991)

[20] Nowostawski, M., Poli, R.: Parallel Genetic Algorithm Taxonomy. In: Proceedings of KES (May 1999)

Distributed Simulated Annealing with MapReduce

Atanas Radenski

Chapman University, Orange 92865, USA
Radenski@Chapman.edu

Abstract. Simulated annealing's high computational intensity has stimulated researchers to experiment with various parallel and distributed simulated annealing algorithms for shared memory, message-passing, and hybrid-parallel platforms. MapReduce is an emerging distributed computing framework for large-scale data processing on clusters of commodity servers; to our knowledge, MapReduce has not been used for simulated annealing yet. In this paper, we investigate the applicability of MapReduce to distributed simulated annealing in general, and to the TSP in particular. We (i) design six algorithmic patterns of distributed simulated annealing with MapReduce, (ii) instantiate the patterns into MR implementations to solve a sample TSP problem, and (iii) evaluate the solution quality and the speedup of the implementations on a cloud computing platform, Amazon's Elastic MapReduce. Some of our patterns integrate simulated annealing with genetic algorithms. The paper can be beneficial for those interested in the potential of MapReduce in computationally intensive nature-inspired methods in general and simulated annealing in particular.

Keywords: simulated annealing, MapReduce, traveling salesperson (TSP).

1   Introduction

Simulated annealing is a metaheuristic that is used to find near-optimal solutions for various hard combinatorial optimization problems; it does so by imitating the physical process by which melted metal is cooling slowly to form a frozen structure with minimal energy. Simulated annealing is computationally intensive and this has stimulated the exploration of a variety of high-performance simulated annealing algorithms based on popular paradigms: shared memory [9], message-passing [6], and hybrid-parallel [2, 4].

MapReduce (MR) is an increasingly popular distributed computing framework for large-scale data processing that is amenable to a variety of data intensive tasks. Users specify serial-only computation in terms of a *map* method and a *reduce* method, and the underlying implementation automatically parallelizes the computation, tends to machine failures, and schedules efficient inter-machine communication [3]. MR was first implemented as a proprietary platform by Google. Soon afterwards, Apache offered Hadoop MR [15] as open source, and cloud computing providers offer MR platforms on a cost-effective pay-per-use basis.

By design, MR supports fault-tolerance, load-balancing, and scalability. This is in contrast to well understood but lower level high-performance frameworks, such as

C. Di Chio et al. (Eds.): EvoApplications 2012, LNCS 7248, pp. 466–476, 2012.
© Springer-Verlag Berlin Heidelberg 2012

MPI and OpenMP, in which users - rather than the frameworks - need to tend to machine failures and scheduling. Such advantages of MR to more traditional frameworks have motivated us to explore its suitability for high-performance simulated annealing and to our knowledge, this is the first study of its kind. MR is known to work well on large datasets. MR's applicability to computationally intensive problem domains with smaller datasets - such as simulated annealing and TSP - poses challenges, primarily because of the lack of direct control over tasks and data allocation. This paper makes contributions towards better understanding of MR's potential in computationally intensive problem domains with smaller datasets in general, and simulated annealing and TSP in particular.

The remainder of the paper is organized as follows. Section 2 introduces MR and then specifies algorithmic patterns for simulated annealing with MR. Section 3 describes a conversion of the patterns into MR implementations for the TSP; it also evaluates the solution quality and performance (execution time and speedup) in the Amazon cloud. Section 4 reviews related work and Section 5 offers conclusions.

2   Placing Simulated Annealing on MapReduce

The MapReduce Framework. Excellent general introductions of the MR framework [3, 8] and its implementation within the Hadoop platform [15] are available to the interested reader. In this paper, we offer only a brief description of MR features needed for the understanding of our simulated annealing algorithm design.

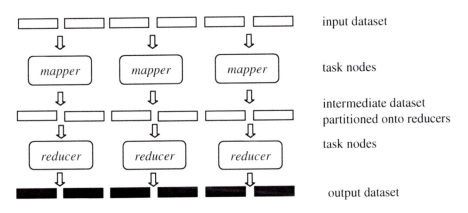

Fig. 1. A simplified representation of an MR job. The size of the output dataset can be different from the size of the input data set. An MR cluster consists of a number of *tasks nodes* and a single *main node* that controls tasks nodes. All mappers and reducers run on task nodes.

The MR framework consists of a programming model and runtime behavior. In the programming model, users specify serial map and reduce methods (one of each kind) that transform key-value records into new key-value records. The run-time environment transforms an input set of records into an output sets in two principal stages. First, a user-defined *map* method is applied over all records from the input dataset - in parallel, in a number of separate map tasks, or simply *mappers* – to produce intermediate outputs from all *map* methods. All intermediate records are then shuffled, sorted, and submitted for final processing by a user-defined *reduce* method. In general, the reduce method can be executed in parallel in several reducer tasks, or simply *reducers*, to produce several output sets of records. MR uses intermediate records' keys to *partition* records between reducers. In that, all intermediate records with the same key are always assigned to the same reducer; yet the same reducer may possibly handle a number of different keys. The MR framework assigns records to mappers and reducers, guided by record keys and without direct user participation.

The map and reduce stages form a single MR *job* (Fig. 1). It is possible to *pipeline* several MR jobs so that the output from one job is used as the input for the next one (Fig. 2). Input and output data sets for MR jobs are stored in a distributed file system.

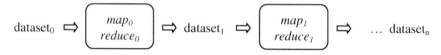

$$dataset_0 \Rightarrow \boxed{\begin{array}{c} map_0 \\ reduce_0 \end{array}} \Rightarrow dataset_1 \Rightarrow \boxed{\begin{array}{c} map_1 \\ reduce_1 \end{array}} \Rightarrow \dots dataset_n$$

Fig. 2. An MR job pipeline. The output of job k becomes the input of job $k+1$. In some MR implementations, such pipelines are referred to as job flows.

We use the following notation for MR pipelines in this paper:

- $A1 + A2 + \dots + Am$ is the pipeline of jobs $A1, A2, \dots Am$
- mA is an abbreviation for a pipeline $A + A + \dots + A$ of length m

In addition to the primary *map* and *reduce* methods, the MR framework includes two methods that can be optionally used to *initialize* and *finalize* mappers and/or reducers. Initialization can create objects that persist during *map (reduce)* invocations within the same mapper (reducer); these objects are also available in finalization.

Pure and Hybrid Annealing Patterns. The rest of this section introduces two MR algorithmic patterns for pure simulated annealing and four algorithmic patterns for hybrid simulated annealing. *Hybrid* simulated annealing patterns use genetic operations, such as crossover, to enhance the annealing process, as opposed to *pure* patterns which employ simulated annealing alone. For readability, we present all patterns in Python-like pseudo code, instead of our actual Java implementations.

Data Representation and MR Tasks. Recall that logically, MR input and output datasets are collections of records. In the general case, an MR record is a key-value pair: *record* = *<key, value>*. *Empty* keys can be used to make records equivalent to values; this option is employed in our simulated annealing algorithmic patterns. Each value represents, in textual form, a possible *solution* to a problem (such as a TSP route, for example). An input/output dataset defines a *population* of candidate-solutions. Our simulated annealing patterns transform *input populations* of candidate-solutions into *output populations* of possibly better candidate solutions.

At the file system level, MR datasets are collections of one or more files. The MR framework uses the number and size of files in the input dataset to determine the number of spawned mappers, without direct user control. In general, each file from a multi-file dataset will be assigned to at least one mapper, with larger files being split by MR and assigned to multiple mappers for the same large file. In particular, a relatively small single-file dataset (such as an input population of candidate solutions for the TSP for example) will be assigned to a single mapper, regardless of the number of available task nodes in the MR cluster; other nodes will remain idle. In contrast to mappers, the number of reducers can be explicitly defined by the user. In particular, it is possible to define a MR job with *zero reducers*, in which case the output dataset is produced by the mappers alone. Without user specification, a default number of reducers are spawned uniformly on each task node.

Single-Job Simulated Annealing Patterns: SA0 & SA. In practice, it is often difficult to assess the accuracy obtained with a single simulated annealing run. In order to find a better solution, a frequently used strategy is to run simulated annealing a number of times and select the best solution from the independent runs [16, 6]. We have adopted this idea in two single-job MR patterns: a *special* simulated annealing pattern, *SA0*, and a *general-purpose* simulated annealing pattern, *SA*.

With the *special pattern*, *SA0*, annealing runs are performed by distributed mappers in a single MR job with zero reducers, in which mappers simply invoke an annealing algorithm over their assigned candidate-solutions (Fig. 3). The *SA0* pattern is termed *special* because it works well only in the special case of single-record input files. Recall that the number of mappers is implicitly determined by MR as a function of the number and sizes of input files. For good *SA0* performance, each candidate-solution must be preloaded in its own file; in this case, each candidate-solution will be assigned by MR to a dedicated mapper. Grouping all candidate-solution in a single file would be detrimental to *SA0*'s performance because all solutions will most likely be assigned by MR to a single mapper and in fact annealed serially.

```
class Mapper:
  method map(key, value):
    solution = parse(value)
    annealer.anneal(solution)
    emit(empty, solution)
```

Fig. 3. Mapper for the *SA0* algorithmic pattern. MR splits the input set of candidate-solutions between mappers and feeds candidate-solutions as "values" into map method invocations. Each map invocation anneals its assigned candidate-solutions then emits the annealed solution with an empty key in the output population. The number of mappers – and the level of parallelism – is determined by MR based on the number and the sizes of input files with candidate-solutions.

With the *general-purpose pattern*, *SA*, annealing runs are simultaneously performed on distributed reducers rather than on mappers (Fig. 4). The mappers themselves are only used to replace default input keys with new uniformly distributed random keys; even a single mapper can be efficient enough in this process. The updated records are then further submitted by MR to available reducers. Note that the initial default MR key for each record is simply the position in the record in its corresponding file. Such default keys may cause MR to partition records onto a small number of reducers and therefore result in non-uniform reducer workloads. By using uniformly distributed random keys, *SA* provides uniform distribution of records onto mappers. This approach provides good performance regardless of the physical representation of the input set of candidate-solution - either as a single file or as a collection of multiple files.

```
class Mapper:                          class Reducer:
  method map(key, value):                method reduce(key, values):
    randomKey = random();                  for value in values:
    emit(randomKey, value)                   solution = parse(value)
                                             annealer.anneal(solution)
                                             emit(empty, solution)
```

Fig. 4. Mapper and reducer for the *SA* algorithmic pattern. MR splits mappers' output onto reducers based on keys emitted by mappers; in this pattern mappers emit random keys to provide uniform distribution onto available reducers and more balanced reducer load. Each reduce invocation anneals all candidate-solutions with the same random key and emits possibly improved solutions in the output population.

Genetic Annealing Pattern: GA+SA. It has been recognized that enhanced initial candidate-solutions for simulated annealing can improve both the quality of the final solution and also the annealing execution time [12]. Such enhanced solutions can be obtained by first applying a genetic algorithm, *GA* (Fig. 5) on a randomly generated initial population of candidate-solutions and then applying simulated annealing, *SA* (Fig. 4) on the genetically evolved population.

```
class Mapper [or Reducer]:               method finalize():
  method initialize: subpopulation = ∅     genetic.evolve(subpopulation)
  method map [or reduce] (key, value):     for solution in subpopulation:
    subpopulation.add(parse(value))          emit(empty,solution)
```

Fig. 5. Mapper and reducer for *GA*, the genetic algorithmic pattern incorporated in the *GA+SA* algorithmic pattern. The *GA*'s mapper and reducer are nearly the same. The initial population of candidate-solutions is split by MR in sub-populations among distributed mappers. Each mapper runs a genetic algorithm on its own subpopulation. All evolved subpopulations are then merged by MR onto a single reducer (in contrast to mappers, the number of reducers can be explicitly controlled programmatically). The reducer then runs the same genetic algorithm to further evolve the entire population. The map/reduce method invocations simply accumulate sub-populations, while the actual genetic computation occurs during finalization.

Thus, *GA+ SA* is a two-job MR pipeline (Fig. 2) in which the first job, *GA* is a basic multi-population genetic algorithm [1]. The second job in *GA+SA* is the general purpose simulated annealing, *SA* (Fig. 4). Note that *SA0*, the special purpose simulated annealing (Fig. 3) must not be used after *GA* because *GA* uses a single reducer and produces a single-file output dataset; recall that *SA0* degrades to serial execution for a single-file input dataset.

Genetic Annealing Pipeline Pattern: m(GA+SA). After a genetic algorithm is trapped in a local minimum, the application of simulated annealing can generate uphill jumps to higher costs solutions thus avoiding premature convergence to a local minimum [5]. This computation can be defined as an MR job pipeline *m(GA+SA)* of *2m* jobs, which consecutive applies *GA+SA* over the dataset produced by the previous application. Again, all intermediate datasets are available, together with the final dataset for the selection of the best solution.

Annealing Genetic Pattern: SA+GA. A genetic algorithm can be used to recombine and possibly improve solutions produced by individual simulated annealing processes [11]. With MR, such computation can be defined in the MR framework as a two-job pipeline in which the first job, *SA* is simulated annealing (Fig. 4) and the second job, *GA* is multi-population genetic evolution (Fig. 5).

Annealing Genetic Pipeline Pattern: m(SA+GA). Genetic recombination can enhance the annealing process by running "simulated annealing followed by genetic recombination" a number of times to gradually obtain a better solution [16]. Such iterative computation can be defined as an MR job pipeline similar to the previously discussed *m(GA+SA)* pipeline, but with *SA* executing before *GA*.

Pipelines in MR. Job pipelines cannot be expressed in the pure MR model proper; such pipelines are often implemented as applications that schedule and run sequences of individual MR jobs. Section 3 contains details on our implementation of pipelines.

3   Implementation and Experimental Evaluation

Annealing the TSP on Amazon's Elastic MapReduce Cloud. The simulated annealing pure and hybrid algorithmic patterns defined in Section 2 can be instantiated to solve specific problems. To instantiate the algorithmic patterns, it suffices to develop *serial-only* domain-specific annealing and genetic algorithms, with no direct involvement of the MR API.

We illustrate this instantiation process with the traveling salesperson problem. We chose TSP because (i) it is known to be computationally intensive for larger problem sizes and because (ii) it is arguably the most popular combinatorial optimization problem that is well studied and well-applied to various specific tasks.

We developed TSP annealing implementations for the Amazon's Elastic MR cloud - a member of Amazon Web Services (AWS). We chose AWS because (i) it is a large and versatile cloud computing platform and because (ii) Amazon supports research through special grants, within its cost-effective pay-per-use business model.

The principal Elastic MR API is for Java. The goal of our proof-of-concept implementation was more to illustrate and evaluate our generic MR algorithmic patterns (Section 2) rather than develop new TSP algorithms. This is why we adopted some features of known serial TSP algorithms to fit the MR Java API.

The TSP aims to find the shortest way to visit each of n points once and return to the initial point. A candidate-solution is an array of different points, referred to as a *tour*. The length of a tour is the sum of the Euclidean distances between its points. In the special case of a *square city grids* of $n = s^2$ points, where s is even, an optimal tour of length n is known to exist [6]. This special case of the TSP problem offers an opportunity to directly assess the solution quality of annealing algorithms.

To instantiate MR algorithmic patterns into TSP implementations, we adapted in Java a proven serial annealing method, originally described by Hansen in SuperPascal [6]. We developed an *Annealer* class with an *anneal* method (Fig. 6) which we plugged into our *SA0* and *SA map* and *reduce* methods (Fig. 3 and Fig. 4).

```
class Annealer:                      method search(tour, temp):
  method anneal(tour):                 na = 0; nc = 0
    temp = tempMax                     while (na < attempts && nc < changes):
    for k = 1, 2, ..., reductions:       tour1 = swap2RandomPoints(tour)
      search(tour, temp)                 if tour1.length() - tour.length < temp:
      temp = alpha * temp                  tour = tour1
```

Fig. 6. Simulated annealing for the TSP problem. Annealing is implemented by swapping two randomly chosen tour points, p and q and reversing the tour path between p and q. The *search* method uses a simple deterministic tour acceptance criterion that has been proven to work just as well as the standard stochastic criterion [10].

For our hybrid annealing TSP implementations, we developed in Java a genetic algorithm with a proven serial crossover method, originally described by Sengoku and Yoshihara [13]. We developed a *Genetic* class with an *evolve* method (Fig. 7) which we plugged into the *GA*'s genetic *map/reduce* methods (Fig. 5).

```
class Genetic:                       method cross(population, m):
  method evolve(population):           parents = roulette(population, m)
    for k = 1, 2, ..., generations:    for i = 1, 3, 5, ..., m:
      select(population, m)              tour1 = crossover(population, i, i+1)
      mutate(population, mutatProb)      population.add(tour1)
      cross(population, m)
```

Fig. 7. Basic genetic algorithm for the TSP problem. Proof-of-concept evolutionary computation involves deleting m tours from the current population, applying mutation stochastically on the remaining tours, selecting $2*m$ parents to crossover, producing a single offspring form every pair of parents, and adding the offspring to the population. The *crossover* method is based on the longest sub-tour crossover operator [13]. Our *mutate* method swaps two random tour points like in simulated annealing.

Pipeline Implementation. AWS's Elastic MR cloud permits the direct implementation of MR job pipelines (Fig. 2) in the form of the so called job flows. At present, to define a job flow in Elastic MR the user must employ Amazon's proprietary lower-level API. We preferred to follow a platform-independent approach, for which we implemented MR job pipelines in Java proper, using reflection: we developed a Java MR utility that reads all classes to be pipelined as command-line arguments then uses a loop to configure and run MR jobs accordingly. In addition, another Java MR utility of ours extracts and sorts all intermediate and final solutions produced by a pipeline and identifies the best solution.

Experimental Evaluation of Solution Quality. We tested experimentally the solution quality of our TSP implementations by means a serial model program; submission and evaluation of Elastic MR jobs is time-consuming and the use of a serial model program helped simplify the evaluation. (We did, however, measure execution times/performance by actually running programs on the Amazon's Elastic MR cloud, as discussed later in this section.) For simulated annealing, we used the same control parameters as in [6]. For genetic computations, we performed $25*sqrt(tour\text{-}size)$ generations with crossover probability of 50% and mutation probability of 20%. The population size for these experiments was 16.

Table 1. Pure simulated annealing *SA/SA0* solution quality

Tour Size	100	400	900	1600
Solution	100	405.30	918.72	1648.48
Error	**0%**	**1.33%**	**2.08%**	**3.03%**
Min error	0%	0.82%	1.84%	2.80%
Max error	0%	1.66%	2.30%	3.31

We tested TSP solution quality with pure simulated annealing, *SA* over *square city grids* of *n* points with known optimal tour length of *n* [6]. Table 1 shows averages of all best solutions obtained in 10 trials over tours of various sizes. The solution quality of *SA0* is the same as the solution quality of *SA* because the two methods differ only in the way they distribute the same annealing process between mappers and reducers. The highest average solution error is about 3% for larger tours.

Table 2. Solution quality of hybrid simulated annealing

Algorithm	SA/SA0	GA+SA	4(GA+SA)	SA+GA	4(SA+GA)
Improvement	Base	**0.28%**	**0.61%**	**0.06%**	**0.94%**
Min improv.	Base	0%	0%	0.01%	0.05%
Max improv.	Base	1.30%	2.01%	0.15%	1.89%

We also tested TSP solution quality with hybrid simulated annealing over *random grids* for which no optimal tours are known a priori. Table 2 shows average solution improvements by each of the hybrid methods, *GA+SA*, *4(GA+SA)*, *SA+GA*, and *4(SA+GA)*, relatively to the best solution obtained by pure annealing methods, *SA/SA0*. Solution improvements were measured in 10 trials over various randomly generated tours of 900 cities and populations of size 16. The hybrid *4(SA+GA)* pipeline (Table 2) provides nearly 1% of improvement compared to pure *SA/SA0* and therefore can reduce almost in half the estimated 2% error of *SA* (Table 1).

Experimental Evaluation of Performance on Elastic MR. On the Elastic MR cloud, we tested experimentally TSP solution performance with pure and hybrid simulated annealing: *SA*, *SA+GA*, *4(SA+GA)*, and *SA0*. We did not test performance of *GA+SA* and *4(GA+SA)* because they are comparable performance-wise to *SA+GA* and *4(SA+GA)* correspondingly. Table 3 shows average execution times $T(p)$ and speedups $S(p)$ on p task nodes as obtained in 3 trials over a randomly generated tour of size 900 and populations of sizes 16 and 32. (In Table 3 population sizes are appended in brackets to algorithm designators.) As nodes, we used AWS 32-bit small instances with 1.7 GB memory, 1 virtual core, and moderate I/O performance.

Table 3 shows that special simulated annealing, *SA0* achieves better speedup than general purpose simulated annealing, *SA*. However, *SA0* achieves this speedup for special single-record input files only, as explained in Section 2. It is an advantage of general purpose *SA* that it can be combined with *GA* to form hybrid pipelines that achieve better quality solutions, while *SA0* cannot be combined with *GA*.

Table 3. Elastic MR execution times (in minutes) and speedup

Algorithm	SA[16]		SA[32]		SA0[16]		SA+GA[16]		4(SA+GA)[16]	
Nodes (n)	T(n)	S(n)	T(n)	S(n)	T(n)	S(n)	T(n)	S(n)	T(n)	S(n)
1	11.7	**1.0**	22.6	**1.0**	11.6	**1.0**	13.8	**1.0**	52.0	**1.0**
8	3.0	**3.9**	5.1	**4.4**	2.3	**5.0**	5.3	**2.6**	20.1	**2.6**
16	2.5	**4.7**	3.7	**6.1**	1.5	**7.7**	4.9	**2.8**	16.9	**3.1**

Despite of the use of random keys in *SA*'s mappers, some reducers are assigned by MR more work than others; such imbalances result in relatively moderate speedups when the population size is equal to the number of task nodes (16 nodes in Table 3). Load imbalances can be reduced by using populations of size $k*nodes$ with $k \geq 2$. In general, the scalability of standalone and pipelined *SA* is limited by the population size.

4   Related Work

The serial components of our implementations are based on work from others [6, 13], as already discussed in the preceding section. To our knowledge, we are the first to parallelize simulated annealing with MR, but there are numerous non-MR parallel simulated annealing algorithms, such as, for example, message passing [6], shared memory [9], message passing combined with shared memory [4], and GPGPU-based [2]. Others have proposed self-contained MR-based genetic algorithms [14, 7] and MR has been used for fitness function calculation in evolutionary algorithms [17]; in contrast, our MR genetic algorithm is not intended as standalone but to be incorporated as a job in hybrid annealing pipelines.

5   Conclusions

In this paper, we investigate the applicability of MapReduce to distributed simulated annealing in general, and to the TSP in particular. The specific technical contributions of this paper are as follows: (i) we propose six MR algorithmic patterns for distributed simulated annealing; (ii) we instantiate the MR patterns into TSP implementations; (iii) we evaluate the MR implementations in cloud computing environment.

A significant advantage of our MR simulated annealing patterns to traditional parallel algorithms is that these patterns provide *fault-tolerant MR parallelism* without user intervention. With the use of MR, we trade some speedup for fault-tolerance and robustness. The lack of direct user control on parallelism however can also be a limitation when the programmer wants to explicitly declare some MR parameters, such as the total number of mappers. A benefit from our annealing MR patterns is that they can be instantiated into MR applications with the addition of serial-only domain code, such as code to represent, anneal, and evolve the TSP for example. Our hybrid annealing patters are slower than the pure annealing patterns but are more precise. In future work, the genetic component of hybrid patterns can be fine-tuned to make then even more precise. The Amazon's Elastic MR cloud offers the advantages of instant cluster provisioning and pay-per-use cost efficiency for users who do not have access to dedicated MR clusters on the premises.

Acknowledgement. This work was supported by an AWS in Education 2011 research grant award from Amazon.

References

[1] Cantú-Paz, E.: Efficient and Accurate Parallel Genetic Algorithms. Kluwer, Boston (2000)

[2] Choong, A., Beidas, R., Zhu, J.: Parallelizing Simulated Annealing-Based Placement using GPGPU. In: Field Programmable Logic and Applications, pp. 31–34. IEEE, New York (2010)

[3] Dean, J., Ghemawat, S.: MapReduce: Simplified Data Processing on Large Clusters. CACM 51(1), 107–113 (2008)

[4] Debudaj-Grabysz, A., Rabenseifner, R.: Nesting OpenMP in MPI to Implement a Hybrid Communication Method of Parallel Simulated Annealing on a Cluster of SMP Nodes. In: Di Martino, B., Kranzlmüller, D., Dongarra, J. (eds.) EuroPVM/MPI 2005. LNCS, vol. 3666, pp. 18–27. Springer, Heidelberg (2005)

[5] Elhaddad, Y., Sallabi, O.: A New Hybrid Genetic and Simulated Annealing Algorithm to Solve the Traveling Salesman Problem. In: World Congress on Engineering (WCE 2010), vol. 1, pp. 11–14. International Association of Engineers, Taipei (2010)

[6] Hansen, P.-B.: Studies in Computational Science. Prentice Hall, Englewood Cliffs (1995)

[7] Huang, D.-W., Lin, J.: Scaling Populations of a Genetic Algorithm for Job Shop Scheduling Problems Using MapReduce. In: 2010 IEEE 2nd International Conference on Cloud Computing Technology and Science, pp. 78–85. IEEE, New York (2010)

[8] Lin, J., Dyer, C.: Data-Intensive Text Processing with MapReduce. Morgan and Claypool, San Francisco Bay Area (2010)

[9] Ma, J., Li, K., Zhang, L.: The Adaptive Parallel Simulated Annealing Algorithm Based on TBB. In: 2nd International Conference on Advanced Computer Control, pp. 611–615. IEEE, New York (2010)

[10] Moscato, P., Fontanari, J.: Stochastic versus Deterministic Update in Simulated Annealing. Physics Letters A 146(4), 204–208 (1990)

[11] Ohlídal, M., Schwarz, J.: Hybrid Parallel Simulated Annealing Using Genetic Operations. In: 10th International Conference on Soft Computing, Mendel 2004, pp. 89–94. University of Technology, Brno (2004)

[12] Ram, J.D., Sreenevas, T.T., Subramaniam, K.G.: Parallel Simulated Annealing Algorithms. J. Par. Distr. Computing 37, 207–212 (1996)

[13] Sengoku, H., Yoshihara, I.: A Fast TSP Solver Using GA on Java. In: 3rd Int. Symp. Artif. Life and Robot., pp. 283–288. Springer, Japan (1998)

[14] Verma, A., Llorà, X., Goldberg, D.E., Campbell, R.H.: Scaling Genetic Algorithms Using MapReduce. In: 9th International Conference on Intelligent Systems Design and Applications, pp. 13–18. IEEE, New York (2009)

[15] White, T.: Hadoop: The Definitive Guide, 2nd edn. O'Reilly Media, Sebastopol (2009)

[16] Yao, X.: Optimization by Genetic Annealing. In: 2nd Australian Conf. Neural Networks, pp. 94–97. Sidney University, Sidney (1991)

[17] Zhou, C.: Fast Parallelization of Differential Evolution Algorithm Using MapReduce. In: 12th Annual Conference on Genetic and Evolutionary Computation, pp. 1113–1114. ACM, New York (2010)

Flex-GP: Genetic Programming on the Cloud

Dylan Sherry, Kalyan Veeramachaneni,
James McDermott, and Una-May O'Reilly

Massachusetts Institute of Technology, USA
{dsherry,kalyan,jmmcd,unamay}@csail.mit.edu

Abstract. We describe Flex-GP, which we believe to be the first large-scale genetic programming cloud computing system. We took advantage of existing software and selected a socket-based, client-server architecture and an island-based distribution model. We developed core components required for deployment on Amazon's EC2. Scaling the system to hundreds of nodes presented several unexpected challenges and required the development of software for automatically managing deployment, reporting, and error handling. The system's performance was evaluated on two metrics, performance and speed, on a difficult symbolic regression problem. Our largest successful Flex-GP runs reached 350 nodes and taught us valuable lessons for the next phase of scaling.

1 Introduction

Cloud computing has emerged as a new paradigm for commercial, scientific, and engineering computation. A cloud allows an organization to own or rent efficient, pooled computer systems instead of acquiring multiple, isolated, large computer systems each commissioned and assigned to particular internal projects [1, 6].

Cloud computing has substantial advantages. First, it offers *elasticity*. Elasticity allows a software application to use as much computational resources as it needs, when it wants. A cloud also offers *redundancy*. If a server fails, a replacement is swiftly available from the resource pool. However it cannot be automatically incorporated into a long-running computation unless the computation is designed to allow this. A cloud also offers *higher utilization*. Utilization refers to the amount of time a pool of resources is in use rather than idle.

For evolutionary algorithms (EA) researchers, the cloud represents both a huge opportunity and a great challenge. Parallelization has been well-studied in the context of EAs, and has been shown to affect population dynamics and diversity, and to improve performance. With the cloud we can aim to run parallel evolutionary algorithms at a scale never before seen, but we must first make our algorithms and implementations cloud-ready.

In the long term we envisage novel refactoring and rethinking of genetic programming (GP) as the cornerstone of a massively scalable cloud-based evolutionary machine learning system. Our immediate goal is to design, implement, deploy and test a cloud-based GP system, which we call Flex-GP.

In this paper, we adopt an island based parallelization model with communication via sockets, and choose Amazon's EC2 as a computational substrate.

C. Di Chio et al. (Eds.): EvoApplications 2012, LNCS 7248, pp. 477–486, 2012.

We choose symbolic regression as an application. We start modestly with a few nodes and scale to tens and then hundreds of nodes. We encounter some unexpected challanges but achieve parallelization of GP up to 350 islands.

In Sect. 2, we begin with a discussion of related work on the parallelization and scaling of EAs. Sect. 3 briefly describes the elastic compute resource provided by Amazon. In Sect. 4 we present the strategies we employed to scale the algorithm and the challenges that arose. We then present a benchmark problem and experimental results in Sect. 5. We present our conclusions and future work in Sect. 6.

2   Related Work

The simplest EC **parallelization models** are the independent parallel runs model and the master-slave fitness evaluation model. Both are useful in some circumstances. Our research interest is in a different model, the *island model*. Multiple populations or islands run independently and asynchronously, with infrequent *migration* of good individuals between islands. The island model has been studied extensively, with surveys by Cantú-Paz [2] and Tomassini [8].

The topology of an island model may be visualised as a network of nodes (each representing a population) and directed edges (representing migration). Island models typically depend on a centralised algorithm to impose the desired neighbourhood structure. Decentralised algorithms have also been studied, where the network structure emerges in a peer-to-peer or bottom-up fashion [4].

Island models may be expected to deliver performance benefits over single-machine EC, due to their larger total populations. As demonstrated by Tomassini [8], island models can in fact do even better. This happens chiefly because a structured population can avoid premature convergence on just one area of the search space. Vanneschi [9] (p. 199) notes that for each problem there is a population size limit beyond which increases are not beneficial. Tomassini found that isolated populations have an advantage in performance over single large populations, where total population size is 2500, and that communicating islands have an advantage over multiple isolated ones.

A key opportunity in cloud computing is its **massive scale**. Most existing research in island model evolutionary algorithms has not used very large numbers of nodes. A typical value in previous experiments is between 5 and 10 nodes [8,9]. Although we note that each of these projects are now quite old, few specific node-counts are available in the recent literature. The most important exception is the 1000-node cluster used by Koza [7] (p. 95).

The *Hadoop* implementation [http://hadoop.apache.org/] of the *MapReduce* framework [3] has been used for genetic algorithms [10]. We started from *ECJ* [5] which is a EC system written in Java. It includes an island model framework which uses sockets for communication in a master-slave arrangement. ECJ is more flexible than MapReduce and can avoid its requirement for a master node which represents a single point of failure and a synchronisation bottleneck; it is actively maintained; and it offers an easy-to-use, but limited, island model. Therefore ECJ was selected for our work.

3   Deploying Flex-GP

We chose to use the Amazon Elastic Compute Cloud (EC2), a versatile cloud computing service. EC2 provides a simple abstraction: the user is granted as many instances (VMs) as needed, but no access to the underlying host machines. Instances can vary in size, with the smallest costing as little as $0.02 per hour. Users can request instances immediately, reserve instances for a set time period, or bid against the current spot price. An instance's software is specified by a VM image called an Amazon Machine Image (AMI). Amazon provides a selection of off-the-shelf AMIs and also allows the definition of custom images. We found the default Amazon Linux AMI to be a suitable platform for our system. We initially chose to use the cheapest instance size, *micro*, on an immediate-request basis. However, *micro* instances are intermittently allocated more processing power for short periods. To more clearly analyze the performance of our system, we transitioned to *small* sized instances, which are granted a fixed amount of processing power.

Table 1. Notations

Name	Notation
Island q	I_q
Number of Neighbors	N_n
Number of islands	Q
Number of instances	n
IP address of the node i	IP_i
Neighbor Destinations	N_{dn}
Time out for replacing instance	T_o

As described in Sect. 2, we chose to use ECJ's island model. It uses a client-server architecture, where each client hosts one island, and a sole server is responsible for setting up the topology, starting and halting computation. In ECJ's off-the-shelf island models, one of the clients doubles up as the server. We chose to configure a separate server with an eye to scaling. Algorithm 1 presents the pseudocode for the n-island model. Parameters like migration size, M_s, rate, F_m, start generation SG are provided by the user.

Each island consists of two Java threads, as shown in Fig. 1. The first is the main thread, which performs evolutionary computation and periodically sends packets of emigrants to neighboring islands. The second acts as a mailbox, and is responsible for receiving packets of incoming immigrants. The main thread periodically fetches newly arrived individuals from the mailbox and mixes them into the population. Note that this architecture implies that if a node crashes, those nodes to which it sends will not be affected, nor will those which send to it. Although the topology of the network will be damaged, all other nodes will continue calculations. This is a limited form of robustness.

Algorithm 1. The socket based n-island model

---

1. Pre-process: Create the necessary *params* files
2. **Initialization**
for $d = 0$ to n **do**
  Initialize I_d
  if $d = 0$ **then**
    $IP_s \leftarrow IP_d$
  end if
  if $d \neq 0$ **then**
    Recv_$I_d(IP_s)$
    Send_$I_0(IP_d)$
  end if
end for
3. **Set up communications**
if $d = 0$ **then**
  for $d = 1$ to n **do**
    Send neighborhood information: Send_$I_d(N_n, N_{dn}, N_{dn}^{id}, IP_{dn})$.
    Send Migration parameters: Send_$I_d(M_s, F_m, SG)$.
    Send GP parameters: Send_$I_d(psize, ngens)$.
  end for
end if
4. **Start computation**
if $d = 0$ **then**
  for $d = 1$ to n **do**
    Instructs I_d to start computation
  end for
end if
5. **Stopping computation**
if $d \neq 0$ **then**
  while I_d did not receive stop signal **do**
    Send_$I_0(O_i)$ where O_i is the fitness at the end of ith generation.
  end while
end if
if $d = 0$ **then**
  Recv(O_i)
  if $O_i \geq O_d$ **then**
    for $d = 1$ to n **do**
      Send_$I_d(stop\ signal)$
    end for
  end if
end if
The server and all islands have exited

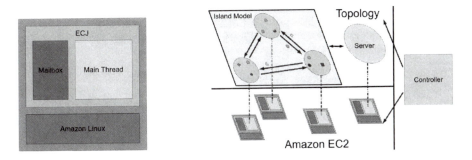

Fig. 1. An ECJ island consists of multiple processes (left). The three-island model on EC2 (right).

4   Scaling Flex-GP

As we proceeded from an initial run towards large numbers of islands, new issues emerged. We present our progress as a series of milestones: $Q = 3, Q = 20$, $Q = 100$ and finally $Q = 350$ islands. In this section our aim is only to consider scaling, and so the details of the problem and the fitness values are achieved are not reported. For completeness, we note that the experimental setup was the same as that in Sect. 5.

Milestone 1: Three Islands. Our initial goal was a proof of concept. We manually constructed a three-island ring topology on three EC2 instances, with the server hosted on a fourth. This enabled us to understand the steps involved in launching, starting and running a basic island-based GP system on EC2. We found that three key ECJ components were easy to use and ready to run on EC2: socket based communication, an evolutionary loop, and experiment setup. The three-island model ran successfully.

Milestone 2: 20 Islands. As soon as more than three islands were required, the overhead required to manually start each island as above became unachievable. We automated the instance requests using EC2's Python API, *boto*. We avoided the transfer of files by creating a custom AMI containing our code-base. With this setup we achieved the 20-island milestone.

Milestone 3: 100 Islands. During the next phase of scaling, several more issues became apparent.

Instance boot was both unreliable (about 1 in 250 requested nodes simply fails to start) and highly variable in the time required (from 15 seconds to several minutes). The time required for instance network connection was also variable, up to 30 seconds. Since the ECJ island computation does not begin until all islands have reported successful startup, it was therefore essential to provide **monitoring and dynamic control of instance startup**. In Algorithm 2 we present a dynamic launch monitor and control process. It has two parameters, the wait time α and the time for timeout, T_o. All EC2 interfaces have latency of a few seconds, so α

must be long enough to allow for that. We set T_o as the mean time required for an instance to launch and connect. Its main function is to make instance requests and take account of requests which are apparently failing.

Algorithm 2. Dynamic logic for instance startup

Generate a cloneable image C
$d \leftarrow 0$
while $d \leq n$ **do**
   Request an instance of C via *Boto*
   while $T_o \neq \gamma$ **do**
      if Instance is running according to API **then**
         if Instance is connected **then**
            $d \leftarrow d + 1$
            $T_o \leftarrow \gamma$
         else
            wait for α
            $T_o \leftarrow T_o + \alpha$
         end if
      else
         wait for α
         $T_o \leftarrow T_o + \alpha$
      end if
   end while
end while

Even after an instance is correctly created, a variety of problems can occur at runtime, including software and configuration bugs, network problems, and other unknown errors. **Error tolerance and reporting** is essential. To better handle debugging and post-run analysis we first coded these messages. Once we resorted to a coded catalogue/dictionary for errors, it enabled us to incorporate more log messages throughout our code-base. The coding of messages helped to reduce the bandwidth when transferring logs from the islands.

Milestone 4: 350 Islands. Amazon limits new EC2 users to 20 concurrent instances. Requests for increases may be placed and are usually fulfilled incrementally some days later. After several requests our limit stands at 400 instances. Our next goal was to approach this limit. The main questions to be considered were: would the socket-based model withstand communication among hundreds of islands? Would the fact that the server is a single point of failure prove problematic?

At this level, two major augmentations were required. We added an additional dedicated instance as a monitoring/log server. To do so, we added a *LogServer*, supported by the *Twisted* open source Python networking engine. The *LogServer*'s role was to aggregate and display information about the current status of computation across the network. Two types of information were transmitted to the *LogServer*: performance and migration tracking. In our larger tests we inadvertently "stress tested" the capacity of this server. During benchmarks with 350 islands, the *LogServer* received and recorded almost 100,000 lines of text over several minutes, successfully receiving 100% of incoming messages.

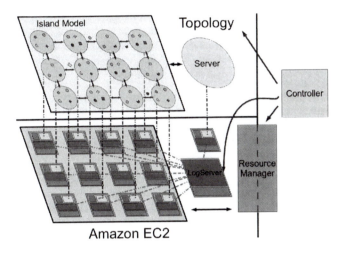

Fig. 2. The 350-Island System

We also wished to take advantage of our large limit of 400 instances by running multiple tests simultaneously, for example almost the entirety of the 1, 2, 4, 16, 128 and 256 island runs mentioned in the next section. We introduce the concept of a *bucket*, that is a set of nodes allocated to a single task. We added a *ResourceManager* to manage instances and buckets, consisting of the following components: a *Connection* class which communicates with EC2; a list of *free instances*; and a set of buckets to which the free instances can be allocated. Each test run requires one bucket. Managing multiple buckets in the ResourceManager means that multiple tests can be run simultaneously. The ResourceManager favoured the serial re-use of buckets to minimise setup time.

5   Experimental Setup

Our next goal was to evaluate our system. The benchmark was a two-variable symbolic regression problem taken from [11]. The aim is to match training data produced from a known target function,

$$f(x,y) = \frac{e^{-(x-1)^2}}{1.2 + (y - 2.5)^2}$$

The training data was 100 points randomly generated in the interval $(x, y) \in [0.3, 4]$, with a different set of points being generated for each island. No separate testing phase was run. Fitness was defined as the mean error over the training points. The function set was $\{x, y, +, *, -, \%, \text{pow}, \exp, \text{square}\}$, where $\%$ indicates protected division (if the denominator is zero, 1 is returned).

The population was initialised using the ramped half-and-half algorithm, with minimum depth 2 and maximum depth 6. Tournament selection was used with tournament size 7. Subtree crossover, biased 90/10 in favour of internal nodes, was used with probability 0.9, and reproduction with probability 0.1. No mutation was used, nor was elitism. The maximum tree depth was 17. The population size at each island was 3000 and the number of generations was 100. The island topology used was a non-toroidal four-neighbor grid. Each island was configured to send 40 emigrants to its destination neighbors every four generations. We ran the benchmark on 1, 2, 4, 16, 64, 128 and 256 islands. For each of these cases we conducted at least 10 runs. We evaluated benchmark performance on two metrics, accuracy and time.

Accuracy is the best fitness achieved, i.e. $1/(1 + e)$ where e is the mean squared error on our benchmark problem. In Fig. 3 we show the improvement in accuracy of the system as a function of number of islands. We plot the average fitness achieved at the end of each generation. We average this number over 10 independent runs. This is shown in Figure 3(right). Fitness generally improves as we add more resources. However, as we add more resources the gain in fitness achieved at the end of 100 generations reduces. For example, the biggest gain in fitness is achieved when we go from 1-island model to 4-islands and the least gain is achieved when we use 256 instead of 128 islands. In the experiment only the number of islands was varied, and all other parameters left fixed. The larger trials thus had a lower *relative* degree of information flow between islands, which may have impaired performance. Finally it is possible that this result reflects Vanneschi's finding that for any problem, there is a limit to total population size beyond which performance does not increase and can even be impaired [9] (p. 199). We also show the variance in the best fitness at the hundred generations for multiple runs as a box plot in Figure 3(left). It is interesting to note that the variance in the best fitness achieved significantly reduces as we add more resources.

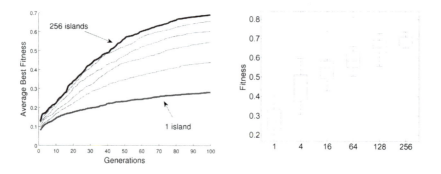

Fig. 3. Results achieved on a benchmark symbolic regression problem

Time is measured as achieving this higher accuracy in the same amount of time that would be used on a single machine. We measure both communications and infrastructure setup time and total compute time, with results as shown in Fig. 4. We plot four different times. First one is the initialization of the *LogServer*. The second is the time taken by our system to set up monitors, and the third is the time taken by the evolutionary server to set up islands and the communications. Both these times see an increase in time from 10 seconds to 180 seconds. Finally we show the actual computation time of the islands. Even though we add 255 nodes the compute time only increases by a factor of 3.

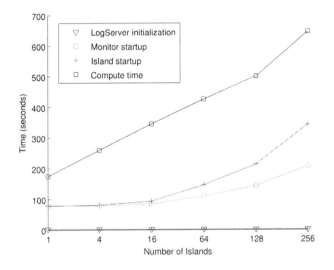

Fig. 4. Time taken by the server to set up island infrastructure and topology for communications

6 Conclusions and Future Work

In this paper we have described development of Flex-GP, which we believe to be the first large-scale cloud GP implementation. This is timely and is made possible by the advances made in virtualization and cloud infrastructure. We chose the Amazon EC2 cloud service. From several pre-existing software packages, we chose *ECJ* which provides a simple off-the-shelf island model with socket communication. We made each island a separate EC2 instance.

In order to scale up to 350 islands we had to develop many additional software features including *cloning*, *dynamic launch*, and a *LogServer*. We made use of some publicly available open source tools like *twisted, boto, nmap*. We encountered and overcame several problems during scaling. We were able to identify when certain features become critical: for example, automatic launch via *cloning* becomes necessary for island numbers above 5.

Our success is demonstrated through Flex-GP performance on a benchmark problem. Increased resources improve performance, with some cost in time.

Our next goal is to scale Flex-GP to at least 1000 islands. From experience in scaling to 350, we expect to require a number of additional infrastructural features such as *distributed startup, visualization tools* and *no single point of failure*. We also aim to modify the Flex-GP island model to introduce features of *elasticity* (add or remove instances when needed) *resiliency* (gracefully handle node failures). After achieving these we aim to examine other distribution models.

Acknowledgements. This work was supported by the GE Global Research center. Any opinions, findings, and conclusions or recommendations expressed in this material are those of the authors and do not necessarily reflect the views of General Electric Company.

References

1. Armbrust, M., Fox, A., Griffith, R., Joseph, A., Katz, R., Konwinski, A., Lee, G., Patterson, D., Rabkin, A., Stoica, I., Zaharia, M.: Above the clouds: A berkeley view of cloud computing. EECS Department, University of California, Berkeley, Tech. Rep. UCB/EECS-2009-28 (2009)
2. Cantú-Paz, E.: A survey of parallel genetic algorithms. Calculateurs Paralleles 10(2) (1998)
3. Dean, J., Ghemawat, S.: MapReduce: Simplified data processing on large clusters. Communications of ACM 51(1), 107–113 (2008)
4. Laredo, J.L.J., Castillo, P.A., Paechter, B., Mora, A.M., Alfaro-Cid, E., Esparcia-Alcázar, A.I., Merelo, J.J.: Empirical Validation of a Gossiping Communication Mechanism for Parallel EAs. In: Giacobini, M. (ed.) EvoWorkshops 2007. LNCS, vol. 4448, pp. 129–136. Springer, Heidelberg (2007)
5. Luke, S., Panait, L., Balan, G., Paus, S., Skolicki, Z., Bassett, J., Hubley, R., Chircop, A.: ECJ: A Java-based evolutionary computation research system (2007), http://cs.gmu.edu/~eclab/projects/ecj/
6. Ograph, B., Morgens, Y.: Cloud computing. Communications of the ACM 51(7) (2008)
7. Poli, R., Langdon, W., McPhee, N.: A field guide to genetic programming. Lulu Enterprises UK Ltd. (2008)
8. Tomassini, M.: Spatially structured evolutionary algorithms. Springer, Heidelberg (2005)
9. Vanneschi, L.: Theory and Practice for Efficient Genetic Programming. Ph.D. thesis, Université de Lausanne (2004)
10. Verma, A., Llora, X., Goldberg, D.E., Campbell, R.H.: Scaling genetic algorithms using MapReduce. In: Proceedings of Intelligent Systems Design and Applications, pp. 13–18 (2009)
11. Vladislavleva, E., Smits, G., Den Hertog, D.: Order of nonlinearity as a complexity measure for models generated by symbolic regression via Pareto genetic programming. IEEE Transactions on Evolutionary Computation 13(2), 333–349 (2009)

Customized Normalcy Profiles
for the Detection of Targeted Attacks

Victor Skormin[1], Tomas Nykodym[1], Andrey Dolgikh[1], and James Antonakos[2]

[1] Binghamton Univeristy, Binghamton, NY, USA
{vskormin,tnykody1,adolgik1}@binghamton.edu
[2] Broome Community College, Binghamton, NY, USA
antonakos_j@sunybroome.edu

Abstract. Functionality is the highest semantic level of the software behavior pyramid that reflects goals of the software rather than its specific implementation. Detection of malicious functionalities presents an effective way to detect malware in behavior-based IDS. A technology for mining system call data, discussed herein, results in the detection of functionalities representing operation of legitimate software within a closed network environment. The set of such functionalities combined with the frequencies of their execution constitutes a normalcy profile typical for this environment. Detection of deviations from this normalcy profile, new functionalities and/or changes in the execution frequencies, provides evidence of abnormal activity in the network caused by malware. This approach could be especially valuable for the detection of targeted zero-day attacks. The paper presents the results of the implementation and testing of the described technology on the computer network testbed.

Keywords: Behavior Based Intrusion Detection, Functionalities, Colored Petri Net, Targeted Attacks.

1 Introduction

The warfare in the cyberspace today is manifested primarily by the deployment of advanced malware and novel IDS technologies. Nevertheless, malware presents a significant threat to national infrastructures. Modern malware demonstrates features of both "carpet bombing" and high precision weapons: information attacks perpetrated by malicious codes are widespread and are commonly used to inflict costly massive disruptions, conduct espionage, and as in the case of the Stuxnet worm, even industrial sabotage [1], [2]. Consequently, Intrusion Detection is a very active area of research that continues evolving as the malware techniques improve to overcome existing defenses. However, the most popular malware detection schemes are still dominated by the binary signature-based approach. Although it has many practical advantages, this technology can be evaded by using automatic tools like code packers and metamorphic engines, and leads to a dead end due to exponentially growing database of binary signatures. In addition, it is inherently incapable of addressing targeted, zero-day malware attacks not represented by a binary sample in a database.

C. Di Chio et al. (Eds.): EvoApplications 2012, LNCS 7248, pp. 487–496, 2012.

Behavioral analysis offers a more promising approach to malware detection since behavioral signatures are more obfuscation resilient than the binary ones. Indeed, changing behavior while preserving the desired (malicious) functions of a program is much harder than changing only the binary structure. More importantly, to achieve its goal, malware usually has to perform some system operations (e.g. registry manipulation). Since system operations can be easily observed and they are difficult to obfuscate or hide, malicious programs are more likely to expose themselves to behavioral detection. Consequently, while database of specific behavioral signatures is still to be utilized, its size and rate of increase are incomparably lower than those in the case of binary signatures. However, the behavioral detector has to be able to distinguish malicious operations from benign (executed by benign program) ones which is often difficult. Moreover, maliciousness of an executed functionality can often be determined only by its context or environment. Therefore the challenge of behavioral detection is in devising a good model of behavior which is descriptive enough to allow for discrimination of benign versus malicious programs and which can be tuned to the target environment.

In principle, there are two kinds of behavior detection mechanisms: misuse detection and anomaly detection. Misuse detection looks for specific behavioral patterns known to be malicious, while the anomaly based approach responds to unusual (unexpected) behavior. The advantage of anomaly based detection is in its ability to protect against previously unseen threats; however, it usually suffers from a high false positive rate. Misuse detection is usually more reliable in terms of detection performance (fewer false positives and often no false negatives) but it has two major drawbacks. First, defining a set of malicious patterns (signatures) is a time consuming and error prone task that calls for periodic updating, similarly to how binary signatures are used today. Second, it cannot detect any malicious code that does not expose known malicious behavior patterns and thus its capabilities to detect a zero day attack are limited. Consequently, it seems logical to combine both detection mechanisms thus resulting in a highly dependable IDS technology.

This paper presents a technology centered on the concept of *functionality*, i.e. such a combination of computer operations that is defined not by its specific implementation but by the result of its realization reflecting the intent (goal) of the developer. First, we discuss the formalization aspects of behavioral signatures representing functionalities, either benign or malicious, in the inclusive and obfuscation resilient form. Then we address the approach enabling the functionality extraction from a Kernel Object Access Graph capturing how kernel objects (objects managed by operating system, e.g. files, processes) are manipulated. Then we discuss the formation of a database containing the functionalities pertaining to the particular network environment that in combination with their frequencies of execution constitutes a customized normalcy profile. The resultant IDS would perform an ongoing task of functionality detection and assess the deviation of the observed network behavior from the earlier established normalcy profile. Finally, the implementation results of the particular components of the described system will be presented.

The described technology is expected to be instrumental in the detection of targeted information attacks against high value targets such as banks, power plants, government installations, etc. Indeed, Stuxnet-type malware is highly tuned to the particular network environment, and execution of its functionalities at the early stage of the attack would cause an anomaly in the network behavior undetectable by traditional means.

2   Modeling Behavior

Behavior analysis can be performed on the basis of system call data. System calls represent the lowest level of behavior semantics. Mere aggregation of system calls has inherent limitations in terms of behavior analysis. Instrumental behavior analysis must involve all levels of the semantics, from its foundation at system call level to API functions and to its highest level – functionalities. Functionality is described as a sequence of operations achieving well recognized results in the programs environment. Being defined not by its specific implementation but by the result of its realization, functionality is the only level of behavior semantics where maliciousness could be recognized. Indeed, there is no such thing as a malicious system call or a malicious API function; however the following is a partial list of commonly recognized malicious functionalities: *Self code injection, Self mailing, Download and Execute, Remote shell, Dll/Thread injection, Self manage cmd script create and execute, Remote hook, Password stealing.*

One can realize that malware inherently implements at least one malicious functionality, otherwise it cannot be qualified as malware. A technology for discovering known functionalities by monitoring computer behavior offers a dependable tool for software classification. It includes the off-line task of representing functionalities by appropriate behavioral signatures defined in the system call domain. At the run time, these pre-defined signatures could be extracted from system call data thus manifesting a benign operation or malware attack. To implement this approach, the following challenges were addressed [3]:

- **Expressiveness of behavioral signatures.** It is crucial for the success of IDS in detecting new realizations of the same malware. Since most malware incidents are derivatives of some original code, a successful signature must capture invariant generic features of the entire malware family, i.e. the signature should be expressive enough to reflect the most possible malware realizations.
- **Behavioral obfuscation.** Malicious software might writers might attempt to hide the malicious behavior of software. It may be implemented through the multipartite attacks perpetrated by coordinated operation of several individually benign codes. This is an emerging threat that, given the extensive development of behavior-based detection, is expected to become a common feature of future information attacks.
- **Run-time signature matching efficiency.** Efficiency determines the share of computer resources consumed by its security systems by the utilization of Colored Petri nets (CPN). Development of the enabling technology for this approach, a general purpose CPN tool, e.g. a unique software module that could be programmed to implement various CPN.

Ideally, malicious functionalities such as infection of executable files as performed by classical viruses are never executed by legitimate programs. However, most functionalities cannot be declared as strictly malicious. For example, code injection is usually a strong sign of malicious behavior, but it can be also used by legitimate programs for debugging or profiling purposes. Similarly, some functionalities such as user interaction are seldom executed by malicious software. Although some of the malware (e.g. rogue antiviruses) can interact with the user, most malicious programs try to stay hidden. Consequently, according to testing results of [3] the detection of pre-defined, either malicious or benign, functionalities shows 100% success rate. The functionality-based attack detection presents a more complex task. It can be accomplished only by the detection of multiple functionalities. One can base this detection on conditional probabilities of occurrence of particular functionalities subject to the execution of malicious and legitimate programs.

3   Dealing with Functionalities

In [3] the functionality specification domain (abstract OS objects) is separated from the detection domain (system calls). Abstract specification domain allows an expert to concentrate on the conceptual realization of a functionality omitting certain implementation details. To achieve higher signature expressiveness, the functionalities of interest (for example, those indicative of malware attack) can be specified by activity diagrams (AD) in terms of both standard system objects and abstract behavioral constructs named functional operations [3].

In the detection domain pre-specified functionalities are to be reconstructed from the system call flow. This computational task can be very efficiently performed by executing respective Colored Petri nets (CPN) obtained from the generalized AD specification [3, 5].

It was found that CPN provide sufficient expressive power and are highly dependable and efficient for recognizing specified functionalities in the flow of system calls as well as utilized data [4, 5]. Depending on the AD specification, the technology results in a coarse-grained detector or fine grained detector. Coarse-grained detectors trace only system call execution discarding information dependencies. Some functionalities are to be specified by AD with informational dependency between the operation attributes. In these cases fine-grained detectors trace information flows using dynamic data tracing techniques such as taint propagation thus potentially providing additional discriminative power.

It could be seen that CPN is the enabling technology for the proposed IDS: it is used to define the functionalities of interests as behavior signatures, and at the same time it provides the mechanism for the signature detection (matching). The authors have developed and offered to the user community a generic CPN simulator suitable for specifying any complex event (such as functionality of interest) as a unique combination of interrelated low-level events. This software presents an *effective handshake mechanism for matching/identification of complex structures* [6].

Most aspects of the technology described herein were implemented in a prototype IDS [6] and validated on dozens of malware and hundreds of legitimate programs by successful detection of pre-defined malicious functionalities employed by network worms and bots, including self-replication engines and various malicious payloads. A series of experiments performed to estimate run-time overhead due to IDS implementation (except taint propagation engine that could be performed by dedicated hardware module) showed only 4% run time increase.

4   Automatic Functionality Extraction

One of the drawbacks of the above technology is the presence of an expert who will be engaged each time the set of functionalities is to be expanded. Automatic functionality extraction would eliminate this drawback, and being enacted on a continuous basis, has a potential for the development of novel computer security mechanisms that is the main target of this paper.

Functionality could be observed as a frequently occurring pattern in kernel object access graphs. Kernel objects such as files or processes represent important objects in the operating system. They are managed by the operating system kernel and so they can be accessed and manipulated only through system calls. Programs access kernel objects via handles, which are identifiers unique in the context of each process. Since all important data structures and resources in the operating system are represented by a kernel object, programs cannot do much without manipulating at least some of them. Kernel object dependencies are difficult to hide and since we ignore any other data dependencies, this model provides a robust, obfuscation resilient tool for malware analysis [6].

The simplest possible functionality is just a single system call. More complex functionalities can have constraints on their arguments and contain several system calls and dependency relations between them. Significant functionalities are those which provide good descriptive power for program classification.

Kernel Object Access Graph Dependencies between kernel objects can be represented by a directed graph, where each vertex represents an access (by system call) of a particular kernel object and edges represent dependency on (or modification of) the object state. Every vertex is labeled by system call number and edges are labeled by the type of access. Being directed and acyclic, the graph defines partial ordering on the system calls observed in the trace. Kernel Objects are tracked by their handles if they are not named, or by their names if they are available.

The following operations are monitored resulting in a kernel object dependency graph [7]: Object Creation, Obtaining a Handle to an Object, Read access, Write/modify access

4.1   Access Graph Compression

Since programs can easily execute several thousand system calls in less than a minute and each system call is represented by a vertex, the size of graphs built from execution traces quickly becomes unmanageable. Moreover, graphs often contain redundant information which obfuscates the graph rather than carries useful information. Therefore, constructed graphs must be compressed by removing redundant information and condensing frequently appearing subgraphs into single vertices.

There are three basic types of repetition: *Sequence, Parallel Sequence,* and *Loop.*

Repetitions can be composed of subgraphs which are isomorphic to each other. Sequence and loop repetitions are replaced by the first member of the component while absorbing the outgoing edges from the last component and introducing a backwards edge, creating a loop. The parallel sequence is simply replaced by only one occurrence.

The reuse of the code and executing the same functionality in multiple contexts results in virtually identical structures appearing in traces of the same or several different programs. This is typical of API functions composed from more than one system call that inherently show up as a frequent subgraph structure. Since they always represent the same function there is no need to keep several copies of them and each instance can be replaced by a single node. This operation further reduces the data size and thus reduces the cost of subsequent processing. Since API functions are executed frequently and some of them translate into relatively large subgraphs, this size reduction can often be significant. Moreover, it also identifies structures which should always remain the same. Any deviation in standard API functions is by itself suspicious.

4.2   Functionality Extraction

Functionalities can be identified as frequent subgraphs of the kernel object access graph. However, subgraph mining is computationally expensive and the graphs, though considerably reduced, are still large in size. Moreover, manual inspection of the obtained graphs indicates that most of the frequent structures seem to be contained together. This is because functionality is usually executed at one time without many additional system calls related to the same Kernel Object. These reasons justify the use of a heuristic technique which first converts graph components into a string and then performs a search for repeated factors [8]. While graphs cannot be expressed as strings without losing information, the string search algorithm cannot be fully relied on to find all functionalities. However, the heuristics [8] works well and computing common string factors is a much more efficient operation than frequent subgraph mining. While the described approach may not work should behavioral obfuscation techniques be applied, a proper subgraph mining algorithm might be required.

4.3   Evaluation of the Functionality Extraction Procedure

The described scheme was evaluated on execution traces obtained from several benign and malicious programs running on Windows XP. System call traces were recorded from our driver which intercepted system calls with their arguments by hooking into the SSDT table. Since we wanted to evaluate our approach in general conditions without any prior knowledge about importance of individual system calls for security, we intercepted all of the calls referenced by the service table, except a few for which we could not find the correct specification of input arguments. We used our driver to obtain execution traces from several malicious and benign applications.

Malicious programs were obtained from the Offensive Computing website [9] and include malware samples of different types and from several families. Benign programs were selected to represent a typical user setup. We joined the obtained samples into three testing traces so that each trace consisted of several malware types and several benign programs.

Table 1. Testing/validation of the functionality extraction procedure

Trace #	Number of system calls	Number of unique graph components	Number of detected functionalities	Number of malicious functionalities detected
1	6927937	1047	341	23
2	3704217	862	307	21
3	20719	217	49	9

Since we monitored all of the system calls, the size of execution traces grew rapidly with time, quickly exceeding 10GB for large traces. Therefore we used traces obtained only for a limited amount of time, ranging from 1 minute to 20 minutes in the case of longest execution trace. Currently, the compression was applied to the entire graph that could have over ten thousand nodes. In the future, some incremental, real time compression schemes could be used allowing the processing of much longer traces. The results could be seen in Table 1. Figure 1 illustrates the functionality extraction process. Numbers in circles correspond to different system calls issued during execution of a process.

5   Customized Normalcy Profile and Its Utilization in an IDS

Proliferation of targeted information attack expected in the nearest future calls for novel IDS approaches capable of protecting "high value targets". Such a target is visualized as one or several networked machines operating in a closed network environment executing a fixed set of approved programs that service industrial systems/processes and/or government facilities. It is known that StuxNet worm, perpetrating such an attack, employs several techniques to conceal its activity and to keep low activity profile. Such techniques include [2], [10]:

- Controlled self-propagation that is achieved by limiting the number of generation of the worm
- Uses multiple attack vectors to self-propagate via USB removable drives and local network
- Uses rootkits digitally signed with stolen valid certificates that hide the worm binaries
- Uses centralized and decentralized mechanisms to update itself on infected machines
- Employs various techniques to precisely identify the target of its attack

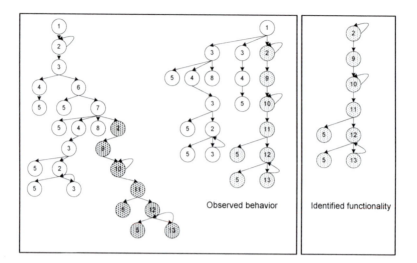

Fig. 1. Functionality extraction

To further our understanding of StuxNet type malware, we evaluated the robustness of modern antiviruses for detecting StuxNet. The following simple experiment was performed on worm W32/Stuxnet.A downloaded from a public malware repository (md5- 74ddc49a7c121a61b8d06c03f92d0c13) [9].

1. We uploaded the malware binary to Virus Total website [11] that utilizes most common anti-viruses to detect the malware. It was found that almost all, 42 out of 43, anti-viruses detected the worm.
2. We modified the original worm by changing the code section without affecting the behavior of the worm. We should point out that OllyDbg dumper added additional sections to the binary, but the executable section of the code remained intact.
3. The modified worm was submitted to the Virus Total website. The scanning results showed that 25 out of 43 anti-viruses, including NOD32, Avast and Kaspersky, failed to detect the modified worm.

Such fragile performance of anti-viruses could be attributed to the nature of the current anti-virus technology utilizing binary patterns that are highly bound to rarely changed portions of the code. Our experiment demonstrates that targeted StuxNet-type attacks can easily overcome current computer defense mechanisms.

The utilization of the earlier described functionality extraction methodology offers a new perspectives to the development of a much more dependable IDS. Such an IDS would utilize a customized normalcy profile representing the normal operation of a network and a generic abnormal behavior profile comprising known malicious functionalities. Both profiles are to be represented by CPN models and implemented on the basis of the CPN tool [6].

The customized normalcy profile is perceived as a set of automatically extracted functionalities, subjected to generalization and augmentation, and accompanied by the frequencies of their execution. Unlike a public network providing services to a wide community of users, network of a "high value facility" is expected to demonstrate a very rigid set of functionalities and their frequencies. The abnormal behavior profile represents typical malicious functionalities by CPN-based behavioral signatures as in [3] and will be expanded as new malicious functionalities be detected.

The initial deployment of the proposed IDS will result in the extraction of the functionalities routinely executed within the network. However, the functionality extraction/recognition will be an on-going task. The detection of one or more malicious functionalities will indicate an attack. The detection of unseen earlier, not necessarily malicious functionalities, and/or changes in the execution frequencies of the functionalities would also indicate an attack. The alarm would prompt the necessary attack mitigation measures. Figure 2 illustrates the described IDS concept. The implementation of the described approach includes the development and periodic updating of the normalcy profile, and the on-going tasks of the functionality extraction, detection of known malicious functionalities, and the anomaly detection in network operation.

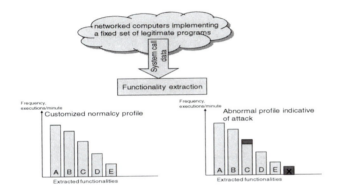

Fig. 2. IDS utilizing a customized normalcy profile

6   Related Work

This paper is a continuation of previous work on malicious functionality specification using CPN technology [3]. The utilization of CPN [4] enables automation of the definition process. An approach to behavioral matching based on dependencies of system call arguments is presented in [12]. Authors in [13] developed a method for constructing near-optimal behavior signatures from the behavior graph and use behavior graphs following data dependencies which capture even complex semantic relationships. As per [14], malware employing advanced behavior obfuscation such as using covert channels to pass the information could escape detection. A system-centric model of benign programs observing sequences of executed system calls is built in [15].

Acknowledgments. This research has been funded by the Air Force Office of Scientific Research (AFOSR). The authors are grateful to Dr. Robert Herklotz of AFOSR for supporting this effort and to Dr. A. Tokhtabayev of George Mason University for valuable suggestions and friendly discussions.

References

1. Percoco, N., Ilyas, J.: Malware Freakshow 2010., White paper for Black Hat USA (2010)
2. Falliere, N., Murchu, L., Chien, E.: W32.Stuxnet Dossier. version 1.4. Symantec Security Response (April 2011)
3. Tokhtabayev, A.G., Skormin, V.A., Dolgikh, A.M.: Expressive, Efficient and Obfuscation Resilient Behavior Based IDS. In: Gritzalis, D., Preneel, B., Theoharidou, M. (eds.) ESORICS 2010. LNCS, vol. 6345, pp. 698–715. Springer, Heidelberg (2010)
4. Jensen, K.: Coloured Petri nets: basic concepts, analysis methods and practical use, 2nd edn., vol. 1. Springer, Berlin (1996)
5. Dolgikh, A., Nykodym, T., Skormin, V., Antonakos, J.: Colored Petri Nets as the Enabling Technology in Intrusion Detection Systems. In: MILCOM 2011, Baltimore, VA (2011)
6. http://apimon.codeplex.com
7. Nykodym, T., Skormin, V., Dolgikh, A., Antonakos, J.: Automatic Functionality Detection in Behavior-Based IDS. In: MILCOM 2011, Baltimore, VA (2011)
8. Melichar, B., Holub, J., Polcar, T.: Text Searching Algorithms. Czech Technical University in Prague Faculty of Electrical Engineering, Department of Computer Science and Engineering (November 2005)
9. http://offensivecomputing.net/ (accessed in April 2011)
10. Matrosov, A., Rodionov, E., Harley, D., Malcho, J.: Stuxnet under the microscope. ESET LLC (September 2010)
11. http://www.virustotal.com/
12. Kolbitsch, C., Comparetti, P.M., Kruegel, C., Kirda, E., Zhou, X., Wang, X.: Effective and efficient malware detection at the end host. In: Proceedings of the 18th USENIX Security Symposium (Security 2009) (August 2009)
13. Fredrikson, M., Jha, S., Christodorescu, M., Sailer, R., Yan, X.: Synthesizing Near-Optimal Malware Specifications from Suspicious Behaviors. In: 2010 IEEE Symposium on Security and Privacy (2010)
14. Cavallaro, L., Saxena, P., Sekar, R.: On the Limits of Information Flow Techniques for Malware Analysis and Containment. In: Zamboni, D. (ed.) DIMVA 2008. LNCS, vol. 5137, pp. 143–163. Springer, Heidelberg (2008)
15. Lanzi, A., Christodorescu, M., Balzarotti, D., Kirda, E., Kruegel, C.: AccessMiner: Using System-Centric Models for Malware Protection. In: Proceedings of the 17th ACM CCS

A Novel Multiobjective Formulation of the Robust Software Project Scheduling Problem

Francisco Chicano[1], Alejandro Cervantes[2],
Francisco Luna[1], and Gustavo Recio[2,*]

[1] University of Málaga, Spain
{chicano,flv}@lcc.uma.es
[2] University Carlos III of Madrid, Spain
{acervant,grecio}@inf.uc3m.es

Abstract. The Software Project Scheduling (SPS) problem refers to the distribution of tasks during a software project lifetime. Software development involves managing human resources and a total budget in an optimal way for a successful project which, in turn, demonstrates the importance of the SPS problem for software companies. This paper proposes a novel formulation for the SPS problem which takes into account actual issues such as the productivity of the employees at performing different tasks. The formulation also provides project managers with robust solutions arising from an analysis of the inaccuracies in task-cost estimations. An experimental study is presented which compares the resulting project plans and analyses the performance of four different well-know evolutionary algorithms over two sets of realistic instances representing the problem. Statistical parameters are also provided in order to help the project manager in the decision process.

Keywords: Software Project Scheduling, Robustness, Multi-objective Optimisation, Evolutionary Algorithms.

1 Introduction

As software projects become larger, the need to control people and processes, and to efficiently allocate resources turn out to be increasingly important. Managing such projects usually involves scheduling, planning, and monitoring tasks. This paper focuses on minimising both, the project cost and its make-span, during the assignment of employees to particular tasks in the context of a software project. The problem studied here is known in the literature as the Software Project Scheduling (SPS) problem [2].

In general, the solution of a multi-objective problem, such as SPS, consists of a set of non-dominated solutions known as the *Pareto optimal set*, which is often called *Pareto border* or *Pareto front* when plotted in the objective space [3].

* This work has been partially funded by the Spanish Ministry of Science and Innovation and FEDER under contract TIN2008-06491-C04. It has also been partially funded by the Andalusian Government under contract P07-TIC-03044.

C. Di Chio et al. (Eds.): EvoApplications 2012, LNCS 7248, pp. 497–507, 2012.

Solutions within this set are optimal in the sense that there are not solutions which are better with regards to one of the objectives without achieving a worse value in at least another one. Particularly, in the context of the SPS problem, it is not possible to reduce the project cost without increasing its make-span (or vice versa). Previous works in the literature have addressed the SPS problem using single-objective and multi-objective formulation with meta-heuristics [2]. The contribution of this research differs from previous works in three ways. First, a new formulation of the problem is presented which involves more realistic assumptions than the previous ones [2], e.g. the need of taking into account several constraints was removed simplifying then the optimisation process. Second, the concept of robustness of a solution was introduced into the formulation of the problem in order to deal with inaccuracies in task-cost estimations. Third, the experimental study carried out here consists on applying four multi-objective evolutionary algorithms to the problem using three different robustness approaches over two different instances derived from a realistic software project. The resulting solutions were analysed using correlation measures between solution features and objective function values.

This paper is organised as follows. The new formulation of the SPS problem and the way robustness is considered are described in Section 2. The experimentation carried out with the corresponding analysis of results is detailed in Section 3. Finally, Section 4 deals with discussion of the main findings and contributions of this research.

2   The SPS Problem

Consider the set of people potentially involved in a software project E, where each person is denoted as $e_i \in E$, with i varying from 1 to $|E|$ (the number of employees), and e_i^s being the salary of an employee. The set of tasks to be performed in the project and each individual task are referred to as T and $t_j \in T$ respectively (with j varying between 1 and $|T|$). The cost in person-hour of task t_j is denoted with t_j^c. The tasks must be performed according to a Task Precedence Graph (TPG) that indicates which tasks must be completed before a new task is started. The TPG is an acyclic directed graph $G(T, A)$ which nodes represent the tasks and an arc $(t_i, t_j) \in A$ exists if task t_i must be completed, with no other intervening tasks, before task t_j can start. Each instance of the problem includes a productivity matrix P of size $|E| \times |T|$ in which element $P_{i,j} \in [0, 1]$ is a positive real value which describes the productivity of employee e_i in task t_j. This productivity value is related to the time required by the employee to finalise the task. If employee e_i is working alone in task t_j then $t_j^c / P_{i,j}$ hours are required to complete the task.

A solution to this problem $x = (d, r, q)$ consists of a real valued vector of employee dedication $d \in \mathbb{R}^{|E|}$, an integer valued vector of task delays $r \in \mathbb{N}^{|T|}$ and an integer valued matrix of priorities $q \in \mathbb{N}^{|E| \times |T|}$. Each component d_i of the dedication vector refers to the percentage of a full working day that employee e_i spends in tasks related to the project. Thus, $d_i = 0.5$ means that half working day is spent in the project by employee e_i. If working alone in task t_j with productivity $\boldsymbol{P}_{i,j} = 1$ then the task takes $2t_j^c$ hours for completion. The component r_j within the vector of task delays refers to the number of hours that task t_j is delayed with respect to the earliest possible starting time, e.g. if task t_j can start at time h, then, applying task delays it will start at time $h + r_j$. Task delays where introduced in the formulation as under certain circumstances they are needed in order to represent optimal solutions in terms of make-span. Without considering delays, the model can only generate solutions with as many tasks as possible processed in parallel, i.e. tasks are started as soon as the TPG allows it. If some of those paralleled tasks are in the critical path (that is, their make-span highly influences the total make-span), a better total make-span can be achieved if the critical tasks are completed as soon as possible. The use of task delays allow the model to represent such solutions: non-critical tasks can start later than allowed by the TPG, so they have less parallelisation than critical tasks. Critical tasks get as much dedication as possible minimising their contribution to the total make-span. The priority matrix $\boldsymbol{q}$ specifies which task is performed by each employee. In the case in which an employee is simultaneously working in several tasks, the matrix also specifies the distribution of employee time between parallel tasks. An employee e_i works in task t_j when $\boldsymbol{q}_{i,j} > 0$ and $\boldsymbol{P}_{i,j} > 0$. If employee e_i is working at a given time τ in tasks $t_{j_1}, t_{j_2}, \ldots, t_{j_l}$ then the amount of time dedicated to task t_{j_m} is given by $\boldsymbol{q}_{i,j_m}/(\sum_{k=1}^{l} \boldsymbol{q}_{i,j_k})$.

Optimising the cost and the make-span of the proposed scheduling of the software project is the aim behind this research. Therefore, the evaluation of cost and make-span becomes highly important. Consider a discrete time where the time variable is represented by τ. The working hours in the software company in which the project are begin developed are represented as finite values of the time variable $\tau = 0, 1, 2, \ldots$ (being $\tau = 0$ the starting time of the project).

Since the employee dedication to a task is time dependent (due to simultaneous tasks), computing the make-span of the project involves an auxiliary time-dependent real valued vector of manpower for each task. Such vector will be denoted as π, where π_j refers to the manpower of the team at performing task t_j, e.g. if $\pi_j(7) = 2$ then at time $\tau = 7$ the remaining cost of task t_j is reduced in 2 persons-hour. The set of finalised and active tasks at time τ based on the values of $\pi(\tau')$ for $\tau' < \tau$ are defined as

$$done(\tau) = \left\{ t_j \in T \left| \sum_{\tau'=0}^{\tau-1} \pi_j(\tau') \geq t_j^c \right. \right\}, \tag{1}$$

$$active(\tau) = \{t_j \in T | \forall t_i, (t_i, t_j) \in A : t_i \in done(\tau - r_j)\} - done(\tau), \tag{2}$$

where A stands for the arc set of the TPG. Notice that the computation of both, the $active(\tau)$ and $done(\tau)$ sets only depends on the values of $\pi(\tau')$ for $\tau' < \tau$. In particular, $done(0)$ and $active(0)$ do not depend on π, as $done(0) = \emptyset$ and $active(0)$ is the set of initial tasks: $active(0) = \{t_j \in T | \nexists t_i \in T, (t_i, t_j) \in A\}$.

The vector $\pi(\tau)$ can be computed for each time step $\tau = 0, 1, \ldots$ in an iterative manner as follows:

$$\pi_j(\tau) = \begin{cases} \displaystyle\sum_{e_i \in E: q_{i,j} > 0} \frac{d_i \cdot P_{i,j} \cdot q_{i,j}}{\sum_{t_k \in active(\tau)} q_{i,k}} & \text{if } t_j \in active(\tau), \\ 0 & \text{otherwise.} \end{cases} \tag{3}$$

The project make-span is: $makespan(x) = \min\{\tau \in \mathbb{N} | done(\tau) = T\}$, that is, the *make-span* refers to the amount of time required to complete all the tasks in the project. A well defined make-span involves that all tasks must be performed by at least one employee with non-zero productivity in the corresponding task. This is the only constraint imposed to the solutions.

The cost is computed by multiplying the salary per hour of each employee, the dedication of the employee and the number of hours dedicated to tasks in the project. Then the salaries for all employees are sum together to compute the total cost of the project, that is:

$$cost(x) = \sum_{e_i \in E} e_i^s \cdot d_i \cdot |\{\tau \in \mathbb{N} | \exists t_j \in active(\tau) : q_{i,j} \cdot P_{i,j} > 0\}| . \tag{4}$$

Considering the expressions for the makespan and the cost, the SPS problem can be modelled as a bi-objective optimisation problem with objective function $f(x) = (cost(x), makespan(x))$.

2.1   Adding Robustness to the Solutions

In a real scenario, task costs are usually estimations made on the basis of previous experiences and they are not accurate. Indeed, a review of studies in estimation accuracy points out that software projects overspend on average 30-40% more effort than estimated [10]. Taking into account these uncertainties in the problem statement allows search algorithms to propose not only good solutions according to the main objectives (cost and make-span), but also to provide robust solutions whose cost and make-span are not sensitive to changes in the the cost of individual tasks due to inaccuracies of the initial estimations. These disturbances in task costs can be modelled by using a multivariate random variable $\mathcal{T}^c = \{t_1^c, \ldots, t_{|T|}^c\}$ following a probability distribution $\mathcal{C}$. Given a solution x, the project cost and make-span are now defined by a bivariate random variable S that can be computed as $S(x) = (makespan(x), cost(x)) = f(x, \mathcal{T}^c)$ where $\mathcal{T}^c$ was explicitly introduced in the notation to clarify that the solution evaluation depends on the varying estimated cost of the tasks (represented here with its random variable).

The average and the standard deviation of each component of $S(x)$ are used as a measure of the quality and the robustness of each objective for a given solution, respectively. These values are computed by sampling over a number of H simulations of $\mathcal{T}^c$. The bi-objective formulation of the problem is, thus, transformed into a four-objective one:

$$f(x) = (makespan_{avg}(x), makespan_{sd}(x), cost_{avg}(x), cost_{sd}(x)), \tag{5}$$

where sub-indices in the original objective functions were used to denote the average and the standard deviation of the sampling performed to compute the robustness. As before, these four objectives are expected to be minimised.

Three different robustness scenarios are being considered. The first one, denoted as NR, assumes perfect knowledge on the task cost. The second one assumes that only one task has been miss-estimated (noted as OTR). The third one assumes that all tasks could have been miss-estimated (noted as STR). The probability distribution used to generate the perturbation in the cost of a task in the two last scenarios is such as that each t_j^c is multiplied by a value uniformly drawn from the interval $[0.5, 2.0]$. That is, each task can be carried out from half to double of its original estimated cost.

2.2   Comparison against Other Scheduling Problems

An analogy can be established between the shop scheduling [6] and SPS problems. The tasks would be the same in both problems, employees in SPS would be analogous to machines in shop scheduling problems, the productivity $P_{i,j}$ of employee e_i in task t_j when considering SPS would be related to the length of task j in machine i for shop scheduling problems. However, in shop scheduling only one machine can perform a task, while in SPS problems tasks are performed by working teams of employees. In addition, the decision variables in SPS determine the dedication of an employee to a software project whereas in the case of shop scheduling the "dedication" or efficiency of a machine cannot be modified.

Another problem related to SPS is the Resource-Constrained Project Scheduling (RCPS) [12]. In RCPS there are several kinds of resources while SPS involves only one: the human resource. Each activity in RCPS requires different amounts of each resource while SPS does not imposes a minimum or maximum number of employees in a working team developing a task.

Two important works that also use modelling of software project scheduling were presented in Gutjahr et al. [7] and Chang et al. [1]. The former includes a model of learning capabilities of employees and a portfolio selection. In the second, a solution accounts for the time variation of the assignment of employees to tasks. The drawback of complex formulations, like the previous ones, is the large number of parameters that the project manager must configure to provide a complete instance of the problem, which in turn increases the chances of miss-estimating such parameters. Hence, the improved accuracy obtained using a more realistic formulation of the problem turns out to be limited by a larger inaccuracy of the problem instance parameters. Then, a new formulation which is a trade-off between realistic (but complex) and simple (and unrealistic) formulations is proposed in this work.

3   Experimental Study

Four meta-heuristics have been used to carry out the experimental study: NSGA-II [5], SPEA2 [14], PAES [9], and MOCell [11]. They all use evolutionary computation which is by far the most popular meta-heuristic technique for solving

MOPs due to their ability in finding a set of trade-off solutions in one single run
[3,4]. Binary tournament selection, two-point crossover and a random mutation
that randomly chooses a value within the range defined for each variable were
used in NSGA-II, SPEA2 and MOCell. On the other hand, the PAES execution
sequence consisted of a single individual population which is iteratively modified
by using only random mutation (no crossover operator was used), pareto front
solutions in this case are obtained by using an external archive where all non-
dominated solutions are stored. Population sizes of 100 individuals were used for
NSGA-II, SPEA2, and MOCell, whereas the pareto front size was limited to 100
solutions in the four approaches. Crossover and mutation rates were $p_c = 0.9$
and $p_m = 1/L$ respectively, where L refers to the length of the tentative solution.
Aiming at performing a fair comparison between different algorithms, the stopping criterion for them all consisted in computing 1000000 function evaluations.
Finally, the size of the Monte Carlo sampling used to evaluate the solutions of
the robust SPS versions was set to a neighbourhood of $H = 100$.

Two quality indicators have been used to measure the performance of the
multi-objective algorithms: the hyper-volume (HV) [15] and the attainment surfaces [8]. The HV is considered as one of the more suitable indicators by the
EMO community since it provides a measure that takes into account both the
convergence and diversity of the obtained approximation set. The empirical attainment surfaces have been defined to be a kind of "average" Pareto front of a
randomised multi-objective algorithm. For each pair of algorithm vs test problem instance, 100 independent runs were carried out. The HV indicator and the
attainment surfaces were then computed. In the case of HV computations, a
multiple comparison test was carried out in order to check if the differences were
statistically significant or not. All the statistical tests were performed with a
confidence level of 95%.

Two realistic instances that are variations of a project scheduling which is
available at the online repository of the MS Project tool will be solved in this
research. The same TPG (see Fig. 1), tasks cost and number of employees as in
the original instance will be used and the values for the employees salary and
the productivity matrix will also be provided. Table 1 summarises the above
information.

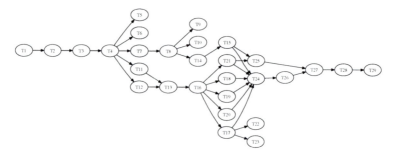

Fig. 1. Task Precedence Graph for the two instances of the SPS problem being solved

Both instances, denoted with ms1 and ms2, differ in the definition of their productivity matrix. In ms1 all the values in the productivity matrix are 0 or 1 and are based on the original assignment of employees to tasks in the sample project (denoted as "base solution"). On the other hand, instance ms2 contains a more flexible resource productivity table, with larger overlap between resources' abilities, and also fractional (not 1.0) productivity in tasks.

Table 1. Productivity matrices $P_{i,j}$, task cost t_j^c and employee salary e_i^s

e_i e_i^s		Task (t_j) 1	2	3	4	5	6	7	8	9	10	11	12	13	14	15	16	17	18	19	20	21	22	23	24	25	26	27	28	29
e_1 50	ms1	1	0	0	0	1	1	0	0	0	0	0	0	0	0	0	0	0	0	0	0	0	0	0	0	0	0	0	0	0
	ms2	1	0	0	0	1	1	0	0	0	0	0	0	0	0	0	0	0	0	0	0	0	0	0	0	0	0	0	0	0
e_2 40	ms1	0	1	1	1	1	1	1	1	1	1	1	1	1	1	1	0	0	0	0	0	0	0	0	0	1	0	0	1	1
	ms2	0	1	1	1	1	1	1	1	1	1	1	1	1	1	1	0	0	0	0	0	0	0	0	0	.5	0	0	1	1
e_3 10	ms1	0	0	0	0	0	0	0	1	1	1	0	0	1	0	0	0	0	1	0	1	0	0	0	1	0	1	0	0	0
	ms2	0	0	0	0	0	0	0	.3	.3	.3	0	0	.5	0	0	0	0	.5	0	.5	0	0	0	.5	0	.5	0	0	0
e_4 15	ms1	0	0	0	0	0	0	0	1	1	1	1	1	1	0	0	0	0	1	0	1	0	0	0	1	1	1	1	1	0
	ms2	0	0	0	0	0	0	0	1	1	1	.5	.5	.5	0	0	0	0	.8	0	.8	0	0	0	.8	.8	.8	.8	.8	0
e_5 20	ms1	0	1	1	1	1	1	0	0	0	0	1	1	1	0	0	1	1	0	0	0	1	1	0	0	0	0	1	0	
	ms2	0	.5	.5	.5	.5	.5	0	0	0	0	1	1	1	0	0	1	1	0	0	0	1	1	1	1	1	0	1	1	0
e_6 30	ms1	0	1	1	1	1	1	1	1	1	1	1	1	1	1	1	0	0	0	0	0	0	0	1	0	0	0	0	0	
	ms2	0	1	1	1	1	1	1	1	1	1	1	1	1	1	1	0	0	0	0	0	0	0	0	.8	0	0	0	.8	0
e_7 30	ms1	0	1	1	1	1	1	1	1	1	1	1	1	1	1	1	0	0	0	0	0	0	0	0	1	1	1	1	1	0
	ms2	0	.7	.7	.7	.7	.7	.7	.7	.7	.7	.7	.7	.7	.7	0	0	0	0	0	0	0	0	0	1	1	1	1	1	0
t_j^c		6	680	408	8	10	10	378	10	10	162	48.6	8.8	720	6	198	180	6	108	6	30	36	36	18	540	120	180	450	3	

3.1 Performance of the Algorithms

A comparison of the performance of the four multi-objective algorithms within the three robustness scenarios is carried out in this Section. The performances have been evaluated using the HV indicator which values are summarised in Table 2. The best performances are highlighted in a dark grey background whereas second to best are shown in light grey. We also mark with * the results having statistically significant differences with the best result. Several conclusions can be drawn from these values. Both NSGA-II and MOCell obtained the best (largest) values for the two instances (as well as many of the second to best values). NSGA-II resulted in the best performance when tackling the robust versions of the instances (in 3 out of the 4 scenarios the approximated Pareto front with best HV indicator was returned). On the other hand, MOCell seems to be specially well suited for the non-robust setting, yielding the higher HV indicator for the two instances. PAES seems to be clearly the worst algorithm with respect to this indicator, specially for the robust versions. The uncertainty in the objective functions could be the main reason behind this fact. Regarding the runtime, all the algorithms require between 2.5 and 5 minutes in the NR scenario, while they require around 5 hours in the OTR and STR scenarios.

Table 2. Median and IQR of the HV value for the two instances

Rob.	NSGAII	SPEA2	PAES	MOCell	NSGAII	SPEA2	PAES	MOCell
		ms1				ms2		
NR	$0.943^{*}_{0.000}$	$0.943^{*}_{0.000}$	$0.518^{*}_{0.065}$	$0.944_{0.000}$	$0.904^{*}_{\pm0.000}$	$0.905^{*}_{\pm0.001}$	$0.543^{*}_{\pm0.031}$	$0.905_{\pm0.000}$
OTR	$0.829^{*}_{0.027}$	$0.807^{*}_{0.030}$	$0.328^{*}_{0.039}$	$0.816_{0.032}$	$0.738_{\pm0.025}$	$0.730_{\pm0.018}$	$0.287^{*}_{\pm0.020}$	$0.695^{*}_{\pm0.043}$
STR	$0.746_{0.028}$	$0.688^{*}_{0.063}$	$0.345^{*}_{0.036}$	$0.742_{0.025}$	$0.764_{\pm0.025}$	$0.717^{*}_{\pm0.030}$	$0.387^{*}_{\pm0.032}$	$0.769_{\pm0.022}$

3.2 Analysis of Solutions

This section focuses on analysing the solutions obtained using the multi-objective algorithms. Figure 2 (left) shows the result of an NSGA-II execution over both instances using the NR approach. The base solution for instance ms1 is close to a minimum-make-span solution, as all available employees are committed to tasks for which they have non-zero productivities. None the less, the algorithm is able to improve this minimum make-span. The Pareto front includes solutions with smaller cost which were obtained by reducing the dedication of the most expensive resources when developing their tasks. In instance ms2 improvements in both, cost and make-span, using NSGA-II with respect to the base solution were also observed.

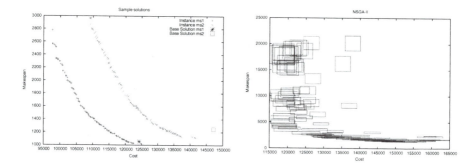

Fig. 2. Pareto front sample and base solution for the two instances (left). 50%-attainment surface for ms1 in the STR robust approach (right). The position of the boxes is determined by the average value and the size by the standard deviation.

Fig. 2 (right) shows the 50%-attainment surface of NSGA-II for ms1 within the STR scenario. A four objective problem requires 4D data to be represented in order to visually inspect the resulting Pareto fronts. In order to show both the quality (average) and the robustness (standard deviation) in the cost and make-span of a project scheduling problem, the approach taken consists on displaying boxes such that the position of the center of a box is defined by the two average values ($cost_{avg}(x)$ and $makespan_{avg}(x)$), whereas the width and the height are proportional to $cost_{sd}(x)$ and $makespan_{sd}(x)$. It is worth mentioning that when the average values of cost and make-span are reduced (bottom left corner of the plot), the standard deviation is increased (larger boxes). It was also observed that high-cost solutions show a low make-span and are quite robust in make-span, whereas low-cost solutions are not robust in make-span or cost. This can be explained by the larger need of average parallelism required by low make-spam solutions, thus, task deviations are distributed among several employees working in the same task.

Consider now the features of the solutions x in the approximated Pareto front. In particular, a detail analysis must be done accounting for the number of employees performing each task t_j (denoted as $t_j^e(x)$) and the average number of tasks that each employee e_i performs in parallel (denoted as $e_i^p(x)$).

Only results from MOCell over the ms2 instance will be analysed due to space constraints. All solutions of the approximated Pareto front obtained in different independent runs of MOCell are being considered. The $e_i^p(x)$ and $t_j^e(x)$ values have been computed for each employee and each task in all the solutions and the Spearman rank correlation coefficients [13] between all the $e_i^p(x)$, $t_j^e(x)$, $makespan(x)$ and $cost(x)$ have been calculated. The correlation coefficients are shown in Fig. 3. An arrow pointing up means positive correlation whereas an arrow pointing down means negative correlation. The absolute value of the correlation is shown in grey scale (the darker the higher).

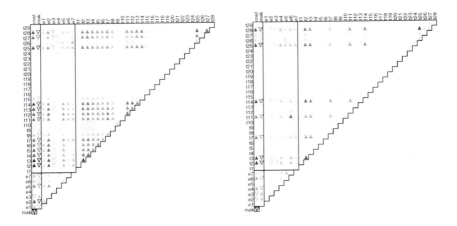

Fig. 3. Correlations between cost, duration, the number of average parallel tasks performed by the employees and the number of employees per task for the NR approach (left) and the STR approach (right) using MOCell. Solutions for ms2 in the approximated Pareto front of all the independent runs.

Regarding the current values of $e_i^p(x)$ and $t_j^e(x)$ in all the solutions of all the independent runs of MOCell, $e_i^p(x)$ ranges between 1.00 and 1.61 with average values around 1.04. On the other hand, $t_j^e(x)$ ranges between 1 and 6 with average values around 1.56. This means that it is not common to have large working teams or a large number of parallel tasks per employee, therefore the communication overhead or the reduction of productivity due to parallel tasks is not high.

Focusing on the correlation between the make-span and the number of parallel tasks performed by the employees, a negative correlation with the exception of e_3 (and e_2 using the STR approach) can be observed. A negative correlation means that in order to reduce the make-span of the project, the employees will have to work in several tasks simultaneously. This seems to agree with common sense. Then, why does a positive correlation between make-span and employee e_3 appear? This employee is the only one able to do some tasks in the critical path of the project. Therefore, such critical tasks are assigned to this employee by the algorithm in order to reduce the execution time of the tasks. The above also explains the negative correlation between the size of the working teams $t_j^e(x)$ and $e_3^e(x)$. It is expected that in order to reduce the make-span the size of working teams must be increased, which also implies an increase in the number of parallel tasks each employee has to develop. This explains the positive correlation between $e_i^p(x)$ and $t_j^e(x)$ for the remaining employees.

Considering now the correlations between the make-span and the number of employees in each task, it is noticed, with no surprise, that reducing the make-span implies that more employees have to work on the tasks. However, some blank cells can be observed for which no correlation is detected. This happens in the tasks of the project for which only one employee has the required skills (non-zero productivity), like task t_1. This is just an illustration on how the analysis of solutions can provide some interesting information for the project manager.

4   Conclusions and Future Work

A new formulation of the Software Project Scheduling problem taking into account the productivity of the employees in developing different tasks of a software project and considering the inaccuracies of task cost estimations was presented. Experimental studies were carried out in order to analyse the performance of four multi-objective algorithms on real-like instances for this problem. Solutions were analysed to illustrate the way project managers can use this tool to improve their decision making. Results show that MOCell is the best algorithm in solving this formulation of the problem, improving even the original solutions proposed by a project manager to the instances used in the experimental section. The analysis of the solutions reveals that the algorithms have been able to identify the tasks in the critical path and the most important employees for the project.

This work can be extended in several ways. An empirical study using real projects and their corresponding scheduling can be done with the help of data provided by software companies. Second, different robustness approaches can be used to take into account the inaccuracies in the productivity values. Third, new operators or search methods can be developed to improve the solutions or the required computational effort.

References

1. Chang, C.K., Yi Jiang, H., Di, Y., Zhu, D., Ge, Y.: Time-line based model for software project scheduling with genetic algorithms. Information and Software Technology 50(11), 1142–1154 (2008)
2. Chicano, F., Luna, F., Nebro, A.J., Alba, E.: Using multi-objective metaheuristics to solve the software project scheduling problem. In: Proceedings of GECCO, pp. 1915–1922 (2011)
3. Coello Coello, C.A., Lamont, G.B., Van Veldhuizen, D.A.: Evolutionary Algorithms for Solving Multi-Objective Problems, 2nd edn. Springer, New York (2007)
4. Deb, K.: Multi-objective optimization using evolutionary algorithms. John Wiley & Sons (2001)
5. Deb, K., Pratap, A., Agarwal, S., Meyarivan, T.: A fast and elitist multiobjective genetic algorithm: NSGA-II. IEEE Trans. on Ev. Comp. 6(2), 182–197 (2002)
6. Garey, M.R., Johson, D.S.: Computers and Intractability. A Guide to the Theory of NP-Completeness. W.H. Freeman and Company (1979)
7. Gutjahr, W., Katzensteiner, S., Reiter, P., Stummer, C., Denk, M.: Competence-driven project portfolio selection, scheduling and staff assignment. Central European Journal of Operations Research 16(3), 281–306 (2008)
8. Knowles, J.: A summary-attainment-surface plotting method for visualizing the performance of stochastic multiobjective optimizers. In: ISDA, pp. 552–557 (2005)
9. Knowles, J., Corne, D.: Approximating the nondominated front using the pareto archived evolution strategy. Evolutionary Computation 8(2), 149–172 (2000)
10. Moløkken, K., Jørgensen, M.: A review of surveys on software effort estimation. In: 2003 Int. Symp. on Empirical Software Engineering, pp. 223–231 (2003)
11. Nebro, A.J., Durillo, J.J., Luna, F., Dorronsoro, B., Alba, E.: A cellular genetic algorithm for multiobjective optimization. In: NICSO 2006, pp. 25–36 (2006)
12. Palpant, M., Artigues, C., Michelon, P.: LSSPER: Solving the resource-constrained project scheduling problem with large neighbourhood search. Annals of Operations Research 131, 237–257 (2004)
13. Sheskin, D.J.: Handbook of Parametric and Nonparametric Statistical Procedures, 4th edn. Chapman & Hall/CRC (2007)
14. Zitzler, E., Laumanns, M., Thiele, L.: SPEA2: Improving the strength Pareto evolutionary algorithms. In: EUROGEN 2001, pp. 95–100 (2002)
15. Zitzler, E., Thiele, L.: Multiobjective evolutionary algorithms: a comparative case study and the strength pareto approach. IEEE TEC 3(4), 257–271 (1999)

Optimizing the Unlimited Shift Generation Problem

Nico Kyngäs[1], Dries Goossens[2], Kimmo Nurmi[1], and Jari Kyngäs[1]

[1] Satakunta University of Applied Sciences, Tiedepuisto 3, 28600 Pori, Finland
{nico.kyngas,jari.kyngas,cimmo.nurmi}@samk.fi
[2] KU Leuven, Naamsestraat 69, 3000 Leuven, Belgium
dries.goossens@econ.kuleuven.be

Abstract. Good rosters have many benefits for an organization, such as lower costs, more effective utilization of resources and fairer workloads. This paper introduces the unlimited shift generation problem. The problem is to construct a set of shifts such that the staff demand at each timeslot is covered by a suitable number of employees. A set of real-world instances derived from the actual problems solved for various companies is presented, along with our results. This research has contributed to better systems for our industry partners.

Keywords: Shift Generation Problem, Workforce Scheduling, Staff Rostering, Real-World Scheduling, Computational Intelligence.

1 Introduction

Workforce scheduling is a difficult and time consuming problem that every company or institution that has employees working in shifts or on irregular working days must solve. The workforce scheduling problem has a fairly broad definition. Most of the studies focus on assigning employees to shifts, determining working days and rest days or constructing flexible shifts and their starting times. Different variations of the problem are NP-hard and NP-complete [1]-[5]. The first mathematical formulation of the workforce scheduling problem based on a generalized set covering model was proposed by Dantzig [6]. Good overviews of staff scheduling are published by Alfares [7], Ernst et al. [8] and Meisels and Schaerf [9].

Shift generation, an important part of the workforce scheduling process, has only received limited attention in the literature. The problem is to construct an optimal shift structure from the staff demand. To the best of our knowledge, there are only a few academic publications concentrating on the shift generation problem [5,10,11] and its applications, such as airport ground [12,13], bank [14] and retail store [15,16]. Furthermore, there are very few cases (such as [5]) where academic researchers have been able to close a contract with such a problem owner.

The aim of this paper is to solve the unlimited shift generation problem as it occurs in various lines of business and industry. Section 2 introduces the workforce scheduling process and necessary terminology. In Section 3 we describe the unlimited shift generation problem. Section 4 gives an outline of our solution method. Section 5 presents a set of real-world instances and our computational results.

C. Di Chio et al. (Eds.): EvoApplications 2012, LNCS 7248, pp. 508–518, 2012.

2   Workforce Scheduling

Workforce scheduling consists of assigning employees to tasks and shifts over a period of time according to a given timetable. The *planning horizon* is the time interval over which the employees have to be scheduled. Each employee has *competences* (qualifications and skills) that enable him or her to carry out certain tasks. Days are divided into *working days* (days-on) and *rest days* (days-off). Each day is divided into timeslots. A *timeslot* is the smallest unit of time. A *shift* is a contiguous set of working hours and is defined by a day and a starting timeslot on that day along with a *shift length* (the number of occupied timeslots). Each shift may be composed of a number of *tasks*. A work schedule for an employee over the planning horizon is called a *roster*. A roster is a combination of shifts and days-off assignments that covers a fixed period of time.

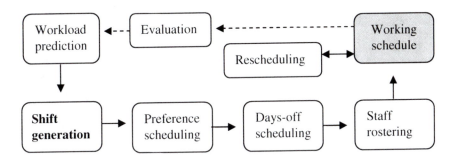

Fig. 1. The real-world workforce scheduling process as described in [17]

We classify the real-world workforce scheduling process as given in Figure 1. *Workload prediction*, also referred to as demand forecasting or demand modeling, is the process of determining the staffing levels - that is, how many employees are needed for each timeslot of the planning horizon. In this presentation, workload prediction also includes determination of planning horizons, competence structures, regulatory requirements and other constraints. *Shift generation* is the process of determining the shift structure, the tasks to be carried out in particular shifts and the competence needed in different shifts. The shifts generated from a solution to the shift generation problem form the input for subsequent phases in workforce scheduling. Another important goal for shift generation is to determine the size of the workforce required to solve the demand. Shifts are created anonymously, so there is no direct link to the employee that will eventually be assigned to the shift.

It is essential to find an accurate match between the predicted workload and the scheduled workforce. Scheduling too few employees can lead to reduced productivity, low-quality service levels, customer dissatisfaction and employee dissatisfaction. On the other hand scheduling more employees than necessary leads to increased costs due to employee salaries.

3   The Unlimited Shift Generation Problem

In workforce scheduling in airlines, railways and buses the demand for employees is quite straightforward because the timetables are known beforehand. In other applications of scheduling, such as call-centers, retail stores and emergency polyclinics, the demand fluctuates. The random arrivals of customers should be forecast using models based on techniques such as queueing theory, simulation and statistics. The result is the number of employees required at each competence level for each timeslot over the planning horizon. It should be noted that sick leaves and other no-shows should be considered when calculating the staff demand.

The *unlimited shift generation problem* is to create a set of shifts that cover the demand as well as possible, while satisfying the large number of constraints arising from regulatory and operational requirements and operational and employees' preferences. The most important goal is to minimize understaffing (shortage on shifts) and overstaffing (surplus on shifts). Low-quality rosters can lead either to an undersupply of employees with a need to hire part-time employees or an oversupply of employees with too much idle time. We define the strict version of the problem such as each timeslot should be exactly covered by the correct number of employees. Furthermore, it is important to have as few shifts as possible. Fewer shifts make schedules easier to read and manage, and this helps to keep teams of people together.

We next give an outline of the optimization criteria of the unlimited shift generation problem. We make no strict distinction between hard and soft constraints; that will be given by the instances themselves. The goal for an instance is to find a feasible solution that is the most acceptable for the problem owner. That is, a solution that has no hard constraint violations and that minimizes the weighted sum of the soft constraint violations. The weights will also be given by the instances themselves. Still, one should bear in mind that an instance is usually just an approximation of practice. In reality, hard constraints can turn out to be soft, if necessary, while giving weights to the soft constraints can be difficult. We classify the criteria into coverage, volume and placement criteria:

Coverage
(C1)   The number of employees at each timeslot over the planning horizon must be exactly as given (strict version).
(C2)   The sum of the excesses of employees at each timeslot over the planning horizon must be minimized.
(C3)   The sum of the shortages of employees at each timeslot over the planning horizon must be minimized.

Volume
(V1)   The number of shifts must be minimized.
(V2)   Shifts of exactly k_1 timeslots in length must be maximized.
(V3)   Shifts of less than k_2 and over k_3 in length must be minimized.
(V4)   The average shift length should be as close to k_4 timeslots as possible.

Placement

   (P1)   Shifts that start between timeslots k_4 and k_5 must be minimized.

   (P2)   Shifts that end between timeslots k_6 and k_7 must be minimized.

   (P3)   Shifts of at least k_8 timeslots in length must include a break (e.g. a lunch) of k_9 timeslots in length, which must be located between $k_{10}\%$ from the beginning of the shift and $k_{11}\%$ from the end of the shift.

The unlimited shift generation problem is related to the minimum shift design problem as discussed in [5] and [11]. In the minimum shift design problem, shifts are limited to a number of types, for which the length and the starting time of the shifts have to be within certain ranges. In our problem, length and starting time of a shift are not strictly limited. Furthermore, each shift corresponds to one worker, whereas in the minimum shift design problem, the number of workers per shift is part of the optimization. Thus, the unlimited shift generation minimizes the number of workers (V1), whereas the minimum shift design problem minimizes the number of different shifts. The minimum shift design problem can be modeled as a network flow problem, namely as the cyclic multi-commodity capacitated fixed-charge min-cost max-flow problem. Di Gaspero et al. [5] show that the minimum shift design problem is NP-hard. Alternatively, the problem can be formulated as the bin-packing problem [12], which is also an NP-hard problem [1].

Some other papers on shift generation (see section 1) also handle the competences required in each timeslot in the shift generation phase. We, on the other hand, consider competences in the days-off scheduling and staff rostering phases.

For academics, publications are usually more important than making business. This has a consequence that the scope of the models academics create seems to be relatively small. Although the workforce scheduling models developed by academics solve the problem instance at hand, they may fall short of meeting the complex needs of the customers. Academic solutions are often not only computer and platform dependent, but also use commercial mathematical programming solvers. This approach, however, may not be able to cope with the size or the complexity of a real-world problem. We believe that the best action plan in order to get the research results implemented into commercial systems is to use computational intelligence heuristics. The next section describes our solution method.

4   Solution Method

Our PEAST algorithm is a population-based local search method. Population-based methods use a population of solutions in each iteration. The outcome of each iteration is also a population of solutions. Population-based methods are a good way to escape from local optima. Our algorithm is based on the cooperative local search method. In a cooperative local search scheme, each individual carries out its own local search, in our case the GHCM heuristic [22]. The pseudo-code of the algorithm is given in Figure 2. The PEAST algorithm has been used to solve real-world school timetabling problems [18], real-world sports scheduling problems [19] and real-world workforce scheduling problems [20].

The reproduction phase of the algorithm is, to a certain extent, based on steady-state reproduction: the new schedule replaces the old one if it has a better or equal objective function value. Furthermore, the least fit is replaced with the best one when *n* better schedules have been found, where *n* is the size of the population. Marriage selection is used to select a schedule from the population of schedules for a single GHCM operation. In the marriage selection we randomly pick a schedule, *S,* and then we try at most *k* – 1 times to randomly pick a better one. We choose the first better schedule, or, if none is found, we choose *S*.

Set the time limit *t*, no_change limit *m* and the population size *n*
Generate a random initial population of schedules
Set *no_change* = 0 and *better_found* = 0
WHILE elapsed_time < *t*
  REPEAT *n* times
    Select a schedule *S* by using a marriage selection with *k* = 3
    (explore promising areas in the search space)
    Apply GHCM to *S* to get a new schedule *S'*
    Calculate the change Δ in objective function value
    IF Δ < = 0 THEN
      Replace S with *S'*
      IF Δ < 0 THEN
        better_found = better_found + 1
        no_change = 0
      END IF
    ELSE
      no_change = no_change + *1*
    END IF
  END REPEAT
  IF *better_found* > *n* THEN
    Replace the worst schedule with the best schedule
    Set *better_found* = 0
  END IF
  IF *no_change* > *m* THEN
    (escape from the local optimum)
    Apply shuffling operators
    Set *no_change* = 0
  END IF
  (avoid staying stuck in the promising search areas too long)
  Update simulated annealing framework
  Update the dynamic weights of the hard constraints (ADAGEN)
END WHILE
Choose the best schedule from the population

Fig. 2. The pseudo-code of the population-based PEAST algorithm

The heart of the GHCM heuristic is based on similar ideas to the Lin-Kernighan procedures [23] and ejection chains [24]. The basic hill-climbing step is extended to generate a sequence of moves in one step, leading from one solution candidate to another. The GHCM heuristic moves an object, o_1, from its old position, p_1, to a new position, p_2, and then moves another object, o_2, from position p_2 to a new position, p_3, and so on, ending up with a sequence of moves.

Picture the positions as cells, as shown in Figure 3. The initial cell selection is random. The cell that receives an object is selected by considering all the possible cells and selecting the one that causes the least increase in the objective function when only considering the relocation cost. Then, another object from that cell is selected by considering all the objects in that cell and picking the one for which the removal causes the biggest decrease in the objective function when only considering the removal cost. Next, a new cell for that object is selected, and so on. The sequence of moves stops if the last move causes an increase in the objective function value and if the value is larger than that of the previous non-improving move. Then, a new sequence of moves is started. The initial solution is randomly generated.

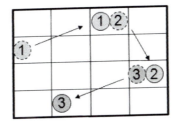

Fig. 3. A sequence of moves in the GHCM heuristic

In the unlimited shift generation problem, each *row* corresponds to a shift, and each *column* to a timeslot. An *object* is a block of a shift that is as long as a timeslot (e.g. if we have 24 timeslots per day, an object is an hour-long piece of a shift). A *move* involves removing an object from a certain timeslot in one shift and inserting it in another shift at the same timeslot. Hence, only vertical moves are allowed.

The decision whether or not to commit to a sequence of moves in the GHCM heuristic is determined by a simulated annealing refinement [21]. This is useful to avoid staying stuck in the promising search areas for too long. The initial temperature T_0 is calculated by

$$T_0 = 1 / \log(1/X_0) . \tag{1}$$

where X_0 is the degree to which we want to accept an increase in the cost function (we use a value of 0.75). The exponential cooling scheme is used to decrement the temperature:

$$T_k = \alpha T_{k-1} , \tag{2}$$

where α is usually chosen between 0.8 and 0.995. We stop the cooling at some predefined temperature. Therefore, after a certain number of iterations, m, we continued to accept an increase in the cost function with some constant probability, p. Using the initial temperature given above and the exponential cooling scheme, we can calculate the value:

$$\alpha = (-1/(T_0 \log p))^{-m} . \tag{3}$$

We choose m equal to the maximum number of iterations with no improvement to the cost function and p equal to 0.0015.

A hyperheuristic [25] is a mechanism that chooses a heuristic from a set of simple heuristics, applies it to the current solution, then chooses another heuristic and applies it, and continues this iterative cycle until the termination criterion is satisfied. We use the same idea, but the other way around. We apply shuffling operators to escape from the local optimum. We introduce a number of simple heuristics that are normally used to improve the current solution but, instead, we use them to shuffle the current solution - that is, we allow worse solution candidates to replace better ones in the current population. In the unlimited shift generation problem the PEAST algorithm uses three shuffling operations:

1) Make a random move from a random shift to another random shift and repeat this l_1 times.
2) Pick a pair of random shifts and make a random move from the first to the second and a random move from the second to the first and repeat this l_2 times.
3) Pick a random shift, S. Shift S consists of little pieces, each of which are the same length as a timeslot. For each such piece in S, move the piece to another random shift with probability p_1.

A random shuffling operation is selected in every $k/20$th iteration of the algorithm, where k equals the maximum number of iterations with no improvement to the cost function. The best results were obtained using the values $l_1 = 8$, $l_2 = 4$ and $p_1 = 0.3$.

We use the weighted-sum approach for multi-objective optimization. A traditional penalty method assigns positive weights (penalties) to the soft constraints and sums the violation scores to the hard constraint values to get a single value to be optimized. We use the ADAGEN method (as described in [21]) which assigns dynamic weights to the hard constraints. The weights are updated in every kth generation using the somewhat complicated formula given in [21].

5   Computational Results

Researchers quite often only solve some special artificial cases or one real-world case. The strength of artificial and random test instances is the ability to produce many problems with many different properties. Still, they should be sufficiently simple for each researcher to be able to use them in their test environment. The strength of real-world instances is self-explanatory. Solving real-world cases is our ultimate goal. However, an algorithm that performs well in one practical instance may not perform well in another practical instance, which is why we present a collection of test instances for both artificial and real-world cases. Both the artificial and the real-world instances can be requested by email from the authors. The artificial instances can be requested by email from the authors along with the detailed data for the real-world instances.

The four real-world instances introduced in this section are based on cases we have solved for our business partner, Numeron. The instances are derived from various lines of business and industry in Finland.

The following hard constraint is used in all the benchmark cases:

V3.   Shifts of less than k_2 and over k_3 in length must not exist.

The soft constraint violations are calculated as follows:

C1.   For each timeslot s, let d_s be the difference between the number of shifts in s and the staff demand for s. The total number of C1 violations is given as $\sum_{s=1,n}(d_s^2)$, where n is the number of timeslots.

V1.   One violation for each shift.

V2.   One violation for each shift whose length is not 8 hours.

V4.   One violation for each minute that the average shift length differs from 6 hours. The difference is calculated as an absolute value and rounded down.

P3.   One violation for each shift that is at least 6 hours long and either contains no lunch break or the lunch break starts before 30% of the shift length or after 70% of the shift length.

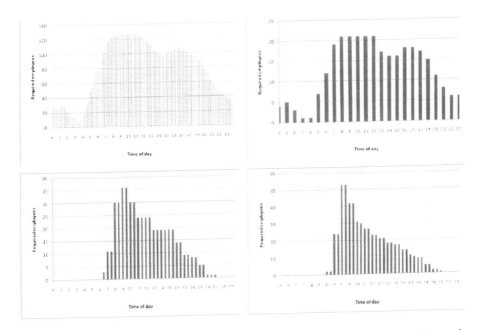

Fig. 4. The staff demand for the instances, from left to right and from top to bottom in numerical order

Table 1. Four unlimited shift generation real-world instances (*Staff demand* = the maximum and average number of employees needed per timeslot over the planning horizon, *V3* = every shift must be between 4 and 8 hours in length, *C1* = attempt to fulfill the staff demand exactly, *V1* = minimize the number of shifts, *V2* = maximize the number of shifts of certain length, *V4* = the average shift length should be as close to a certain length as possible, *P3* = minimum shift length to grant a lunch break; length of said lunch break; the limits of the position of the break within the shift)

ID	Staff demand	V3	C1	V1	V2	V4	P3
R1	Max 126; avg 78	4h – 8h	Yes	No	No	6h	No
R2	Max 21; avg 12.7	5h – 7h	Yes	Yes	No	No	No
R3	Max 36; avg 16.7	4h – 8h	Yes	Yes	8h	No	6h; 30 min; 30%-70%
R4	Max 53; avg 19.3	4h – 8h	Yes	Yes	8h	No	6h; 30 min; 30%-70%

Table 1 shows four unlimited shift-generation instances based on real-world cases. The staff demand for each instance is visualized in Figure 4. Cases R3 and R4 are essentially different from the other two as employees are not needed during the night. That makes their shift structures independent of the adjacent days. They also have two additional constraints compared to the other cases. First, the number of shifts of maximum length (8 hours) is to be maximized. Second, each shift that is at least 6 hours long must have a 30-minute lunch break at some time between 30% and 70% of the shift length.

For cases R1 and R2, in which employees are required 24 hours per day, we assume a contiguous sequence of similar days, so that for all instances the optimal shift structure is only dependent on the data shown in Figure 4. In R2 the shift length is five to seven hours. In R1, the number of shifts is not minimized. Instead, the average shift length should be as close to 6 hours as possible.

Our results are shown in Table 2. The lengths of all the shifts are acceptable (hard constraint V3). The number of employees at each timeslot over the planning horizon is exactly as given for instances R1, R2 and R3 (soft constraint C1). Each shift in instances R3 and R4 includes an acceptable lunch break (soft constraint P3). Instance R4 is interesting in that at least 70 shifts are required to fulfill constraints V3 (shift length) and C1 (staff demand) exactly. However, due to the extreme peak in the morning, it proves beneficial to have significantly fewer shifts than that (soft constraint V1). This is the main reason we use the quadratic violation function for C1.

Table 2. Our results for the instances ($V3$ = number of shifts whose length is not between 4 and 8 hours, $C1$ = total sum of workforce excess and shortage over all the timeslots, $V2$ = number of 8-hour shifts, $V4$ = the absolute difference between the sought average shift length and the actual average shift length, $P3$ = number of misplaced lunch breaks, sol = value of the objective function, *Running Time* = the approximate time our algorithm ran on a machine with Intel Core i7-980X Extreme Edition 3.33GHz and 6GB of RAM running Windows 7 Professional Edition)

ID	Avg length of shift	V3	C1	V1 (# of shifts)	V2 (# of 8-hour shifts)	V4	P3	sol	Running Time
R1	359 min	0	0	313		0		0	40h
R2	398 min	0	0	46				46	1h
R3	327 min	0	0	46	17		0	75	7h
R4	314 min	0	34	57	12		0	260	7h

Our business partner was satisfied with the results for the artificial and real-world instances. They are seeking to integrate our algorithm into their workforce management software, which is the market leader in Finland. The software will

1) allow users to specify the importance of the optimization criteria,
2) minimize the scheduling time required by personal managers,
3) run on virtually any modern desktop computer,
4) generate different solutions to choose from, and
5) be helpful as a planning tool for future scenarios.

6   Conclusions

We have introduced the unlimited shift generation problem. We believe that a considerable number of real-world scenarios can be modeled using the model presented in this paper. This research has contributed to better systems for our industry partner Numeron. A set of real-world instances derived from the actual problems solved for various companies were presented. We have published the best solutions we have found. We invite the workforce scheduling community to challenge our results. We believe that the instances will help researchers to test the implementation value of their solution methods.

References

1. Garey, M.R., Johnson, D.S.: Computers and Intractability: A Guide to the Theory of NP-Completeness. Freeman (1979)
2. Tien, J., Kamiyama, A.: On Manpower Scheduling Algorithms. SIAM Rev. 24(3), 275–287 (1982)
3. Lau, H.C.: On the Complexity of Manpower Shift Scheduling. Computers and Operations Research 23(1), 93–102 (1996)
4. Marx, D.: Graph coloring problems and their applications in scheduling. Periodica Polytechnica Ser. El. Eng. 48, 5–10 (2004)
5. Di Gaspero, L., Gärtner, J., Kortsarz, G., Musliu, N., Schaerf, A., Slany, W.: The minimum shift design problem. Annals of Operations Research 155(1), 79–105 (2007)

6. Dantzig, G.B.: A comment on Edie's traffic delays at toll booths. Operations Research 2, 339–341 (1954)
7. Alfares, H.K.: Survey, categorization and comparison of recent tour scheduling literature. Annals of Operations Research 127, 145–175 (2004)
8. Ernst, A.T., Jiang, H., Krishnamoorthy, M., Sier, D.: Staff scheduling and rostering: A review of applications, methods and models. European Journal of Operational Research 153(1), 3–27 (2004)
9. Meisels, A., Schaerf, A.: Modelling and solving employee timetabling problems. Annals of Mathematics and Artificial Intelligence 39, 41–59 (2003)
10. Herbers, J.: Models and Algorithms for Ground Staff Scheduling On Airports, Dissertation, Rheinisch-Westfälische Technische Hochschule Aachen, Faculty of Mathematics, Computer Science and Natural Sciences (2005)
11. Musliu, N., Schaerf, A., Slany, W.: Local search for shift design. European Journal of Operational Research 153(1), 51–64 (2004)
12. Draghici, C., Hennet, J.C.: Generation of shift schedules - a time slot approach. In: International Conference on Industrial Engineering and Systems Management, pp.653–672 (2005)
13. Clausen, T.: Airport Ground Staff Scheduling, Dissertation, Technical University of Denmark (2011)
14. Demassey, S., Pesant, G., Rousseau, L.-M.: A Cost-Regular Based Hybrid Column Generation Approach. Constraints 11(4), 315–333 (2006)
15. Zolfaghari, S., El-Bouri, A., Namiranian, B., Quan, V.: Heuristics for Large Scale Labour Scheduling Problems in Retail Sector. INFOR 45(3), 111–122 (2007)
16. Chapados, N., Joliveau, M., Rousseau, L.-M.: Retail Store Workforce Scheduling by Expected Operating Income Maximization. In: Achterberg, T., Beck, J.C. (eds.) CPAIOR 2011. LNCS, vol. 6697, pp. 53–58. Springer, Heidelberg (2011)
17. Ásgeirsson, E.I., Kyngäs, J., Nurmi, K., Stølevik, M.: A Framework for Implementation-Oriented Staff Scheduling. In: Proc of the 5th Multidisciplinary Int. Scheduling Conf.: Theory and Applications (MISTA), Phoenix, USA (2011) (submitted for publication)
18. Nurmi, K., Kyngäs, J.: A Framework for School Timetabling Problem. In: Proc. of the 3rd Multidisciplinary Int. Scheduling Conf.: Theory and Applications (MISTA), Paris, France, pp. 386–393 (2007)
19. Kyngäs, J., Nurmi, K.: Scheduling the Finnish Major Ice Hockey League. In: Proc. of the IEEE Symposium on Computational Intelligence in Scheduling (CISCHED), Nashville, USA (2009)
20. Kyngäs, J., Nurmi, K.: Shift Scheduling for a Large Haulage Company. In: Proc. of the 2011 International Conference on Network and Computational Intelligence (ICNCI), Zhengzhou, China (2011)
21. Nurmi, K.: Genetic Algorithms for Timetabling and Traveling Salesman Problems, Dissertation, Dept. of Applied Math., University of Turku, Finland, (1998), http://www.bit.spt.fi/cimmo.nurmi/
22. Ross, P., Ballinger, G.H.: PGA - Parallel Genetic Algorithm Testbed, Department of Articial Intelligence, University of Edinburgh, England (1993)
23. Lin, S., Kernighan, B.W.: An effective heuristic for the traveling salesman problem. Operations Research 21, 498–516 (1973)
24. Glover, F.: New ejection chain and alternating path methods for traveling salesman problems. In: Sharda, Balci, Zenios (eds.) Computer Science and Operations Research: New Developments in Their Interfaces, pp. 449–509. Elsevier (1992)
25. Cowling, P., Kendall, G., Soubeiga, E.: A Hyperheuristic Approach to Scheduling a Sales Summit. In: Burke, E., Erben, W. (eds.) PATAT 2000. LNCS, vol. 2079, pp. 176–190. Springer, Heidelberg (2001)

Ant Colony Optimization
with Immigrants Schemes
for the Dynamic Vehicle Routing Problem

Michalis Mavrovouniotis[1] and Shengxiang Yang[2]

[1] Department of Computer Science, University of Leicester
University Road, Leicester LE1 7RH, United Kingdom
mm251@mcs.le.ac.uk
[2] Department of Information Systems and Computing, Brunel University
Uxbridge, Middlesex UB8 3PH, United Kingdom
shengxiang.yang@brunel.ac.uk

Abstract. Ant colony optimization (ACO) algorithms have proved to be able to adapt to dynamic optimization problems (DOPs) when they are enhanced to maintain diversity and transfer knowledge. Several approaches have been integrated with ACO to improve its performance for DOPs. Among these integrations, the ACO algorithm with immigrants schemes has shown good results on the dynamic travelling salesman problem. In this paper, we investigate ACO algorithms to solve a more realistic DOP, the dynamic vehicle routing problem (DVRP) with traffic factors. Random immigrants and elitism-based immigrants are applied to ACO algorithms, which are then investigated on different DVRP test cases. The results show that the proposed ACO algorithms achieve promising results, especially when elitism-based immigrants are used.

1 Introduction

In the vehicle routing problem (VRP), a number of vehicles with limited capacity are routed in order to satisfy the demand of all customers at a minimum cost (usually the total travel time). Ant colony optimization (ACO) algorithms have shown good performance for the VRP, where a population of ants cooperate and construct vehicle routes [5]. The cooperation mechanism of ants is achieved via their pheromone trails, where each ant deposits pheromone to its trails and the remaining ants can exploit it [2].

The dynamic VRP (DVRP) is closer to a real-world application since the traffic jams in the road system are considered. As a result, the travel time between customers may change depending on the time of the day. In dynamic optimization problems (DOPs) the moving optimum needs to be tracked over time. ACO algorithms can adapt to dynamic changes since they are inspired from nature, which is a continuous adaptation process [9]. In practice, they can adapt by transferring knowledge from past environments [1]. The challenge of such algorithms is how quickly they can react to dynamic changes in order to maintain the high quality of output instead of premature convergence.

C. Di Chio et al. (Eds.): EvoApplications 2012, LNCS 7248, pp. 519–528, 2012.
© Springer-Verlag Berlin Heidelberg 2012

Developing strategies for ACO algorithms to deal with premature convergence and address DOPs has attracted a lot of attention, which includes local and global restart strategies [7], memory-based approaches [6], pheromone manipulation schemes to maintain diversity [4], and immigrants schemes to increase diversity [11,12]. These approaches have been applied to the dynamic travelling salesman problem (DTSP), which is the simplest case of a DVRP, i.e., only one vehicle is used. The ACO algorithms that are integrated with immigrants schemes have shown promising results on the DTSP where immigrant ants replace the worst ants in the population every iteration [11].

In this paper, we integrate two immigrants schemes, i.e., random immigrants and elitism-based immigrants, to ACO algorithms and apply them to the DVRP with traffic factor. The aim of random immigrants ACO (RIACO) is to increase the diversity in order to adapt well in DOPs, and the aim of elitism-based immigrants ACO (EIACO) is to generate guided diversity to avoid randomization.

The rest of the paper is organized as follows. Section 2 describes the problem we try to solve, i.e., the DVRP with traffic factors. Section 3 describes the ant colony system (ACS), which is one of the best performing algorithms for the VRP. Section 4 describes our proposed approaches where we incorporate immigrants schemes with ACO. Section 5 describes the experiments carried out by comparing RIACO and EIACO with ACS. Finally, Section 6 concludes this paper with directions for future work.

2   The DVRP with Traffic Jams

The VRP has become one of the most popular combinatorial optimization problems, due to its similarities with many real-world applications. The VRP is classified as *NP*-hard [10]. The basic VRP can be described as follows: a number of vehicles with a fixed capacity need to satisfy the demand of all the customers, starting from and returning to the depot.

Usually, the VRP is represented by a complete weighted graph $G = (V, E)$, with $n + 1$ nodes, where $V = \{u_0, \ldots, u_n\}$ is a set of vertices corresponding to the customers (or delivery points) u_i $(i = 1, \cdots, n)$ and the depot u_0 and $E = \{(u_i, u_j) : i \neq j\}$ is a set of edges. Each edge (u_i, u_j) is associated with a non-negative d_{ij} which represents the distance (or travel time) between u_i and u_j. For each customer u_i, a non-negative demand D_i is given. For the depot u_0, a zero demand is associated, i.e., $D_0 = 0$.

The aim of the VRP is to find the route (or a set of routes) with the lowest cost without violating the following constraints: (1) every customer is visited exactly once by only one vehicle; (2) every vehicle starts and finishes at the depot; and (3) the total demand of every vehicle route must not exceed the vehicle capacity Q. The number of routes identifies the corresponding number of vehicles used to generate one VRP solution, which is not fixed but chosen by the algorithm.

The VRP becomes more challenging if it is subject to a dynamic environment. There are many variations of the DVRP, such as the DVRP with dynamic demand [14]. In this paper, we generate a DVRP with traffic factors, where each edge (u_i, u_j) is associated with a traffic factor t_{ij}. Therefore, the cost to travel from u_i to u_j is $c_{ij} = d_{ij} \times t_{ij}$. Furthermore, the cost to travel from u_j to u_i may differ due to different traffic factor. For example, one road may have more traffic in one direction and less traffic in the opposite direction.

Every f iterations a random number $R \in [F_L, F_U]$ is generated to represent potential traffic jams, where F_L and F_U are the lower and upper bounds of the traffic factor, respectively. Each edge has a probability m to have a traffic factor, by generating a different R to represent high and low traffic jams on different roads, i.e., $t_{ij} = 1 + R$, where the traffic factor of the remaining edges is set to 1 (indicates no traffic). Note that f and m represent the frequency and magnitude of changes in the DVRP, respectively.

3   ACO for the DVRP

The ACO metaheuristic consists of a population of μ ants where they construct solutions and share their information with the others via their pheromone trails. The first ACO algorithm developed is the Ant System (AS) [2]. Many variations of the AS have been developed over the years and applied to difficult optimization problems [3].

The best performing ACO algorithm for the DVRP is the ACS [13]. There is a multi-colony variation of this algorithm applied to the VRP with time windows [5]. However, in this paper we consider the single colony which has been applied to the DVRP [13]. Initially, all the ants are placed on the depot and all pheromone trails are initialized with an equal amount. With a probability $1 - q_0$, where $0 \le q_0 \le 1$ is a parameter of the *pseudo-random* proportional decision rule (usually 0.9 for ACS), an ant k chooses the next customer j from customer i, as follows:

$$p_{ij}^k = \begin{cases} \dfrac{[\tau_{ij}]^\alpha [\eta_{ij}]^\beta}{\sum_{l \in N_i^k} [\tau_{il}]^\alpha [\eta_{il}]^\beta}, \text{if } j \in N_i^k, \\ 0, \qquad\qquad\quad \text{otherwise,} \end{cases} \tag{1}$$

where τ_{ij} is the existing pheromone trail between customers i and j, η_{ij} is the heuristic information available a priori, which is defined as $1/c_{ij}$, where c_{ij} is the distance travelled (as calculated in Section 2) between customers i and j, N_i^k denotes the neighbourhood of unvisited customers of ant k when its current customer is i, and α and β are the two parameters that determine the relative influence of pheromone trail and heuristic information, respectively. With the probability q_0, the ant k chooses the next customer with the maximum probability, i.e., $[\tau]^\alpha [\eta]^\beta$, and not probabilistically as in Eq. (1). However, if the choice of the next customer leads to an infeasible solution, i.e., exceed the maximum capacity Q of the vehicle, the depot is chosen and a new vehicle route starts.

When all ants construct their solutions, the best ant retraces the solution and deposits pheromone globally according to its solution quality on the corresponding trails, as follows:

$$\tau_{ij} \leftarrow (1 - \rho)\tau_{ij} + \rho \Delta \tau_{ij}^{best}, \forall (i, j) \in T^{best}, \tag{2}$$

where $0 < \rho \leq 1$ is the pheromone evaporation rate and $\Delta\tau_{ij}^{best} = 1/C^{best}$, where C^{best} is the total cost of the T^{best} tour. Moreover, a local pheromone update is performed every time an ant chooses another customer j from customer i as follows:

$$\tau_{ij} \leftarrow (1 - \rho)\tau_{ij} + \rho\tau_0, \tag{3}$$

where ρ is defined as in Eq. (2) and τ_0 is the initial pheromone value.

The pheromone evaporation is the mechanism that eliminates the areas with high intensity of pheromones that are generate by ants, due to stagnation behaviour[1], in order to adapt well to the new environment. The recovery time depends on the size of the problem and magnitude of change.

4 ACO with Immigrants Schemes for the DVRP

4.1 Framework

The framework of the proposed algorithms is based on the ACO algorithms that were used for the DTSP [11,12]. It will be interesting to observe if the framework based on immigrants schemes is beneficial for more realistic problems, such as the DVRP with traffic factors, as described in Section 2.

The initial phase of the algorithm and the solution construction of the ants are the same with the ACS; see Eq. (1). The difference of the proposed framework is that it uses a short-term memory every iteration t, denoted as $k_{short}(t)$, of limited size, i.e., K_s, which is associated with the pheromone matrix. Initially, $k_{short}(0)$ is empty where at the end of the iteration the K_s best ants will be added to $k_{short}(t)$. Each ant k that enters $k_{short}(t)$ deposits a constant amount of pheromone to the corresponding trails, as follows:

$$\tau_{ij} \leftarrow \tau_{ij} + \Delta\tau_{ij}^k, \forall \ (i,j) \in T^k, \tag{4}$$

where $\Delta\tau_{ij}^k = (\tau_{max} - \tau_0)/K_s$ and T^k is the tour of ant k. Here, τ_{max} and τ_0 are the maximum and initial pheromone value, respectively.

Every iteration the ants from $k_{short}(t - 1)$ are replaced with the K_s best ants from iteration t, a negative update is performed to their pheromone trails, as follows:

$$\tau_{ij} \leftarrow \tau_{ij} - \Delta\tau_{ij}^k, \forall \ (i,j) \in T^k, \tag{5}$$

where $\Delta\tau_{ij}$ and T^k are defined as in Eq. (4). This is because no ants can survive in more than one iteration because of the dynamic environment.

In addition, immigrant ants replace the worst ants in $k_{short}(t)$ every iteration and further adjustments are performed to the pheromone trails since $k_{short}(t)$ changes. The main concern when dealing with immigrants schemes is how to generate immigrant ants, that represent feasible solutions.

---

[1] A term used when all ants follow the same path and construct the same solution.

4.2   Random Immigrants ACO (RIACO)

Traditionally, the immigrants are randomly generated and replace other ants in the population to increase the diversity. A random immigrant ant for the DVRP is generated as follows. First, the depot is added as the starting point; then, an unvisited customer is randomly selected as the next point. This process is repeated until the first segment (starting from the most recent visit to the depot) of customers do not violate the capacity constraint. When the capacity constraint is violated the depot is added and another segment of customers starts. When all customers are visited the solution will represent one feasible VRP solution.

Considering the proposed framework described above, before the pheromone trails are updated, a set S_{ri} of $r \times K_s$ immigrants are generated to replace the worst ants in $k_{short}(t)$, where r is the replacement rate.

RIACO has been found to perform better in fast and significantly changing environments for the DTSP [11]. This is because when the changing environments are not similar it is better to randomly increase the diversity instead of knowledge transfer. Moreover, when the environmental changes are fast the time is not enough to gain useful knowledge in order to transfer it. However, there is a high risk of randomization with RIACO that may disturb the optimization process. A similar behaviour is expected for the DVRP.

4.3   Elitism-Based Immigrants ACO (EIACO)

Differently from RIACO, which generates diversity randomly with the immigrants, EIACO generates guided diversity by the knowledge transferred from the best ant of the previous environment. An elitism-based immigrant ant for the DVRP is generated as follows. The best ant of the previous environment is selected in order to use it as the base to generate elitism-based immigrants. The depots of the best ant are removed and adaptive inversion is performed based on the inver-over operator [8]. When the inversion operator finishes, the depots are added so that the capacity constraint is satisfied in order to represent one feasible VRP solution.

Considering the proposed framework above, on iteration t, the elite ant from $k_{short}(t-1)$ is used as the base to generate a set S_{ei} of $r \times K_s$ immigrants, where r is the replacement rate. The elitism-based immigrants replace the worst ants in $k_{short}(t)$ before the pheromone trails are updated.

The EIACO has been found to perform better in slowly and slightly changing environments for the DTSP [11]. This is because the knowledge transferred when the changing environments are similar will be more useful. However, there is a risk to transfer too much knowledge and start the optimization process from a local optimum and get stuck there. A similar behaviour is expected for the DVRP.

5   Simulation Experiments

5.1   Experimental Setup

In the experiments, we compare the proposed RIACO and EIACO with the existing ACS, described in Section 3. All the algorithms have been applied to the vrp45, vrp72, and vrp135 problem instances[2].

To achieve a good balance between exploration and exploitation, most of the parameters have been obtained from our preliminary experiments where others have been inspired from literature [11]. For all algorithms, $\mu = 50$ ants are used, $\alpha = 1$, $\beta = 5$, and $\tau_0 = 1/n$. For ACS, $q_0 = 0.9$, and $\rho = 0.7$. Note that a lower evaporation rate has been used for ACS, i.e. $\rho = 0.1$, with similar or worse results. For the proposed algorithms, $q_0 = 0.0$, $K_s = 10$, $\tau_{max} = 1.0$ and $r = 0.4$.

For each algorithm on a DVRP instance, $N = 30$ independent runs were executed on the same environmental changes. The algorithms were executed for $G = 1000$ iterations and the overall offline performance is calculated as follows:

$$P_{offline} = \frac{1}{G} \sum_{i=1}^{G} \left(\frac{1}{N} \sum_{j=1}^{N} P_{ij}^* \right) \tag{6}$$

where P_{ij}^* defines the tour cost of the best ant since the last dynamic change of iteration i of run j [9].

The value of f was set to 10 and 100, which indicate fast and slowly changing environments, respectively. The value of m was set to 0.1, 0.25, 0.5, and 0.75, which indicate the degree of environmental changes from small, to medium, to large, respectively. The bounds of the traffic factor are set as $F_L = 0$ and $F_U = 5$. As a result, eight dynamic environments, i.e., 2 values of $f \times 4$ values of m, were generated from each stationary VRP instance, as described in Section 2, to systematically analyze the adaptation and searching capability of each algorithm on the DVRP.

5.2   Experimental Results and Analysis

The experimental results regarding the offline performance of the algorithms are presented in Table 1 and the corresponding statistical results of Wilcoxon rank-sum test, at the 0.05 level of significance are presented in Table 2. Moreover, to better understand the dynamic behaviour of the algorithms, the results of the largest problem instance, i.e., vrp135, are plotted in Fig. 1 with $f = 10$, $m = 0.1$ and $m = 0.75$, and $f = 100$, $m = 0.1$ and $m = 0.75$, for the first 500 iterations. From the experimental results, several observations can be made by comparing the behaviour of the algorithms.

First, RIACO outperforms ACS in all the dynamic test cases; see the results of RIACO $\Leftrightarrow$ ACS in Table 2. This validates our expectation that ACS need

---

[2] Taken from the Fisher benchmark instances available at
http://neo.lcc.uma.es/radi-aeb/WebVRP/

Table 1. Comparison of algorithms regarding the results of the offline performance

	$f = 10$				$f = 100$			
$m \Rightarrow$	0.1	0.25	0.5	0.75	0.1	0.25	0.5	0.75
Alg. & Inst.				vrp45				
ACS	897.5	972.5	1205.6	1648.0	883.4	929.1	1120.2	1536.9
RIACO	841.2	902.4	1089.5	1482.9	834.9	867.5	1016.1	1375.1
EIACO	840.1	899.8	1083.8	1473.5	839.8	860.6	1009.1	1355.5
Alg. & Inst.				vrp72				
ACS	305.3	338.6	426.2	596.2	297.3	324.6	412.7	547.9
RIACO	294.4	322.8	401.7	562.5	280.6	303.5	375.2	489.6
EIACO	289.9	319.4	397.8	557.0	276.2	298.5	366.7	476.5
Alg. & Inst.				vrp135				
ACS	1427.7	1567.3	1967.4	2745.7	1383.7	1519.4	1820.5	2536.2
RIACO	1417.8	1554.2	1922.1	2676.0	1353.1	1457.2	1698.6	2358.4
EIACO	1401.3	1542.1	1907.6	2663.1	1329.1	1444.3	1668.5	2293.8

Table 2. Statistical tests of comparing algorithms regarding the offline performance, where "+" or "−" means that the first algorithm is significantly better or the second algorithm is significantly better

Alg. & Inst.	vrp45				vrp72				vrp135			
$f = 10, m \Rightarrow$	0.1	0.25	0.5	0.75	0.1	0.25	0.5	0.75	0.1	0.25	0.5	0.75
RIACO ⇔ ACS	+	+	+	+	+	+	+	+	+	+	+	+
EIACO ⇔ ACS	+	+	+	+	+	+	+	+	+	+	+	+
EIACO ⇔ RIACO	+	+	+	+	+	+	+	+	+	+	+	+
$f = 100, m \Rightarrow$	0.1	0.25	0.5	0.75	0.1	0.25	0.5	0.75	0.1	0.25	0.5	0.75
RIACO ⇔ ACS	+	+	+	+	+	+	+	+	+	+	+	+
EIACO ⇔ ACS	+	+	+	+	+	+	+	+	+	+	+	+
EIACO ⇔ RIACO	−	+	+	+	+	+	+	+	+	+	+	+

sufficient time to recover when a dynamic change occurs, which can be also observed from Fig. 1 in the environmental case with $f = 100$. This is because the pheromone evaporation is the only mechanism used to eliminate pheromone trails that are not useful to the new environment, and may bias the population to areas that are not near the new optimum. On the other hand, RIACO uses the proposed framework where the pheromone trails exist only in one iteration.

Second, EIACO outperforms ACS in all the dynamic test cases as the RI-ACO; see the results EIACO ⇔ ACS in Table 2. This is due to the same reasons RIACO outperforms the traditional ACS. However, EIACO outperforms RI-ACO in almost all dynamic test cases; see the results of EIACO ⇔ RIACO in Table 2. In slowly and slightly changing environments EIACO has sufficient time to gain knowledge from the previous environment, and the knowledge transferred has more chances to help when the changing environments are similar. However, on the smallest problem instance, i.e., vrp45, with $f = 100$ and $m = 0.1$ RIACO performs better than EIACO. This validates our expectation where too much

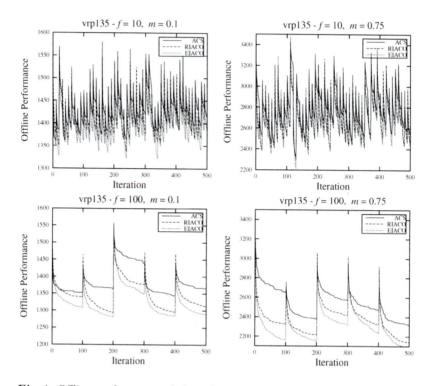

Fig. 1. Offline performance of algorithms for different dynamic test problems

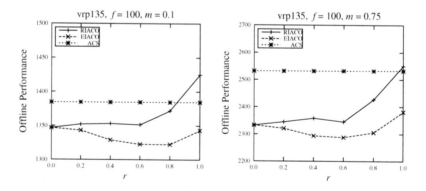

Fig. 2. Offline performance of RIACO and EIACO with different replacement rates against the performance of ACS in slowly changing environments

knowledge transferred does not always mean better results in dynamic environments. On the other hand RIACO, was expected to perform better than EIACO in fast and significantly changing environments, since the random immigrants only increase the diversity, but that it is not the case. This may be possibly because of too much randomization that may disturb the optimization process and requires further investigation regarding the effect of the immigrant ants.

Third, in order to investigate the effectiveness of the immigrants schemes, further experiments have been performed on the same problem instances with the same parameters used before but with different immigrant replacement rates, i.e., $r \in \{0.0, 0.2, 0.4, 0.6, 0.8, 1.0\}$. In Fig. 2 the offline performance of RIACO and EIACO with the varying replacement rates are presented[3], against the ACS performance, where $r = 0.0$ means that no immigrants are generated to replace ants in the $k_{short}(t)$. The results confirm our expectation above, where the random immigrants in RIACO sometimes may disturb the optimization and degrade the performance. On the other hand, elitism-based immigrants in EIACO improve the performance, especially in slightly changing environments.

Finally, the proposed framework performs better than ACS, even if no immigrants are generated; see Fig. 2. The RIACO with $r = 1.0$ performs worse than the ACS, whereas the EIACO with $r = 1.0$ better than ACS. This is because RIACO destroys all the knowledge transferred to the $k_{short}(t)$ from the ants of the previous iteration with random immigrants, whereas EIACO destroys that knowledge but transfers new knowledge using the best ant from the previous iteration.

6   Conclusions

Different immigrants schemes have been successfully applied to evolutionary algorithms and ACO algorithms to address different DOPs [11,16]. ACO-based algorithms with immigrants, i.e., RIACO and EIACO, have shown good performance on different variations of the DTSP [11,12]. In this paper, we modify and apply such algorithms to address the DVRP with traffic factors, which is closer to a real-world application. The immigrant ants are generated either randomly or using the previous best ant as the base and replace the worst ones in the population. The aim is to maintain the diversity of solutions and transfer knowledge from previous environments in order to adapt well in DOPs.

Comparing RIACO and EIACO with ACS, one of the best performing ACO algorithms for VRP, on different test cases of DVRPs, the following concluding remarks can be drawn. First, the proposed framework used to integrate ACO with immigrants schemes, performs better than the traditional framework, even when immigrant ants are not generated. Second, EIACO is significantly better than RIACO and ACS in almost all dynamic test cases. Third, RIACO is significantly better than ACS in all dynamic test cases. Finally, the random immigrants may disturb the optimization process with a result to degrade the performance, whereas elitism-based immigrants transfers knowledge with a result to improves the performance for the DVRP with traffic factor.

An obvious direction for future work is to hybridize the two immigrants schemes. However, from our preliminary results the performance of the hybrid scheme is better than RIACO but worse than EIACO in all dynamic test cases. Therefore, to find another way to achieve a good balance between the knowledge

---

[3] The experimental results of the remaining problem instances and dynamic test cases are similar for EIACO, whereas for RIACO there is an improvement when $r > 0.0$ on the smallest problem instance.

transferred and the diversity generated would be interesting for future work. Another future work is to integrate memory-based immigrants with ACO, which have also performed well on the DTSP [12], to the DVRP with traffic factors.

References

1. Bonabeau, E., Dorigo, M., Theraulaz, G.: Swarm Intelligence: From Natural to Artificial Systems. Oxford University Press, New York (1999)
2. Dorigo, M., Maniezzo, V., Colorni, A.: Ant system: optimization by a colony of cooperating agents. IEEE Trans. on Syst. Man and Cybern. Part B: Cybern. 26(1), 29–41 (1996)
3. Dorigo, M., Stützle, T.: Ant Colony Optimization. The MIT Press, London (2004)
4. Eyckelhof, C.J., Snoek, M.: Ant Systems for a Dynamic TSP. In: ANTS 2002: Proc. of the 3rd Int. Workshop on Ant Algorithms, pp. 88–99 (2002)
5. Gambardella, L.M., Taillard, E., Agazzi, G.: MACS-VRPTW: A multiple ant colony system for vehicle routing problems with time windows. In: Corne, D., et al. (eds.) New Ideas in Optimization, pp. 63–76 (1999)
6. Guntsch, M., Middendorf, M.: Applying Population Based ACO to Dynamic Optimization Problems. In: Dorigo, M., Di Caro, G.A., Sampels, M. (eds.) Ant Algorithms 2002. LNCS, vol. 2463, pp. 111–122. Springer, Heidelberg (2002)
7. Guntsch, M., Middendorf, M.: Pheromone Modification Strategies for Ant Algorithms Applied to Dynamic TSP. In: Boers, E.J.W., Gottlieb, J., Lanzi, P.L., Smith, R.E., Cagnoni, S., Hart, E., Raidl, G.R., Tijink, H. (eds.) EvoIASP 2001, EvoWorkshops 2001, EvoFlight 2001, EvoSTIM 2001, EvoCOP 2001, and EvoLearn 2001. LNCS, vol. 2037, pp. 213–222. Springer, Heidelberg (2001)
8. Tao, G., Michalewicz, Z.: Inver-over Operator for the TSP. In: Eiben, A.E., Bäck, T., Schoenauer, M., Schwefel, H.-P. (eds.) PPSN 1998. LNCS, vol. 1498, pp. 803–812. Springer, Heidelberg (1998)
9. Jin, Y., Branke, J.: Evolutionary optimization in uncertain environments - a survey. IEEE Trans. on Evol. Comput. 9(3), 303–317 (2005)
10. Labbe, M., Laporte, G., Mercure, H.: Capacitated vehicle routing on trees. Operations Research 39(4), 616–622 (1991)
11. Mavrovouniotis, M., Yang, S.: Ant Colony Optimization with Immigrants Schemes in Dynamic Environments. In: Schaefer, R., Cotta, C., Kołodziej, J., Rudolph, G. (eds.) PPSN XI. LNCS, vol. 6239, pp. 371–380. Springer, Heidelberg (2010)
12. Mavrovouniotis, M., Yang, S.: Memory-Based Immigrants for Ant Colony Optimization in Changing Environments. In: Di Chio, C., Cagnoni, S., Cotta, C., Ebner, M., Ekárt, A., Esparcia-Alcázar, A.I., Merelo, J.J., Neri, F., Preuss, M., Richter, H., Togelius, J., Yannakakis, G.N. (eds.) EvoApplications 2011, Part I. LNCS, vol. 6624, pp. 324–333. Springer, Heidelberg (2011)
13. Montemanni, R., Gambardella, L., Rizzoli, A., Donati, A.: Ant colony system for a dynamic vehicle routing problem. Journal of Combinatorial Optimization 10(4), 327–343 (2005)
14. Psaraftis, H.: Dynamic vehicle routing: status and prospects. Annals of Operations Research 61, 143–164 (1995)
15. Rizzoli, A.E., Montemanni, R., Lucibello, E., Gambardella, L.M.: Ant colony optimization for real-world vehicle routing problems - from theory to applications. Swarm Intelli. 1(2), 135–151 (2007)
16. Yang, S.: Genetic algorithms with memory and elitism based immigrants in dynamic environments. Evol. Comput. 16(3), 385–416 (2008)

Evolving Communication in Robotic Swarms Using On-Line, On-Board, Distributed Evolutionary Algorithms

Luis E. Pineda[1], A.E. Eiben[2], and Marteen van Steen[2]

[1] Instituto de Cálculo Aplicado, Universidad del Zulia, Maracaibo, Venezuela
[2] Dept. of Computer Science, Vrije Universiteit Amsterdam, The Netherlands
lpineda@ica.luz.edu.ve

Abstract. Robotic swarms offer flexibility, robustness, and scalability. For successful operation they need appropriate communication strategies that should be dynamically adaptable to possibly changing environmental requirements. In this paper we try to achieve this through evolving communication on-the-fly. As a test case we use a scenario where robots need to cooperate to gather energy and the necessity to cooperate is scalable. We implement an evolutionary algorithm that works during the actual operation of the robots (on-line), where evolutionary operators are performed by the robots themselves (on-board) and robots exchange genomes with other robots for reproduction (distributed). We perform experiments with different cooperation pressures and observe that communication strategies can be successfully adapted to the particular demands of the environment.

Keywords: swarm robotics, communication, on-line, on-board, distributed.

1 Introduction

Swarm robotics has emerged in recent years as an important field of research. Drawing inspiration from the behavior of social insects, the main idea behind swarm robotics is that a group of simple robots, by means of cooperation, are able to perform tasks beyond the capabilities of a single individual. The motivations for this approach are increased robustness, flexibility, and scalability [5].

For robotic swarms to be successful, a key component is the development of appropriate communication strategies, particularly due to the requirement that robots operate in a decentralized manner. Furthermore, robotic swarms are expected to operate in dynamic environments for which a high degree of flexibility and adaptation is required. Thus, instead of using fixed communication policies, it is better to equip robots with the ability to adapt their communication strategies to environmental requirements.

A promising way to achieve this is through the use of an evolutionary robotics (ER) approach, i.e., using evolutionary algorithms to evolve the robots' controllers [12]. ER techniques have been applied to diverse problems such as gait control for legged robots [16], and navigation for aerial vehicles [2]. The taxonomy offered by Eiben et al. classifies ER techniques according to *when* evolution happens (off-line vs. on-line), *where*

C. Di Chio et al. (Eds.): EvoApplications 2012, LNCS 7248, pp. 529–538, 2012.
© Springer-Verlag Berlin Heidelberg 2012

it takes place (on-board vs. off-board), and *how* it happens (encapsulated/centralized, distributed, or a hybrid of these two) [7]. The huge majority of work in ER is based on off-line, off-board evolution, assuming the presence of an omniscient master.

In this work we study the evolution of communication in robotic swarms using on-line, on-board, and distributed evolutionary algorithms. This means that evolution takes place during the actual operation of the robots (on-line), evolutionary operators are performed exclusively inside each robot (on-board), and robots exchange genomes with other robots instead of maintaining purely local pools of genomes (distributed). In particular, the evolutionary algorithm (EA) used in this work, Hybrid EvAg, is a hybrid between a purely distributed evolutionary algorithm and a purely local one [10]. In Hybrid EvAg, each robot maintains both a local pool of genomes and a cache of robot neighbors for periodical exchange of genomes.

We study a group of robots that require cooperation to gather energy sources randomly distributed in a rectangular arena. Our experiments draw ideas from the work of Buzing et al. [4], the main one being that communication arises as a means to facilitate cooperation, and thus no fitness is explicitly given to robots for communicating. We study the effect of different cooperation pressures in the communication preferences evolved and, as in [4], we draw a distinction between talking and listening behaviors.

2   Related Work

Many authors have used computer simulations to study the environmental and evolutionary conditions conducive to communication. According to Perfors [14], work in this area can be divided in two categories: the evolution of syntax [3,17,15] and the evolution of communication and coordination [13,4,9].

One key difference between this and other existing work is that we do not intend to establish conclusions about the emergence of communication as an evolutionary construct. Our question is more practical: can we use on-line, on-board, distributed EAs as a tool to allow robotic swarms to develop appropriate communication strategies on their own? While several previous works have studied solutions to the problem of evolving appropriate communication strategies for swarms of robots (e.g., [1,9,11,6]), to the best of our knowledge, no on-line, on-board solutions have been proposed. Nevertheless, the work of Buzing et al. [4] and Floreano et al. [9] are particularly relevant to our research. Our experimental setting, as well as the idea of varying degrees of environmental pressure, is directly based on [4]. On the other hand, our neural network-based controllers are similar to those used in [9]. A comparison between the present work, [4], and [9] is shown in Table 1, and a more detailed description of their work is discussed next.

Buzing et al. [4] studied the evolution of communication within what they named the VUSCAPE model. This model, based on SUGARSCAPE [8], consists of a discrete landscape in which sugar seeds are periodically redistributed and agents need to collect them in order to survive. In addition, pressure towards cooperation is introduced in the form of a limit to the amount of sugar agents can collect on their own. In order to facilitate cooperation, agents have a hard-wired ability to communicate (using messages with fixed syntax and semantics), but their attitude towards using communication is not fixed and evolves over time. The authors used this model to study how communication

Table 1. Comparison between Buzing et al.[4] , Floreano et al.[9], and the present work

	Buzing et al. [4]	Floreano et al. [9]	This work
Dynamic Environment	YES (energy redistributed)	NO	YES (energy redistributed)
Hard-wired semantics	YES	NO	YES
Varying cooperation pressure	YES	NO	YES
Means of communication	Message board. Messages only travel parallel to the axes	Emitting blue light	Broadcasting within a certain circular range
2 agents on 1 location	YES	NO	NO
Agents die	YES	NO	NO
Controller	Rule set	Neural network	Neural network
Actions	2 behavior macros: go to largest sugar seed or random move. Talk / Listen with a certain probability	Spin left/right wheel. Turn on/off blue light	3 behavior macros: random move, avoid obstacle, go to largest energy source. Talk / Listen with a probability
Fitness function	Environmental fitness based on energy	Number of cycles stepping on the energy source minus number of cycles stepping on the poison source	Energy gained
On-line	YES	NO	YES
On-board	YES	NO	YES
Distributed	YES	NO	YES
Selection	No parent selection. Agents mate when at the same location. Environmental survivor selection (agents that run out of sugar die)	Individual and colony-level	Global parent selection, local survivor selection

evolves under different levels of cooperation pressure, and concluded that higher levels of cooperation pressure translate into increased attitudes towards communication.

On the other hand, Floreano et al. [9] studied the evolutionary conditions that facilitate the emergence of communication. Their setting investigated colonies of robots that could forage in an environment with food and poison sources (one of each), and in which robots could use a blue light to (possibly) signal about the location of the food/poison sources. In contrast to Buzing et al., the semantics of the messages were not hard-wired into the system, and they found that different communication strategies evolved depending on the kin structure and selection level of the population (individual-versus colony-level).

3 Problem Description

The test scenario proposed is directly based on the VUSCAPE model developed by Buzing et al. [4]. Our scenario consists of a number of robots set in a rectangular arena in which several energy sources (corresponding to sugar in VUSCAPE) are randomly distributed (according to a uniform distribution). Each robot's fitness is determined by how much energy it is able to collect over a certain period of time. However, collecting energy is made difficult by the following factors:

- Robots constantly lose energy over time. Whenever a robot's energy counter reaches zero, the robot is immediately switched off for the rest of an evaluation period, thus receiving minimal fitness.
- The environment requires that robots cooperate in order to successfully collect energy. In order to study different levels of cooperation pressure, we add an experimental parameter, the cooperation threshold (CT), specifying how much energy a robot can collect from a single source on its own. Specifically, a source carrying an amount of energy higher than the CT must be collected by two or more robots, in which case the energy is distributed equally among the collecting robots.
- The only way for a robot to gather knowledge (on its own) about the location of an energy source is through a fixed set of sensors of limited range.
- Energy sources are relocated once they are collected, thus increasing the need for robots to have an exploratory behavior. Whenever a robot collects an energy source, this source is instantly relocated to a randomly drawn position (uniform distribution).

In order to surmount these difficulties, robots are able to facilitate cooperation and exploration through a hardwired ability to communicate. In particular, robots can use (with a certain probability) information given by other robots about the location and size of energy sources (i.e., listening), and multicast (with a certain probability) the size and location of energy sources they are not able to collect on their own (i.e., talking). Notice that while robots possess an innate ability to communicate, the extent to which they are willing to do so is not fixed; we deliberately leave it subject to adaptation through evolution.

Note that the problem described is not dynamic from the evolutionary algorithm's perspective (once the proper behavior is learned it remains valid throughout a robot's operation). Nevertheless, the problem is dynamic from the point of view of the robots, since the environment is constantly changing in a way that is unpredictable to them. Furthermore, from the evolutionary algorithm's perspective, the fitness function is stochastic.

3.1 Controller

Each robot is controlled through a neural network that decides between different pre-programmed control policies. The twelve (12) inputs of the neural network are: measurements from eight (8) distance sensors that detect obstacles and other robots in the vicinity, angle to the largest energy source the robot has knowledge of, distance to the largest energy source the robot has knowledge of, current energy level, and bias node.

The five (5) outputs of the neural network are: three (3) outputs corresponding to different actions (the highest valued output determines the next action of the robot), talk preference, (i.e., the probability that the robot multicasts information about an energy source when it needs to cooperate), and listen preference, (i.e., the probability that the robot incorporates knowledge about energy sources seen by other robots).

The robots' actions are implemented as follows:

Random Walk. The robot chooses a random direction and moves as far as it can in a straight line in the chosen direction.

Avoid Obstacles. The robot moves straight in the direction it is currently facing until its sensors detect an obstacle. It then rotates away from the obstacle and moves in a straight line again.

Go to Largest Energy Source. The robot rotates so that it faces the largest energy source it is aware of and moves towards this source as fast as it can.

3.2 Evolutionary Algorithm

The controllers in our experiments (i.e., neural networks) were adapted using Hybrid EvAg, a variant of the on-line, on-board, distributed evolutionary algorithm for robotics described in [10]. In Hybrid EvAg, in addition to a local cache of neighbors (other robots) for genome exchange, each robot maintains a local pool of $\mu+1$ genomes (μ stored in the internal population plus one active controller). Parental selection is performed by selecting two neighbors from the cache (i.e., the external population) and using their current genomes (active controllers) as parents. If, after evaluation, the new genome turns out to be better than the worst one in the local pool of μ genomes, the worst one is replaced by the new. This local pool of genomes is used to randomly choose genomes for reevaluation. Thus, in Hybrid EvAg survival selection is local while parental selection is (approximately) global.

The cache of neighbors in Hybrid EvAg is maintained using the Newscast gossiping protocol as explained in [10]. We compared the performance of the Newscast-based Hybrid EvAg with that of a panmictic variant in which each agent has access to the local pools of all the other agents for parent selection. This allows us to study the effect that the lack of information about the true global genome pool has on gossiping-based distributed evolutionary algorithms.

The genome representation of the neural network was a real-valued vector consisting of the neural network's weights and a mutation step size for every weight. Mutation was performed using Gaussian perturbation, and the recombination operator was standard two-parent arithmetic crossover. Binary tournament was used for parent selection. The following evolutionary parameters were used in our experiments: $\mu=10$ (size of the local pool of genomes), $\sigma=1$ (initial mutation step size), crossover rate = 0.5, re-evaluation rate = 0.2, mutation rate = 1, and Newscast cache size = 20.

4   Experiments

4.1   Experimental Details and Performance Measures

Our experiments were run using the RoboRobo simulator developed by Nicholas Bredeche, a fast and simple 2D robot simulator built in C++. We used a group of 20 robots and performed 56 different simulations to account for the stochasticity in the evolutionary algorithms. Each simulation ran for 2,000,000 steps, with a new generation of controllers being evaluated each 1,000 steps. After each evaluation period, the controller's fitness was calculated and the evolutionary algorithm described in Sec. 3.2 was carried to select a new controller. Each robot's energy counter was then reset to its initial value and the robot was allowed to move randomly for 250 steps in order to avoid difficult conditions inherited from the previous evaluation.

The performance of the evolutionary algorithms was evaluated in terms of the performance metrics described next. Note that the values reported in Sec. 4.2 correspond to these measures averaged over the 56 experiments.

- Fitness: the median fitness of the group of robots for each generation.
- Talk/listen preferences: the median average talk/listen preference during 250 controller steps (i.e., not counting the random relocation steps).
- Frequency of controller actions: the median frequency of controller actions during 250 controller steps.

As we are interested in assessing whether robots can develop appropriate strategies for different environmental demands, we study the effect of the cooperation threshold (CT), and thus environmental pressure, on the evolved strategies; for this, two values of the CT were considered (CT = 1 and CT = 5). In one case (LOWCT) the CT was set so that robots needed cooperation to collect any of the energy points in the arena; in the other (HIGHCT), cooperation was not required for any of the energy points. While in our experiments the CT remained fixed throughout the simulation, these settings allow us to evaluate how well the robots adapt to unforeseen environments of different nature.

4.2   Experimental Results

For the two CT values considered, both the Newscast-based and panmictic variants of Hybrid EvAg were able to improve (Wilcoxon rank-sum test, $p < 0.00001$ for both CT values) the average fitness of the robots over time (see Figs. 1a and 1b). Interestingly, although the mating pool for each robot was smaller in the Newscast-based variant, it showed much quicker convergence than the panmictic variant. For the HIGHCT case this resulted in the Newscast-based variant having a somewhat better fitness at the end of the simulation (not statistically significant - Wilcoxon rank-sum test, $p=0.064$). However, the panmictic variant showed a better final performance (not statistically significant - Wilcoxon rank-sum test, $p=0.104$) in the LOWCT case (see Table 2).

Evolved talking and listening preferences were very high in the LOWCT case (see Table 2), which indicates that communication evolved as a response to the environmental pressure to cooperate (see Fig. 2a). With the panmictic variant of Hybrid EvAg, the

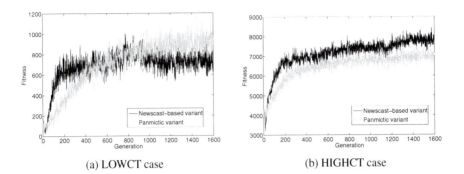

(a) LOWCT case   (b) HIGHCT case

Fig. 1. Fitness vs. Number of generations for CT = 1 (LOWCT implying high pressure to cooperate) and CT = 5 (HIGHCT implying low pressure to cooperate). Mind the different scales on the Fitness axes.

average talking and listening probabilities converged to close to 100% after approximately 600,000 controller steps (600 generations). On the other hand, although the Newscast-based variant quickly reached high talking/listening probabilities (approximately 90% in less than 200 generations), the final values were considerable lower (Wilcoxon rank-sum test, p<0.0001 both for talking and listening) than those obtained with the panmictic variant (see Table 2); in fact, talk/listen probabilities show a decreasing trend over time. This partially explains why the fitness was lower for the Newscast-based variant in the LOWCT case, as a lower preference for communication was detrimental to the robots' capacity to cooperate.

In the HIGHCT case the talk/listen preferences were considerably lower (Wilcoxon rank-sum test, p<0.007 both for panmictic and Newscast-based variants) than in the LOWCT case (see Fig. 2b and Table 2). This is not surprising since cooperation was not required in order for robots to succeed in this arena and, due to cooperation involving a split of the resources among cooperating robots, it would have only resulted in less fitness overall. However, one interesting observation is the different talking/listening evolution trends obtained with the Newscast-based and the panmictic variants. The Newscast-based variant's evolution history was highly irregular and showed no sign of convergence, in contrast to the typical evolution pattern observed with the panmictic variant; the reason for these differences requires further investigation. Nevertheless, it is worth noting that the difference in the final talking/listening probabilities between the Newscast-based and the panmictic variants was not statistically significant (Wilcoxon rank-sum test, p=0.27 and p=0.12 for talking and listening, respectively).

Finally, regarding the frequency of controller actions, there are significant differences between the strategies evolved using the panmictic and Newscast-based variants of Hybrid EvAg. Both in the LOWCT and HIGHCT cases the controllers evolved using the Newscast-based variant showed a much higher preference for the "Avoid Obstacles" action than those evolved using the panmictic variant (see Figs. 3 and 4). Significant differences can also be observed in the preferences for the "Go to Largest Energy Source" action in the LOWCT case (see Fig. 3), with the panmictic variant converging to a higher value than the Newscast-based variant.

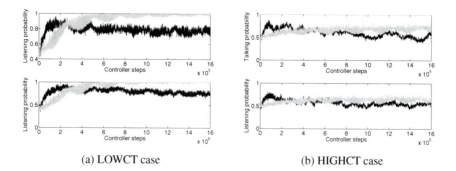

(a) LOWCT case   (b) HIGHCT case

Fig. 2. Talking (upper) and Listening (lower) probabilities vs. Controller steps. Dark line: Newscast-based variant. Light line: Panmictic variant.

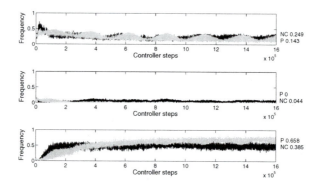

Fig. 3. Frequency of controller actions: Random (upper), Avoid Obstacles (middle), and Go to Largest Energy Source (lower). Dark line: Newscast-based variant. Light line: Panmictic variant (LOWCT case).

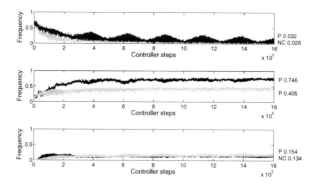

Fig. 4. Frequency of controller actions: Random (upper), Avoid Obstacles (middle), and Go to Largest Energy Source (lower). Dark line: Newscast-based variant. Light line: Panmictic variant (HIGHCT case).

Table 2. Performance (mean and standard deviation) of the Newscast-based (NC) and panmictic (P) variants at the end of the simulation (LOWCT and HIGHCT cases)

	LOWCT		HIGHCT	
	NC	P	NC	P
Fitness	786.6(740.3)	937.1(676.3)	7489.6(1858.8)	7030.5(1482.5)
Talk preference	0.81(0.32)	0.99(0.01)	0.59(0.41)	0.72(0.33)
Listen preference	0.83(0.35)	1.00(0)	0.56(0.43)	0.70(0.35)

5   Conclusions and Future Work

In this paper we presented an initial study on the applicability of on-line, on-board, distributed evolutionary algorithms (e.g., Hybrid EvAg) for evolving communication in robotic swarms. For this first study we assumed robots possessed the ability to communicate using messages with fixed semantics, and focused on studying the communication strategies evolved under different degrees of cooperation pressure. We also draw a distinction between the preference for sending messages (i.e., talking) and that for receiving messages (i.e., listening).

The results show that our on-line, on-board, distributed evolutionary mechanism enabled robots to develop appropriate communication attitudes: a high communication preference when the environmental pressure to cooperate is large, and a low preference when the environmental pressure to cooperate is low. However, we observed a distinction between the communication preferences evolved using a distributed algorithm with full information of the global genome pool (panmictic variant), versus one in which each robot only has a local approximation of the genome pool (Newscast-based variant). The reason for these differences require further investigation, but it is probably related to the information loss inherent to the Newscast-based variant. Note that in some cases (e.g, HIGHCT case) the Newscast-based variant can offer a higher performance than the panmictic variant.

In future work we aim to study the evolution of communication on groups of robots having a lesser degree of hard-wired abilities (such as the current fixed controller actions and semantics). Also, we are currently studying larger groups of robots (e.g., 500 robots) since the computational advantages of the Hybrid EvAg algorithm are more relevant in such a context, and different types of communication behavior may emerge.

References

1. Ampatzis, C., Tuci, E., Trianni, V., Dorigo, M.: Evolution of signaling in a multi-robot system: Categorization and communication. Adaptive Behavior 16(1), 5–26 (2008)
2. Barlow, G.J.: Autonomous controller design for unmanned aerial vehicles using multi-objective genetic programming. In: Proceedings of the Graduate Student Workshop at the 2004 Genetic and Evolutionary Computation Conference (GECCO 2004, Seattle, WA (June 2004); Winner of Best Paper award at the Graduate Student Workshop at the 2004 Genetic and Evolutionary Computation Conference (GECCO 2004)

3. Briscoe, E.J.: Grammatical acquisition and linguistic selection. In: Linguistic Evolution through Language Acquisition: Formal and Computational Models, ch. 9. Cambridge University Press (2002)
4. Buzing, P.C., Eiben, A.E., Schut, M.C.: Emerging communication and cooperation in evolving agent societies. Journal of Artificial Societies and Social Simulation 8(1), 1–16 (2005)
5. Şahin, E.: Swarm robotics: From sources of inspiration to domains of application. Technical Report METU-CENG-TR-2005-01, Department of Computer Engineering, Middle East Technical University (January 2005)
6. de Greeff, J., Nolfi, S.: Evolution of implicit and explicit communication in mobile robots. In: Evolution of Communication and Language in Embodied Agents, pp. 179–214. Springer, Heidelberg (2010)
7. Eiben, A.E., Haasdijk, E., Bredeche, N.: Embodied, on-line, on-board evolution for autonomous robotics. In: Levi, P., Kernbach, S. (eds.) Symbiotic Multi-Robot Organisms: Reliability, Adaptability, Evolution, ch. 5.2, pp. 361–382. Springer, Heidelberg (2010)
8. Epstein, J.M., Axtell, R.: Growing Artificial Societies: Social Sciences from Bottom Up. Brooking Institution Press and The MIT Press (1996)
9. Floreano, D., Mitri, S., Magnenat, S., Keller, L.: Evolutionary conditions for the emergence of communication in robots. Current Biology: CB 17(6), 514–519 (2007)
10. Huijsman, R., Haasdijk, E., Eiben, A.E.: An On-line On-board Distributed Algorithm for Evolutionary Robotics (2011)
11. Montes-Gonzalez, F., Aldana-Franco, F.: The Evolution of Signal Communication for the e-puck Robot. In: Batyrshin, I., Sidorov, G. (eds.) MICAI 2011, Part I. LNCS, vol. 7094, pp. 466–477. Springer, Heidelberg (2011)
12. Nolfi, S., Floreano, D.: Evolutionary Robotics: The Biology, Intelligence, and Technology of Self-Organizing Machines. MIT Press, Cambridge (2000)
13. Oliphant, M., Batali, J.: Learning and the emergence of coordinated communication. The Newsletter of the Center for Research in Language 11(1) (1997)
14. Perfors, A.: Simulated evolution of language: a review of the field. Journal of Artificial Societies and Social Simulation 5(2), 1–62 (2002)
15. Steels, L.: Modeling the formation of language: Embodied experiments. In: Nolfi, S., Mirolli, M. (eds.) Evolution of Communication and Language in Embodied Agents, pp. 235–262. Springer, Berlin (2010)
16. Teo, J.: Darwin + robots=evolutionary robotics: Challenges in automatic robot synthesis. In: 2nd International Conference on Artificial Intelligence in Engineering and Technology (ICAIET 2004), Kota Kinabalu, Sabah, Malaysia, pp. 7–13 (August 2004)
17. Vogt, P.: The emergence of compositional structures in perceptually grounded language games. Artificial Intelligence 167(1-2), 206–242 (2005); Connecting Language to the World

Virtual Loser Genetic Algorithm for Dynamic Environments

Anabela Simões[1,2] and Ernesto Costa[2]

[1] Coimbra Institute of Engineering, Polytechnic Institute of Coimbra
[2] Centre for Informatics and Systems of the University of Coimbra
abs@isec.pt, ernesto@dei.uc.pt

Abstract. Memory-based Evolutionary Algorithms in Dynamic Optimization Problems (DOPs) store the best solutions in order to reuse them in future situations. The memorization of the best solutions can be direct (the best individual of the current population is stored) or associative (additional information from the current population is also stored). This paper explores a different type of associative memory to use in Evolutionary Algorithms for DOPs. The memory stores the current best individual and a vector of inhibitions that reflect past errors performed during the evolutionary process. When a change is detected in the environment the best solution is retrieved from memory and the vector of inhibitions associated to this individual is used to create new solutions avoiding the repetition of past errors. This algorithm is called Virtual Loser Genetic Algorithm and was tested in different dynamic environments created using the XOR DOP generator. The results show that the proposed memory scheme significantly enhances the Evolutionary Algorithms in cyclic dynamic environments.

1 Introduction

The use of Evolutionary Algorithms (EA) in dynamic optimization problems (DOPs) has been widely explored in the last decades. In order to make EAs robust to dynamic problems, several enhancements have been proposed: mechanisms to promote diversity when a change is detected [1], methods to maintain the diversity through the entire run [2], the incorporation of memory [3], [4], the use of multi-populations [5] or the anticipation of the change [6].

Memory-based EAs are beneficial when past situations reappear in the future. This way, memorized solutions can help the EA to readapt to the new conditions [5]. Different types of memory have been investigated: direct memory approaches, associative memory and also memory schemes using immigrants. This paper is centered in associative memory, which stores the best individual from the population and additional information about the environment. This information is used to create new individuals every time a change occurs [4]. Different types of associative memory have been investigated in the past: Trojanowski et al. [7] introduced an EA which memorizes information associated to the individual's ancestors. Yang [4] introduced an associative memory

C. Di Chio et al. (Eds.): EvoApplications 2012, LNCS 7248, pp. 539–548, 2012.

scheme inspired in Baluja's population-based incremental learning (PBIL) algorithms. In this memory scheme, the best solution in the population together with the probability vector, which represents the current environment, is stored in the memory. When a change is detected, the probability vector associated with the best memory sample is used to create new individuals. Other PBIL's inspired associative memory scheme were investigated in [8] and [9]. Barlow [10] investigated a memory-enhanced EA to the dynamic job shop scheduling problem. This memory-based EA uses a classifier-based memory for abstracting and storing information about schedules, that is used to build similar schedules at future times.

This paper studies a new form of associative memory. The main idea is to create a vector of inhibitions during the evolutionary process, which contains knowledge about errors that were performed. The stored vector, called *the virtual loser*, is a template of the unfit individuals of the population evolved between two consecutive changes. This knowledge about previous errors is used when a change is detected to create new individuals to reintroduce into the population. The underlying metaphor is that the new solutions should be as different as possible from the virtual loser, in order to avoid repetition of past errors. The virtual loser scheme was proposed by Sebag et al. [11] and was tested in static function optimization, using a $(\mu + \lambda)$-ES, assuming that the fitness landscape was fixed and that the fitness of an individual was independent from the other individuals in the population. The knowledge stored in the virtual loser vector was used by a mutation operator, called flee-mutation, that created offspring derived from the parents. The virtual loser was used to compute different mutation probabilities to a selected number of bits in such a way that the 'bad' mutations performed in the past were avoided.

As far as we know this idea of memorizing errors was never explored in the context of dynamic optimization. This paper investigates the potentialities of an associative memory that stores the best individual and the virtual loser in EAs for dynamic optimization. The proposed algorithm is called *Virtual Loser Genetic Algorithm* (VLGA) and is tested in three different problems using several instances of cyclic and random environments.

The rest of the paper is organized as follows: the next section describes the *Virtual Loser Genetic Algorithm* for DOPs. Section 3 details the experimental setup used in this work. The experimental results and analysis are presented in Section 4. Section 5 concludes the paper and some considerations are made about future work.

2   Virtual Loser Genetic Algorithm

The VLGA is a standard GA using a different type of associative memory. The proposed mechanism is denoted by *the virtual loser* and is based on the memorization of past errors of evolution in order to avoid (inhibit) their future occurrence. This idea resembles to Tabu Search, but instead of memorizing a list of past individuals to avoid, which would only reflect a small part of the search space, uses a tractable and general description of past errors. The virtual loser vector (VL) is initialized at the beginning of the run (or whenever a change happens) and is updated at every generation until a change is detected in the environment. The virtual loser is initialized by analyzing the bits of the best individual of the population and the corresponding bits in the worst individuals of the population. For instance, in Table 1, A is the best individual, B, C and D, the worst individuals (adapted from [11]). The VL calculates, for each position, the percentage of bits of the worst individuals that differ from the corresponding bit in the best individual. For instance, bit 1 was always different in the unfit solutions when compared with the best one. This means that the high fitness of A can be due to bit $1 = 1$ and this value should be preserved. Contrariwise, nothing can be said about the influence of bit 3 in the fitness of A, since this bit has the same value for all individuals. In the future, when applying mutation to solution A, bit 3 should be more affected than bit 1. So, for the first bit the mutation should be strongly inhibited and the opposite for the third bit. At every generation, the VL is updated by relaxation from a fraction η of the worst individuals in the current population. The relaxation factor γ is used to update VL as follows (eq. 1): $VL(t+1) = (1-\gamma)VL(t) + \gamma \times dVL$, where dVL is the average of the η worst individuals. The memory is updated whenever a change occurs. The relevant information stored in memory refers to the population before the change. This information consists of the pair $< B^{p-1}, VL^{p-1} >$, where B^{p-1} is the best individual of the population before change and VL^{p-1} is the VL vector evolved until the generation before change. When the environment changes, the memory is reevaluated according to the new environment and the best memory point $< B^m, VL^m >$ is retrieved. The solution B^m and the associated VL^m vector are used to create a set of $\alpha \times n$ new individuals, that will replace the worst individuals in the population. The parameter $\alpha \in [0, 1]$ is called associative factor and determines the number of individuals created from the memory when a change happens. To create those new individuals, each bit of B^m is mutated according to a certain probability, calculated using VL^m and the value of the gene. The probability to mutate a bit B_i^m is given by: $P_i = |VL_i - B_i^m|$ (eq. 2). Two mechanisms for creating the new individuals I from B^m are investigated: probabilistically and tournament-based. Let X be the number of potential bits to mutate in B^m. In the probabilistic VLGA (VLGAp), X bits are randomly

Table 1. Creating the virtual loser

bit	1	2	3	4	5	6	Fitness
A	1	1	1	1	1	1	high
B	0	0	1	0	0	1	low
C	0	0	1	1	1	1	low
D	0	1	1	0	0	0	low
VL	1.00	0.67	0.00	0.67	0.67	0.33	

selected and each bit i $(i = k_1, k_2, ..., k_X)$ is mutated according the value of P_i, and using the ith allele from B^m: if $rand(0, 1) < P_i$ then $I_i = 1 - B_i^m$. In the tournament-based VLGA (VLGAt), three bits are randomly selected and the one with the highest value of P_i is the winner of the tournament. The winner bit is always mutated. This process is repeated X times. Moreover, after a change, the VL is reinitialized using the procedure illustrated in Table 1. Then, at every evolutionary step, after selection, crossover and mutation, the VL is updated using eq. 1 until the occurrence of the next change. Fig. 1 presents the pseudo code of VLGA.

Function $VLGA$

    L: chromosome length p: population size
    m: memory size n: n. of individuals
    X: n. of bits to mutate α: associative factor
    γ: relaxation factor η: fraction of worst

$t = 0$;
Initialize and Evaluate($P(0)$)
Initialize and Evaluate($M(0)$)
$VL(t) = initVL(P(t))$
repeat
    Preserve best individual from $P(t-1)$
    if change is detected then
    //Update Memory
    $B' =$ best from $P(t-1)$
    $VL' = updateVL(VL^p(t-1), P(t-1), \gamma, \eta)$
    $M(t) = UpdateMemory(B', VL')$
    //Retrieve Memory
    Retrieve $B^m(t)$ and $VL^m(t)$ from memory
    $I(t) = CreateIndividuals(B^m(t), VL^m(t), X, \alpha)$
    $P(t) = ReplaceWorst(P(t), I(t), \alpha)$
    // Reinitialize VL
    $VL^p(t) = initVL(P(t))$
  $P'(t) = Selection(P(t))$
  $Crossover(P'(t))$
  $Mutation(P'(t))$
  $P(t+1) = P'(t)$
  $Evaluate(P(t+1))$
  $Evaluate(M(t+1))$
  $VL^p(t+1) = updateVL(VL^p(t), P(t+1), \gamma, \eta)$ //eq. 1
  $t = t+1$
until stop_condition

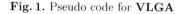

Fig. 1. Pseudo code for **VLGA**

3   Experimental Design

3.1   Dynamic Test Environments

The dynamic environments to carry out our experimentation were created using Yang's Dynamic Optimization Problems (DOP) generator [4]. This generator allows constructing different dynamic environments from any binary-encoded stationary function using the bitwise exclusive-or (XOR) operator. The characteristics of the change are controlled by two parameters: the speed of the change, r, which is the number of function evaluations between two changes, and the magnitude of the change, ρ, that controls how different is the new environment from the previous one. The DOP generator can construct three types of dynamic environments: cyclic, cyclic with noise and random. In this work we constructed two types of environments: cyclic and random. For each type of DOP, the parameter r was used with 1000 and 5000. The ratio ρ was set to different values in order to test different levels of change: 0.1 (a light shifting) 0.2, 0.5 and 1.0 (severe change). A random change was also tested concerning severity. For this case, the new environment has no relation with the previous one. Three binary-encoded problems were selected as the stationary functions: the 0-1 knapsack , the Royal Road F1 [12] and the Onemax function. Those benchmark problems were transformed from static to dynamic using the DOP generator, described before.

3.2   Parameters Setting

In the experiments, four GAs were investigated using the above constructed DOPs: the standard GA (SGA), the Hypermutation GA (HGA) proposed by [1], the Associative Memory Genetic Algorithm (AMGA) investigated in [4] and the two versions of the Virtual Loser Genetic Algorithm (VLGAp and VLGAt) detailed in the previous section. Standard parameters were used for all GAs: generational replacement with elitism of size one, tournament selection with tournament of size two, uniform crossover, with probability $p_c = 0.7$ and flip mutation with probability $p_m = 0.01$. Binary representation was used with chromosomes of size 100 for the Knapsack, 64 for the Royal Road F1 and 300 for the Onemax. The associative factor α for AMGA, VLGAp and VLGAt was set to 0.5. The hypermutation rate used in HMGA was 0.2. All algorithms used a global number of individuals $n = 100$. The memory size for AMGA and VLGA was set to $m = 0.2 \times n$. For VLGA, the relaxation factor μ was set to 0.02, the fraction η of the worst individuals was 0.5 and the number of potential mutations per individual was $M = L$ (L is the chromosome length). The *similar* replacing strategy proposed by [5] was used to replace a memory solution when the maximum capacity for the memory is attained. A change in the environment was detected if at least one individual in the memory changed its fitness. For each experiment of an algorithm, 50 runs were executed. Each algorithm was run for a number of generations corresponding to 200 environmental changes. The overall performance used to compare the algorithms was the offline performance [5] averaged over 50 independent runs, executed with the same random seeds.

4   Experimental Results

4.1   Analysis of the Results

Fig. 2 and Table 2 show the results obtained for the Knapsack problem in cyclic and random environments. Fig. 3 and Table. 3 present the results obtained for the Royal Road F1 problem. The statistical validation was made using the non-parametric Friedman's ANOVA test at a 0.01 level of significance. After this test, the multiple pair wised comparisons were performed using the Nemenyi procedure with Bonferroni correction. The notation used in the statistical results tables is $s+$, $s-$ and $\sim$, when the first algorithm is significantly better than, significantly worse than, or without statistical evidence with the second algorithm, respectively. The statistical results refer only to the comparison of the proposed method with the peer algorithms. No comparisons between SGA, HGA and AMGA are presented. From the plots and statistical analysis result that, for cyclic environments, the performance of VLGA is significantly better than SGA, HGA and AMGA. The few exceptions occur in the Knapsack problem, in environments with light severity of change ($\rho = 0.1$). This is understandable, as for light changes, keeping individuals from previous environments can be advantageous. Comparing the two versions of VLGA, the results show that, in cyclic environments, VLGAt outperforms VLGAp. The justification for this may be related to the number of genes mutated when a new individual is created. In VLGAt the bits to mutate are selected using tournaments of size three and the same gene can be selected in different tournaments. This can have an effect in the total number of mutated bits that can be lower than X. If the same bit is mutated an odd number of times, it returns to its initial value (no mutation). Inversely, if it is selected an even number of times, the process of mutation is repeated X times, but the effective number of mutations is below X. This indicates that mutating the bits with higher P can be more effective than randomly selection the bits, as used in VLGAp. For random environments, as expected, the use of memory is not beneficial. The performance of VLGA is better than SGA, HGA and AMGA when $\rho = 1.0$ and for some $\rho = rnd$, and similar or worse in the remaining cases. It is important to observe that the random environment is cyclic for $\rho = 1$ (as every bit is changed in each period, i.e., the environment is repeated every two periods), what explains the obtained results. The differences in the performances are more related to the diversity of the population promoted by the algorithms than the use of memory. The results for the Onemax problem were similar to the ones obtained for the Knapsack and the Royal Road F1.

4.2   Analysis of the Effect of α

In this work the influence of the associative factor α is also analyzed. Besides $\alpha = 0.5$, AMGA and VLGA were run using $\alpha = 1.0$. This means that when a change happens all the individuals of the population are replaced by new ones. Table 4 compares the results for AMGA, VLGAp and VLGAt using $\alpha = 0.5$ and $\alpha = 1.0$. The results marked in bold indicate that, for that value of α, the results

Table 2. Statistical Results for the Dynamic Knapsack

$\alpha = 0.5$	Cyclic		Random	
	$r = 1000$	$r = 5000$	$r = 1000$	$r = 5000$
$\rho \Rightarrow$	0.1 0.2 0.5 1.0 rnd	0.1 0.2 0.5 1.0 rnd	0.1 0.2 0.5 1.0 rnd	0.1 0.2 0.5 1.0 rnd
VLGAp - SGA	s+ s+ s+ s+ s+	~ s+ s+ s+ s+	s− ~ ~ ~ s+	s− s− s+ s+ s+
VLGAp - HGA	~ ~ s+ s+ s+	s+ s+ s+ s+ s+	~ s− s− s+ ~	~ ~ ~ s+ ~
VLGAp - AMGA	~ s+ s+ s+ s+	~ s+ s+ s+ s+	s− ~ ~ s+ s+	~ ~ ~ s+ s+
VLGAp - VLGAt	s− s− s− s− s−	s− s− s− s− s−	~ ~ ~ s+ ~	s+ s+ s+ ~ ~
VLGAt - SGA	s+ s+ s+ s+ s+	s+ s+ s+ s+ s+	s− s− s+ s+ s+	s− s− ~ s+ s+
VLGAt - HGA	s+ s+ s+ s+ s+	s+ s+ s+ s+ s+	s− s− s− s+ ~	~ s− s− s+ ~
VLGAt - AMGA	s+ s+ s+ s+ s+	s+ s+ s+ s+ s+	s− s− s− s+ s+	s− s− ~ s+ ~

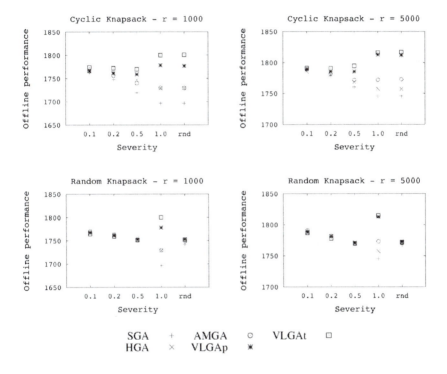

Fig. 2. Offline performance for the dynamic Knapsack

are significantly better. The results refer to the Knapsack problem, but the same conclusions were obtained for Royal Road F1 and Onemax. From the analysis of the results we conclude that, for cyclic environments, the performance of AMGA is significantly decreased for $\alpha = 1.0$. Contrarily, for VLGAp and VLGAt, in general, $\alpha = 1.0$ improves the algorithms' performance. The exception occurs for light shifts in the environment, where the performances are, in general, similar. Analyzing the diversity of the population in cyclic environments, for $\alpha = 0.5$ and $\alpha = 1.0$, we see that the diversity decreases when α is higher. This means

Table 3. Statistical Results for the Dynamic Royal Road F1

$\alpha = 0.5$	Cyclic										Random									
	$r = 1000$					$r = 5000$					$r = 1000$					$r = 5000$				
$\rho \Rightarrow$	0.1	0.2	0.5	1.0	rnd	0.1	0.2	0.5	1.0	rnd	0.1	0.2	0.5	1.0	rnd	0.1	0.2	0.5	1.0	rnd
VLGAp - SGA	$s+$	$s+$	$s+$	$s+$	$s+$	$s+$	$s+$	$s+$	$s+$	$s+$	$s-$	$\sim$	$\sim$	$s+$	$\sim$	$s-$	$\sim$	$s+$	$s+$	$s+$
VLGAp - HGA	$s+$	$s+$	$s+$	$s+$	$s+$	$s+$	$s+$	$s+$	$s+$	$s+$	$\sim$	$\sim$	$\sim$	$s+$	$\sim$	$\sim$	$\sim$	$\sim$	$s+$	$\sim$
VLGAp - AMGA	$s+$	$s+$	$s+$	$s+$	$s+$	$\sim$	$s+$	$s+$	$s+$	$s+$	$\sim$	$\sim$	$\sim$	$s+$	$\sim$	$\sim$	$\sim$	$\sim$	$s+$	$\sim$
VLGAp - VLGAt	$s-$	$s-$	$s-$	$s-$	$s-$	$\sim$	$s-$	$s-$	$s-$	$s-$	$s-$	$\sim$	$\sim$	$s-$	$\sim$	$\sim$	$\sim$	$\sim$	$s-$	$\sim$
VLGAt - SGA	$s+$	$s+$	$s+$	$s+$	$s+$	$s+$	$s+$	$s+$	$s+$	$s+$	$s-$	$s-$	$\sim$	$s+$	$\sim$	$\sim$	$s-$	$\sim$	$s+$	$\sim$
VLGAt - HGA	$s+$	$s+$	$s+$	$s+$	$s+$	$s+$	$s+$	$s+$	$s+$	$s+$	$s-$	$s-$	$\sim$	$s+$	$s-$	$\sim$	$s-$	$\sim$	$s+$	$s-$
VLGAt - AMGA	$s+$	$s+$	$s+$	$s+$	$s+$	$\sim$	$s+$	$s+$	$s+$	$s+$	$s-$	$\sim$	$\sim$	$s+$	$\sim$	$\sim$	$\sim$	$\sim$	$s+$	$\sim$

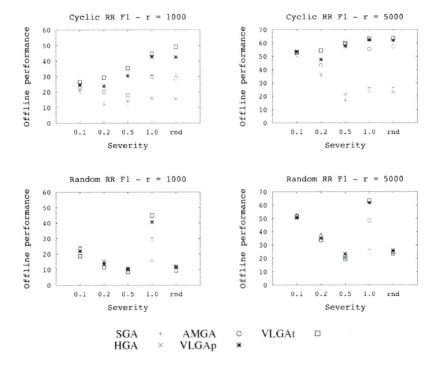

Fig. 3. Offline performance for the dynamic Royal Road F1

that, although a larger number of new individuals is being created, they are more similar with each other. This indicates that the new individuals created in AMGA are unfit to the new environments, and the new individuals formed using the proposed VL mechanism are fitter than the individuals preserved from the previous population. These results show that, for cyclic environments, the memorization of information about the errors performed in the evolutionary process can successfully be used to guide the creation of new individuals suitable to the new environment. For random environments, the loss of diversity observed when $\alpha = 1.0$ is detrimental. As before, the memory doesn't help the GA when the environment changes and the performance is better if the GA is able to promote a higher diversity.

Table 4. Effect of α in the Dynamic Knapsack

	r	ρ	AMGA0.5	AMGA1.0	VLGAp0.5	VLGAp1.0	VLGAt0.5	VLGAt1.0
CYCLIC	1000	0.1	**1766.16**	1585.24	**1766.55**	1759.20	**1773.98**	1763.32
		0.2	**1756.29**	1590.01	1761.62	**1770.77**	1772.26	**1778.29**
		0.5	**1740.01**	1596.44	1758.85	**1776.52**	1769.69	**1792.93**
		1.0	**1729.74**	1600.50	1778.40	**1784.90**	1799.92	**1803.80**
		rnd	**1729.05**	1601.25	1776.88	**1782.75**	1800.76	**1803.55**
	5000	0.1	**1789.32**	1753.07	1789.21	1789.34	1791.29	1790.96
		0.2	**1781.93**	1760.83	1785.54	**1803.00**	1790.94	**1804.92**
		0.5	**1771.78**	1766.31	1785.68	**1810.59**	1794.65	**1813.22**
		1.0	**1772.06**	1768.37	1811.73	**1814.59**	1815.79	1817.58
		rnd	**1772.76**	1769.00	1811.96	**1814.69**	1816.60	1816.90
RANDOM	1000	0.1	**1769.91**	1582.15	**1767.09**	1748.42	**1764.45**	1746.99
		0.2	**1762.71**	1580.04	**1760.93**	1749.53	**1758.99**	1744.37
		0.5	**1751.22**	1579.46	**1752.03**	1747.77	**1750.96**	1744.78
		1.0	**1729.22**	1601.71	1777.64	**1785.82**	1799.05	**1803.20**
		rnd	**1748.57**	1579.43	**1752.85**	1748.88	**1751.59**	1744.47
	5000	0.1	**1790.80**	1742.98	**1788.54**	1775.94	**1786.68**	1776.52
		0.2	**1781.93**	1738.15	**1781.23**	1772.95	**1777.90**	1770.38
		0.5	**1769.35**	1735.32	1770.66	1770.91	**1769.67**	1766.38
		1.0	**1773.08**	1770.04	1811.70	**1814.81**	1815.97	1817.27
		rnd	**1770.61**	1736.74	1772.54	1771.97	**1772.72**	1768.37

5   Conclusions and Future Work

This paper proposes a different type of associative memory for GAs addressing DOPs. The new mechanism, denoted by the virtual loser, memorizes the best individual and additional knowledge about the 'bad' mutations performed during the evolutionary process. The memorized information is used to create new individuals when a change is detected. The proposed algorithm, called VLGA, creates new solutions by mutating the best memory solution, avoiding the bad mutations applied in the past. Two mechanisms of generating new individuals were investigated: probabilistic (VLGAp) and tournament-based (VLGAt). Moreover, the influence of the number of new individuals to create (α) was analyzed. From the experimental results on a series of dynamic benchmark problems, the following conclusions can be drawn. First, VLGA outperforms the peer algorithms for almost all cyclic dynamic environments. For random environments, the memory is, in general, detrimental to the performance of the GA. Second, for cyclic environments, the proposed VLGAt obtains better performances than VLGAp, indicating that the choosing the bits with higher VL values is better than a random choice. Third, in cyclic environments, the associative factor α significantly influences the performance of VLGA. For higher values of α the performance is better in VLGA and worse in AMGA. As the diversity decreases with $\alpha = 1.0$, the improvement on the performance of VLGA proves that the proposed VL mechanism, although creating more similar individuals, they have higher fitnesses than the individuals from the previous population. The same does not happen with the new individuals created using the AMGA mechanism. For future work we intend to study and analyze the influence of the other parameters used in VLGA - X, μ and η. Moreover, for X, instead of using a fixed value, it would be interesting to analyze the benefits of a mechanism for adjusting the value of X during the run.

References

1. Cobb, H.G.: An investigation into the use of hypermutation as an adaptive operator in genetic algorithms having continuous, time-dependent nonstationary environments. Technical Report TR AIC-90-001, Naval Research Laboratory (1990)
2. Grefenstette, J.J.: Genetic algorithms for changing environments. In: Männer, R., Manderick, B. (eds.) Proceedings of PPSN II, pp. 137–144 (1992)
3. Simões, A., Costa, E.: Variable-Size Memory Evolutionary Algorithm to Deal with Dynamic Environments. In: Giacobini, M. (ed.) EvoWorkshops 2007. LNCS, vol. 4448, pp. 617–626. Springer, Heidelberg (2007)
4. Yang, S.: Explicit memory schemes for evolutionary algorithms in dynamic environments. In: Yang, S., Ong, Y.S., Jin, Y. (eds.) Evolutionary Computation in Dynamic and Uncertain Environments. SCI, vol. 51, pp. 3–28. Springer, Heidelberg (2007)
5. Branke, J.: Evolutionary Optimization in Dynamic Environments. Kluwer Academic Publishers (2002)
6. Simões, A., Costa, E.: Prediction in evolutionary algorithms for dynamic environments using markov chains and nonlinear regression. In: Proceedings of GECCO 2009, pp. 883–890. ACM Press (2009)
7. Trojanowski, K., Michalewicz, Z.: Searching for optima in nonstationary environments. In: Proceedings of the IEEE Congress on Evolutionary Computation (CEC 1999), pp. 1843–1850. IEEE Press (1999)
8. Yang, S., Yao, X.: Population-based incremental learning with associative memory for dynamic environments. IEEE Transactions on Evolutionary Computation 5(12), 542–561 (2008)
9. Yang, S., Richter, H.: Hyper-learning for population-based incremental learning in dynamic environments. In: IEEE Congress on Evolutionary Computation, CEC 2009, pp. 682–689 (May 2009)
10. Barlow, G.J., Smith, S.F.: A Memory Enhanced Evolutionary Algorithm for Dynamic Scheduling Problems. In: Giacobini, M., Brabazon, A., Cagnoni, S., Di Caro, G.A., Drechsler, R., Ekárt, A., Esparcia-Alcázar, A.I., Farooq, M., Fink, A., McCormack, J., O'Neill, M., Romero, J., Rothlauf, F., Squillero, G., Uyar, A.Ş., Yang, S. (eds.) EvoWorkshops 2008. LNCS, vol. 4974, pp. 606–615. Springer, Heidelberg (2008)
11. Sebag, M., Schoenauer, M., Ravisé, C.: Toward civilized evolution: Developing inhibitions. In: Bäck, T. (ed.) Proceedings of the 7th Int. Conference on Genetic Algorithms (ICGA 1997), pp. 291–298. Morgan Kaufmann, San Francisco (1997)
12. Mitchell, M., Forrest, S., Holland, J.: The royal road for genetic algorithms: fitness landscape and GA performance. In: Varela, F.J., Bourgine, P. (eds.) Proceedings of the First European Conference on Artificial Life, pp. 245–254. MIT Press (1992)

Author Index